Tactile Internet
with Human-in-the-Loop

Tactile Internet
with Human-in-the-Loop

Edited by

Frank H.P. Fitzek
Technische Universität Dresden
Dresden, Germany

Shu-Chen Li
Technische Universität Dresden
Dresden, Germany

Stefanie Speidel
National Center for Tumor Diseases, Partner Site Dresden
Division of Translational Surgical Oncology
Dresden, Germany

Thorsten Strufe
Karlsruhe Institute of Technology
Karlsruhe, Germany

Meryem Simsek
University of California, Berkeley
International Computer Science Institute
Berkeley, CA, United States

Martin Reisslein
Arizona State University
School of Electrical, Computer and Energy Engineering
Tempe, AZ, United States

ACADEMIC PRESS
An imprint of Elsevier

Library of Congress Cataloging-in-Publication Data
A catalog record for this book is available from the Library of Congress

British Library Cataloguing-in-Publication Data
A catalogue record for this book is available from the British Library

ISBN: 978-0-12-821343-8

For information on all Academic Press publications
visit our website at https://www.elsevier.com/books-and-journals

Publisher: Mara Conner
Acquisitions Editor: Tim Pitts
Editorial Project Manager: Andrea Gallego Ortiz
Production Project Manager: Nirmala Arumugam
Designer: Matthew Limbert

Typeset by VTeX

Contents

CHAPTER 1   **Tactile Internet with Human-in-the-Loop:**
                New frontiers of transdisciplinary research **1**
                Frank H.P. Fitzek, Shu-Chen Li, Stefanie Speidel, and
                Thorsten Strufe

PART 1 Domains of applications

CHAPTER 2   **Surgical assistance and training** **23**
                Stefanie Speidel, Sebastian Bodenstedt,
                Felix von Bechtolsheim, Dominik Rivoir, Isabel Funke,
                Eva Goebel, Annett Mitschick, Raimund Dachselt, and
                Jürgen Weitz

PART 2 Key technology breakthroughs

Eckehard Steinbach, Shu-Chen Li, Başak Güleçyüz,
Rania Hassen, Thomas Hulin, Lars Johannsmeier,
Evelyn Muschter, Andreas Noll, Michael Panzirsch,
Harsimran Singh, and Xiao Xu

PART 4 Technological standards and the public

CHAPTER 15   Tactile Internet standards of the IEEE P1918.1 Working Group **351**
Meryem Simsek, Sharief Oteafy, Zaher Dawy,
Mohamad Eid, Oliver Holland, and
Eckehard Steinbach

CHAPTER 16   Public opinion and the Tactile Internet **375**
Sven Engesser, Lisa Weidmüller, and Lutz M. Hagen

List of contributors

Gökhan Akgün
Technische Universität Dresden, Dresden, Germany

Ercan Altinsoy
Technische Universität Dresden, Dresden, Germany

Uwe Aßmann
Technische Universität Dresden, Dresden, Germany

Christel Baier
Technische Universität Dresden, Dresden, Germany

Tina Bobbe
Technische Universität Dresden, Dresden, Germany

Karlheinz Bock
Technische Universität Dresden, Dresden, Germany

Sebastian Bodenstedt
National Center for Tumor Diseases, Partner Site Dresden, Dresden, Germany

Juan A. Cabrera G.
Technische Universität Dresden, Dresden, Germany

Lingyun Chen
Technical University of Munich, Munich, Germany

Chokri Cherif
Technische Universität Dresden, Dresden, Germany

Darío Cuevas Rivera
Technische Universität Dresden, Dresden, Germany

Raimund Dachselt
Technische Universität Dresden, Dresden, Germany

Zaher Dawy
American University of Beirut, Beirut, Lebanon

Annika Dix
Technische Universität Dresden, Dresden, Germany

Clemens Dubslaff
Technische Universität Dresden, Dresden, Germany

Sebastian Ebert
Technische Universität Dresden, Dresden, Germany

Mohamad Eid
New York University Abu Dhabi, Abu Dhabi, United Arab Emirates

Frank Ellinger
Technische Universität Dresden, Dresden, Germany

Sven Engesser
Technische Universität Dresden, Dresden, Germany

Gerhard P. Fettweis
Technische Universität Dresden, Dresden, Germany

Christof W. Fetzer
Technische Universität Dresden, Dresden, Germany

Frank H.P. Fitzek
Technische Universität Dresden, Dresden, Germany

Norman Franchi
Technische Universität Dresden, Dresden, Germany

Isabel Funke
National Center for Tumor Diseases, Partner Site Dresden, Dresden, Germany

Eva Goebel
Technische Universität Dresden, Dresden, Germany

Diana Göhringer
Technische Universität Dresden, Dresden, Germany

Dominik Grzelak
Technische Universität Dresden, Dresden, Germany

Başak Güleçyüz
Technical University of Munich, Munich, Germany

Sami Haddadin
Technical University of Munich, Munich, Germany

Lutz M. Hagen
Technische Universität Dresden, Dresden, Germany

Simon Hanisch
Technische Universität Dresden, Dresden, Germany

Ardhi Putra Pratama Hartono
Technische Universität Dresden, Dresden, Germany

Rania Hassen
Technical University of Munich, Munich, Germany

Adamantini Hatzipanayioti
Technische Universität Dresden, Dresden, Germany

Jens R. Helmert
Technische Universität Dresden, Dresden, Germany

Diego Hidalgo
Technical University of Munich, Munich, Germany

Oliver Holland
Advanced Wireless Technology Group, Ltd., London, United Kingdom

Thomas Hulin
German Aerospace Center (DLR), Oberpfaffenhofen, Germany

Sebastian A.W. Itting
Technische Universität Dresden, Dresden, Germany

Lars Johannsmeier
Technical University of Munich, Munich, Germany

Stefan J. Kiebel
Technische Universität Dresden, Dresden, Germany

Konstantin Klamka
Technische Universität Dresden, Dresden, Germany

Stefan Köpsell
Technische Universität Dresden, Dresden, Germany

Jens Krzywinski
Technische Universität Dresden, Dresden, Germany

Vincent Latzko
Technische Universität Dresden, Dresden, Germany

Simone Lenk
Technische Universität Dresden, Dresden, Germany
Fraunhofer-Gesellschaft, Dresden, Germany

Shu-Chen Li
Technische Universität Dresden, Dresden, Germany

Jakub Limanowski
Technische Universität Dresden, Dresden, Germany

Tianfang Lin
Technische Universität Dresden, Dresden, Germany

Yun Lu
Technische Universität Dresden, Dresden, Germany

Lisa-Marie Lüneburg
Technische Universität Dresden, Dresden, Germany

Christian Mayr
Technische Universität Dresden, Dresden, Germany

Sebastian Merchel
Technische Universität Dresden, Dresden, Germany

Johannes Mey
Technische Universität Dresden, Dresden, Germany

Annett Mitschick
Technische Universität Dresden, Dresden, Germany

Jens Müller
Technische Universität Dresden, Dresden, Germany

Evelyn Muschter
Technische Universität Dresden, Dresden, Germany

Susanne Narciss
Technische Universität Dresden, Dresden, Germany

Krzysztof Nieweglowski
Technische Universität Dresden, Dresden, Germany

Andreas Nocke
Technische Universität Dresden, Dresden, Germany

Andreas Noll
Technical University of Munich, Munich, Germany

Luca Oppici
Technische Universität Dresden, Dresden, Germany

Sharief Oteafy
DePaul University, Chicago, IL, United States

Sebastian Pannasch
Technische Universität Dresden, Dresden, Germany

Michael Panzirsch
German Aerospace Center (DLR), Oberpfaffenhofen, Germany

Johannes Partzsch
Technische Universität Dresden, Dresden, Germany

Dirk Plettemeier
Technische Universität Dresden, Dresden, Germany

Ariel Podlubne
Technische Universität Dresden, Dresden, Germany

Martin Reisslein
Arizona State University, Tempe, AZ, United States

Dominik Rivoir
National Center for Tumor Diseases, Partner Site Dresden, Dresden, Germany

Christian Scheunert
Technische Universität Dresden, Dresden, Germany

René Schilling
Technische Universität Dresden, Dresden, Germany

Anna Schwendicke
Technische Universität Dresden, Dresden, Germany

Patrick Seeling
Central Michigan University, Mount Pleasant, MI, United States

Merve Sefunç
Technische Universität Dresden, Dresden, Germany

Meryem Simsek
International Computer Science Institute, Berkeley, CA, United States

Harsimran Singh
German Aerospace Center (DLR), Oberpfaffenhofen, Germany

Stefanie Speidel
National Center for Tumor Diseases, Partner Site Dresden, Dresden, Germany

Eckehard Steinbach
Technical University of Munich, Munich, Germany

Thorsten Strufe
Karlsruhe Institute of Technology, Karlsruhe, Germany

Ronald Tetzlaff
Technische Universität Dresden, Dresden, Germany

Andreas Traßl
Technische Universität Dresden, Dresden, Germany

Andrés Villamil
Technische Universität Dresden, Dresden, Germany

Uwe Vogel
Fraunhofer-Gesellschaft, Dresden, Germany

Felix von Bechtolsheim
Technische Universität Dresden, Dresden, Germany

Jens Wagner
Technische Universität Dresden, Dresden, Germany

Lisa Weidmüller
Technische Universität Dresden, Dresden, Germany

Jürgen Weitz
Technische Universität Dresden, Dresden, Germany

Hans Winger
Technische Universität Dresden, Dresden, Germany

Xiao Xu
Technical University of Munich, Munich, Germany

Jiajing Zhang
Technische Universität Dresden, Dresden, Germany

Sandra Zimmermann
Technische Universität Dresden, Dresden, Germany

About the editors

Frank H.P. Fitzek is a Professor and head of the *Deutsche Telekom Chair of Communication Networks* at Technische Universität Dresden (TUD) coordinating the 5G Lab Germany since 2014. Since 2019 he is a speaker of the German Research Foundation (DFG, Deutsche Forschungsgemeinschaft) Cluster of Excellence *Centre for Tactile Internet with Human-in-the-Loop (CeTI)*. He received his diploma (Dipl.-Ing.) degree in EE from RWTH Aachen, Germany, in 1997 and his Ph.D. (Dr.-Ing.) in EE from the Technical University Berlin, Germany in 2002 and became Adjunct Professor at the University of Ferrara, Italy in the same year. In 2003, he joined Aalborg University as Professor. He has visited various research institutes, including Massachusetts Institute of Technology (MIT), VTT, and Arizona State University. He cofounded several start-up companies since 1999. He received several awards, such as the NOKIA Champion Award and the Nokia Achievement Award. In 2011, he received the SAPERE AUDE research grant from the Danish government, and in 2012 the Vodafone Innovation prize. In 2015, he was awarded the honorary degree *Doctor Honoris Causa* from Budapest University of Technology and Economics (BUTE).

Shu-Chen Li is a Professor and head of the Chair of Lifespan Developmental Neuroscience at TUD since 2012. She is a speaker of the DFG Cluster of Excellence CeTI since 2019. She received her Ph.D. degree in cognitive psychology from the University of Oklahoma in the USA in 1994. After working as a postdoc at the McGill University in Canada, she continued her research career at the Max Planck Institute for Human Development in Germany for 16 years until she took up the professorship at TUD. From 2006 to 2008, she was also an adjunct professor of the Brain Research Center in the College of Electrical and Computer Engineering at the National Chiao-Tung University in Taiwan. A key aspect of her research focuses on understanding brain mechanisms of neuronal gain control and their implications on age-related differences in perception and cognition across the human life span. For several years she served as the associated editor of *Developmental Psychology*, one of the flagship journals of the American Psychological Association. She is currently a member of the editorial board of *Neuroscience and Biobehavioral Reviews*.

Stefanie Speidel is a Professor for *Translational Surgical Oncology* at the National Center for Tumor Diseases (NCT), Partner Site Dresden, since 2017 and speaker of the DFG Cluster of Excellence CeTI since 2019. She received her Ph.D. (Dr.-Ing.) from Karlsruhe Institute of Technology (KIT) with distinction in 2009 in the context of the research training group *Intelligent Surgery* (KIT, University of Heidelberg, DKFZ), and led a junior research group *Computer-Assisted Surgery* from 2012–2016 at KIT. She has been (co)-authoring more than 100 publications and regularly organizes workshops and challenges, including the Endoscopic Vision Challenge@MICCAI as well as the Surgical Data Science workshop. She has been general chair and program chair for a number of international events, including IPCAI and MICCAI conference.

Thorsten Strufe is a Professor for *IT Security* at Karlsruhe Institute of Technology (KIT), Adjunct Professor for Privacy and Network Security at TUD, a speaker of the DFG Cluster of Excellence CeTI, and director of the Helmholtz Security Labs KAS-TEL at KIT. His research interests lie in the areas of large distributed systems and social media, with a focus on privacy and resilience. More recently, he has focused on studying user behavior and security in social media, and on ways to provide privacy-friendly and secure social networking services; he is fascinated by protection through decentralization. One of the challenges that drives him is how to create competitive web services and mobile apps without extensive collection of personal information, thus respecting the privacy of their users. To this end, his group measures and analyzes behavioral data on a large scale, develops algorithms and protocols to improve privacy and security, and formally analyzes anonymization networks for making their actual protection against new attacks formally verifiable.

Meryem Simsek is a Senior Research Scientist at the International Computer Science Institute, UC Berkeley, USA. She received her Ph.D. (Dr.-Ing.) from University of Duisburg-Essen on Learning-Based Techniques for Intercell-Interference Coordination in LTE-Advanced Heterogeneous Networks in 2013. Dr. Simsek has initiated and is currently chairing the IEEE Tactile Internet Technical Committee and serves as the Vice Chair for the IEEE P1918.1 Standardization Working Group, which she co-initiated. On the basis of her roles at IEEE, she disseminates and standardizes the achievements of CeTI. She was a recipient of the IEEE Communications Society Fred W. Ellersick Prize in 2015 and the Rising Star in Computer Networking and Communications by N2Women in 2019.

Martin Reisslein is a Professor in the School of Electrical, Computer, and Energy Engineering at Arizona State University (ASU), Tempe, and an external associated investigator with the DFG Cluster of Excellence CeTI, TUD, Germany. He received the Ph.D. in systems engineering from the University of Pennsylvania, Philadelphia, in 1998. He was a post-doctoral researcher with the Fraunhofer FOKUS institute and the Technical University Berlin from 1998 to 2000, when he joined ASU as Assistant Professor.

Preface

This book is a result of intensive collaborations among the contributors during the period of applying for a grant from the German Research Foundation (DFG, Deutsche Forschungsgemeinschaft) to establish a Cluster of Excellence and from the ongoing research activities during the first year after the Cluster had been successfully established at Technische Universität Dresden (TUD) in 2019. Together with researchers from other participating institutions, including Technical University of Munich (TUM), the Fraunhofer Institutes and the German Aerospace Center (Deutsches Zentrum für Luft- und Raumfahrt), the National Center for Tumor Diseases (Partner Site Dresden) and several international partners, a core team of researchers from five faculties (Electrical Engineering, Mechanical Engineering, Computer Science, Psychology, and Medicine) at TUD launched the Centre for Tactile Internet with Human-in-the-Loop (CeTI) to pursue new frontiers of research to promote disruptive innovations for digitally transmitted human–machine interactions that may revolutionize many aspects of our lives. This new field of transdisciplinary research will tackle a broad spectrum of theoretical, methodological, and technological challenges. In doing so, the emerging research on Tactile Internet with Human-in-the-Loop (TaHiL) will chart new frontiers for basic and applied research in human and engineering sciences to yield breakthroughs for next-generation multimodal, quasi-real-time human–machine interactions in real, virtual, mixed, and remote environments with broad applications in medicine, industry, and digital transformation technologies for daily-life usages.

Advancing the frontiers of science and technology relies on intensive collaborations among established fields of research that in the end may yield transdisciplinary breakthroughs of much broader impacts than the sum of the outputs from the individual disciplines involved. For a large number of researchers from several disciplines to join forces in embarking on disruptive, transdisciplinary research, as in the case of the research on TaHiL, basic understandings about the key principles of the involved fields, common languages, and shared visions need to be developed across the disciplines. Furthermore, synergistic research needs to be systematically structured and interconnected. This book is conceived as a handbook for the research on TaHiL to serve exactly these purposes. The book follows the structure of a synergistic research program with twelve research building blocks that have been established in the Cluster of Excellence CeTI at TUD. The building blocks are hierarchically interconnected, such that together they form a research pyramid (see the synergistic research program introduced in Chapter 1 for details). The respective research aims, approaches, and activities of the twelve building blocks are each covered by a chapter in this book. With the aim to serve as a handbook, representative work in the relevant areas beyond the research and technologies currently pursued in CeTI are also reviewed in the respective chapters.

Following the introduction (Chapter 1), which provides an overview of the research on TaHiL, the twelve chapters are divided into three parts, proceeding from the top to the base of the pyramid of the synergistic research structure. The first part showcases three selected domains of applications, which are robotic-assisted surgery (Chapter 2), human–robot cohabitation in industrial settings (Chapter 3), and Internet of Skills for other daily applications (Chapter 4). These use-cases presented in Part 1 require the key technologies and methods—in particular haptic codecs (Chapter 5), intelligent networks (Chapter 6), augmented perception and interaction (Chapter 7), as well as human-inspired models and computing (Chapter 8)—that are presented in Part 2. The challenges combined have to be tackled by systematically organized integrative research from several disciplines. These target primary research fields are presented in Part 3, which cover basic research on human multisensory perception (Chapter 9), sensors and actuators (Chapter 10), communications and control (Chapter 11), electronics for textile integration (Chapter 12), and tactile computing (Chapter 13). The last part of the book extends to cover cross-cutting topics, such as a digital trace library (Chapter 14) and standardization (Chapter 15) as well as technology transfer and communication to the public (Chapter 16).

This volume can serve as a handbook for the research on TaHiL for students and researchers from several contributing disciplines. For readers who would like to find out what TaHiL is and what applications the research in this new frontier may have, we recommend surveying Chapter 1 and the chapters in Part 1. For researchers who are already acquainted with topics about some aspects of TaHiL, we recommend sampling chapters from Part 1 for the specific applications of interests, and reading through chapters in Part 2, which highlight key areas of technological breakthroughs that require interdisciplinary research. Furthermore, to foster interdisciplinary understanding, the chapters in Part 3 are recommended for students and researchers to gain knowledge about fundamental questions and methods that are important for the research on TaHiL from the perspectives of other disciplines. Last but not least, the chapters in Part 4 address topics on technological standards and public communication, which are also crucial for the success of developing new technologies to serve better human–machine interactions.

This book marks a beginning. We hope it will kindle more interest and attract intensive research attention for the emerging transdisciplinary field of TaHiL. Interested readers are also referred to the CeTI webpage (ceti.one) for research updates.

Frank H.P. Fitzek, Shu-Chen Li, Stefanie Speidel, Thorsten Strufe,
Meryem Simsek, and Martin Reisslein
Dresden, Germany
2021

Acknowledgments

First of all, we would like to thank the German Research Foundation (DFG, Deutsche Forschungsgemeinschaft).[1] Many results reported in this book are funded by the DFG as part of Germany's Excellence Strategy[2] in support of the Cluster of Excellence *Centre for Tactile Internet with Human-in-the-Loop* (CeTI) established at Technische Universität Dresden (TUD).

The editors would like to thank all the authors who have contributed to the different chapters collected in this book. Most of the authors are members of CeTI, including many current CeTI Ph.D. students and postdocs, who have invested significant amounts of time and effort in addition to their regular duties to make this book possible.

Many thanks also go to our international research partners and consultants, such as (*i*) Muriel Médard from Massachusetts Institute of Technology (MIT), (*ii*) Adam Gazzaley from the University of California San Francisco (UCSF), (*iii*) Uta Noppeney from the Radboud University, and (*iv*) Gene Tsudik from University of California, Irvine (UCI), who supported us while we applied for the excellence initiative funding to establish CeTI or support us in CeTI's Advisory Board.

In alphabetical order, we express deep gratitude to our industrial partners, such as Atlantic Labs, CampusGenius, Deutsche Telekom, Mimetik, and Wandelbots.

We thank our design team, Jens Krzywinski, Tina Bobbe, Lisa Lüneburg, and their associates, for the support in creating designs and graphics for several demonstrators as well as the illustrations presented in the book. Their work not only gives this book a nice touch, but has also helped us to convey CeTI's main ideas of future communication systems to the public over the last years.

We are deeply thankful to Christian Scheunert and Hrjehor Mark for their support in managing the LATEX sources and their patience over the last months. It is their achievement to have all the sources of this book pulled together.

The work presented in this book would not have been possible without the endless support of our universities, i.e., Technische Universität Dresden, Technical University

---

[1] Funded by the German Research Foundation (DFG, Deutsche Forschungsgemeinschaft) as part of Germany's Excellence Strategy – EXC 2050/1 – Project ID 390696704 – Cluster of Excellence *Centre for Tactile Internet with Human-in-the-Loop* (CeTI) of Technische Universität Dresden.

[2] A funding program of the Federal Government and the states to strengthen cutting-edge research at universities.

of Munich, as well as the Fraunhofer Institutes, the German Aerospace Center (DLR), and the National Center for Tumor Diseases, Partner Site Dresden (NCT).

Frank H.P. Fitzek, Shu-Chen Li, Stefanie Speidel, Thorsten Strufe,
Meryem Simsek, and Martin Reisslein
Dresden, Germany
2021

Acronyms

3D	Three-dimensional
3GPP	3rd-Generation Partnership Project
5G	Fifth Generation
ADC	Analog-to-Digital Converter
AFE	Analogue Frontend
AG	Actor Gateway
AI	Artificial Intelligence
AM	Additive Manufacturing
AOP	Aspect-Oriented Programming
API	Application Programmer Interface
AR	Augmented Reality
ARQ	Automatic Repeat Requests
ASF	Acceleration Sensitivity Function
ASIC	Application-Specific Integrated Circuit
ASP	Application Service Provider
ASQ	Action Sequence
AVB	Audio-Video Bridging
BAN	Body Area Network
BCH	Body Computing Hub
BDD	Binary Decision Diagram
BFGS	Broyden-Fletcher-Goldfarb-Shanno
CACC	Cooperative Adaptive Cruise Control
CATI	Computer-Assisted Telephone Interviews
CBR	Constant Bit Rate
CBSE	Component-based Software Engineering
CELP	Code-Excited Linear Prediction
CeTI	Centre for Tactile Internet with Human-in-the-Loop
CEW	Communication and Early Warning
CI	Communication Interruption
CMD	Command
CMOS	Complementary Metal Oxide Semiconductor

CNN	Convolutional Neural Network
CNS	Central Nervous System
COM/SDB	ComSoc Standards Development Board
COP	Context-Oriented Programming
CORA	Core Ontologies for Robotics and Automation
CPE	Control Planc Entity
CPS	Cyber-Physical System
CPU	Central Processing Unit
CR	Compression Ratio
CT	Computed Tomography
DARPP-32	Dopamine- and cAMP-Regulated Neuronal Phosphoprotein
DB	Deadband
DC	Direct current
DCT	Discrete Cosine Transform
DDoS	Distributed Denial of Service
DDS	Data Distribution Service
DNF	Disjunctive Normal Form
DNN	Deep Neural Network
DoF	Degrees of Freedom
DPMC	Dorsal Premotor Cortex
DPR	Dynamic Partial Reconfiguration
DSL	Domain-Specific Language
DT	Detection Threshold
DTAG	Deutsche Telekom
DTLS	Datagram Transport Layer Security
DVFS	Dynamic Voltage and Frequency Scaling
DWT	Discrete Wavelet Transform
E2E	End-to-End
ECU	Electronic Control Units
EEG	Electroencephalography
eMBB	Enhanced Mobile Broadband
eSAP	External Service Access Point
ESE	Energy Storage Element
ETSI	European Telecommunications Standards Institute
FB	Feedback
FDX	Fully Depleted Silicon-on-Insulator

fMRI	Functional Magnetic Resonance Imaging
FPGA	Field Programmable Gate Array
F-RAN	Fog Computing based Radio Access Network
FSK	Frequency Shift Keying
FSM	Finite State Machine
FUNc	Network Functional Compression
GN	Gateway Node
GNC	Gateway Node Controller
GPU	Graphics Processing Unit
GRAND	Guessing Random Additive Noise Decoding
HCI	Human–Computer Interaction
HCTG	Haptic Codecs Task Group
HDL	Hardware Description Language
HIC	Haptic Interpersonal Communication
HLS	High Level Synthesis
HO	Human Operator
HPC	High Performance Computing
HPD	High Performance Demonstrator
HRTF	Head-Related Transfer Function
HSI	Human–System Interface
IAT	Inter-Arrival Time
IC	Integrated Circuit
ICN	Information Centric Networks
IEEE	Institute of Electrical and Electronics Engineers
IEEE-SA	IEEE Standards Association
IMU	Inertial Measurement Unit
IoS	Internet of Skills
IoT	Internet of Things
IP	Internet Protocol
IPL	Inferior Parietal Lobe
IPS	Intraparietal Sulcus
ISDN	Integrated Services Digital Network
ISI	Inter-Stimulus Interval
ISO	International Standards Organization
ISS	Input-to-State Stability
ITU-T	International Telecommunication Union Standardization Sector

IVR	Immersive Virtual Reality
JIGSAWS	JHU-ISI Gesture and Skill Assessment Working Set
JND	Just Noticeable Difference
JSON	Javascript Object Notation
JVM	Java Virtual Machine
K	Key Technologies and Methods
KPI	Key Performance Indicator
LAN	Local Area Network
LED	Light-Emitting Diode
LGN	Lateral Geniculate Nucleus
LNA	Low Noise Amplifier
LTE	Long Term Evolution
MAC	Medium Access Control
MAD	Maximally Allowable Delay
MATI	Maximum Allowable Transmission Interval
MBCC	Model-Based Cobotic Cell
MDE	Model-Driven Engineering
MDP	Markov Decision Process
MEC	Mobile Edge Cloud
MGC	Medial Geniculate Complex
MGD	Mini-Batch Stochastic Gradient Descent
MIMO	Multiple-Input Multiple-Output
ML	Machine Learning
MMT	Model Mediated Teleoperation
mMTC	Massive Machine Type Communication
MRI	Magnetic Resonance Imaging
NAcc	Nucleus Accumbens
NC	Network Controller
NCS	Networked Control System
NesCom	New Standards Committee
NFV	Network Function Virtualization
NR	New Radio
NS	Network Slicing
NUI	Natural User Interface

OBG	Observer-Based Gradient method
OFDM	Orthogonal Frequency Division Multiplexing
OLED	Organic Light-Emitting Diode
OOK	On-Off Keying
OR	Operating Room
OS	Operating System
OSATS	Objective Structured Assessment of Technical Skills
OSM	Orthographic Software Modeling
PA	Power Amplifier
PAR	Project Authorization Request
PC	Passivity Controller
PCTL	Probabilistic Computation Tree Logic
PDMS	Polydimethylsiloxane
PDU	Protocol Data Unit
PE	Processing Element
PHY	Physical Layer
PL	Programmable Logic
PLL	Phase-Locked Loop
pRRH	pico-Remote-Radio-Head
PSNR	Peak Signal to Noise Ratio
PTP	Precision Time Protocol
PU	Polyurethane
QAM	Quadrature Amplitude Modulation
QoC	Quality-of-Control
QoE	Quality-of-Experience
QoP	Quality-of-Performance
QoS	Quality-of-Service
QPSK	Quadrature Phase-Shift Keying
RAG	Reference Attribute Grammar
RAM	Random-Access Memory
RAN	Radio Access Network
RC	Reflection Coefficient
RCS	Reconfigurable Computing System
RGB	Red Green Blue
RGB-D	Red Green Blue and Depth

RL	Reinforcement Learning
RLNC	Random Linear Network Coding
RO	Robot Operator
ROP	Role-Oriented Programming
ROS	Robot Operating System
RQs	Research Questions
RRI	Responsible Research and Innovation
RRM	Radio Resource Management
RRSI	Rapid Reaction Standardization Initiative
RT	Reaction Time
RTOS	Real-Time Operating System
SAP	Service Access Point
SAS	Self-Adaptive System
SAW	Spatial Audio Workstation
SDC	Silent Data Corruption
SDN	Software Defined Network
SE	Subjective Equality
SFA	Successive Force Augmentation
SFC	Service Function Chaining
SGX	Software Guard eXtension
SLP	Sparse Linear Prediction
SMPTE	Society of Motion Picture and Television Engineers
SMR	Signal-to-Mask Ratio
SNc	Substantial Nigra Parc Compacta
SNR	Signal-to-Noise Ratio
SoA	Service-oriented Architecture
SoC	Systems-on-Chip
SPIHT	Set Partitioning In Hierarchical Trees
SPL	Software Product Line
SR	Stimulus Response
SR-ARQ	Selective Repeat ARQ
STS	Superior Temporal Sulcus
SW-ARQ	Stop and Wait ARQ
TA	Technology Assessment
TADF	Thermally Activated Delayed Fluorescence
TaHiL	Tactile Internet with Human-in-the-Loop
TAM	Technology Acceptance Model

TD	Tactile Device
TDPA	Time Domain Passivity Approach
TDPA-ER	Time Domain Passivity Approach Energy Reflection
TE	Tactile Edge
TEE	Trusted Execution Environment
TFT	Thin Film Transistor
TI	Tactile Internet
TIM	Tactile Internet Metadata
TLS	Transport Layer Security
TNM	Tactile Network Manager
ToF	Time of Flight
TP	Talent Pool
TPU	Thermoplastic Polyurethane
TSM	Tactile Service Manager
TSN	Time Sensitive Network
TSX	Transactional Synchronization eXtension
TT	Tactile Traces
TUD	Technische Universität Dresden
TUM	Technical University of Munich
U	Use Cases
UE	User Equipment
UML	Unified Modeling Language
UPE	User Plane Entity
URLLC	Ultra-Reliable Low-Latency Communication
UTAUT	Unified Theory of Acceptance and Use of Technology
V2V/V2I	Vehicle-to-Vehicle/Vehicle-to-Infrastructure
V2X	Vehicle-to-Any
VCO	Voltage Controlled Oscillator
VPL	Ventral Posterolateral Nucleus
VPM	Ventral Posteromedia Nucleus
VPMC	Ventral Premotor Cortex
VR	Virtual Reality
VTA	Ventral Tegmental Area
WAN	Wide Area Network
WFS	Wave Field Synthesis
WG	Working Group

WiFi   Wireless Fidelity

XML   eXtensible Markup Language

ZOH   Zero-Order Hold

Tactile Internet with Human-in-the-Loop: New frontiers of transdisciplinary research

1

Frank H.P. Fitzek[a], Shu-Chen Li[a], Stefanie Speidel[b], and Thorsten Strufe[c]

[a]*Technische Universität Dresden, Dresden, Germany*
[b]*National Center for Tumor Diseases, Partner Site Dresden, Dresden, Germany*
[c]*Karlsruhe Institute of Technology, Karlsruhe, Germany*

What seemingly was often overlooked... is that the human brain itself... is something that is co-shaped by experience..., something that does not operate in an environmental vacuum, but at any moment is subject to environmental constraints and affordances.
– Paul B. Baltes[☆]

Engineering is a living branch of human activity and its frontiers are by no means exhausted.
– Igor Sikorsky

Above all things expand the frontiers of science: without this the rest counts for nothing.
– Georg C. Lichtenberg

1.1 Motivation and vision of TaHiL

In the summer of 1969 the Internet was created by coupling a small number of computer nodes to share files across different locations. Back then a small number of services was available to a small number of experts. Fifty years later the Internet is among the most important global infrastructures worldwide. It has become a key infrastructure that is used by everyone and touching almost every aspect of human daily lives. The current Internet provides the service of democratizing access to information for everybody, independent of location or time (as illustrated in Fig. 1.1). One

---

[☆] P. B. Baltes, P. A. Reuter-Lorenz, and F. Rösler (Eds.), *Lifespan Development and the Brain: The Perspective of Biocultural Co-constructivism*, p. 4, Cambridge University Press, 2006.

Tactile Internet. https://doi.org/10.1016/B978-0-12-821343-8.00010-1

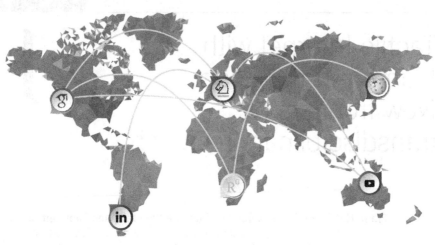

Fig. 1.1

Nowadays Internet: Democratizing access to information for everybody regardless of location or time.

important enabler for this function of global information access was the introduction of the World Wide Web, which allowed laypersons and computer scientists alike to create contents for distribution and to consume information through and from the Internet, respectively. Focusing on characteristics of the most popular services, such as video streaming, social networking, or web browsing, the current Internet has been optimized for high data rates to facilitate quick access and live consumption during the download. The next-generation Internet, the Tactile Internet (TI), takes these ideas one big step further. It envisions new opportunities and is faced by entirely novel challenges. The Institute of Electrical and Electronics Engineers (IEEE) P1918.1 Tactile Internet Standardization Working Group (http://ti.committees.comsoc.org/) defines the TI communication platform as: *A network or network of networks for remotely accessing, perceiving, manipulating or controlling real, or virtual objects, or processes in perceived real time by humans or machines* [1,2]. To transcend the possibilities of just gaining access to information, as we use the Internet today (compare Fig. 1.1), the Human-in-the-Loop approach [3] needs to be thoroughly realized en route to further technological advancements in the TI. The resulting new field of Tactile Internet with Human-in-the-Loop (TaHiL) research aims at democratizing access to skills and expertise to promote equity for people of different age, genders, cultural backgrounds, or physical limitations (see Fig. 1.2). To reach such breakthroughs for bringing digitally transmitted human–machine interactions to a new era, it is indispensable that transdisciplinary research involving researchers from several fields—ranging from psychology, cognitive neuroscience, and medicine to the fields of computer science, electrical, mechanical, and material engineering—needs to be conducted. One such transdisciplinary research center, the Centre for Tactile Internet with Human-in-the-Loop (CeTI), has been recently established at Technische Universität Dresden (TUD).

Fig. 1.2

TaHiL: Democratizing access to skills and expertise to promote equity for people of different ages, genders, cultural backgrounds, or physical limitations.

The Internet today, which is mainly optimized for increased throughput, cannot support TI applications. In this overview chapter, we will highlight several new challenges that need to be overcome to achieve the goals of TaHiL. The subsequent chapters in the book then present exemplified domains of applications that can be enabled through tackling challenges in a structured array of basic research and technological innovations. Here we highlight three frontiers in engineering and human research that still need to be explored and established: (*i*) a communication network that is optimized for skill (beyond information) transfer and hence supports extremely low latencies and different Quality-of-Service (QoS) support for modalities, such as video, audio, and haptics, (*ii*) novel human–machine interfaces that utilize a large array of sensors and actuators, and (*iii*) systematic understanding of goal-directed human multisensory perception and action, and the impacts of lifespan development and learning on these processes. Albeit establishing the TI as the next-generation infrastructure for global skill transfer is one important factor, the communication network alone falls short of several other challenges. Advanced wearable and adaptive sensors as well as actuators need to be developed as new types of interfaces for communications between humans and machines via the TI. Furthermore, the research on TaHiL has to consider the principles and mechanisms of human goal-oriented multisensory perception and action in people with different ages, learning experiences, and skill levels. Only then will the TI allow broad populations of human users to immerse themselves into virtual, remote, or inaccessible real environments, to exchange skills and expertise; thus create new opportunities and novel ways for people to learn, to work, and to interact (as illustrated in Fig. 1.3).

Fig. 1.3

(left) How may we learn in the future? (middle) How may our work change due to robots? (right) How may TaHiL technologies help the old and the oldest-old in the future?

1.1.1 Skill transfer from humans to machines

Different paths can be chosen to exchange skills among humans and machines. Here we consider a leading example of skill transfer from a human expert to a machine, in this case a standard industry robot (see Fig. 1.4). One way for this type of skill transfer would be to equip a human expert with any kind of human–machine interface, such as a simple remote control (e.g., a game console controller). This solution, albeit simple, has a couple of caveats. Not all human experts are able to operate such a controller, nor do they understand the relationship between the movements and accuracy of the robot.

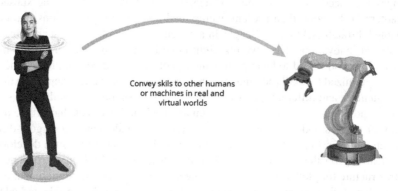

Convey skils to other humans
or machines in real and
virtual worlds

Fig. 1.4

Skill transfer from humans to machines.

Even replacing the simplistic controller with a more advanced human–machine interface, such as wearables that track and map human behavior directly to the robot, does not solve another issue regarding the scalability of direct remote control. A single machine still needs to be controlled by one human. In scenarios with millions of consumers demand a specific robot skill, the control has to be scaled up and the skill itself has to be transferred to millions of robots to meet the demand of the consumers. Industry has long recognized this problem, and there are standard processes for conveying skills and expertise to industrial robots (as illustrated in Fig. 1.5).

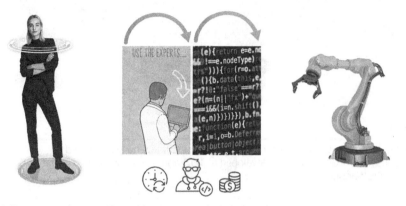

Fig. 1.5

Common industrial view on skill transfer.

A domain expert in such industry use cases explains the necessary expertise and actions a robot has to perform to a computer scientist. This programming expert will then, to the best of his/her understanding, convey these descriptions into the software that is subsequently executed on the industrial robot. Once successful, the software can be deployed to several robots. While this approach is more scalable than remote control, it still has several problems. First, the communication between the human expert and the computer scientist is error prone and often only a best-effort service. Second, the cost factor of the computer scientist is two to six times higher than the cost of the industrial robot. In light of decreasing prices for robots, this ratio will become even higher. Third, completing the skill transfer, potentially in a sequence of several trial-and-error cycles, is rather time-consuming.

To overcome these limitations, a novel approach developed in the field of TaHiL research at TUD has been proposed and implemented (see Fig. 1.6). This innovation builds specifically on the idea of using wearable clothing that is instrumented by sensors, and worn by domain experts. By performing the action routine several times, the activity of the expert is used to train the robot by demonstration through natural human movements. The training sequences recorded by the sensors are evaluated by machine-learning algorithms, which then output the software for the robot automatically. In other words, by combining sensor recordings of human behavior with machine learning, a direct form of human-to-machine demonstration teaching can be established. Such an approach is one of the promising avenues for further research on TaHiL. Indeed, this solution has been spun off in 2018, leading to a start-up company, Wandelbots, that has now developed several products for industry (for example, see Fig. 1.7).

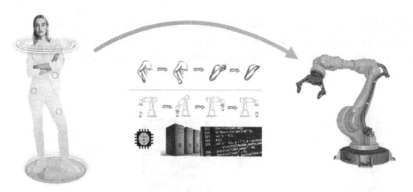

Fig. 1.6

One of the TaHiL approaches for skill transfer using machine learning and demonstration-based teaching developed at TUD and now implemented in the start-up company, Wandelbots.

Fig. 1.7

The CEO of the Volkswagen group testing the demonstration-based robot teaching developed by the TUD spin-off Wandelbots.

This natural demonstration-based teaching is promising. However, conveying the sensor data from the wearables over a communication network to remote robots in real-time is still a big challenge in the research on TaHiL. Future developments along this novel approach may meet the challenges of specific tasks that would require global exchanges of skills, for instance an expert in Europe trains a robot in Tokyo, Japan (see Fig. 1.8). In this scenario, the task of providing the expert with timely multimodal feedback, as it is required for efficient remote training and for giving the expert the feeling of virtually being right next to the remote robot, is particularly challenging. In fact, this is not possible with the Internet technology as it stands today, and derives several challenges ahead of us to realize the TI.

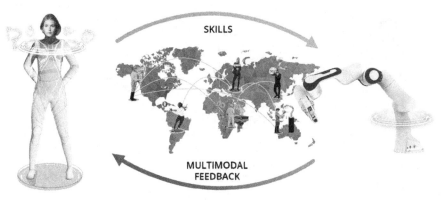

Fig. 1.8

Future scenarios of global skill exchange in the TI.

An overview of the manifold challenges in different research building blocks of TaHiL is shown in Fig. 1.9. The multimodal feedback needs to comprise haptic, in addition to video and audio, information. Each of these information modalities differs in its requirements of bandwidth and latency. It is already clear that video requires more bandwidth at relaxed delay constraints as compared to audio interaction. Furthermore, the properties and requirements for haptic information, so far, have not yet been extensively investigated and are far from been understood. Thus further basic research on haptic information processing in humans and applied research on haptic technologies would be necessary. With respect to latency and reliability of human sensory and perceptual processes, the latencies range from several milliseconds for video, over around 3 ms for audio, to only about 1 ms for haptic information [4] (see also Chapters 5 and 9). Such strict latency requirements create novel challenges, particularly given the laws of physics. Since the speed of light becomes a limiting factor for the possible distance between the human expert and robot for timely real-time remote interaction, as illustrated in the scenarios above. Even without considering the time needed for sensing, encoding, and processing, light travels at around 300 kilometers per millisecond, which limits the distance between the expert and robot to a range that falls substantially short of the requirements of global skill transfer and space communications.

Furthermore, human–machine real-time interactions in the form of coworking or training over longer distances will require local predictions of the remote behaviors, which would then also need to be corrected upon reception of the actual remote updates. While modeling a robot in a well-described physical environment without potential error sources is easy and follows the technical specifications of the machine, modeling and predicting human behaviors are significantly harder tasks that depend on many more parameters and are characterized by a huge number of degrees of freedom (see Chapters 9 and 11). These challenges require the Human-in-the-Loop approach [3] to be thoroughly realized on the foundation of human goal-directed perception and action in all aspects and steps of new technological developments.

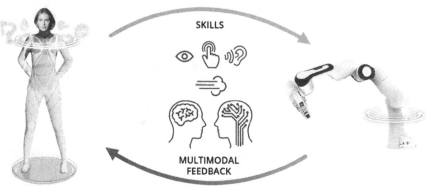

Fig. 1.9

Challenges in global skill exchange via the TI.

1.1.2 Skill transfer from machines to humans

So far we have considered one direction of skill transfer, i.e., having the human to teach a machine. But the reverse direction covers scenarios that could be applicable in other use cases. Assuming that the wearables are not only equipped with sensors but also with actuators, learning signals can either be generated live or in advance, and then conveyed to the human user. Taking physical rehabilitation of the elderly as an example, the movements of a remote physiotherapist could be generated online and transmitted to either wearables equipped with actuators that the elderly person wears, or a Cyber-Physical System (CPS) to help performing physiotherapy exercises at home (see Fig. 1.10). The potential application domains for such skill transfer from machines to humans are not limited to health and nursing care, they also cover teaching new skills in schools, at work, or of personal interests, i.e., the broad domain of Internet of Skills (IoS) [5]. Fig. 1.11 depicts two specific examples of this last class of applications that involve training rowing and climbing with specific wearables for detecting and correcting inefficient or potentially harmful movements. These technologies are currently under development in CeTI at TUD.

1.1.3 Skill transfer in holistic settings

There are several other application domains, for which digital skill transfer may also be valuable. Research on TaHiL aims to enable humans and machines to work collaboratively together in multiple learning activities in the future. Besides the aforementioned scenarios, skill transfer among robots of different manufacturers in completely different environments is a further potential application field. Furthermore, digitally mediated learning from human-to-human over long distances or between restrained environments will open disruptive new opportunities for the democratization of expertise and learning opportunities for acquiring various skills. Holistically, global

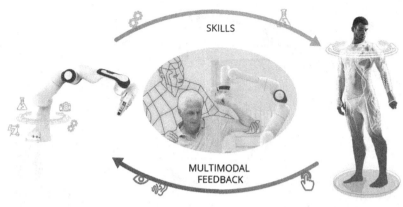

Fig. 1.10

A scenario of reverse skill transfer from machines to humans: The case of remote health care.

Fig. 1.11

Scenarios of teaching and training humans (see Chapter 4) in the specific cases of (left) rowing and (right) climbing.

digitally mediated skill exchanges can involve multiple combinations of human-to-machine, machine-to-human, and human-to-human interactions (see Fig. 1.12).

1.2 Research objectives to meet the challenges of TaHiL

There are several fundamental objectives for the research on TaHiL, which all revolve around the main building blocks of next generation multimodal closed-loop human–machine interactions that take place in the TI in perceived real-time (see Fig. 1.13): the human, who is augmented by large numbers of sensors and actuators that are connected through an intelligent network, cooperates with CPS (e.g.,

Fig. 1.12

Scenarios of the holistic human–machine skill transfer via the TI.

robots and other virtual- or mixed-reality entities) that are equipped with inherent sensors and actuators as well as adaptive learning mechanisms. Such quasi-real-time closed-loop interactions lead to a plethora of multisensory feedback information that has to be conveyed from the human to the machine and back over the same intelligent network, but with different communication characteristics in terms of latency and resilience. Note that the closed-loop human–machine interaction is not limited to one human, robot, or other CPS; instead, the TaHiL concept generalizes to other combinations and extensions that include an arbitrary number of these components in holistic settings of applications.

This section describes six key research objectives of TaHiL in a logical order. As shown in Fig. 1.13, we start with the Human-in-the-Closed-Loop system reflecting the first objective on *human perception and action*. In particular, the first objective mainly concerns the modeling and prediction of human goal-directed multisensory perception and action. The second objective focused on *human–machine coaugmentation* and addresses the novel bendable electronics, sensors, and actuators required for TaHiL. The aforementioned intelligent network is reflected by the third objective, which focuses on developing *human–machine networks* that can assure real-time communication, storage, and computing for all involved communication elements. The fourth objective concerns learning strategies for humans and machines to learn from and adapt to each other and is therefore called *human–machine learning*. The fifth objective of *human–machine computation* targets the computing infrastructure that is necessary for human–machine interactions. The sixth objective, *human–machine communication*, aims to develop new information theoretical approaches for communication, compression, coding, and control. (See Fig. 1.14.)

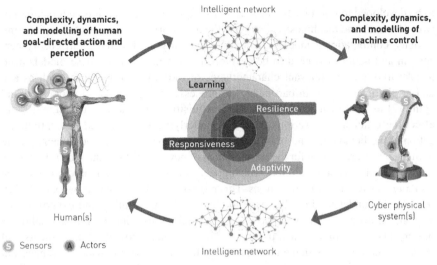

Fig. 1.13

Conceptual representation of the TaHiL.

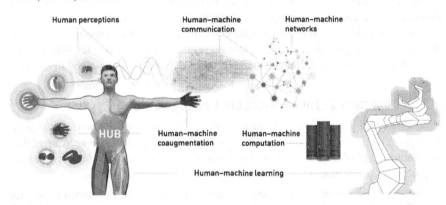

Fig. 1.14

Logical derivation of key objectives of the research on TaHiL.

1.2.1 **Objective 1: Human perception and action**

Model and predict human goal-directed behavior, which entails flexible and dynamic interactions between sensation, multisensory perception, cognition, and action in contexts.

The novel technologies to be developed for human–machine interactions via the TI will create new digital environments for humans to interact with a wide range of CPS, with substantially, if not completely, changed hardware and software interfaces that require extensive multisensory information processing (see Chapter 9, Fig. 9.1). Thus innovative approaches for system and interface designs would need to be developed to optimize the new digitally transmitted closed-loop interaction between humans and machines. To establish the necessary requirements for engineering designs, computational models of flexible, human goal-directed multisensory perception and action will need to be developed (see Chapter 13). These models need to take into account relevant characteristics of individual differences, such as age and levels of expertise. In particular, the processes of human development [6] and aging [7] as well as mechanisms of skill acquisition and mastery can significantly affect the efficiency of various processes underlying goal-directed perception and action at the behavioral and brain levels. Thus models characterized by appropriate human factors are crucial for the development of new algorithms and technologies for human–machine coadaptation, in which goal awareness and action prediction are the prerequisites for smooth interactions. This requires research to go far beyond the current state of understanding. We need to characterize and understand expertise- and age-related differences in key parameters of multisensory integration and delay requirements. Methodologically, psychophysical, and neurocognitive experiments will need to be conducted with large samples of individuals covering wide age ranges and expertise levels. The experiments will need to encompass different sensory modalities (e.g., auditory, visual, and haptic) across an array of perceptual decision and sensorimotor tasks that entail flexible switching between goal sequences or task contexts. Psychophysical and Bayesian active inference models will need to be developed to model and predict expertise- and age-related differences in the complexity and constraints of human goal-directed perception and action (see Chapter 9 for details about research on these topics and experimental studies currently been pursued in CeTI).

1.2.2 **Objective 2: Human–machine coaugmentation**

Produce wearable peripherals for fast sensing and actuating with multimodal feedback for human perception, cognition, and action based on ultra-small, bendable, stretchable, and ultra-low-power electronic circuits that precisely localize humans and objects in real-time.

New fast and flexible sensors and actuators will need to be developed to provide plausible multimodal feedback, such as soft exoskeletons (e.g., eGloves or eBody-Suits) that go beyond existing products. A high-quality multimodal feedback system should recognize and interpret the inputs from different modalities to provide multimodal outputs for human multisensory processing. Intelligent adaptive sensors are required for these purposes. Human psychophysical parameters provide the requirements for the actuators and multimodal interfaces. Psychophysical thresholds (e.g., tactile acuity or just noticeable level of difference in sound frequency) define the necessary information for developing and designing interfaces with realistic and compelling multisensory feedback that also entail tactile and kinesthetic signals. Latency sensitive haptic and visual codec applications will be required in addition to the feedback design. As a concrete scenario, novel robotic hand–arm systems will need to be developed in TaHiL that utilize haptic feedback to enable new control designs and methods for digitally transmitted remote physical manipulation and learning. The manipulation aspect of human-oriented feedback information is comparable to the approach in MP3 to avoid unnecessary information in audio. The haptic feedback will also be used to connect to smart wearables to immerse the human user in virtual or augmented reality. To achieve fast sensing and actuating for multimodal feedback, a new generation of electronics will need to be developed. It has to provide real-time operation, while consuming very low energy. Moreover, to facilitate the natural interaction between the human and the CPS, which are both equipped with high number of sensors and actuators, the electronic hardware has to be ultracompact, bendable, and stretchable. The transceiver dimensions are strongly determined by the antenna size, which massively decreases with increasing frequency. In this regard, by using very high frequencies, around 100 GHz, the antenna area can be decreased by several orders of magnitude allowing on-chip integration. To minimize the average power consumption of these wireless transceivers to below 1 mW and to allow real-time operation, aggressive duty cycling with record circuit reaction time in the nanosecond regime, allowing very efficient sleep and wake-up modes, would be necessary. The huge amount of sensor data has to be preprocessed adaptively and locally to guarantee low latency. For this, a mobile high-performance body computing platform will need to be designed with significantly reduced power consumption (20 TOPS/W). To allow mechanical flexibility, the chips will need to be thinned and integrated on stretchable substrates. Many of these new technologies are being used and developed in CeTI (see Chapters 10 and 12 for details).

1.2.3 Objective 3: Human–machine networks

Develop completely softwarized network solutions for wireless and wired communication that provide low latency, resilience, and security to enable human–machine co-operation.

To realize the vision of TaHiL, it is necessary to develop novel communication and network solutions for the body area, local area, and wide area networks that go beyond the mere relaying of information in the wireless and wired domain. New concepts of softwarization, including software defined networks, network function virtualization, and information-centric networking, will need to be exploited with respect to low latency and learning capabilities [8]. The latter is needed to allow the network to learn about different multimodal information flows (i.e., audio, video, and haptic) to adapt the network capabilities (e.g., network slicing). Moreover, the intelligent network will need to provide the means for learning from human behavior. For instance, when the connection to the human gets lost, the intelligent network should predict action courses during the period of interrupted communication (e.g., in a mobile edge cloud). Furthermore, the human's high-level but abstract mental models of behavior as well as low-level sensorimotor programs that make it possible for the human to cope with delays, ambiguities, and disruptions of the technical world will need to be better understood when designing and implementing communication networks. Such knowledge is a prerequisite for the design of efficient human–technology feedback strategies on all levels and provides an end-user-oriented direction for solving technical conflicts between latency, bandwidth, and resilience. Resolving this challenge has the potential to tremendously accelerate the progress in this field (for details see Chapters 6 and 11 for research activities on these topics that are currently being pursued in CeTI for details).

1.2.4 Objective 4: Human–machine learning

Provide an integrated framework that leverages the effects of continuous, mutual adaptive learning between humans and machines. Tune explanation facilities towards the demands and objectives of the human user. Assess boundary conditions and benefits for skill acquisition and training.

To achieve this objective, it is necessary to develop a portfolio of models, methods, and tools for automated human-inspired computing, to provide the Human-in-the-Closed-Loop interactive system with explanation and learning assistive equipment. In our view, the main innovation will be a new model-based approach that incorporates recent neuroscientific achievements on nonlinear dynamic models for human decision-making. To ensure smooth closed-loop human–machine interaction, it is important to enable human and machine to learn to predict and support each other's actions online and thereby bringing interaction to a new era of immersed closed-loop human–machine cooperation and coaugmentation. On the one hand, human actions can be augmented by machines; on the other hand, machines can learn from human behavior to represent and generate expert knowledge with various methods. The software systems for TaHiL applications have to provide self-explanation techniques to help end-users to understand the correct function and the rationale of decisions that are taken by machines. Towards an integrated framework, model-based explanations for machine behaviors, explanations for machine-learning results, and their appropriate visualization to facilitate human understanding will need to be developed. To this end, the TaHiL framework (see Figs. 1.13 and 9.1) can be applied to different domains of competency acquisition to explore the interplay of individual and situational factors to identify the conditions under which best benefits can be obtained for human–machine learning and coadaptation (see Chapter 8 for details).

1.2.5 Objective 5: Human–machine computation

Deliver a secure and scalable computing infrastructure that enables intuitive haptic interaction and automatically adapts to changes in task contexts and world models.

The softwarization of the TI leads to highly immersive software and services. The software infrastructure not only needs to perform fast, but also needs to be highly resilient to failures and attacks. Thus safety, security, privacy, and scalability are prerequisites for the tactile computing infrastructure serving TaHiL applications. The software for these applications, being naturally embedded into the physical environments, will need to rely on plausible world and context models that can also adapt to changes (see Chapter 13 for details).

1.2.6 **Objective 6: Human–machine communication**

Provide novel coding and compression methods, such as haptic codecs, compressed sensing, and network coding that take into consideration human factors to enable a combined control and communication system.

The TaHiL approach will produce a massive amount of sensor and actuator information to be exchanged among multiple communication and control nodes to facilitate human–machine cohabitation. Currently, control systems employ almost exclusively wired communication, because wireless solutions suffer from the traditional hard trade-off between latency, resilience, and throughput. However, the TaHiL applications need to rely on wireless communication to enable higher degrees of freedom for humans and machines. The massive amounts of sensor and actuator data that will be produced thus calls for a novel compression method—haptic codecs [9], which is akin to source compression for audio and video, but this time for all possible types of haptic information. Beside this end-to-end approach, new distributed approaches, using network coding and compressed sensing will soften the aforementioned trade-off, allowing for more flexible optimization strategies. Both approaches need to be tailored to the software-defined networks and network virtualization solutions that have to be developed. Furthermore, the amount of sensing and feedback data will need to be reduced significantly through learning strategies and optimizations of the communication network and control loop. The amount of traffic needed for closed-loop human–machine interactions in the TI will need to be reduced by at least one, but potentially up to two, orders of magnitude compared to the current standard. In addition, haptic communication requires qualitative and quantitative assessments of the user's quality-of-experience. In contrast to time-consuming subjective tests, reliable automated quality metrics for the evaluation of human-in-the-loop systems with haptic feedback will need to be developed. A range of research activities is currently being undertaken in CeTI to resolve these challenges (see Chapters 5, 6, and 11 for details).

1.3 **A synergistic research program**

To achieve the six research objectives and tackle the challenges presented above, we suggest a research program to facilitate synergies between the different fields for transdisciplinary research (see Fig. 1.15). Specifically, we identified five Talent Pools (TPs) that together build the foundation for the research on TaHiL. They span research activities on (i) human perception and action, (ii) novel sensors and actuators, (iii) core networking technologies, (iv) flexible electronics for on-body computation, as well as (v) computation and computing infrastructure for the TI.

Through close collaborations, researchers from the disciplines of the TPs engage in developing Key Technologies and Methods (K), particularly with regards to (i) codecs for tactile and haptic communication, (ii) secure, intelligent networks,

Fig. 1.15

A synergistic research program involving 12 interlinked research groups for the research on TaHiL.

(iii) novel user interfaces, and (iv) human–machine mutual learning techniques. Outcomes from basic research conducted by the research groups at the TP-level and novel technologies developed by research groups at the K-level can, in turn, be integrated and tested in research groups at the Use Cases (U)-level (see Fig. 1.15). We have identified three broad use cases (U) involving application in (i) medicine (see left panel of Fig. 1.16), (ii) industry (see middle panel of Fig. 1.16), and (iii) the Internet of Skills (see right panel of Fig. 1.16). Other application domains can be further integrated into the research program. However, the three domains we focus on here are complete in the sense that requirements for future research in the field of TaHiL for other applications can be derived from them.

An instantiation of this suggested research program is CeTI, which has been established at TUD since 2019. The structure of the research program (Fig. 1.15) illustrates the dependencies and synergistic interplays among the various groups of experts to progressively enable solutions from highly specialized to increasingly interdisciplinary teams. Outcomes of the solutions are then evaluated in different application domains, through which new assumptions and requirements are derived and transmitted back to the lower levels of the research structure for further research and technological refinements. Here we also use this structure to organize the chapters in this book. Key themes and issues indicated in the first three parts of the book and tackled by each of the twelve U, K, and TP research groups are presented in detail in corresponding chapters. Furthermore, topics on standardization, digital trace data library and communicating technological developments to the public are also treated in separate chapters in the last part of the book.

Fig. 1.16

(left) The vision for medicine and health. (middle) The vision for industry 4.0. (right) The vision for internet of skills.

1.4 Research outreaches and societal impacts

Other than scientific and technological impacts, the public may perceive the new era of digital communication and human–machine interaction to be brought about by breakthroughs in the research on TaHiL as a double-edged sword. Whereas technological advancements can bring new horizons on the one hand, the public may also have well-justified concerns about new technologies and their societal implications on the other hand. It is therefore very important to communicate with and engage the general public about the scientific and technological developments en route the research on TaHiL. Furthermore, research on societal and ethical implications of new human–machine interaction technologies is also increasingly gaining importance [10–12]. We highlight here a few research outreach activities that support different facets of public life by using selected examples from CeTI, the newly established TaHiL research program that is based at TUD.

In the current public discourse about digitalization, there is a common narrative of concerns about being replaced by machines or artificial intelligence. To address this issue, CeTI helps by conveying the TaHiL concept, which puts the human at the center of all developments, and hence focuses on technologies which serve as enablers for humans in different facets of life. The developed systems will open new chances and opportunities for humans to optimize and expand the ranges of their functions and performances. In the applied research towards the IoS (see Chapter 4) and as part of outreach activities and impact assessments, CeTI organizes public events to reach a wide spectrum of individuals in the society and uses existing public media channels, such as newspapers, TV programs, and social media to publicize and disseminate information about the new technologies and their implications. Above and beyond, to involve a broader audience from the precollege population, CeTI consolidates key research and technological developments in the research on TaHiL in the *CeTI truck*, with several demonstrations for human–robot skill transfer being installed on the cargo area of a truck. A first realization of the *CeTI truck* was available in summer 2019; over 5000 people visited the truck in the first week, including several school classes (see Fig. 1.17).

Mobile labs, such as the *CeTI truck*, can also visit the secondary schools and bring outcomes of current states of TaHiL research to the youth. School visits of this type

Fig. 1.17

CeTI truck and some activities. (top) *CeTI truck* on Postplatz Dresden as a mobile lab for the public. (left) A demonstration showing the principle of demonstrated-based skill transfer from human to robot. (right) Pupils learning about TaHiL research in the *CeTI truck*.

are particularly important for the society's goal towards gender equality, as it opens up possibilities to engage more women into science and technology at younger ages. Such earlier exposures in schools are likely to yield impacts on equal opportunity for their later choices of topics for academic pursuits and career developments.

The research program of TaHiL could also bring information about new technologies into local high school classrooms. An example of such outreach undertaken by CeTI is the established formal cooperation with a newly founded high school in the city, a university preparatory secondary school that has been established to become the first of a novel school type that has a special focus on information technology, computer and media literacy, and digital learning. In this case, CeTI and Gymnasium

Dresden-Pieschen mutually benefit from the cooperation in several ways. On the one hand, the high school students (age 10 to 18 years) have the opportunity to become acquainted with the fields of robotics, computer science, psychology, and medicine in selected teaching subjects at school. This serves as a head-start program for the youth to develop interests in science and technology. On the other hand, teaching in the high school can be enriched by hands-on activities offered by researchers in school classrooms or in the CeTI laboratories, allowing researchers to gain insights from the perspective of users.

1.5 Conclusion and outlook

In summary, the frontiers of transdisciplinary research on TaHiL will create multi-faceted impacts on the involved scientific and technological fields. As will be detailed in the remaining chapters of the book, there are still many research challenges that need to be solved on the way to realize the next-generation human–machine interaction in TI. These challenges can only be overcome through synergistic research and methodological exchanges across several fields. Although the intellectual and financial investments for such transdisciplinary research are high, research on TaHiL promises a high return. Outcomes that will flourish from the research on TaHiL are bound to bring human–machine interactions into a new era and to revolutionize numerous facets of human life (see Chapters 2, 3, and 4), including the domains of medicine (e.g., context-aware robotic assisted surgical or rehabilitation systems [13]), industry (e.g., human–robot co-working industrial space [14]) and Internet of Skills [5] (e.g., education and skill acquisition for the general public).

Domains of applications

Outline

We commence the book by introducing three application domains, namely medicine, Industry 4.0, and the Internet of Skills with one chapter each. Note that the Tactile Internet is not limited only to applications in these three domains. Other applications, such as nursing care, architecture and urban design, space communication as well as entertainment and product marketing could easily be added. We focus on the three mentioned domains, as we regard them as spanning a complete set of requirements for future research in the field of TaHiL, and allow deriving a broad array of contemporary applications.

Surgical assistance and training

Stefanie Speidel[a], Sebastian Bodenstedt[a], Felix von Bechtolsheim[b], Dominik Rivoir[a], Isabel Funke[a], Eva Goebel[b], Annett Mitschick[b], Raimund Dachselt[b], and Jürgen Weitz[b]

[a]*National Center for Tumor Diseases, Partner Site Dresden, Dresden, Germany*
[b]*Technische Universität Dresden, Dresden, Germany*

> *A theory must be tempered with reality.*
> — Jawaharlal Nehru

2.1 Introduction

In general, research in the context of Tactile Internet with Human-in-the-Loop (TaHiL) is driven from the use cases, which establish assumptions as well as requirements and demonstrate the research results in realistic scenarios. Following the *Theory that matters!* approach, three use cases, which have different requirements regarding latency, responsiveness, reliability, and security, taking into account the human's neurocognitive capacities as well as the machine's computational power and algorithmic complexity are demonstrated within this book. This chapter introduces the medical use case along with specific research questions and challenges regarding TaHiL.

Crucial for the success of medical treatments is the individual expertise of the clinician. While computer- and robotic-assistance in medical practices has emerged, the current systems do not incorporate a broad expert knowledge base and often do not exceed the capabilities of mechanical solutions. To go beyond current technologies, the goal of this use case is to develop a context-aware, real-time assistant with human-in-the-loop. This continuously learning- and knowledge-accumulating system acquires expert knowledge by demonstration of a certain type of treatment procedure and supports the clinician based on an interpretation of sensor data. The aim is an enhanced surgical environment through the system's situation awareness, which provides the appropriate assistance at the right time, thereby minimizing the complexity of the clinician's cognitive workload.

TaHiL addresses these aspects for intraoperative as well as training applications in highly dynamic sensor/actuator-enhanced environments, such as the Operating Room (OR) and surgical training settings. Therefore novel sensor-processing devices and data-driven methods for capturing, modeling, and transferring surgical expert skills, based on multimodal and temporal data, have to be exploited. To enable an intuitive

Tactile Internet. https://doi.org/10.1016/B978-0-12-821343-8.00012-5

human–machine collaboration, novel real-time interaction strategies with low-latency visual and haptic feedback have to be researched. Furthermore, new possibilities for immersive Virtual Reality (VR) and Augmented Reality (AR) training environments need to be investigated.

Towards this end, the medical use case integrates three different objectives in the context of exemplary surgical applications. The first objective *human-to-machine* focuses on the challenge of capturing and modeling surgical skills. The second objective *machine-to-human* investigates different learning and training concepts in multi-user VR/AR environments to teach previously modeled expert skills. The third objective *human–machine collaboration* aims for a context-aware real-time assistance based on the modeled expertise and the interpretation of intraoperative sensor data.

2.2 Human-to-machine: Modeling surgical skills

Generally, surgical skills and expertise can be divided into two main categories: technical, such as manual motions and handling of tools, and nontechnical, such as error management strategies, communication and teamwork [15]. While this seems like a clear-cut distinction, it has been acknowledged that skills and expertise pertaining to surgery are not completely technical [16]. Instead, a performance framework of five interrelated domains has been proposed by [17]: Psychomotor skills, declarative knowledge, interpersonal skills, personal resources, and advanced cognitive skills. Here the focus is mostly on psychomotor skills, also known as perceptual-motor skills [18], which can be directly observed from clinical performances. The other four types of skills are more difficult to capture directly. Here, thinking aloud protocols [19] or clinical logs [20] could be used.

2.2.1 State of the art

Data-driven modeling of surgical skills and expertise generally encompasses three steps: (*i*) the performance of one or more clinicians of certain tasks is captured using a multitude of sensors (Fig. 2.1) and (*ii*) the resulting data is annotated, preferably by clinical experts; eventually, (*iii*) the annotated data serves as basis for machine-learning methods for (workflow) analysis and modeling to make the captured information available to machines.

In what follows, we will elaborate on the state of the art in all of these areas.

2.2.1.1 Capturing surgical performance

Determining how to acquire psychomotor skills is the first step towards making them available to other clinicians and machines. For this, they first have to be demonstrated by surgeons, either in a clinical or a lab environment. This demonstration can then be captured and transformed into a digital format. Capturing digital psychomotor skills and expertise in a clinical setting is a data-driven process, meaning that the process relies on data collected from one or multiple sensors that observe performances from

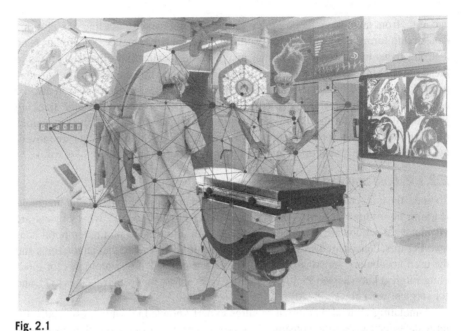

Fig. 2.1

Capturing surgical skills in sensor-enhanced connected environment, e.g., an operating room (Sensor-OR).

clinicians [15]. These collected performances could then serve as basis for modeling surgical psychomotor skills.

The types of sensors that can be used for adequate observations of clinical performances rely heavily on the clinical context. In the surgical domain, for example, the type of surgery being performed would decide which sensor modalities should be used and what they should focus on. In conventional open surgery, surgical tools and soft-tissues are handled directly by the surgeons, making it reasonable to capture skills through direct observation of the surgeon. To this end, work has mostly been done by using video streams from a single or from multiple Red Green Blue (RGB) cameras for capturing and analyzing surgical performance data [21,22] and for pose estimation [23]. Further work has extended the standard camera to include depth by relying on RGB-D camera systems, allowing for more accurate results in pose estimation [24,25] and for analyzing spatio-temporal features [26]. A further extension to camera-based data acquisition is the usage of a full motion-capture system for marker-based tracking of surgical personnel and their poses [27].

In types of minimally invasive surgery, such as laparoscopic surgery, retinal surgery, or microscopic surgery, the surgical performance is limited to a small site, which is only indirectly manipulated, often with long tools. Due to this and in contrast to open surgery, the focus when examining surgical skills is generally not directly on the surgeon's pose, but instead on the surgical tools and on their movements and tissue contacts. As these types of surgery are generally performed via a camera image, e.g., from an endoscope or a microscope, and not using direct line of sight, this data stream forms an ideal basis for capturing surgical performances. There has been a strong focus in the last years on the image-based localization of surgical tools in these types of image streams [28], which is an important milestone to acquiring surgical skills. Instead of image-based, the motions of surgical instruments can also be acquired via accelerometers [29], optical markers [30] or electromagnetic tracking [31]. Additionally, during robot-assisted minimally invasive surgery, the kinematics of the robot provide information not only on how the surgical instruments are moved, but also on the movement of the manipulators of the surgeon [32–34].

2.2.1.2 Data annotation

To make the collected data accessible to automated processing, in particular machine-learning algorithms, it has to be enriched with knowledge, e.g., what objects are visible, what actions are being performed, or with information on the temporal context. For this, consistent data annotations are required. To ensure that annotations and the used vocabulary are consistent, ontologies are utilized, e.g., ontologies for modeling surgical knowledge [35,36], to identify risks across medical processes [37], and for surgical workflows [38–40].

Annotating clinical data often requires expert knowledge, which, especially for large datasets, can be time-consuming and expensive to muster and is thereby often a bottleneck for clinical applications. Different approaches to assist during annotation and to reduce the annotation effort have been proposed in literature. Active learning approaches [41,42] and error detection methods [43], which reduce the annotation effort by selecting only the most informative or only erroneous data points for annotation, have been suggested. Furthermore, approaches for crowd-sourcing clinical annotation tasks [44–47] have been explored.

2.2.1.3 Workflow analysis and modeling

Capturing and analyzing clinical skills and expertise requires an understanding of the temporal context in which they are performed. To this end, approaches for workflow analysis, generally based on machine-learning algorithms, are utilized [48]. Commonly, clinical interventions are subdivided into phases, which describe different parts of a procedure and their clinical contexts. Methods in literature for automatically segmenting procedures into phases generally rely on information on tool usage [49,50] or directly infer the phases from visual data, such as from a surgical endoscope [51,52].

Actions, such as holding, cutting or suturing, are a more fine-grained division of surgical phases and directly relate to surgical skills and expertise. Many approaches for detecting clinical actions in data from a variety of sensors exist in literature. In open surgery, spatio-temporal features extracted from RGB-D data have been found useful for recognizing actions [26,53]. In laparoscopic surgery, approaches generally rely solely on the video stream [53] or on tool usage [54], while in robotic laparoscopic surgery the kinematic data is often used by itself [55,56] or in combination with the video stream [34,57,58].

The state of the start is limited with regard to modeling and transferring skills and expertise to machines. While methods for capturing data from a clinical setting and analyzing them in regards to temporal context exist, the skills and expertise are often not modeled. However, some work in laparoscopic robotic surgery has investigated the automation of certain subtasks [33] and the partial automation of certain movements [32].

2.2.2 Capturing surgical skills in the Sensor-OR

A primary goal for TaHiL-related research is the systematic capturing, modeling, and analysis of multimodal data in a sensor-equipped OR, which can be investigated in both a clinical and an experimental setting (Sensor-OR, Fig. 2.1).

In these environments, we aim to capture surgical skills and expertise (psychomotor) by using a multitude of sensors and devices, e.g., medical devices, camera arrays, optical tracking, robot kinematics, and novel real-time sensors in combination with approaches, such as eye tracking and data gloves (see Chapter 10), to collect performances from clinicians. Additionally, to enable the transfer of skills, codecs for transferring relevant sensor and device data, developed by K1, will be incorporated and evaluated. A slimmed-down version of the experimental OR setup is used by us to assess and evaluate surgical skills (see Section 2.3). In addition, temporal and spatial relations between instruments, anatomical structures as well as haptic and visual interaction will be acquired in a highly dynamic environment (phantom/ex-vivo/in-vivo), especially for robot-assisted procedures. This requires fast and novel sensors (visual, haptic, audio, wearables) to capture activities of medical experts, procedural progress, and patient status, which will be semiautomatically annotated (see Chapter 10).

To make the collected data accessible to further processing steps via machine learning, expert annotations are required. To facilitate the data annotation process, we are exploring methods for active learning [41] to limit efforts only to the data points that contain the most relevant information. We are also working on methods to further streamline the annotation process by developing methods that provide experts with suggestions from machine-learning algorithms that only have to be corrected. Furthermore, we are examining methods for transfer learning and for using unlabeled data to reduce the amount of data annotations that is required to reach peak performances [51]. In addition, a promising thread of research is to generate realistically looking synthetic images based on images from a simple simulation by using generative adversarial networks to overcome the lack of annotated training data [59] (Fig. 2.2).

Using machine-learning approaches, we aim to model surgical skills and expertise from the annotated sensor data, resulting, for the first time, in a holistic machine-interpretable acquisition of clinical expertise. This shall pave the way for generating a library of medical skills like knot-tying, endoscope guidance, caring procedures, such as repositioning of patients or sonography tasks. Furthermore, using methods for workflow analysis, we aim to acquire and analyze the clinical skills and expertise

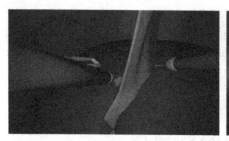 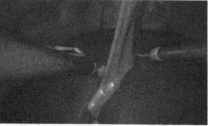

Fig. 2.2

Unpaired image-to-image translation to generate realistic laparoscopic images from simulations. (left) Laparoscopic simulation. (right) Generative adversarial networks translate the simulated images to look like real laparoscopic images. These images, along with their generated labels, can be used without further annotation effort.

temporally, which, combined with a sensor-based modeling of the clinical environment, e.g., patient anatomy, will allow us to view the skills and expertise in the context that they occur in, a huge benefit when trying to make these available to robotic assistants, e.g., for skill automation (see Section 2.4).

2.3 Machine-to-human: Surgical training

Constant training of crucial tasks is an indispensable part of education [60], particularly in high-risk domains, such as surgery (Fig. 2.3). Surgery is one of the most difficult psychomotor activities and a certain amount of first-hand experiences is necessary to obtain proficiency. Similar to musical and athletic skills, most surgical skills develop through training, though a certain amount of talent might influence the steepness of the learning curve. With evidence showing that a surgeon's lack of individual expertise as well as poor surgical skill can cause severe complications in patients [61,62], it is well conceivable that adequate surgical training systems could considerably improve surgical outcome and patient safety. To ensure high-quality patient care, it is therefore crucial to train surgical skills effectively and efficiently [15].

In general, surgical skills can be divided into technical and nontechnical skills [15] (see Section 2.2). Technical skills address competences, such as handling of instruments and bimanual working, while nontechnical skills require a high level of cognitive and social competence, involving teamwork, situation awareness or communication aspects [63].

Besides the basic skills for open surgery, e.g., stitching, knot tying, and cutting, modern medicine additionally requires the understanding and application of more complex devices and techniques, e.g., for minimally invasive and robotic surgery. Furthermore, surgical training should not be limited to surgical beginners, but should apply also to advanced surgeons and should adapt to different levels of expertise. Another critical issue is the transferability of trained skills to the real surgical procedure

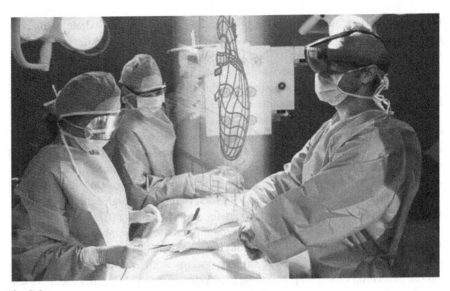

Fig. 2.3

AR-based surgical training setting.

as common tasks are usually trained in an unrealistic setting. In addition, individual feedback strategies that aim to shorten the trainee's learning curve are missing.

2.3.1 State of the art

In surgery, the official role of granting allowance to operate independently falls to certain state certification boards. The actual practical skills of a surgeon applying for certification are usually not part of the evaluation. Therefore surgical training of practical skills is heterogeneous and differs widely from hospital to hospital. Despite surgery being a high-risk domain, adequate training opportunities for surgical skill education are limited, especially for more demanding techniques, such as minimally invasive and robotic surgery [64]. These techniques pose increased demands through technology, which became so extensive in terms of handling and interaction among surgeon, device, and patient, that learning-by-doing is not a safe and effective way to train. The following subsections review the state of the art regarding skill training for minimally invasive and robot-assisted surgery.

2.3.1.1 Training systems for minimally invasive and robot-assisted surgery

Skills for minimally invasive and robot-assisted surgery can be trained on a phantom or box trainers, on animal models or on VR training systems [65,66]. In general, a distinction can be made between dry lab and wet lab for surgical training. Dry lab embraces training devices using specially designed practice tasks or organ dummies.

On the other hand, surgical training on animal or human organs, whole animals or even human corpses is summarized as wet lab. Most VR simulators offer a simulation of dry and wet lab tasks. Box trainers (or pelvi trainers) are a common and effective way to especially train the basic skills for minimally invasive surgery. Nevertheless, the disadvantage is the limited immersion due to the lack of a realistic setting and haptic sensation of artificial organs. This can be enhanced by using perfused animal organs, but acquisition and storage of perfusable organs is very costly.

VR simulators offer a wide range of exercises, including basic instrument handling tasks or complex operations, mainly focusing on dexterity or workflow-related tasks. Modern VR simulators can pretend haptic feedback, but nevertheless the immersion can vary, depending on graphics and tissue simulation. Despite the debate about the effectiveness of stand-alone VR-based training and high acquisition prices, such trainers can be found in many centers for surgical training [67–70].

Regarding robot-assisted surgery, the training platform resembles the intraoperative control console in terms of layout and function, which enhances immersion. The intuitive controls of today's robotic surgery systems seem to result in better learning curves for certain tasks compared to laparoscopic surgery [71,72], but there are still drawbacks, such as realistic tissue behavior and appearance.

2.3.1.2 Skill assessment and feedback

The importance and positive effects of feedback, such as an improved learning curve and a reduced error rate, have been shown on simulators [73] as well as in the OR [74]. Here, feedback can be defined as *the provision of information pertaining to the operator's surgical performance with the aim of improving subsequent performance* [75].

A prerequisite for providing feedback is the assessment of trainees. The current gold standard for skill assessment is rating based on task-specific checklists or global rating scales [76], such as Objective Structured Assessment of Technical Skills (OSATS) [77], which include criteria, such as efficiency, tissue handling, or bimanual dexterity. A common approach to automatic surgical skill assessment [15,78] is to (*i*) record sensor signals during the performance of a task and (*ii*) to analyze the data to assess performance (see Section 2.2). Various sensor types have been used for skill assessment, such as cameras, force sensors, or tracking systems [15]. Few approaches use sensors to measure aspects concerning the trainee, such as their posture [79], eye gaze [80,81], or cognitive load. So far, algorithms for skill assessment mainly focus on motion or dexterity analysis [78]. For example, a popular approach [82,83] is to calculate descriptive metrics, such as time to completion or path length based on instrument trajectories. These metrics are fed into machine-learning classifiers to categorize surgeons according to skill level. Recent studies [84,85] propose using deep learning-based methods to learn features directly from robot kinematic data.

A broader approach to providing feedback is coaching, which emphasizes goal setting and strategies towards reaching these goals [86]. In this context, the trainee should be instructed on how to practice specific skills in a repetitive and mindful manner (*deliberate practice* [60]). The positive effect of coaching on skill acquisition has been demonstrated in simulated [87,88] and real [89] surgical settings. In the real setting, coaching interventions were based on the recorded intraoperative video. However, manual coaching by an expert surgeon is expensive in terms of time and resources and limited regarding feasibility and availability. Automatic assessment and feedback functions, which could enhance training significantly, have yet to be established.

The most feasible automatic feedback strategy is the provision of general, implicit feedback, e.g., visualizations regarding the proximity of a surgical instrument to risk structures [90] or the magnitude of the force being applied [91]. Besides visual feedback, auditory [92], haptic, or multimodal feedback can be used [93]. For explicit feedback, automatic methods usually rely on predefined features computed from the available sensor data, which are assumed to be descriptive of surgical performance. Typical examples are basic motion-based metrics, which are presented, without further explanation, at the end of a training task [94,95]. In contrast, some approaches compare the trainee's performance to a given optimal performance to provide feedback, mostly using a set of manually engineered rules [96–98]. Finally, there are some approaches to automatically provide guidance during training. If an optimal trajectory is known, the deviation of the currently observed trajectory can be indicated using visual [96], auditory, or haptic cues. Also step-by-step walkthroughs for complete training scenarios were implemented in VR training systems [96,99]. However, they are usually defined in a static way, only provide limited predefined user interaction, and do not adapt to the individual performance of the trainee.

2.3.2 Data-driven surgical training

The fields of interest in surgical training are especially the mechanisms of learning surgical skills taking data-driven methods and psychological expertise into account. A special focus lies on how training, in general, and different training methods, in particular, affect the learning curve. TaHiL-related research regarding training includes automatic assessment of surgical skills based on sensor data as well as feedback strategies to improve the learning curve, especially haptic feedback. Multiuser VR as well as AR training scenarios will be investigated to teach skills to novices. The VR/AR systems incorporate multiple sensor devices that enhance immersion, such as gloves for haptic feedback and eye tracking.

The training is based on a multidimensional-structured curriculum developed at the Department of Visceral, Thoracic and Vascular Surgery, University Hospital Dresden. In addition to surgical training in open, laparoscopic, and robotic surgery in a dry and wet lab setup, the curriculum compromises intraoperative teaching as well as clinical and scientific education [100].

To enable data-driven assessment and feedback strategies, novices and experts are recorded during completion of surgical skills in specific training scenarios using different kinds of sensor devices to capture the performance as described in Section 2.2. First attempts demonstrate a sensor-enhanced box trainer (Fig. 2.4) that includes cameras to track instrument motion. The box trainer will be gradually enhanced by

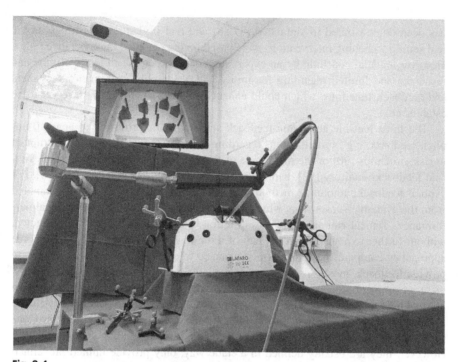

Fig. 2.4

Laparoscopic setup for capturing and evaluating surgical skills at the National Center for Tumor Diseases, Partner Site Dresden.

additional sensor devices, such as eye trackers, haptic sensors, and data gloves. The recorded sensor stream is rated by experts using a scale based on OSATS [77].

The recorded video can be used to assess the skill level automatically based on machine-learning approaches. A major challenge is finding a meaningful feature representation of highly complex video data. In [101], the authors have demonstrated that spatio-temporal features for skill assessment can be learned from raw video using 3D Convolutional Neural Networks (CNNs). The performance of trainees will be compared to expert models, which are used to generate training instructions and to objectively score the performance by different metrics.

It still remains an open research question regarding which way instructions and feedback shall be provided to trainees. Therefore more foundational research has to be conducted on how to actually design effective feedback strategies for people with varying skill levels. Incorporating the sensor-enhanced box trainer, novel strategies have to be designed and implemented that utilize several levels of visual feedback, most notably provided by means of Augmented Reality, haptic as well as acoustic feedback. The investigated feedback modalities might also be used synergetically and should be adapted to the current skill level and personal state of trainees. Comparison studies have to be conducted to compare the appropriateness and effectiveness of the developed approaches.

Further research aims on the perception of stress during open, laparoscopic, and robotic surgery, trying to identify the specific stressors in each approach. For this approach, a combination of subjective and objective values is collected, using questionnaires and measuring physiological parameters. A standardized experimental setup within a real operation theater using perfused animal organs to operate on increases the immersion while at the same time providing comparability. The aim of this research is the understanding of effective surgical training as well as identifying potential confounders and influencing factors.

2.4 Human–machine collaboration: Context-aware assistance

The human capability to manage and process different types of assistance functions that are not integrated in the surgical workflow can be easily exceeded [102]. An integral part of human–machine collaboration in surgery is to provide the right assistance at the right time, similar to an experienced human assistant (Fig. 2.5). This refers to the term *context-aware assistance* [50,103], which avoids an information overflow and decreases the cognitive load, in particular in an already stressful and complex environment, such as the OR. Examples of context-aware assistance during surgery range from visualization of risk and target structures or the prediction of adverse events to semiautomation of specific tasks on a robotic platform [104]. A prerequisite for context-awareness is to automatically and continuously perceive, analyze, and predict the context and progress of the surgical workflow in real-time. In addition, new strategies for man–machine interaction in the OR have to be exploited to enable an intuitive collaboration that does not distract from the actual task, namely taking care of the patient.

2.4.1 State of the art

Context-aware real-time assistance during surgical procedures is still an open challenge. Research ranges from visualization of additional information [105], automation of robot-assisted tasks [106] to workflow analysis based on sensor data [41] (Fig. 2.6), but the challenging, dynamic environment and the highly procedure-specific complexity still pose many open research questions.

The following subsections provide an overview of the state of the art regarding context-aware surgical assistance systems for the aforementioned examples, taking into account Human–Computer Interaction (HCI) strategies.

2.4.1.1 Visualization of navigation information

Intraoperative visualization of hidden subsurface structures, such as tumors or risk structures, is an essential component to assist surgical navigation. The approach is

Fig. 2.5

Different examples for context-aware human–machine collaboration. (left) AR-based visualization of risk and target structure during surgery. (right) Cooperation between surgeon and robotic assistant.

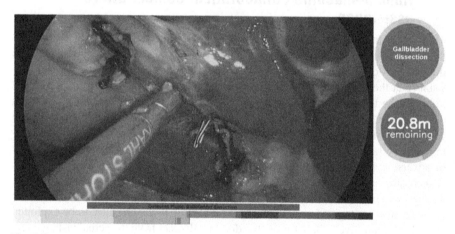

Fig. 2.6

Analyzing the surgical workflow, e.g., determining the current surgical phase, is a prerequisite for context-aware surgical assistance.

based on preoperative patient planning data, such as Computed Tomography (CT) or Magnetic Resonance Imaging (MRI), which is registered with intraoperative imaging data, e.g., from endoscopes or ultrasound. Challenges arise due to soft-tissue deformation, real-time requirements, and integration into the surgical workflow.

Augmented Reality is a promising technology for visualizing navigation information. One of the first studies on using AR for computer-assisted surgery introduced concepts, such as augmenting the endoscope, slice view projection, or a virtual mirror [107]. [108] discussed the potential advantages and disadvantages of 3D imaging techniques, including surface rendering, volume rendering, D3D, VR, and AR in diagnostic imaging. They concluded that AR has the potential to improve diagnosis and patient care, while even saving costs.

In some disciplines, such as neurosurgery, computer-assisted navigation is common practice [109]. In other disciplines, like liver surgery, appropriate and feasible assistance is still on research level. First commercial solutions exist, but do not consider the soft tissue deformation [110,111]. A state-of-the-art review on the clinical use of VR and AR in liver surgery is given in [105,112]. Other publications focus on the use of AR for nephrectomy and urological [113,114] or colorectal [115] interventions.

A recent comprehensive review of AR in robot-assisted surgery by [116] reveals that most of the related work is focused on laparoscopic interventions, and the main visual source for surgical guidance is stereo laparoscopy. It also reveals that most approaches use the surgeon's console or separate monitors as AR interfaces instead of *full AR* head-mounted displays.

In contrast to laparoscopic surgery, open surgery poses different challenges as discussed in [117]. They present a review, including diverse output techniques for open surgery, such as head-mounted displays, 2D/3D displays or handhelds.

The opportunities of AR in medical surgeries can support avoiding cognitive overloads of the surgeons that lead to medical errors [118]. Hence, the surgeon's workload should best stay in the area of optimal performance, which is located between high and low workload [119]. To monitor the surgeon's workload during the surgery, real-time tools are required. A systematic review on measurement tools developed in [120] shows that most studies use heart rate variability or eye tracking for this purpose.

A prerequisite for visualizing navigation information is registering the preoperative patient data with the intraoperative scene. One of the most challenging aspects regarding registration are soft-tissue deformations due to the dynamic behavior of tissue. A comprehensive review regarding registration approaches for minimally invasive surgery can be found in [121,122]. In addition, data-driven methods to learn soft-tissue deformation by combining deep learning and simulation methods are a promising approach [123,124].

2.4.1.2 Input modalities and interaction techniques

Though the listed research examples have shown the diversity of assistant systems, the field of Human Computer Interaction still offers a lot of potential that can be tapped. 2D displays, such as screens displaying the endoscopic video stream, could be extended with three-dimensional content using an approach, such as the augmented displays demonstrated in [125]. Not only presentation but also interaction can be enhanced by integrating innovative user interaction techniques. [126] use tangible windows to explore in three-dimensional space, while in [127] the authors go even further by using spatial interaction with mobile devices for this purpose.

Different input modalities have been considered for interacting with digital content in the OR. [128] introduced hands-free navigation of larger information spaces through gaze-supported foot input. [129] already identified the opportunities of hands-free interaction and adopted the system to the OR. To maintain sterility in the OR, [130] produced an application of Natural User Interface (NUI) devices for touch-free control of radiological images during surgery. In this context, they compared the accuracy, utility, and usability of Leap Motion and Microsoft Kinect v2 controllers with computer mouse, according to surgeons' and radiologists' estimates.

To make interaction more tangible, the authors of [131] developed Tangible Organs. They evaluated *the possibilities of 3D-printed organ models for interaction in VR for the use case of surgery planning*. The Tangible Organs can be used to assist the surgical planning phase. On the other side of the spectrum, *telemanipulation* has been explored in [132]: in a field study, they investigated the impacts of telemanipulated surgical robots on the work practices of surgical teams, concluding that *physically separating surgeons from their teams makes them more autonomous, shifts their use of perceptual senses, and turns the surgeon's assistant into the robot's assistant.*

2.4.1.3 Transfer of surgical skills to robotic platforms

Robotics has great potential to improve surgical outcome by assisting surgeons in high-dexterity and high-precision tasks or reducing workload by automating whole subtasks.

Robotic assistance targets specific human limitations, while leaving high-level control to the surgeon. For example, vitreoretinal microsurgery is extremely challenging since it requires high-precision operation on a microscopic scale. [133,134] propose a cooperatively-controlled robot assistant. The robot replicates the surgeon's movements, while actively compensating physiological hand tremor. [135] extend this idea using a handheld manipulator with incorporated tremor removal, which eliminates the need for a static robotic platform. Furthermore, virtual fixtures are incorporated, which constrain the instrument to stay within safe positions, motions, and velocities [136].

Another application for enhancing the surgeon's precision can be found in prostate cancer treatment, where a needle is inserted into the prostate to deposit radioactive seeds on precise targets. [137] propose a robotic assistant, which leaves the pace of insertion to the surgeon, but corrects the needle trajectory to enable more precise targeting.

During heart surgery, motion caused by heartbeat and respiration poses a great challenge. Several robotic platforms have been proposed for compensating these motions and giving the surgeon the perception of a stable environment [138–140]. Heart and robot motion are automatically synchronized, while the surgeon manually controls the distance between tool and tissue. On tool-tissue contact, forces can be controlled according to physiological motions to ensure safe operating [140]. As an alternative to active motion compensation, [141] propose a teleoperated, organ-mounted robot to eliminate the need for tracking physiological motion.

Robot assistance has further been used to enhance ultrasound imaging. For applications in ultrasound-aided breast cancer diagnosis, [142] minimized the force applied to the tissue by a manually operated robotic ultrasound probe and control the probe to remain perpendicular to the tissue. This enhances the repeatability of scannings. [143] equip the daVinci robotic system with an ultrasound probe for intraoperative elastography. The scanning trajectory is teleoperated by the surgeon, while the vertical palpation motion is automatically overlaid by the robot. Elasticity images compensate the loss of haptic information in laparoscopic and robotic surgery.

In contrast to robotic assistance, *task autonomy* aims at fully automating surgical subtasks. Envisioned benefits are the reduction of the surgeon's workload or enhanced repeatability of certain substeps of a procedure.

Autonomous endoscopic camera guidance has been investigated in several studies to provide the surgeon with better vision during laparoscopic surgery. In current approaches, a robotic arm fully automatically controls the camera position based on instrument positions, the current surgical phase or previously recorded instrument trajectories [144–146].

Knot-tying is a frequently needed surgical subtask, but is tedious in minimally invasive procedures. Therefore it has been a popular problem for task automation and has been approached by learning from recorded demonstrations [147–151]. [147] use concepts from fluid dynamics for collision avoidance, while [150] exploit the robotic setting to perform knot-tying at superhuman speed.

Similarly, [152] and [153] learn scanning trajectories from demonstrations and transfer them to a robotic arm to increase repeatability during ultrasound or microscopic scanning.

Further preliminary research on task autonomy in surgery includes cutting and removing tissue from the patient [154,155] or intraoperative blood suction [156] using reinforcement learning. However, these learning-based approaches were applied to simple environments, where target locations were either known or easy to identify. In the approaches, which learn from demonstrations, instrument trajectories are merely imitated and likely not transferable to slightly different settings. Aiming at task autonomy in more realistic, dynamic environments would be a huge step towards usage in the OR.

2.4.2 Human–machine collaboration in the Sensor-OR

TaHiL-related research aims at providing computer and robot assistance, which is able to adapt to the complex and dynamic environment during a surgical procedure, i.e., providing information when it is most useful to the surgeon and planning safe robotic intervention with regard to the nonrigid, vulnerable surroundings. This requires real-time control and low-latency communication networks, online analysis of sensor data with machine-learning methods as well as near-to-eye information display with appropriate visualizations. The approaches here build upon the captured skills from the skill library outlined in Section 2.2.

For visualization of navigation information and skill automation, several subtasks have to be solved to assure effective assistance, which does not disrupt the surgeon or harm tissue. Advances in computer vision and deep learning are promising for solving many of these problems using image data from the laparoscope. Determining when robotic assistance is required can be approached by learning from recorded surgeries. Object detection and segmentation methods can help identifying targets or relevant structures. Methods for 3D reconstruction in images as well as registration are a prerequisite for safe trajectory planning and soft-tissue navigation.

Regarding soft-tissue navigation, we aim to learn how tissue deforms by combining deep learning and simulation approaches. The result is a data-driven deformation model that estimates a displacement field of all points inside an organ when given only the displacement of a part of the organ's surface [123]. This is a prerequisite for registration, an essential part of soft-tissue navigation.

A prerequisite for a task-autonomous system is to recognize at what time during surgery robotic assistance is required. We aim to learn the timing of certain tasks from recorded surgeries and use this information to trigger the robotic system to intervene. Specifically, we learn to anticipate the usage of surgical instruments, which are dedicated to specific tasks. Example, by learning to anticipate the usage of the irrigator, which is often used to remove blood from the intraoperative scene, we hope to learn visual features of strong bleedings, which indicate that robotic assistance is required.

For the application of autonomous blood suction, we further require visual recognition of hemorrhages and bloody regions in the intraoperative scene. We investigate segmentation methods for identifying bloody regions with a focus on methods which require less data annotation. Since blood often has homogeneous visual appearance, this task has great potential for learning from incomplete or high-level annotations, such as scribbles or class labels [157,158].

Planning safe and nondisruptive robot trajectories to a target requires knowledge about the intraoperative 3D environment. Hence, depth estimation in laparoscopic images is essential for navigation [159–162]. We can further benefit from visual recognition tasks. The robotic system requires awareness for vulnerable tissue and the positions of other surgical instruments, which can be identified using image segmentation [163] or object detection [164] methods.

Effective human–machine collaboration and context-aware assistance requires novel means of user interfaces within an OR as well as novel AR visualization approaches. Based on our advances in developing task-autonomous systems and robot assistance, the interface between human operators and technical OR systems has to be designed. Though some research has already been conducted on novel input modalities during surgery (including gestures, controllers, gaze, or foot input), it still remains an open question, which one to use in a specific situation and how to adapt them to the current workflow and stress level of the surgeon. We will also investigate the usage of wearable and tactile input devices in this context.

To keep the surgeon in the loop, decision made by the robot assistance needs to be appropriately communicated to the people in the OR. We have to design effectively when, how, and what level of detail information has to be presented to surgeons to not distract them, yet provide sufficient details about automated decisions.

For the visualization side, we want to investigate effective means of providing near-to-eye information, which is augmented to the surgeon. This includes the comparison of pure AR displays as opposed to alternatively or synergetically using existing OR displays (such as for displaying medical imagery) as a primary step. Novel wearable smart textiles (such as the digitally enhanced doctor's white coat mentioned in [165]) shall also be considered as possible solutions. Secondly, the way how the information—such as surgical navigation information—is provided to the surgeon as required by them also needs to be carefully designed from a visualization perspective. Here, the usage of gaze data gathered from augmented reality glasses might help to better determine information needs and to optimize the layout.

2.5 Conclusion and outlook

The medical use case aims to improve the safety, quality, and efficiency of patient care by capturing clinical expertise and augmenting clinical performance in the context of computer- and robot-assisted medical therapy and training. The use case builds upon several key technology breakthroughs and primary fundamental research in the context of TaHiL that are related to all three objectives. This includes knowledge of psychological- and age-related processes relating to multisensory integration, learning, cognitive control and action (see Chapter 9) that not only enable the modeling and understanding of skills, but also ensures an intuitive human–machine collaboration by incorporating human factors. Sensor-based modeling and teaching of surgical skills in clinical as well as VR/AR training environments requires novel sensor-actuator approaches, e.g., wearables for natural touch, coherent optical feedback, and realistic sound taking low latency, low weight, and high quality of user experience into account (see Chapter 10). In addition, multisensor acquisition (haptic, vision, audio) leads to a vast amount of sensor data that requires efficient solutions for data compression, especially for haptic data (see Chapter 5) as well as data transfer regarding robot-assisted tasks (see Chapter 6). To support perception and processing of provided information by the human, multimodal interfaces can be a promising solution (see Chapter 7). To complement the machine-learning-based approaches to predict surgical actions, novel reasoning techniques that compete with latency constraints and feedback requirements and mimic human decision-making are investigated (see Chapter 8).

Human–robot cohabitation in industry

3

Uwe Aßmann[a], Lingyun Chen[b], Sebastian Ebert[a], Diana Göhringer[a], Dominik Grzelak[a], Diego Hidalgo[b], Lars Johannsmeier[b], Sami Haddadin[b], Johannes Mey[a], and Ariel Podlubne[a]

[a]*Technische Universität Dresden, Dresden, Germany*
[b]*Technical University of Munich, Munich, Germany*

Be self aware, rather than a repetitious robot.
— Bruce Lee

3.1 Introduction

The emergence of safe and collaborative robots is currently changing the way our workplaces are designed and structured. State-of-the-art robotic systems are capable of complex manipulation tasks and intuitive human–robot interaction making them potential coworkers (*cobots*) in many manufacturing scenarios *(human–robotic cohabitation in industry)*. This new trend of collaborative industrial workspaces (*cobotic cells*) is motivated by various factors, such as the growing general durability and precision of robots, the cost-reduction potential for assembly and production, and the discovery of new application areas, such as remote work in dangerous areas.

Although there has already been much research on the design of collaborative workcells and robots, there are still major challenges to overcome. One such challenge is to give robots all the required skills to be useful in a given workplace (*skill learning*). The way state-of-the-art robots are programmed is very different from the tedious and time-consuming processes of earlier systems. For example, *kinesthetic teaching* is a more and more adopted approach to program collaborative robots, enabling much more intuitive and efficient programming schemes by directly guiding the robots. Another challenge is to integrate all the required software and hardware into small mobile and wearable components that are, nevertheless, performant and energy-efficient. Furthermore, in the Tactile Internet, the robots may be connected with each other via high-speed communication. Perhaps the most crucial factor is safety and reliance when robots are deployed in collaborative scenarios. This means not only the safety of all humans in immediate proximity of robots, but also the safety of the robot itself and its environment. The development of novel reference architectures and design paradigms is making progress with the aim of making modern robots inherently safe.

Tactile Internet. https://doi.org/10.1016/B978-0-12-821343-8.00013-7

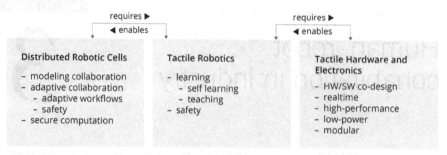

Fig. 3.1

Software- and hardware-related building blocks of distributed adaptive cobotic cells with avatar stations.

3.1.1 Use-case scenario: Distributed cobotic cells

In this chapter, we discuss a specific use-case of tactile robotics that is based on the current research on collaborative robotics, telepresence, and miniaturized mobile hardware. We consider the case that a group of human workers in a factory requires help from an expert to solve a specific problem, yet that expert is at a different location (called *distributed cobotic cell*). For example, pandemic situations can prevent physical contact between the expert and the workers in the factory, so that the expert has to work from remote.

To this end, the expert should be able to log into one of the robots at the factory and take full control of it. Here, *robotic immersion (robotic telepresence)* is required, the ability to interact with the workcell via an avatar through the Tactile Internet [166]. With robotic immersion, the expert can send commands directly to the remote robotic avatar and perceive the robot's sensor input, i.e., through audiovisual as well as tactile feedback. Ideally, this avatar would give the illusion of being directly at the remote site *(immersive cobotic cell)*. Using the avatar body, the expert can then support the factory workers remotely. This scenario consists of several components and challenges that are discussed in the following sections.

The essential building blocks to move from the above specification to an actual implementation are illustrated in Fig. 3.1. For their explanation, we first discuss the architectural design of distributed cobotic cells from a software engineering perspective (Section 3.2), which will impose new challenges owed to the collaboration application scenario. Specifically, modeling adaptive behavior with respect to collaboration in distributed environments is far from trivial [167]. Furthermore, Tactile Robotics platforms are required (see Section 3.3), which in turn require modular hardware-software codesign with real-time capabilities (Section 3.4). The Tactile Internet provides the underpinning for Tactile Robots, which are characterized by enhanced abilities to perceive tactile information, thus allowing human–robot interactions in near real-time at another level, compared to today's industrial robots fenced in safety-cages.

Before we continue, we shall give some clarification of both the terms cobotic cell and avatar station that represent the main physical entities being realized for the implementation of our use case.

3.1.2 Categorization of cobotic cells

Not surprisingly, we find that the definition of classic robotic assembly cells commonly used in the robotic literature [168,169] to be slightly in disagreement with cells defined in more human–robot interaction/collaboration-related literature [170–173]. The former tend to regard robotic assembly cells in the sense of static entities assigned to one specific task, more or less located within a cage, where humans are absent. The latter, the collaborative robotic cell, is a product which operates with a given feature set to provide value to its customers [168] and is thus considered a "value-adding" process [168].

Cobotic cells as collaborative robotic cells

In accordance to the literature, we assume human interaction to be unconditionally involved in the manufacturing process in a collaborative robotic cell. Though a distinction can be made between a *conventional collaborative manufacturing cell* and a *hybrid manufacturing cell*, both involve human–robot interaction in a shared space to carry out a collaborative task. For example, as stated in [170–172], a hybrid manufacturing cell has the primary motivation to minimize the cost of resources by adopting the human flexibility to carry out specific tasks that are too difficult for a robot to perform. In hybrid cells, the workspace between robots and humans is explicitly shared, and direct interaction with each other is desired. To give an example, the hybrid assembly cell is regarded as a workshop "where the semi-finished products are assembled by humans and robots together" [171, p. 1067]. In contrast, in a conventional collaborative manufacturing cell, the human and robot work alternatively [171], hence, "each cell [is] responsible for a complete unit of work" [174, p. 477]. This implies also a distinct separation of the workspace between both coworkers, robots, and humans. We can observe that this strict separation may also be nullified to consider disruptive nonpredictable events, such as path intercepts as investigated by Unhelkar et al. [173] in the course of performing a logistic-related task (e.g., collecting items from a depot, which the robot fills). Such work infers the inclusion of human-aware motion planners solely for safety reasons, but not for interactive collaboration.

Nevertheless, in both forms of manufacturing cells, the robot is aware of the human to varying extents, whether it be simple collision detection and path replanning, or direct interaction, adaption, and optimal rescheduling (more sophisticated approaches concerning interaction) to adapt to the continuously changing human behavior and performance, or unexpected internal/external events.

In conclusion, we can give an informal definition of a cobot and a cobotic cell with reference to [175]. *Cobot* are robots tightly collaborating with humans; this is realized in a *cobotic cell*, which is a hybrid cell without a safety cage. Therefore these robots must be aware of human movements and autonomously adapt their behavior to prevent accidents with humans or other robots. In our use-case a cobot could be either a human-controlled or an autonomic working robot.

Distributed cobotic cells and avatar stations

In addition to cobotic cells situated in one location, the Tactile Internet also enables their distribution. Two stages of distributed human–robot coworking are discussed, the spacial distribution of sections of a cobotic cell as well as immersive control of a remote robot using avatar stations.

Distributed cobotic cells A cobotic cell can be local, so that workers can see all robots and work pieces. However, there are industrial scenarios for *remote work*, in which experts in remote locations have to inspect a cobotic cell *(remote inspection)* or certify the quality of a product from remote *(remote certification)*, and other experts have to maintain a machine in the remote cell *(remote maintenance)*, or in which they have to repair a machine remotely *(remote repair)*. Furthermore, more complex scenarios with both experts and robots spread across different locations are possible. In all these situations, a *distributed cobotic cell* is needed, which couples at least two subcells in different locations over the Tactile Internet. Such distributed cobotic cells must guarantee the same functional and nonfunctional properties as local cobotic cells. Due to the increased complexity of a distributed system, more advanced concepts for design, programming, and error handling must be used.

Avatar stations in immersive cobotic cells As outlined in the introduction, a cobotic cell may incorporate avatar stations. Then, we speak of an *immersive cobotic cell*. An avatar station provides both motion-tracking and manipulation mechanisms to control a remote robot avatar and sensory feedback to the operator. Thus the avatar station is used as an entry point for a human user to *immerse into* the remote robotic body located in the cobotic cell. The aim is *to completely immerse* the users, i.e., give them the illusion as if they were physically at the remote location. In Section 3.2.3, we will present a reference architecture for immersive cobotic cells.

With the development of an anthropomorphic robotic hand–arm system, we are building on novel concepts and aiming to fully project the user to a remote location, at an appropriate level for industrial use cases. To immerse the user, the robotic system has to be able to provide proper tactile and audiovisual feedback to the user and must possess real-time responsiveness. Such specifications present not only a major challenge to the communication infrastructure, but also to the technologies used to establish the impression of remote presence. High-performance wearable technologies, capable of exerting forces, pressure, and temperature as well as measuring the user's state (e.g., the arm pose), will be needed.

3.1.3 Applications

This section discusses concrete industrial applications of distributed cobotic cells, which are variants of *remote work*.

In chip factories, cleanrooms are evacuated of all dust particles to create favorable conditions for wafer etching, layering, and cleaning. Often, wafers must be transported between wafer-processing machines. When wafers cannot be unloaded from a machine, for instance, because a wafer sits in a somewhat incorrect position, the wafer machine should be inspected from outside the cleanroom, to find the flaw. If, alternatively, the cleanroom is opened, it gets dirty and has to be decontaminated, which means to produce thousands of wafers in vain until the last dust particle has vanished from the wafer-processing machines. Opening a cleanroom is very expensive, so *immersive inspection* is desirable. In the same scenario, besides an immersive inspection robot, a remote repair robot moving the faultily-positioned wafer into a correct position, would be desirable. This is *immersive remote repair*, and because the required movements are quite delicate, immersion into the scene is indispensable.

These examples of remote work need distribution and immersion, but not coworking. In chemical factories, dangerous work situations appear, when dangerous chemicals have to be handled. Some processes are more safely operated by robots than by humans. Others can be operated by humans, but need remote inspections from experts that are not physically present. Remote experts might even, under immersion, teleoperate a robot in a lab. Such a robot should protect all other chemists in the room, requiring a distributed cobotic cell with immersion.

Many European companies from the mechanical industry deliver complex machines to foreign markets. Such machines usually underlie a maintenance contract, and, regularly, inspection and maintenance teams visit the remote places to control and repair the electronics, the hydraulics, and mechanics. Whereas, for example, the aircraft industries have managed to measure, trace, and control aircraft turbines during their entire lifetimes, the mechanical industry is far away from this sophisticated level of operation. However, every complex machine in a remote country could form a distributed cobotic cell, together with a maintenance and control subcell in Europe. Such distributed cobotic cells need to trace the movements of all involved humans, offer immersion for inspection and certification, offer remote robotic operation, as well as safety mechanisms so that nobody is harmed. Because maintenance is a major cross-cutting business through many industries, remote maintenance offers a potentially huge market for distributed and immersive cobotic cells.

3.1.4 Outline

The remainder of this chapter discusses research challenges and direction from different perspectives. First, an architectural framework for human–robot cohabitation is proposed using techniques and approaches from software engineering, *model-based cobotic cells*. Then, requirements and challenges of tactile robots are discussed, with a focus on both the specific hardware requirements for a tactile robot avatar and the software requirements to control such hardware. Finally, since the computations required for tactile robotics and avatar stations have highly demanding time and performance requirements, hardware acceleration for the cobotic architecture and use case is introduced.

3.2 Model-based cobotic cells

This section discusses how a model-based approach can facilitate the construction and operation of cobotic cells, considering aspects such has safety, adaptivity, and extensibility. After an introduction to the state of the art, research challenges are defined and several directions to go beyond the state of the art are presented.

3.2.1 State of the art for cobotic cell architectures

Models and components fundamentally simplify the development of systems, especially also of cobotic applications.

Models, abstraction, and model-driven engineering (MDE)

A model is an abstraction of a system or of the real world concentrating on specific structural or behavioral properties and representing them in a syntactically and semantically defined language [176, pp. 107–108]. Systems and therefore cobots can be developed and tested based on such models exploiting their ability to abstract. By abstracting away some details of a system in its model, the complexity of the modeled system is hidden, thus enhancing the understanding of the system. For example, verifying the state space of a cobotic cell on a micro level seems impossible because of state space explosion. However, focusing the verification on selected abstractions is possible [177]. In this respect, a suitable level of abstraction, an appropriate *view* on the system must be determined. A too detailed model may not only be limited to a single use case, but also be hard to construct, because of time constraints or high complexity [178]. However, a too generic model may not provide the required expressiveness. Thus the selection of the right level of abstraction is essential. In the following, several modeling techniques relevant for cobotic cells are introduced.

Object-oriented modeling Object-orientation has been used as a programming paradigm for several decades, but has also evolved over time into a modeling paradigm [179]. Examples for object-oriented modeling languages are UML [180] and SysML [181]. The Unified Modeling Language (UML) is a modeling language, which provides a standardized way to visualize the design of a software. SysML is a UML-extension for the modeling of systems, including their hardware and software [176, pp. 107–108]. Object-oriented modeling can be regarded as a specialized graph-based modeling approach, since models are graphs, comprising classes as nodes with some attributes, connected with inheritance and other relations. For a set of models, a *metamodel* can be given, a graph of metaclasses and their relationships. Also, the *instance model* of a model is simply an object graph. Whereas this approach offers great flexibility, the lack of restrictions on the instance level can be problematic, particularly because it complicates modulation, analysis, and verification (see Section 3.2.2) of context-dependent behavior [182]. This is due to the fact that object-oriented models lack features for context-adaptivity [183], which is important for modeling cobotic cells.

Process-based modeling Another category of models situated at a more formal level needs to be employed to prove custom- and domain-specific properties regarding the operational behavior of cobotic cells. For this purpose, semantic models of distributed and parallel systems, such as *Petri nets* and *bigraphs*, have proven to be valuable frameworks. A *Petri net* [184] can be used to formally model concurrent, asynchronous, or nondeterministic properties of a system. Process calculi, such as *bigraphs*, are employed to model reactive systems [185,186], offering a very high level of abstraction by taking a process-only view and enabling the reasoning about interacting processes [186]. Both forms of models are thus suitable to describe and prove certain properties of safety models for cobots and telepresence.

Hybrid models Additionally, many aspects, such as safety or quality, cannot be sufficiently described with the aforementioned discrete models, but require other approaches, such as continuous, dynamic models. Thus some modeling approaches combine dynamic, discrete, and continuous modeling languages, for instance, hybrid Petri nets [187]. For robotics, specific modeling languages exist that further integrate real-time constraints (real-time modeling), or enable traces between specification models and implementations (traceability).

Model-Driven Engineering (MDE) MDE aims to automatically generate executable software from the aforementioned models [188]. Two techniques are important, *model transformation* and *code generation*. For model transformation, several variants of graph transformation techniques are used, such as Triple Graph Grammars [189], ATL [190] or Epsilon [191]. The second technique, code generation, uses template engines or code snippet composition [192]. For the implementation of Domain-Specific Languages (DSLs), *language workbenches* are used, such as Xtext [193], Spoofax [194], or MontiCore [195].

Multimodeling, dimensional, and aspect-oriented modeling For large software specifications, such as for cobotic cells, several structural and behavioral models have to be specified being related to each other (*multimodel*). Often, the models of a multimodel overlap in their semantics and are hard to maintain, because the overlaps easily lead to inconsistencies. If a multimodel maintains consistency constraints specifying how its models should be synchronized, it is called *macromodel* [196]. Macromodels are very important for the specification of complex cyber-physical systems, also robotic systems [197], because they can handle complex aspects of the system uniformly and support their consistent evolution.

One important macromodeling technique is *dimensional modeling*. Approaches, such as Orthographic Software Modeling (OSM) [198], divide the system model, the *single underlying model*, into a number of *dimensions*, each abstracting on a specific *concern*. Usually, the *view in the specific dimension* is much simpler to understand, because it is *viewpoint*-related and problem-specific. An OSM environment keeps all dimensional models consistent with all others and the single underlying model. These techniques offer huge opportunities to evolve complex software systems.

A similar approach is *Aspect-Oriented Modeling* [199]. In this approach, large models can be decomposed into *core models* and *aspect models*, so that aspect models can be reused many times, like a style sheet, and their definitions may crosscut many core models. This, the relationship of the core and the aspect models is not symmetric, as in dimensional modeling.

Dimensional- and aspect-oriented modeling can be extended to runtime models *(Models at runtime, MaRT)* [200]. Then, several models are maintained, inspected and analyzed at runtime of the software system, and their integration is managed via crosslinks of the models. We speak of a *runtime macromodel* if a macromodel is maintained at runtime based on its crosslinks and synchronization constraints.

Components and CBSE

With Component-based Software Engineering (CBSE), systems are assembled from decoupled, independently developed components [201]. Components are encapsulating sets of related services (functionalities or ports) and data, hiding their contents behind *provided* and *required* interfaces. These interfaces decouple components from each other so that they can be easily reused, customized, and exchanged. Components can encapsulate any kind of software package reaching from web services, to source or binary components, as well as to robotic software modules. Therefore CBSE can reuse components on different levels of abstraction, improving the reuse of already tested or verified components and thereby reducing software development costs. Also, variants of components can be developed and assembled, resulting in different variants of the product *(software product-line engineering, SPLE)*. Furthermore, components are able to separate different software concerns, so that they support dimensional and aspect-oriented development. Using a model-based approach on cobotic cells in combination with component-based techniques is, nevertheless, a huge challenge.

The principles of CBSE are already used in many systems today. Firstly, this concerns applications that are not designed for a specific domain. Microservice architectures are an example of this. Microservices are independently deployable and loosely coupled services, each focusing on a specific task [202]. Thus they are components, communicating based on network protocols and composed based on a business-logic. This also affects systems from the field of robotics. Secondly, in a component-based system for cobotic cells, components may comprise a hardware part, such as a component of a robot or a tracking device. The Robot Operating System (ROS) is a software framework, which facilitates the development of robotic applications. It comprises both components to access and control robotic hardware and pure software components, thus bridging the gap between the software and the physical layer of a cobotic cell. Section 3.4 introduces techniques to construct ROS components using reconfigurable hardware.

Self-adaptive techniques for cobotic cells

A cobot can work in different contexts, for example, with or without humans, which increases a system's complexity and decreases its extensibility. Thus context-adaptive behavior is required on modeling, programming, and on the architectural level.

Adaptive models A runtime model is called *(self-)adaptive*, if it contains several variants that can be alternated at the runtime of the software incorporating the model. There exist adaptive extensions for process models, e.g., Petri nets, so that they can describe reconfigurable software and hardware structures [203]. Adaptive models are important for safety-critical systems, because different basic modes of the software, such as safe and nonsafe modes, can be realized by alternating the variants of an adaptive model.

Context-oriented modeling and programming Consider, e.g., a situation in which a human operator is approaching a cobot; then, its behavior must be adapted according to the changed context, so that the operator is not endangered. Such context-adaptive changes can be realized using various techniques. In Context-Oriented Programming (COP) [204], the execution of a specific part of a program does not only depend on the name of a method, but also on its sender, receiver, and an execution context. COP uses a very general definition for contexts, i.e., a context is everything that is computationally accessible. To use a more structured notion of context, Role-Oriented Programming (ROP) [205] can be used. ROP is a programming paradigm, in which objects can play *roles* to adapt their properties and their behavior dynamically. This is achieved by setting up and detaching relations between objects and roles at runtime, depending on the presence of specific *context objects* [206]. This means that roles are able to adapt sets of related objects to context changes.

Self-adaptive systems, autonomic systems, and their architectures Both COP and ROP allow adaptations on a programming-language level. In contrast, a Self-Adaptive System (SAS) provides an architecture adapting an application to context changes. A self-adaptive system may maintain so-called *models at runtime* to decide on adaptations [207]. If a complex runtime macromodel is involved, we call it an *autonomic system*.

The *MAPE-K* pattern is an architectural pattern for controlling self-adaptive and autonomic systems. An *autonomic controller* is based on a control loop with four phases *monitor*, *analyze*, *plan*, and *execute* as well as a shared *knowledge* base [208]. In the *monitor* phase, information about the environment or the current context is obtained from sensors or connected systems. The second *analyze* phase processes and analyzes the obtained information. A plan for required actions, such as moving a robot, is computed in the *plan* phase and executed in the *execute* phase. Therefore the autonomic controller is able to adapt to context changes. Systems based on the MAPE-K pattern are examples for self-aware computing systems [209], due to the fact that they reflect about models representing themselves and their environment.

Runtime macromodeling for cobotics In software engineering for cobotics, models and components should be combined to harvest the advantages of MDE and CBSE. The reference architecture proposed in Section 3.2.3 builds on a runtime macromodel to represent all necessary details for a cobotic cell, as well as its distributed and immersive variants.

3.2.2 Research challenges

In this section, we discuss the research questions and related issues for the design of cobotic cells. We identify the types of models being suitable for cobotic cells, distributed and immersive cobotic cells, as required by our use-case.

For each concern in a cobotic cell, there exists a distinct category of models particularly effective in addressing that specific concern. Hence, *multimodeling* seems to be a viable approach, the seamless combination of various models to resolve the disadvantages of individual ones. In a multimodel, different views on models or systems are associated with each other. This raises a research question concerning the sort of *multimodel* to choose:

RQ 1.1: What are suitable models to collectively describe the structure and behavior of a cobotic cell at design and particularly at runtime? How do they differ in their content, context, and modeling language, and how are they integrated into a multimodel?

If humans and cobots shall collaborate, models must possess a degree of dynamism to adapt to the changing requirements at runtime. In this context, the communication between models is essential because, although they can represent different aspects, these aspects refer to the same world and must therefore relate to each other to guarantee the coevolution of models. In line with this, *macromodels* can support their consistent coevolution. This leads us to the following research question:

RQ 1.2: What specific form of macromodeling is well qualified to support dynamic model evolution for cobotic cells?

Analogous to the requirements for modeling, modularity is important for cobotic cells. First, there are different variants both of software and hardware components that share a common structure. Secondly, components should be *context-dependent* as they exhibit different behavior based on the context they are in. Finally, in a *distributed* cobotic cell, components with adequate distribution and synchronization mechanisms must be used that allow efficient communication. This leads to the next research question:

RQ 1.3: How can the macromodel of a cobotic cell be constructed in components, so that adaptivity, reusability, and efficient distribution of both software and hardware components is supported?

To program a cobotic cell, the application, the coordination, and the robot control code has to be produced, either manually or using code generation with model-driven engineering (MDE). However, whereas macromodeling supports views for different software development phases, such as the specification of requirements, the design, the test cases, the documentation, as well as continuous redeployments in the dev-ops life cycle, MDE for macromodels needs to be developed first. This leads to the next research question concerning the software development process:

RQ 1.4: How to employ model-driven engineering (MDE) for the dynamic macromodels of cobotic cells that supports all standard phases of a software development process (requirement analysis, design, test, verification, and implementation) as well as runtime and redeployment (dev-ops)?

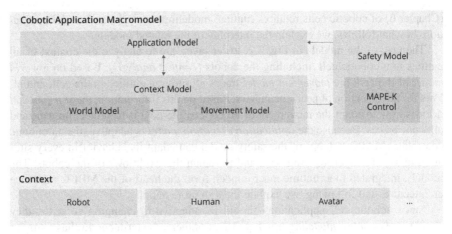

Fig. 3.2

An architecture model for cobots, including data flows.

Though these research questions are still open, we can already present a reference architecture for cobotic cells building on the above-mentioned principles.

3.2.3 Research directions

The cooperation between humans and cobots requires a reference architecture for cobotic applications, supporting the requirements of Section 3.1 and answering major aspects of RQ 1–4. This section discusses a new approach, the Model-Based Cobotic Cell (MBCC) (see Fig. 3.2). Also, there are promising research directions for a component-based and model-driven cobotic architecture, which will be presented afterwards.

A software reference architecture of model-based cobotic cells

The MBCC reference architecture consists of a runtime macromodeling approach with (*i*) a runtime multimodel for context analysis of the robots (world model, movement model), the *afferent model layer*, (*ii*) a context-adaptive architecture, possibly with multiple layers of adaptation, the *efferent model layer*, and (*iii*) a central controller of adaptation (safety model), belonging to both layers.

We suggest a macromodeling approach, because the software of cobotic cells is complex to understand and manage. The collaboration of humans and robots within cobotic cells not only carries the individual inherent complexity of modern robots [210], but also adds the complexity of a distributed, even immersive system. Macromodeling for cobotic cells, however, must rely on runtime macromodels, runtime models integrated by crosslinks. The consistent treatment of critical aspects, such as safety and security (Chapter 13), throughout the software-level to the hardware-levels, of sensors/actuators (Chapter 10), haptic codecs (Chapter 5), and networking

(Chapter 6) of cobotic cells requires runtime modeling, analysis, and model integration. In what follows, we explain the interplay of the related models.

The task of the models in Fig. 3.2, lower layer, are to analyze the context of all entities in a cobotic cell, including the cobots *(context analysis)*. Based on a *world model* of the cell, the *movement model* traces all moving entities in the cell, and allows for the detection of near-encounters. Based on this information, the *safety model* decides how to adapt the movement of all cobots; the goal is to protect humans and all work pieces. Because the *application execution model* is adaptive, the autonomic controller (MAPE-K loop) of the safety model can adapt its behavior in every situation to avoid dangerous situations, and to replan the workflows of the cobots. The models, integrated to a runtime macromodel, form the heart of the MBCC reference architecture, and following, we explain them in more detail.

Since cobotic cell applications should be adaptive to changing contexts, they should apply *context modeling* (see Fig. 3.2). For them, the term *context* has a similar meaning as in COP (Section 3.2.1). In general, a *context* covers any information that allows the characterization of an entity's situation [211]. In the case of cobotic cells, an entity is everything that is considered to be relevant to the interaction between humans and robots, including cobotic applications themselves. Hence, analyzing the space–time information about these entities, *context analysis*, is needed for an appropriate reaction to guarantee safety conditions.

The *world model* describes a global model of the environment, a basic prerequisite for context analysis. A cobot needs a representation of the surrounding entities, thus a cobot is able to reason about and to interact with these entities. The world model describes the states of the world, in which the robots and humans are co-working. Additionally, all states of the relationships between physical entities and their contexts are described. Thus aspects, such as the current performed actions, as well as the position of an entity, can be described. A world model can be realized based on concepts for safety zones, such as voxels or scene graphs [175]. In our use-case, for example, the cobots must know where humans and other robots are situated, how far they are from each other, or whether they are in the same zone of the world model. Especially techniques from the essential building blocks for Tactile Internet applications discussed in Chapter 13 will be used to capture the cobot's surroundings and reason about them. In case a cobotic cell is distributed, the world model is divided into several interlinked world submodels of the distributed subcells. In particular for immersive cobotic cells, the operator cell of the avatar must be modeled as separate submodel, too.

The *movement model* is related to the world model and allows for the computation of future positions of entities, a basic prerequisite to forecast the *change of contexts*. To this end, the world model has to provide the inputs, such as current positions for the movement model, as well as possible and impossible further movements. The movement model may contain a set of continuous equation systems for computing the dynamics of entities, for example, their velocity, direction, and future position. The movement model may combine this with a process model taking up signals from the world model, so that, for example, movements of entities are described by hybrid specifications, such as hybrid automata or hybrid Petri nets [212]. Thus in our use-case a remote controlled cobot is aware *if* a human or other cobot is approaching him, and *when* the encounter will happen so that collision avoidance can be initiated.

Next, a *safety model* describes how cobots achieve protecting humans or other robots. This can be specified by *safety state machines*, i.e., state machines that contain dedicated safety states and transitions between them [175]. An example for a safety model is the safety automaton by Haddadin [213], which contains the basic states of the robot's behavior *autonomous working*, *carefully working in human presence*, *collaborating with human*, and *safety stop*. Based on such a safety model, a cobot can be forced to change to a more careful work mode, depending on how near a human gets to it—eventually, even stopping it. Because humans rely on safety, the safety model must account all possible emerging interactions within a cobotic cell. Indeed, both planned interactions and unforeseen interruptions between many cobots and workers lead to an increased behavioral complexity in cobotic cells that must be properly understood at the design phase prior to the implementation to alleviate as many risks as possible. This can be done based on appropriate testing and verification technologies presented in Section 3.2.1. However, many challenges remain for safety models. For instance, in an environment, in which cobots are mobile and humans can move freely, it is a challenging task to trace the movements of humans and robots. Furthermore, in an immersive cobotic cell, a remote-controlled cobot should be slowed down or stopped if a human or another cobot is approaching it. This requires that the entire macromodel is distributed and treats the immersion.

A safety automaton, however, is only a specific instance of a safety model. In general, the safety of all entities in the cobotic cell should be maintained by an autonomic control loop (MAPE-K loop). To this end, the MAPE-K loop measures (M) and analyzes (A) the context of all entities by querying the world model and the movement model. Additionally, it plans adaptation actions (P) and executes the adaptations (E), while storing insights about the world in its Knowledge base (K). Many other modeling techniques can be used for such autonomic safety controllers.

Adaptive software layer. The top layer in the MBCC application architecture is occupied by the *cobotic adaptive application model*. This model typically consists of a structural model of the application data, as well as a process model of a collaborative and autonomous workflows for human–robot collaboration. To realize adaptation, the application model comprises several variants, which can be activated and configured by the planner of the MAPE-K loop of the safety model. Also, the safety model has to provide the interfaces and communication capabilities to escalate error events to the application layer for adaption. To this end, the safety model is linked with the models in the context analysis layer.

For cobotic applications, the MBCC approach masters their complexity based on the technologies presented in Section 3.2.1. First, the use of a macromodel supports separation of concerns. Second, using a runtime macromodel, the runtime state space is clearly decomposed into several dimensions. Third, the macromodel specifies by crosslinks how the models interact and interplay, so that dependencies and inconsistencies can be detected and cared for at runtime. Fourth, because MDE generates the source code of a control software from formally defined models [188], the level of abstraction can be appropriately chosen for testing, verification, and quality assurance, which decreases the development effort and maintainability of cobotic applications. Fifth, most recent approaches for application architectures of cobotic applications do not treat the software architecture of a cobotic cell in the required generality, but only consider very concrete tasks or use cases. For example, Huang et al. [214] show an anticipatory control method for robots. This method enables robots to prepare and execute actions for a shared goal, based on anticipation and understanding of a human coworker (human–machine coadaption, Chapter 8). However, most actions of humans in a cobotic cell cannot be anticipated. Also, distributed and immersive cobotic cells pose many more challenges, in particular, they enforce that the application consists of several layers.

Self-adaptive multilayer applications for distributed and immersive cobotic cells

If a cobotic application runs in a distributed or immersive cell, it will need more application layers to manage distribution and immersion. Nevertheless, these layers need to be coordinated and consistently adapted *(cross-layer adaptation)*. To this end, the design pattern *multilayer autonomic system (MuLAS)* [167] can be employed. A multilayered autonomic system is self-adaptive, i.e., underlies an autonomic controller in form of a global MAPE-K loop, but is structured in many layers or even hierarchies of autonomic components, which all react consistently if the global MAPE-K loop replans the operational behavior of the system.

In the MBCC reference architecture, the application layer may consist of such a *multilayer autonomic system*, a layered self-adaptive system that can act based on its *context*, e.g., the different operational modes of a cobot in a cobotic cell. Because all layers of the MuLAS application react consistently to replannings of the global MAPE-K controller, safety of human–robotic coworking can be enforced over *all* application components, and even over several applications running in parallel [167]. For our industrial use case, this means that all software components of a cobot react consistently on replannings of the central autonomic controller of the safety model. Therefore the cobot can work autonomously, remotely controlled, or in cooperation with a human, and, nevertheless, will change to a safe mode appropriately [167].

Therefore in combination with context-adaptive modeling (Section 3.2.1), MAPE-K loops are able to activate or deactivate different variants of a software as action in the execution-phase. In our use-case of cobotic cells, they can be used to switch the functional variants for *remotely controlled*, *autonomous*, or *immersive* operation of a cobot. Therefore the MBCC architecture supports the transition between all forms of cobotic cells discussed in this paper.

Research directions for model-based cobotic cells

The presented concept of a reference architecture for MBCC provides various challenging new research directions. In what follows, some of these directions are presented.

Runtime-extensible plug-and-play component architectures A reference architecture of MBCC must be instantiated to concrete application architectures. Components for specific tactile robots have to be integrated into a MBCC architecture,

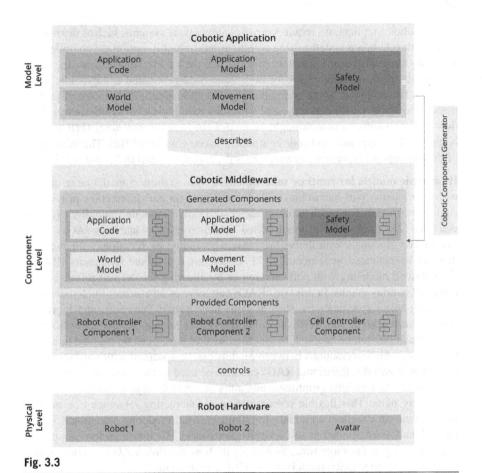

Fig. 3.3

The three layers of a cobotic application, with some software components generated from models and hardware components generated by high-level synthesis.

extending the models of MBCC, in particular, the safety model. It has to be investigated how to enable plug-and-play of robots in a cell, so that additional robots can be set up in a cobotic cell easily. This includes adding new models to the MBCC macromodel and initializing their connections. The question of the geographical distribution of cobotic cells plays a huge role in this respect, particularly concerning the intelligent network approaches of K2 in Chapter 6.

Cobotic model components and their implementation For MBCC, a suitable component technology for models and MDE is required. Existing component systems, such as ROS [210] and ROS 2 [215], only support code composition. Thus techniques to build and compose self-aware adaptive model components and integrate them into existing or novel robot component systems must be investigated. In particular, the challenge to build adaptive low-latency, high throughput components required in industrial robotic applications requires novel context-aware systems, such as distributed context- or role-based systems using combinations of highly-optimized compile time and low-overhead runtime adaptation. Thus it must be investigated how the techniques presented in Section 3.2.1 for both static code generation and dynamic modeling can be combined. Some components in Fig. 3.3 should, for the sake of performance, real-time, and energy-efficiency, be subject to high-level synthesis (see Section 3.4). A suitable approach might be *model-integrated components (MIC)*, combining model component and code component in one container [216]. The integration of these components into the architecture is discussed in detail in Section 3.4.3.

Hierarchic models for runtime macromodels Tree grammars model large objects with many parts (part hierarchies). Modeling based on part hierarchies provides a slightly restricted, but much more efficient alternative modeling paradigm (*grammarware* [217]) than classic graph-based modeling. Actor Gateways (AGs) [218] extend tree grammars with derived *attributes* computed by functions (*attributions*) that can be evaluated incrementally [219]. Reference Attribute Grammars (RAGs) [220] extend modeling with part hierarchies by allowing for derived references, to represent overlay graphs. In this way, RAGs can serve as an efficient implementation for graph-based specifications, such as Petri nets, hybrid Petri nets, and bigraphs. Also, *relations* can be added (Relational RAGs) [221], so that hierarchic models with linking relations can be handled. In this way, several RAGs can be coupled to a multimodel, in which interactions between models can be specified with attributions on relational crosslinks. Relational RAGs can also be used to engineer runtime models, because they can maintain runtime constraints on several, large hierarchical objects with many parts. This flexible representation of interacting reference-tree models makes them an excellent candidate for runtime macromodeling. Relational RAGs have already been applied to self-adaptive systems [222]. It is therefore important, and promising at the same time, to investigate how multiple RAG can consistently support the runtime macromodels of the MBCC architecture.

Bigraphs for context-aware systems Bigraphs [223] explicitly enable specifying the two relevant dimensions for ubiquitous systems, namely, mobility and communication (Bruni et al. [224]). They have already been successfully applied for the formalization of context-aware systems [225], multiagent systems [226], or actor-based models [227], making them promising object of investigation. Furthermore, Relational RAGs can implement bigraphs of arbitrary depth, so that the formalization of context-aware adaptation can be reinterpreted in grammarware.

Verification of models As mentioned in Section 3.2.2, model verification is important, particularly to guarantee safety properties. To improve the verification process for the envisioned component-based architecture, the capabilities of *compositional model checking (model checking by composition)* can be investigated. The correct behavior of the various components of a cobotic cell architecture may be, to some extent, verified independently. However, it is a difficult task to afterwards prove that the product of all components is *safe* according to a specification with respect to human–robot interaction. Thus one promising approach is to exploit the compositional feature of bigraphs and definition of bigraphical subsystems as discussed by Debois and Perrone [228,229], and then, to employ the compositionality of the proofs.

3.2.4 Summary

As can be observed, software engineering of distributed cobotic cells will be complex to understand and manage, especially in environments where mobile cobots and humans can move freely (human–robot cohabitation). To model an architecture of an industrial cobotic cell efficiently, a suitable approach is required combining MDE and CBSE. Models need to be structured into components, views, multimodels and macromodels, providing appropriate abstractions of the cobotic cell by concentrating on specific structural or behavioral properties.

A far from trivial task is to guarantee the safety of multiple cobots and humans performing collaborative tasks. For this purpose, we have proposed the Model-Based Cobotic Cell (MBCC) reference architecture, a fully adaptive runtime macromodeling approach for a cobotic cell, in which the distributed, parallel hybrid system is modeled by a collection of Petri nets, hybrid Petri nets, and bigraphs, integrated into a runtime macromodel, employed as suitable analysis, testing, and verification framework for reasoning about interacting humans, processes, and cobots, in particular, about their safety.

3.3 Tactile robots in the Tactile Internet

We envision tactile robots to become a central physical embodiment of the future Tactile Internet. In contrast to current purely kinesthetic systems, these next-generation devices are characterized by their significantly enhanced ability to perceive tactile information also via tactile sensors, such as artificial skins. With this modality they will be able to detect and measure force and pressure with high accuracy along their entire body, similar to humans. These novel robots will play a significant role in future cobotic workcells since they will be explicitly designed for interaction with humans in unstructured, dynamic environments.

3.3.1 State of the art

Our use case as described in Section 3.1 requires collaborative robots that are able to work directly with humans. Current state-of-the-art robotic platforms intended for interaction with the environment and humans are developed as light-weight and sensitive systems. First examples of this paradigm are [230,231]. More recent platforms are the Sawyer arm from Rethink Robotics, the Universal Robot systems, and the Franka Emika Panda arm. The properties that distinguish these systems from earlier iterations or robots outside the context of human–robot collaboration are their control methodologies, their inherent safety mechanisms, their mechatronic design and the way they are programmed. In a cobotic workcell such robots will be the basic tools for the human workers.

One of the most important developments in this context is the use of torque control with the aid of joint-torque sensors as proposed by [232]. Based on proper estimation methods for the external torque and wrench, this allows for more elaborate control schemes, e.g., the well-known impedance control proposed in [233,234] or force control in [235]. These approaches can also be unified as shown in [236]. Another important control scheme in this context is redundancy control [233], which benefits from the kinematic redundancy of novel robotic platforms.

Since these robots are deployed in collaborative scenarios, safety is of the utmost importance. This means at first the safety of any humans in the vicinity and then the safety of the robot itself and its environment. Especially, in the last decade, there has been extensive research, such as [237], about this topic to determine requirements of safe robots and studies regarding injuries, which resulted in the development of novel design paradigms that aim to make modern robots inherently safe. In particular, some works, such as the one proposed by Sami Haddadin [238], address safety issues and led to new safety standards, such as ISO 13482-3. Joint torque sensors with suitable disturbance observers are used for human–robot contact handling and, more general, for unified collision handling and reflex reaction [239]. To prevent injuries of human coworkers in unintended collisions, safe motion control methods were developed by [240], determining the maximum allowed velocity for ensuring human safety by means of an injury database and the current robot configuration, an essential component to let humans and robots share physical spaces and seamlessly interact. Other works, such as [241], developed methods towards safety-aware robot and task design.

The way state-of-the-art robots are programmed is very different from the tedious and time-consuming processes in earlier systems. Today's technological level allows for much more intuitive and efficient programming schemes [242]. Kinesthetic teaching is a very common way to program collaborative robots and involves a human user directly guiding the robot and teaching it the intended motions and skills. The intuitive and human-centered design of user interfaces enables nonprofessionals to learn interactions with robots in very little time [243]. Furthermore, there has been extensive work on autonomous task planning, which also considers the human coworker and the interactions between human and robot [244].

In collaborative environments, in which humans work alongside robots, it is crucial to have robotic systems that are not only safe to work with, but also have sufficient manipulation capabilities, especially regarding dexterous tasks. Only then, an intuitive and fluent human–robot cooperation is feasible. Clearly, what allows humans to have outstanding manipulation abilities are their hands. This is the reason why robotics researchers have attempted to replicate human hand's behavior for the last decades. Although several endeavors have been made to attain human-like dexterity and capabilities, current designs are still far from reaching human performance. Initially, robotic hand designs were based on rigid joints, rigid actuation, and fully actuated transmissions. However, nowadays they tend to include soft continuous and flexible joints. An example of this are compliant and under actuated robotic hands, such as the one shown in [245]. Nevertheless, robotic hands with rigid joints are still the most widely developed, due to their robustness. However, their actuation and transmission methods have shifted. Coupled transmissions, in which the number of joints is higher than the degrees of freedom, and underactuated transmissions, which allow passive movements between degrees of freedom to adapt the hand to a grasped object, have been included in recent developments. Examples of these hands include the Pisa/IIT SoftHand, DLR Hand I, DLR Hand II. The number of designed hands has increased substantially during the last decade. A surge in the number of prototypes has occurred, in part, due to the pervasive use of rapid prototyping technologies. Another factor, that has promoted the increasing number of robotic hands developments is the growing number of open-source initiatives, such as the Yale OpenHand Project [246], the Soft hands platform, the OpenBionics initiative, among others. A thorough summary on the robotic hands, which have been developed during the last century has been presented in [247]. It is well known that certain particular patterns of muscular activities represent basic postures which, when combined, enable most of the grasping capacities of the human hand, as it was summarized by [248]. These combinations are called synergies. Recent developments, such as [249], have tried to include this concept on the design and control of robotic hands. This will allow future robotic hands to have a more natural human-like behavior, which could eventually emulate and perhaps surpass a human hand's grasping capabilities. To allow robotic hands to work, inevitably, space constraints for its actuation mechanisms need to be considered. In the current use case, this space has been conceived with the shape of a human forearm. Nowadays, a combination of electric motors and speed reducers is widely used as actuators in robotic arm systems, for instance [250]. Especially for commercial off-the-shell products, this combination dominates the market. Other than electric motors, hydraulic actuation is the most widely used alternative for robot manipulators. The high power density and control bandwidth makes hydraulic actuators more suitable for heavy load and tough applications. Within the realm of robotic arms, well-known hydraulic manipulators include HyArm from IIT [251], RL-H1L from Ritsumeikan University [252], the hydraulic one limb model from Tokyo Institute of Technology [250], and HYDRA-MP from KNR System. There are also pneumatic actuators in robotic arms, such as the BionicCobot developed by Festo with flexibility as a major advantage. Clearly, there is vast existing knowledge regarding traditional actuation in robotic platforms, which will be considered when developing the robotic forearm for the robotic hand–arm system.

The hand–arm system aims to mirror the human lower arm, human wrist, and human hand on a robot. Good examples of interfaces, such as the proposed one, can be found in the area of prosthesis, where robots-parts are explicitly used for substituting partial or total loss of a limb. In the work developed by [253], the command input is locally provided by electrodes placed over the arm of the patient, in the work by [254], a solution for targeted reinnervation is proposed to create an artificial sense of touch on a prosthetic limb. Both works rely on locally placed inputs (negligible latency to control), and on the importance of the sense of touch in the closed loop system.

Finally, to fully realize our use case, robotic telepresence will be one of the main components providing the theoretical basis to allow remote workers to interact with the workcell via an avatar through the Tactile Internet [166]. Research outcomes on this topic, such as [255], have already started decades ago and have progressed significantly since then. Major breakthroughs were the application of wave variables, such as in [256,257] and the introduction of PD-like controller schemes by [258]. These approaches were at first proven to be stable for constant delays only, but were later extended to variable delays and packet loss, as stated in [259]. At that point telepresence could be applied via mass-communication infrastructures, such as the Internet. Recent research endeavors in this field, such as [259–261], aim to be as transparent as possible, i.e., they attempt to maintain amplitude and phase of the force feedback as much as possible.

3.3.2 Challenges of current technology

The use case of a cobotic workcell poses various challenges for current robotics technology. The most significant ones are (i) how to design a robotic manipulator such that it integrates optimally into a human work environment, (ii) what technologies to use for an avatar station that is both immersive and intuitive to use for the user and optimally interacts with the robot platform, (iii) what control methodologies to use for robot and avatar station to fulfill the requirements of autonomy, transparent telepresence, overall stability and safety, and (iv) how to keep the Human-in-the-Loop, i.e., immersed in the remote robot platform without breaking immersion.

Design

The design of traditional robotic arms is mostly functional and adheres only to the requirements of a single task. Even many of today's collaborative systems do not really consider the collaboration in their actual design. This is often only achieved through applications and controller paradigms. However, to work efficiently with humans in, in what for robots would be unstructured environments and also serve as a remote body for immersive teleoperation, it is necessary to adapt the design of the robotic arm to the environment to avoid the need for creating highly specialized workspaces. The challenge here is to make the arm to operate both energy efficient and with high performance. This is mostly achieved through highly integrated designs. Moreover, the unstructured nature of a typical human workplace requires the arm to have a capable and robust visual perception as well as a hand to freely manipulate objects. Furthermore, the development of such a hand–arm system poses significant challenges to actuators and sensors as stated by [166].

Based on the remote manipulation and tactile sensing applications, the design of the robotic arm should keep the capability targets, such as maximum joint velocity, payload and flexibility comparable to human arms. The motion capability mismatch would otherwise cause lag and discontinuity in remote manipulation. Also, the overall size and weight should be similar to minimize the efforts on motion and sensor mapping between the robotic arm and the human arm. These requirements indicate that actuators with high power density are necessary as mentioned by [262].

Electric motors are the most common actuators in robotic arms. However, to achieve the peak and average power or torque demonstrated by human arms, the motors tend to be heavier and larger. In addition, transmissions with high gear reductions are required to increase low speed torque performance, which also add mass and size. Hydraulic actuators have much higher power density than electric ones. However, despite the advantages of hydraulic actuators, two fundamental challenges exist. The highly nonlinear dynamic behavior in the articulated robotic arms makes their closed-loop control design and stability analysis an extremely challenging task. Beyond that, traditional hydraulic closed-loop systems are not energy efficient, as shown by [263].

Apart from the actuators, regarding the development of the robotic hand–arm system, its kinematic, dynamic, and actuation architectures also exhibit significant challenges. One is the translation of information from a given set of tasks, which need to be accomplished, to actual design parameters, which can be used to optimize the mechanical properties of the hand–arm system. Another challenge is the inclusion of sensors on the robotic hand–arm system. Robotic hands in general have a limited space for its actuation and transmission components. The inclusion of sensors to measure internal signals and also external forces or torques might prove difficult, as the space which sensors need, might alter the expected behavior of the robotic hand–arm system. Furthermore, to provide tactile perception the hand–arm system needs to be equipped with proper sensors, such as artificial skin. Although, these sensor types are usually lightweight, they need to be fitted to the mechatronic design of the system. This is the kind of challenge which the research group for sensors and actuators (TP2) focuses on. Their outcomes will not only be beneficial to the implementation of robotic systems, but also to other research groups, such as the one involved with communication, compression, and control (TP3).

Avatar

The avatar station is the entry point for a human user into the remote robotic body in the workcell. The aim is to completely immerse the user, i.e., give them the illusion as if they were physically at the remote location. This requires the highest level of transparency as mentioned by [264] in terms of tactile and audiovisual feedback as well as responsiveness of the remote robotic arm. These requirements pose not only a significant challenge to communication bandwidth and speed, but also to the technology used to create the illusion of remote presence. The user would require high-performance wearable technology that is capable of exerting forces (normal and shear), pressure (also based on specific geometric shapes) and temperature. The remote arm would have to be equipped with proper sensors and actuators, which is connected to the system design challenge.

Control methodologies

A human-like collaborative robot arm that at the same time serves as a remote body has to fulfill multiple levels of requirement in terms of control. Firstly, the arm must employ compliance control to safely and robustly interact with its environment, other robots, and humans. Secondly, it must be capable of connecting to an avatar station and provide transparency and responsiveness to the station's user. Thirdly, it should be able to handle any mix between the two extremes. This sliding autonomy would offer a varying granularity of telepresence to the human, i.e., from direct control over high-level control to complete autonomy. The challenge arises from the necessity to prove the stability and performance of the entirety of controller modes.

Maintaining the loop

Assuming that the required technology is available to completely immerse the human user into the remote body and ensure a stable control of the remote robotic system, there are still challenges to overcome. For example, the information density of highly dynamic assembly tasks may exceed the capacities of the robot's tactile perception, or the communication with the user, which could decrease the transparency of teleoperation and thus break immersion to a degree. This could lead to human errors caused by a falsified perception of the environment and must be compensated by local control directly at the remote robot. Moreover, the local control must ensure the safety of the system, other humans in the vicinity, and the avatar user itself, even in the case of critical communication interruptions. Another source of error may be the fusion of multiple perception modalities. In the case that audio, video, and tactile feedback do not match, the human user would be irritated and possibly disoriented. Thus we require algorithms to properly match the different channels in real time.

3.3.3 Future research direction

To overcome the challenges outlined in Section 3.3.2, we will focus our effort on the following points: (*i*) We will develop an anthropomorphic hand–arm system utilizing novel sensor and actuation technology, such as artificial skin and muscles. (*ii*) We will build a complete avatar station as a test bed for various technologies in the context of immersive telepresence. (*iii*) We will research novel controller paradigms in order to unify all components in a safe and robust framework.

Hand–arm system

The most suitable robotic arm design in a human work environment would certainly be one that mimics the human arm perfectly. However, reaching this level of performance and efficiency is one of the most challenging problems in current robotics research. The research directions cover every aspect to improve the performance from higher power density actuators, optimized mechanical structures, and complex and stable controller schemes. Especially the complexity of the hand poses a significant challenge. Our approach will be to develop a highly integrated mechatronic system using novel sensor and actuation technologies developed in the other TaHiL's research groups, such as TP2 and TP4. Examples are artificial skin and muscles, communication devices, and flexible electronics. Integration of these components in

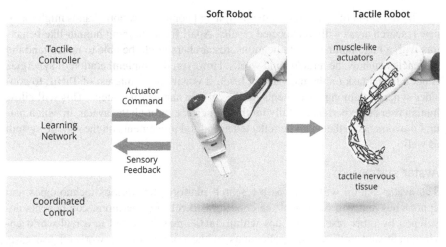

Fig. 3.4

Evolution towards an anthropomorphic hand–arm system.

a single modular package could also provide higher efficiency, higher power density, and higher torque density as stated by [265].

First steps in this direction are the definition of requirements in terms of capabilities necessary for usage in a cobotic workcell, both autonomously as well as a remote body in a telepresence setting. We see the development of such a manipulator as the next logical step in the evolution of collaborative robots so far, see Fig. 3.4.

Most of the current research literature focuses on the optimization of physical features of a robotic hand–arm system, taking into account only its real human hand–arm counterpart. However, most of these approaches do not consider the tasks in which these robotic systems will be used. It would be more beneficial to guarantee that a robotic hand–arm is optimized for a range of specific tasks. By doing so, robustness and repeatability can be guaranteed when the system is being operated by humans through avatars. With new approaches based on the works of [266,267], this design process of robotic hand can be automated and generalized, so that the resulting robotic hand–arm systems can be as optimal as possible. This is beneficial for designers, as it narrows the amount of decisions they need to make, before being able to guarantee the robustness of the robotic system. One of the research directions will be to aim for a hand-designing guideline.

Initially, the experience and widespread information of the design and functionality of a wide variety of hands, which have been developed over the last decades, will be used. The work presented in [247] shows a summary regarding the type of joints, actuation, and transmission of robotic hands. As it can be seen here, over the last decades, there has been a noticeable increase on the use of coupled, underactuated transmissions, and synchronized motions. Examples of this are [245,249,268]. These trends will be exploited during the development of the robotic hand–arm system to avoid design mistakes, which might have already proven trivial.

The inclusion of synergies on the design and control of robotic hands might open new research areas with unexpected results. Apart from attaining human-like behaviors with a huge number of applications, researchers might be able to understand the intricacies on how the human hand works. However, the implementation of synergies is not a trivial task on the mechanical level. It is within the interest of TaHiL to consider synergies among movements on the robotic hand–arm system. This will allow human users to experience a fully intuitive interface through the avatar, in which natural movements to them can be reflected as natural movements on the robotic system as well.

Avatar station

The avatar station will serve as a research platform for various technologies and approaches and will be part of the cobotic workcell. Furthermore, components developed by other research groups within TaHiL may be tested in a real-world environment and with actual applications in the context of the workcell use case. Among these technologies are smart wearables, high-speed communication, exoskeletons, Virtual Reality (VR)/Augmented Reality (AR) devices, data gloves, and more. Fig. 3.5 shows the principle vision behind this station. The user is connected via the Tactile Internet to a remote anthropomorphic hand–arm system in the workcell, controlling it by use of wearable technology. In addition to the bidirectional connection, both user and avatar are able to access the Tactile Internet and make use of cloud-based skills and services. This platform will allow us to conduct research directly in the context of real-world applications and find ways to answer the question of how to keep the Human-in-the-Loop of teleoperation. Other research groups, such

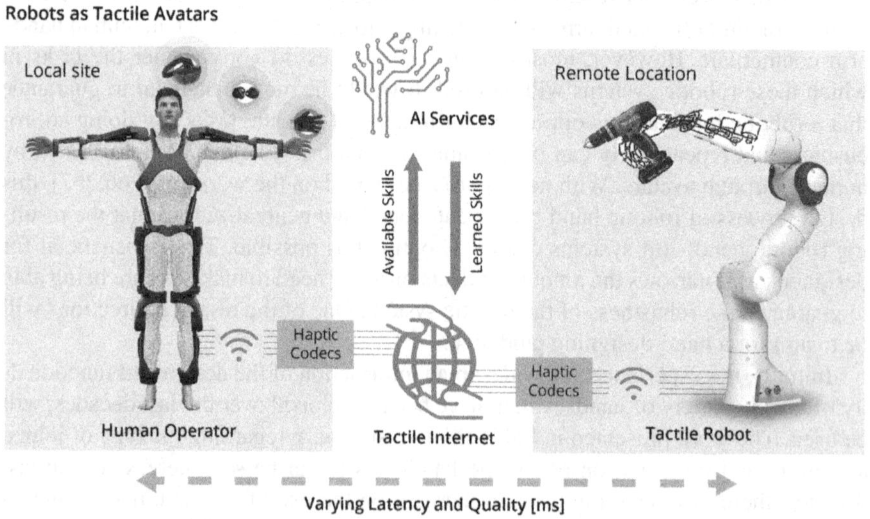

Fig. 3.5

Principle idea of the avatar station.

as Haptic codecs for the Tactile Internet (K1), Augmented perception and interaction (K3), Human–machine co-adaptation (K4), Human-in-the-Loop (TP1), Sensors and actuators (TP2), Communication, compression, and control (TP3) and Flexible electronics (TP4) will also benefit from this, since they can test and demonstrate their approaches in a field test. Thus the avatar station will function as a platform that drives collaborative research within TaHiL.

Controller

The hand–arm system and the avatar station will offer a multitude of novel research possibilities. However, they also require control methodologies that comply to all the requirements, i.e., full autonomy of the hand–arm system, varying degrees of telepresence, meaningful interaction capabilities of any exoskeleton part on the avatar, fusion of numerous sensor modalities, etc. Moreover, these controllers must be stable and safe at all times to avoid harm to human user and coworkers, the robots themselves, or the environment. To meet all these requirements, we will build upon existing work, such as impedance control and extend it first with additional sensor inputs, such as tactile perception. In this context, we will collaborate with other research groups, such as TP1, TP5, and K4. In conjunction with the design of the hand–arm system and the avatar station, we will explore potentially completely new controller paradigms.

3.3.4 Summary

Future cobotic workcells will require robot systems that can safely and efficiently interact with humans and environments made for humans. The design for such platforms will be heavily inspired by the human hand–arm system itself. One of the grand challenges in this context is the development of an optimal design that will not only achieve human-level performance, but will also comply with novel hardware and software control architectures.

Furthermore, the integration into the Tactile Internet requires next-generation robotic systems to fulfill requirements regarding multilevel human–robot interaction and telepresence. Research will focus on the design of an anthropomorphic hand–arm system concurrently with the development of matching control and interaction methodologies. To fully approach this overall challenge, the other research groups within TaHiL will be involved.

3.4 Embedded hardware for robotics

Heterogeneous platforms, such as robots, include multiple types of different sensors, such as cameras, lasers, and Inertial Measurement Units (IMUs). This implies that a large amount of heterogeneous data is generated constantly and needs to be processed, most likely, in real-time. This imposes two challenges. The first one is that every sensor produces different types of data, rising the need for a generic platform able to handle them with ease. The second one is related to meet real-time constraints implying the need for parallel processing. Reconfigurable Computing Systems (RCSs) are an ideal candidate due to their intrinsic characteristic of parallel processing. They can preprocess data very close to sensors [269] and can be, from the performance per watt, close to Application-Specific Integrated Circuit (ASIC) [270,271], but they are more flexible as they are reprogrammable. They provide versatility to specifically design hardware accordingly to the needs and provide results faster than Central Processing Units (CPUs) [272]. Moreover, this implies low latency, which is required to meet real-time constraints. Lastly, they are a good fit to the *self-adaptive techniques for cobotic cells* mentioned in Section 3.2.1 as they offer the possibility for self-adaptation, as some hardware blocks can be modified dynamically at runtime.

Embedded hardware refers to a custom design for a specific application. They are expected to be capable to handling demanding computational operations and efficient in terms of energy consumption. The main focus of this section is around RCSs, as they comply with these two characteristics mentioned before, and they are versatile for custom applications. Moreover, they have called the attention from the robotics community lately due to High Level Synthesis (HLS) techniques [273]. They allow designers with basic hardware knowledge to reuse software applications with minimal changes to comply with some hardware constraints. The main one to consider is data movement between different components (calling functions in terms of software). Therefore they are not limited anymore to experienced hardware developers with knowledge in Hardware Description Languages (HDLs). Besides, RCSs are now even more appealing, as they provide better performance per Watt than standard CPU-based architectures [274]. Furthermore, Xilinx (one of the largest Field Programmable Gate Arrays (FPGAs) manufacturer) integrates in their Systems-on-Chips (SoCs) the software programmability of an ARM-based processor with the hardware programmability of an RCS. This feature allows combining the capabilities of RCSs with the ones from CPUs, where an Operating System (OS) can handle transactions between software and hardware.

3.4.1 State of the art for embedded hardware in robotics

Robotic platforms are a combination of software and hardware. Ideally, they should complement each other, and tools should be provided for designers to mainly focus on their applications. This section intends to show the software and hardware state of the art, and how they can be combined to obtain efficient embedded hardware platforms for robotic applications.

Robotics middlewares

Middlewares provide services to software applications beyond those available in OSs. They simplify the design process providing the infrastructure for communication, letting the designers focus on their specific application, and could be considered as *the software glue*. Robotics middlewares are an abstraction layer that resides between the OSs and software applications. They are specifically designed to manage the complexity and heterogeneity of the hardware and the applications. Their aim is to hide the complexity of low-level communication and sensor's heterogeneity, reuse software infrastructure and production costs [275]. Some of the multiple options that have been proposed over the years are introduced below.

The Orocos project   It develops a general-purpose modular framework for robot and machine control [276]. The Real-Time Toolkit (RTT) and Orocos Component Library (OCL) establish a component-based infrastructure and a library ready-to-use components, providing the high level management of interactions within an application.

YARP   The goal of Yet Another Robot Platform (YARP) is to minimize the effort devoted to infrastructure-level software development by facilitating code reuse, modularity, and so maximize research-level development and collaboration. It supports building a robot control system as a collection of programs communicating in a peer-to-peer way, with an extensible family of connection types (tcp, upd, multicast, local, MPI, etc.) that can be swapped depending on the needs of the developer [277].

NVIDIA Isaac SDK   It is a robotic Artificial Intelligence (AI) development platform with simulation, navigation and manipulation. Its main three components are: (i) Graphics Processing Unit (GPU)-accelerated algorithms and Deep Neural Networks (DNNs) for perception and planning, and Machine Learning (ML) for supervised learning as well as reinforcement learning. (ii) Modular robotic algorithms that provide sensing, planning or actuation for both navigation and manipulation. (iii) Simulation to accelerate robot development and deployment. Isaac SDK runs and it is optimized for the NVIDIA Jetson AGX Xavier system.

ROS and ROS2   The open-source middleware ROS is a software solution that eases the building process of robotic applications. It runs on top of Linux and has been gaining popularity in the robotics community over the last years (see Fig. 3.6). Besides, not only the research community has shown its interest, but industry as well. A consortium integrated by worldwide companies from multiple sectors, such as automotive or aerospace, has been growing over the years to extend the advanced capabilities of ROS software to manufacturing. Currently, there is a lot of effort focused on a new version of the middleware, such as ROS2, due to the fact that the first version does not satisfy real-time requirements. Therefore ROS2 is based on Data Distribution Service (DDS), which is a standard protocol used in industry that meets real-time constraints due to its various transport configurations (e.g., deadlines and fault-tolerance).

Embedded systems used in robotics

There is a wide range of embedded systems to be used in robotics. Below is an introduction to them, depicting their main characteristics.

Embedded PCs   They are small, integrated single-board computers with embedded CPUs (e.g., Intel I5, I7, etc.), GPUs, memory, and several peripherals. Therefore they are quite similar to PCs. The difference is the integration of all the components in a compact design, making them smaller for an embedded system, such as a robot. Like

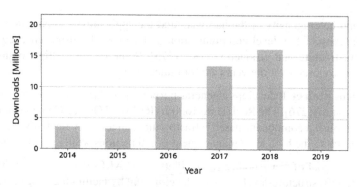

Fig. 3.6

Total ROS packages downloaded (data based on http://wiki.ros.org/Metrics).

this, it is possible to run an OS to control all sensors and actuators, as it is the case for Festo robots.

NVIDIA Tegra SoC and Jetson Platforms Tegra is a SoC series that includes an ARM CPU, a GPU, and memory controller. Jetson is a series of embedded computing boards carrying a Tegra SoC. They include a series of peripherals and interfaces (USB, HDMI, GPIOs, I^2C, I^2S, SPI, UART). They provide the performance and power efficiency to run autonomous machines software faster and with less power. It includes supports for CUDA, libraries for deep learning, computer vision, accelerated computers, and multimedia. The platform also supports drivers for a wide range of sensors.

Reconfigurable computing systems Besides the fact that they provide CPUs capable of running an OS (such as Linux) as the two previous ones, they allow designers to build custom systems according to their needs. This makes them the right choice for self-adaptive computing systems, such as the ones described in Section 3.2.1.

Hardware/software codesign

Hardware/software codesign aims to exploit the synergy between RCSs and CPUs to obtain optimized designs in terms of power consumption, timing, reliability, and costs [278,279]. Most approaches have been focused on ROS as it became the main option for robotics in terms of middlewares. By doing so, common grounds from both fields can be understood from different perspectives to establish the requirements needed to enhance one another. This provides better comprehension of why a combination is possible and how the concepts can be applied for new generation of robots (i.e., cobotics) and the upcoming Tactile Internet.

As mentioned before, ROS runs on top of Linux, which needs a CPU capable for it. Therefore it is easy to think that the solution to integrate ROS and RCSs would be to rely on a SoC that combines a CPU (where Linux can be installed to run ROS) with an RCS and take advantage of the provided mechanisms to establish communication between them. However, this approach increases the complexity of the design, and an entire OS is not justified just to have the possibility to run ROS on top of it. Furthermore, the amount of Random-Access Memory (RAM) on embedded systems (2 to 8 GB) is usually smaller than in regular computers. Hence, it should be used efficiently.

An integration of RCSs to ROS is proposed in the research by the authors in [280] by means of *ROS-Compliant RCS component*, wrapping the hardware design in the Programmable Logic (PL) with software. ROS runs on the CPU along with Xilinux (Xilinx's Linux distribution with RCS-specific functions), which allows establishing communication between CPU and RCS effortlessly by reading/writing from/to a file descriptor from the software side, and a FIFO on the hardware side. This makes the data exchange process quite straightforward. In this case, the embedded hardware from the RCS acts as an accelerator, like traditional hardware/software codesign techniques.

High development costs and time that comes with the design of HDL-based hardware IPs yields to appeal to different techniques, such as HLS as mentioned before. Consequently, the authors in [281] propose a tool that generates the interface circuit (hardware as HDL files) and software interface (C++ files). However, this approach is only bounded to HLS and relies on Linux running on the CPU to provide ROS support.

Running ROS on the CPU not only increases the complexity of the system, but also implies a delay in the communication introduced by transferring data between CPU and RCS. Therefore the authors in [282] implement a TCP/IP stack in the PL to offload the communication handled by ROS to exchange data with other components of a ROS system. However, the limitation comes with the implementation only being able to handle one connection at a time. As a consequence, only data is exchanged directly from the PL as it is a one-to-one communication. The handshaking protocol needed to establish communication between different ROS nodes remains on the CPU as it requires multiple connections at a time.

The authors in [283] propose a generic architecture to have a full hardware implementation without the need of any CPU. It is generic so it can incorporate IPs designed in HDL or HLS. It circumvents previous communication limitations by using an external device capable of handling multiple connections simultaneously. Moreover, it leaves the possibility to replace the corresponding IP block for communications in case a different device is used, or the communications are handled by the CPU if available. In case ROS is not the chosen middleware and a different one is needed, the modular architecture allows replacing the specific-related middleware IP block, thanks to the plug-and-play design. Lastly, it opens the possibility for robotic applications to use Dynamic Partial Reconfiguration (DPR). This is a technique that allows changing parts of the hardware at runtime. Hence, it is not necessary to include all hardware functionalities at the same time as they can be introduced when needed. For example, different image processing filters [273] might be needed in a given application, but not all at the same time. Considering that each hardware block consumes a given amount of power, by dynamically introducing them in the design when needed, it will imply a power reduction.

3.4.2 Research questions to enhance cobotic cells with embedded hardware

Nowadays, modern robots tend to be heterogeneous and complex systems [284]. Moreover, the new generation of robots used for cobotic cells need to be aware of their surroundings and act accordingly. This usually includes a wide range of sensors and actuators that need to be controlled and manipulated by powerful hardware and efficient software. Therefore synergy between these two disciplines is needed to keep up with the increasing complexity of robotic platforms. Furthermore, information from the models described in Section 3.2 is needed to generate complementary hardware components.

RQ 3.1: How to enhance cobotic cells in terms of computational performance with a low-energy budget? Why are RCSs a good option?

One aspect to keep in mind is that the development of software is coupled with CPUs that have been normally designed to fulfill generic operations, such as addition or subtraction. Therefore to obtain any given result, multiple operations have to be performed. A second aspect is that complex operations can be realized by designing a specific hardware for it. For example, multiple functions for image processing can be implemented in an RCS, combined and produce powerful algorithms [273]. Nevertheless, an integration with the software architecture remains an open question.

A third aspect concerns the fact that these complex robots have to perform many computations. Reaching the desired performance represents a challenge, especially in the case where they are equipped with embedded computers consuming a lot of energy, which is limited in a mobile system that is usually battery operated. It follows that other computational resources, such as RCSs or GPUs, which are more power-efficient should be considered for robotics [285]. The first option is advantageous due to their versatility to obtain custom designs capable of performing highly demanding computations with reduced power consumption. Moreover, they have been proven to be more efficient in terms of energy compared to general purpose processors [286]. Besides, RCSs handle parallel and logic operations better than CPUs.

RQ 3.2: How can the models of the cobotic cells be used to generate embedded reconfigurable hardware to enhance their computational power?

Cobotic cells will be mainly modeled following software engineering concepts, as mentioned previously in Section 3.2. Therefore the first challenge, from the embedded hardware side, is how those models can be useful to obtain hardware models that will dictate the norms of meaningful hardware architecture for Tactile Internet applications. This will impose the requirements from the software side that the upcoming hardware architecture will have to comply to fulfill the desired integration between these two research fields.

RQ 3.3: How to obtain hardware components to guarantee adaptivity and reusability? Which techniques are needed?

Considering the fact that models for cobotic cells are well-defined, as described in Section 3.2.2, a similar approach based on MDE is to be followed to generate hardware components. However, embedded hardware components need to consider extra aspects as it is the closest abstraction level to tactile robots in the scheme shown in Fig. 3.1. This means that each hardware component needs to be adapted to different requirements accordingly. Besides, hardware platforms are usually heterogeneous systems using different middlewares or communication protocols. Therefore their characteristics need to be reflected when following the MDE approach. In this way, adaptivity and reusability will be guaranteed.

RQ 3.4: How can hardware components adapt dynamically to cope with security to guarantee the interaction between robots and humans (self-adaptiveness)?

It was mentioned previously that cobotic cells will have to consider the world model describing where the human–robot or robot-robot interaction occurs and safety states to avoid accidents that could potentially injure humans. As a result, some elements of the hardware architecture will also have to adapt dynamically depending on the current safety state. In this case, the design has to be allowed to face a minimal reconfiguration time, making that hardware component nonfunctional for that time. Consequently, this raises the challenge of how to incorporate in higher abstraction levels (*distributed cobotic cells* and *tactile robotics* shown in Fig. 3.1) models those blocks that will be potentially reconfigured dynamically.

3.4.3 Research directions

The state of the art from Section 3.4.1 highlights the convenience of RCSs being versatile and advantageous for power-demanding applications as well as being part of a complex robotic system. The proposed work [283], focused on ROS, serves as the base for the upcoming research directions. However, there are open questions to be further explored to have generic (not only for ROS) and MDE approach.

Modeling Modeling cobotic cells has to leverage the MDE approach. This means that the abstraction provided by modeling has to be maintained regardless of the middleware (ROS, ROS2, etc.) used by designers. Special focus will be on how hardware robotic architectures can be modeled to comply with the requirements dictated by software models of cobotic cells. Besides, concepts for dynamic hardware adaptation (DPR) have to be studied, in addition to how these should be incorporated in cobotic models to follow the safety states proposed previously. Moreover, a direct collaboration to integrate the concepts shown in Section 3.2 is to be exploited. The aim is to obtain hardware models from software models specifying requirements for cobotic cells. This will set up the constraints in terms of concurrency and real-time.

Automation techniques Code generation and automation techniques have to be developed to provide effortless deployment. This is to circumvent the regular cumbersome hardware design work flow. Currently, the methodology demands to do the design manually. Based on the modeling concepts described in Section 3.2, research oriented towards hardware components/accelerators is going to focus on how robotic architectures/applications can be modeled and which modeling techniques can be useful to portray them in a hardware design. Additionally, a set of rules are needed to establish a workflow to follow. This will become imperative in the exploration of toolchains to retrieve the information of higher abstraction levels of cobotic models to autogenerate hardware accelerators as needed.

Energy efficiency   Besides the modeling concepts and how to extract information from cobotic models to generate hardware components, a detailed study regarding the most efficient design in terms of low power consumption and efficiency in computation will be carried out. This will include a software architecture for the embedded hardware platform developed in Chapter 12, targeting a total power consumption of less than 10 mW. Therefore an integration of *embedded software* has to be studied to be included as part of the model.

Adaptivity and reusability   Lastly, the concepts developed in U2 must be connected with other research groups from the project. It is in particular the interest of taking advantage of the Body Computing Hub (BCH) developed in Chapter 12 to build a human avatar. This will become part of cobotic cells in order for humans to teach robots different tasks leading to advanced human–robot interaction. Relying on a MDE approach and automation tools is important to ensure reusability. They will be beneficial not only for the chosen platform of U2 (Franka Emika Panda arm), but any other one. Platform-specific models or specifications will only be required to be incorporated into any cobotic cell.

3.4.4 Summary

RCSs are very versatile, efficient in terms of computation, and have low power consumption. The disadvantage is that achieving good designs is cumbersome due to their complexity and usually hardware knowledge is needed. Cobotic cells can also become quite complex, and it is vital to rely on modeling techniques, such as MDE, to enhance them with proper hardware accelerators. Due to their complexity, giving rise to the complexity to understand and manage them, the realization of embedded hardware components has to rely on cobotic models to autogenerate hardware models to circumvent the cumbersome workflow that hardware design entails. Additionally, cobotic cells have to rely on RCSs to keep up with the demands that the upcoming Tactile Internet will impose.

3.5 Synergistic links

The presented research directions for the industrial use case will rely on many other primary research directions and key technologies in the area of TaHiL.

The consistent treatment of critical architectural aspects throughout all software and hardware levels is required, such as safety and security, further the integration of haptic codecs (Chapter 5), and networking (Chapter 6) to practically employ cobotic cells in industrial environments. For hardware development, especially electronics, actuators, sensors, and wearables, we will collaborate closely with TP2 and TP4. In terms of communication required for telepresence applications, we build on the experience in TP1, TP3, and K1. K2, K3, and K4 extend our state of the art on topics such as human–robot interaction and networks. To prepare and execute actions for a shared goal, based on anticipation and understanding of a human coworker, we will collaborate with K4 in light of human–machine coadaption.

3.6 **Conclusion and outlook**

This chapter illustrated the use of robots in the workplace with a collaborative telepresence use case. Three aspects of the operation of the scenario were described by deriving research questions, surveying the state of the art, and finally highlighting promising research directions that need to be addressed within the TaHiL research field.

Starting top-down, the first aspect is concerned with a software architecture for distributed cobotic cells. Several research challenges are discussed, such as the design of a model-driven engineering approach, the verification of cobotic cells to guarantee safety, and the programming of context-adaptive software.

Secondly, the actual design of a *tactile* robot use-case is discussed. Since the cooperation between humans and the aforementioned robotic systems will first and foremost be applied in industrial environments, one of the crucial factors to consider is the mechatronic design of the robot arm–hand system itself. It is through this mechanism that a human will be able to interact with the robot's environment. Finally, another specific aspect essential to a distributed cobotic system is discussed, hardware accelerated control. The integrated usage of robots in distributed systems poses new challenges to performance, responsiveness, and efficiency of many aspects of a cobotic system. Thus RCSs are proposed as a very versatile approach that is efficient in terms of computation and does not have a large power consumption. Therefore it is vital for cobotic cells to rely on them to keep up with the demands that the upcoming Tactile Internet will impose.

In conclusion, this chapter gives an overview of the essential aspects of human–robot cohabitation in an industrial context, highlighting the challenges and strategies how these will be addressed.

Internet of Skills

4

Luca Oppici[a], **Tina Bobbe**[a], **Lisa-Marie Lüneburg**[a], **Andreas Nocke**[a],
Anna Schwendicke[a], **Hans Winger**[a], **Jens Krzywinski**[a], **Chokri Cherif**[a],
Thorsten Strufe[b], **and Susanne Narciss**[a]

[a]*Technische Universität Dresden, Dresden, Germany*
[b]*Karlsruhe Institute of Technology, Karlsruhe, Germany*

*Learning is not attained by chance, it must be sought for with ardor and
attended to with diligence.*
– Abigail Adams

4.1 Aims of the Internet of Skills

The Internet of Skills aims at investigating how anyone, even at the most remote and
diverse geographical regions, can be provided with a suitable training to improve
their skills and capabilities. It will democratize the access to abilities and expertise
the same way the Internet democratized the access to information and knowledge.
Demographic challenges of an extremely urbanized and aging society require new
technological systems that can assist the population to proficiently perform skills
in daily-life contexts. The Internet of Skills will combine latest scientific advances
in the motor learning field and Tactile Internet with Human-in-the-Loop (TaHiL) to
develop training strategies that foster the acquisition and performance of skills. To
demonstrate that research matters, TaHiL technology will be applied and assessed in
(at least) three exemplary scenarios of motor learning.

The Internet of Skills is classified into three learning strategies: (*i*) machine to human, (*ii*) human to machine, and (*iii*) unsupervised human–machine interaction. In
the first category, TaHiL devices will be used to promote a human's training of relevant skills, e.g., specified individual training activities supervised by a remote human
teacher in quasi-real-time or augmented feedback in real-time. In the second category,
machines will be taught or assisted by humans to approach their level of proficiency
in skilled operation or assistance capabilities. In the third category, the machine itself becomes an unsupervised teacher with pretrained skills. Furthermore, two main
approaches for providing support with regard to skilled behavior and performance
are considered: the first approach provides humans with just the right amount of assistance to produce a skilled performance; the second approach includes all types of
teaching, training, and learning scenarios that promote the acquisition of skills. Importantly, both approaches are primarily concerned with the correct (or functional to

Tactile Internet. https://doi.org/10.1016/B978-0-12-821343-8.00014-9

a task) execution of sensorimotor actions. Correct or functional execution of skilled motor behavior is important for various reasons: (i) to prevent deleterious effects of incorrect movements, particularly in scenarios with a high physical workload (preventive measures related to occupational protection), (ii) to learn or improve motor skills for professional, fitness, or rehabilitation reasons, and (iii) to be able to adjust quickly and flexibly to changes in task requirements. Whereas these three strategies will shape the scope of the Internet of Skills in the long term, this chapter primarily focuses on the first strategy—machine to human, which represents the work currently being performed. Technically challenging scenarios that involve a multimodal interaction with a closed-loop structure will be chosen to demonstrate the innovative hard- and software TaHiL technology. All scenarios will be conceptualized in detail, established, and evaluated with respect to the technical and behavioral parameters as well as their relationships. The various scenarios will allow investigations of several fundamental issues of teaching, assistance, and learning from new perspectives, e.g., (i) how to tailor the level of guidance, assistance, and information provided to a learner given their characteristics, particularly skill level, (ii) characterizing, understanding, and supporting the learning process, (iii) characterizing and understanding how humans gather and integrate task-relevant and multimodal sensory information, and (iv) investigating the benefits and constraints of various types of machine–human mutual adaptation strategies, not only in laboratory tasks, but also in authentic task contexts. This can be achieved through collaboration and joint research of people with different expertise, e.g., movement specialists, psychologists, and engineers. Importantly, finding a suitable balance between what is required from a learning perspective and what is possible from a technical perspective is at the core of successful research in the context of the Internet of Skills, to ensure fruitful collaboration with insightful results. What is ideal from a learning perspective might be hardly achievable from a technical perspective and vice-versa, and the Internet of Skills strives for fine-tuning this balance through a continuous and productive interdisciplinary collaboration. This collaboration, however, will pose another challenge as different fields may use different terminology and different approaches to an issue. This chapter discusses skill learning and how technology can improve the process, and outlines how the Internet of Skills combines interdisciplinary expertise to best integrate technology into effective learning interventions.

4.2 State-of-the-art research: Skill learning and technology

Skill learning requires practice, and various factors can be manipulated to promote the process. Understanding the learning process provides practitioners, such as therapists, coaches, and teachers, with a suitable framework that can guide the design of their intervention programs. In this sense, technology can help applying skill learning principles and in turn promote learning of specific skills.

4.2.1 Skill, skilled behavior, and skill learning

A variety of skills are frequently performed in our daily life (e.g., cooking and cycling), in the workplace (e.g., typing and lifting boxes), and in sport (e.g., swimming and kicking a ball) or music (e.g., playing piano or drums) contexts. Skill generally refers to a goal-directed and voluntary movement that is learnt with practice and couples action with active perception of task/environment information [287]. In the literature, it is often referred to as perceptual-motor skill, given that perception and action are tightly coupled: we perceive environmental information to move, we move to inspect the information flow and perceive key information, and our movement creates new information flow and cues to perceive [288]. Importantly, we (humans) move in accordance with the opportunities for action—affordances—that our environment offers, and the perception of information that specifies affordances guides skill execution [289]. Furthermore, cognitive processes, such as working memory and cognitive flexibility, influence our intentions to move and in turn support perceptual-motor skills. Perception, cognition, and action are intertwined and skill emerges from the interplay of these processes [290].

Skilled behavior is the ability to bring about a specific movement-related result (i.e., achieve the goal of a task) with high certainty, while minimizing energy and time costs, and skilled performers share common characteristics: they consistently achieve successful performance, they can adapt their skill to changes in task demands and environment, they perceive key information to guide their action and they have an extensive repertoire of actions that can be used to perform a task [291]. For example, an expert football player perceives key information from teammates' and opponents' behaviors, makes quick and accurate decisions, and has a repertoire of passing actions, which allows them to consistently perform accurate passes in a variety of conditions and environments.

Skill learning is widely defined as a relatively permanent change in a person's capacity to perform a skill due to practice or experience [292,293]. This definition includes two important aspects that are relevant for developing and investigating training devices for skill learning. Firstly, it emphasizes that learning results in a relatively permanent change of behavior. Secondly, it implies that skill learning requires practice and experience, and in fact the so-called *law of practice* indicates that performance improves and learning progresses as a function of practiced trials [294]. Though the mathematical function that best fits the law is debated (e.g., [295]), it is relevant to highlight that practice is fundamental to promote learning. As a result of practice, perceptual, cognitive, neural, and motor coordinative processes underpin and promote learning [296]. Importantly, from an ecological approach to perception and action [288], the perception of key information that specifies affordances and the calibration of action on that information drives learning [297]. Given the two core aspects that characterize skill learning, several issues arise regarding the assessment of learning processes and products. The next section discusses these issues.

4.2.2 Learning vs. performance: Why this distinction is important

An important distinction that warrants special consideration when learning is assessed is between skill performance and learning. Skill performance refers to the execution of a skill under specific circumstances—at a specific time and in a specific situation, whereas skill learning is a change in a person's capacity to perform a skill. Importantly, performance is observable, while learning is abstract and cannot be directly observed, but must be inferred from changes in performance [293]. For example, you can observe how a kick is performed today at the training center, but you cannot observe an abstract change in a person's capacity to perform a kick. When assessing performance to infer learning, five characteristics indicate whether learning has occurred over time: improvement, consistency, stability, adaptability, and persistence (for further details see [296]). The ideal design to capture these five characteristics includes pre, post (improvement and consistency), transfer (stability and adaptability), and retention (persistence) tests. However, it is practically challenging to implement all those tests, and it is general practice to assess at least three characteristics.

The performance-learning distinction is very important as performance can be a temporary state, whereas learning results in a relatively stable level of performance. Mixing the two concepts can lead to incorrect interpretation of a learning intervention: learning can occur without improvement during practice, and improvement during practice does not necessarily indicate that learning is occurring (for an extensive review see [298]). A good example in the motor learning literature stems from feedback research using Knowledge of Results (KR) to promote learning. Early studies showed that constantly providing KR (i.e., information on the outcome of a movement) during the execution of a movement improved performance, and it was inferred that learning occurred (e.g., [299]). However, Salmoni, Schmidt, and Walter [300] conducted an extensive review and showed that the KR effect was temporary, and performance decreased when KR was removed (i.e., transfer test). These findings show how critical it is to assess at least three performance characteristics to properly infer that learning has occurred.

4.2.3 Approaches to promote skill learning

As previously discussed, skill learning results from practice and experience. However, not all practice conditions have the same effect on the learning process and some influential factors can be manipulated to maximize the efficacy of a training intervention. This section presents instructions, augmented feedback, and the manipulation of task constraints as the main factors that influence learning, given that technology can be used to manipulate them.

Verbal and visual (demonstration) instructions

Practitioners, such as coaches, teachers, and therapists can use instructions to guide an individual executing or learning a skill. Information can be provided through verbal instructions—telling what to do—or visual instructions, showing what to do. For example, a coach can tell a golfer how to handle the club, how to position their feet and how the movement sequence should look like, or can demonstrate the movement and show how the movement looks like. These two approaches have different advantages (and disadvantages), and previous research has indicated how the focus of attention elicited by verbal instructions and how selecting a demonstrator are key factors influencing the performance and learning processes.

The instructional content can direct a learner's focus of attention internally or externally, in reference to one's own body. An internal focus of attention is promoted when the wording of instruction refers to a body part involved in the movement, whereas an external focus of attention is facilitated when an instruction focuses on elements outside of a performer's body (typically the outcome of a movement) [301]. For example, telling a performer what angle a joint should bend at would direct one's attention internally to this specific body part, while telling to throw an object as far as possible would direct the attention externally. Previous research has shown that an external focus of attention is more effective than an internal focus of attention in promoting skill learning and performance, resulting in higher accuracy, quicker learning, higher transfer, and more robust performance under pressure [302,303]. These findings seem to generalize to simple and complex skills and across the expertise continuum (i.e., from novice to expert) [301,304].

Instructions can also be provided visually with a person demonstrating the skill to learn. This instructional approach is generally easier to grasp than verbal instructions, as one can directly see how a skill looks like. For example, a coach can perform a basketball free throw, and learners can see the basic structure of the throwing skill. Two types of models (i.e., person showing the movement) have been examined in previous research: learning model, whereby a learner watches another learner learning a skill, and expert model, where an expert shows the skill. The learning model is typically beneficial for novices as they can observe and learn from common errors, while the expert model is beneficial for skilled performers as they can observe how a successful movement looks like and learn from it [296]. It has been shown that a mix of both models enhances learning as a learner can observe errors and the "optimal" movement at the same time [305]. Furthermore, cues (verbal or visual) can be provided during observation to direct one's attention to the key feature(s) of a movement, which might be difficult to detect due to the numerous degrees of freedom involved in a movement [306].

Augmented feedback

Augmented feedback, also called external or extrinsic feedback, is widely considered a critical component of skill performance and learning [293,307]. It refers to feedback provided by an external source (e.g., coach, trainer or technical device), as opposed to intrinsic feedback, which is naturally available to the senses. Augmented feedback can be classified according to (i) timing, (ii) frequency, (iii) presentation modality, and (iv) content. The effect of augmented feedback on skill learning depends on the interaction between those factors along with skill complexity and a learner's skill level (for an extensive review see [93]). While it is quite difficult to separate the influence of each feedback component in this complex multifactorial process, a brief discussion on each factor is presented next.

Feedback can be presented during a movement (concurrent), after a movement has been executed (terminal), or with a delay after movement execution (delayed). It is generally accepted that concurrent feedback facilitates skill performance during the acquisition phase, but does not promote learning in simple skills (i.e., guidance hypothesis [300]), though it seems to be effective for learning complex skills (e.g., [308,309]). An important component of learning is error recognition, and delayed feedback can facilitate error detection and, in turn, support learning [310]. The effectiveness of feedback timing is strictly related to feedback frequency and different strategies have been examined to counteract the guidance effect.

Frequent concurrent and terminal feedback have typically led to the guidance effect (in simple skills) [300] and strategies—fading, bandwidth, and self-controlled—have been implemented to avoid such effect. A reduction of feedback frequency through fading (i.e., the frequency is reduced over time), bandwidth (i.e., feedback is provided only when an error exceeds a set threshold), and self-controlled (i.e., learner requests feedback) strategies have been shown to be effective and promote skill learning (e.g., [311,312]). Different fading rates have been proposed and it is unclear which rate is optimal for promoting learning; given the multifactorial process, it is unlikely that one rate would fit all. When using the bandwidth strategy, the selection of error threshold is critical and providing a nonoptimal threshold (especially, mixing inherent movement variability with task-related errors) would encourage learning of nonoptimal movement patterns. Self-controlled feedback has received an increased attention lately, and research has shown that it promotes skill learning via different mechanisms, such as strengthen of *correct* movement patterns after successful trials and enhanced self-efficacy [313].

From a content perspective, augmented feedback can be primarily distinguished in Knowledge of Results (i.e., related to a movement outcome, KR [299]) and Knowledge of Performance (i.e., related to the movement itself, KP [314]). For example, in a throwing task, KR would provide information on throwing accuracy, while KP would focus for instance on elbow angle during the throw. KP can be further categorized in kinematic—movement coordination patterns—and kinetic (forces produced during the movement) feedback [315]. As previously mentioned, one size does not fit all and feedback content has a different effect depending on the type of task performed. As a general rule, augmented feedback improves the specified parameter and, consequently, it should specify the criterion of a task to improve performance and learning [316]. In relatively simple lab-based tasks with one degree of freedom (e.g., lever positioning task), the spatial or temporal characteristic of a movement typically corresponds to the outcome/criterion of a task and feedback specifying that parameter promotes learning [317]. However, in complex skills with several degrees of freedom, such as hitting a tennis ball, the selection of a key parameter or task criterion is challenging and highly depends on the type of task performed. Fowler and Turvey [318] argued that the degrees of freedom of feedback should reflect the degrees of freedom of the movement of interest. Accordingly, a combination of KR and KP might be necessary or video replay of a movement can be used to capture its degrees of freedom, and research has shown benefits of such strategy (e.g., [319]).

Lastly, augmented feedback can be presented using different modalities—visual, auditory, haptic, or a combination of them. For example, visual feedback is typically presented on a screen or through augmented reality, and visualizations include showing movement errors using plots, bars or lines, and showing animated or real videos. Considering that vision is highly engaged during skill execution and visual feedback might cognitively overload a learner, auditory and haptic feedback modalities are gaining interest. Auditory feedback can be provided in the form of melody, pitch tones or music, and movement sonification, whereby a movement parameter or error is sonified, is a promising avenue for skill learning [320]. Haptic feedback can guide/assist a movement or amplify a movement error through vibration. Multimodal feedback has been suggested to promote learning; specifically, audiovisual feedback can enhance perception, and visuohaptic feedback can improve spatiotemporal property of a movement [93].

Manipulation of task constraints

The constraints-led approach, grounded in dynamical systems theory, provides a suitable framework for manipulating characteristics of a task or environment to promote skill learning [321]. In brief, this theory contends that movement (and learning) emerges from the self-organization process under the interaction of constraints [322], which have been classified in individual (i.e., characteristics of a performer), task (i.e., requirements of a task), and environmental (i.e., properties of the environment). A change in one of the constraints acting on a movement modifies how a skill emerges, and practitioners can purposefully manipulate constraints to facilitate the emergence of skills that are functional to achieving the goal of a task [323].

Here, we focus on how manipulating task constraints influences skill learning, considering that practitioners can readily modify this category of constraints and previous research has shown its benefits. Equipment modification is common practice (especially in youth sport), and research has shown that scaling equipment to match the developmental stage and skill level of learners facilitates the execution of a skill and promotes learning [324]. For example, the size of racquets and net can be scaled to children's age and in turn facilitate learning of tennis skills [325,326], and football ball stiffness can be reduced and in turn promote learning of a kicking skill [327]. Furthermore, modifying the rule of a task, such as game area and type of movement allowed influences how skill emerges [328]. For example, reducing the game area and number of players typically results in enhanced skill involvement and accuracy in sport (e.g., [329,330]).

4.2.4 TaHiL technology and skilled behavior and skill learning

Technology can be used to manipulate instructions, augmented feedback, and task constraints and create a learning environment that fosters people's skills. For example, in physiotherapy, sensors can detect how a person is moving a certain limb after surgery and provide feedback in the form of sound or vibration directing the person to a more desirable movement. The same strategy could be adopted to teach workers how to safely lift and position heavy boxes on shelves, or correct movement form of expert athletes. Similarly, real-time videos of trainers providing instructions and feedback can be displayed to learners, and actuators placed on a learner's body could direct a learner's attention to specific body parts (using light, audio or vibration) according to a trainer's instructions. Previous research has primarily focused on how technology can be used to provide augmented feedback in a variety of contexts. For instance, real-time video of an expert's performance, sonification of a learner's movement in relation to an expert's, tactile feedback guiding to a desirable movement in rowing [331,332]; auditory feedback to guide alignment of the rifle to the target trajectory in rifle shooting [333], and values of exerted forces in a surgery training [334].

Considering the distinction between skill performance and learning, it is important to differentiate technology devices that aim to assist performance from the devices that aim to promote learning. A device that assists performance would primarily guide an individual performing a movement accurately, such as reaching for a cup or moving in space. In short, it would promote skilled behavior without considering how it may impact learning. For example, an exoskeleton can guide and assist the movement of an upper limb, whereby the device applies forces on the limb segments guiding one's movement towards a target. On the other hand, devices that promote learning would encourage learners perceiving and acting on key information (specifying affordances), which will result in a relatively stable improvement in performance (i.e., learning). For example, sonifying a movement parameter in rowing (e.g., timing and spatial accuracy) directs a learner's attention to that parameter and enhances their perception, which will in turn promote skill learning. This distinction should be accounted for when a technology device is designed, as its characteristics and functions will depend on its application. Accordingly, the efficiency of newly developed technological devices should be assessed considering the distinction between performance and learning described above.

TaHiL technology can improve the human–machine interaction and can further enhance how technology assists and promotes skills. TaHiL technology refers to cooperating cyber-physical systems (CPS) containing adaptive learning mechanisms, numerous sensors and actuators, connected through an intelligent network with humans or other CPSs, which enable fluid interactions (see Chapter 1 for further details).

To facilitate learning, novel cooperating CPS need to be developed with characteristics adapted to the specifics of the respective learning scenarios. Therefore the different components of TaHiL technologies have to be evolved accordingly: (*i*) Human perception: new algorithms and protocols for human–machine coadaption, in which goal awareness and action prediction are prerequisites for fluid interaction and integration based on models characterized by appropriate human factors, (*ii*) Human–machine augmentation: wearable peripherals for fast sensing and actuating with multimodal haptic feedback for human perception, cognition, and action, (*iii*) Human–machine networks: completely softwarized network solutions for wireless and wired communication that provide low latency, resilience, and security to enable human–machine cooperation, (*iv*) Human–machine computation: secure and scalable computing infrastructure that enables intuitive haptic interaction and automatically adapts to changes in task contexts and world models, and (*v*) Human–machine communication: novel coding and compression methods, such as haptic codecs that take into consideration human factors, compressed sensing, and network coding to enable a combined control and communication system. Certainly, these technologies can introduce a change in learning by intervening in human behavior in an unprecedented way.

TaHiL technology can enhance how devices collect, process, and deliver information to a learner, improving the learning process. Most skills involve multiple joints, body segments, and muscles, which produce forces that drive movement. TaHiL can allow collecting information on the most important muscles and joints involved, processing information, selecting the most relevant parameter and deliver it to a learner, in real-time. Furthermore, it is well known how the coach's and teacher's experience in providing instructions and leading a training session influences learning, i.e., typically the higher the experience the better the teaching. TaHiL may allow practitioners to run a training session in remote with reliable transmission of video and information. This would benefit disadvantaged areas or contexts that lack qualified and experienced personnel. For example, experienced music teachers located in a certain location can provide instructions and visual demonstrations on how to play/learn piano to individuals located in another location that lacks music schools. TaHiL would provide real-time information transmission, which ensures quality videos with no delay.

4.3 Key requirements and challenges in designing skill learning with TaHiL technology

The idea of using TaHiL technology to promote learning and performance of important skills, such as handling an object, catching, throwing, and lifting, in contexts that are of utmost importance for society (e.g., occupational sector, rehabilitation, sport, and music) is certainly appealing. However, there are several challenges that need to be considered from a learning, technical, and design perspective. An example to highlight the challenges from both perspectives: a volleyball player is receiving an opponent's serve with the ball traveling at 80 km/h and we want to assist the receiving player preparing for ball contact, making a decision on which teammate to pass the ball to, and impacting the ball to proficiently pass the ball to the intended teammate. From a learning perspective: What, when, and how information on the player's body posture, coordination, teammate's position and movement, and ball's trajectory should be provided? From a technical perspective: What are the relevant physical parameters that determine the player's, teammates' and ball behavior, how to detect them and process the detected information computing only the key parameters, model the optimal response, and deliver the information to the performer in a meaningful form in such a short period of time?

Placing humans in the TaHiL technology loop also poses the challenge of privacy and data security [335,336]. TaHiL devices and the Internet of Skills will closely sensor users' physiological and behavioral data, which contains extremely sensitive and personal information [337]. Considering realistic applications, whereby the devices will be provided by commercial entities rather than public universities (or schools and associations), preventing the abuse of this data is a vital aspect for the public's acceptance of TaHiL as an actual democratizing technology. Deployed TaHiL applications will capture extensive datasets of movement and behavior traces from large groups of users to extract descriptions of actions at different skill levels and potentially prescribe intervention techniques. As such, the Internet of Skills users will wear comprehensive sensor kits and will perform their activities in front of cameras and location trackers. Potential disclosures in these databases and traces are twofold: Identity disclosures will allow for the identification of the individuals who volunteered their skilled behavior to be used for training the models, or the individuals using the TaHiL applications to improve their skills; Attribute disclosures in addition will allow for the inference of certain characteristics of the respective users [338].

With regards to the identification, extensive work has aimed at identifying individuals, based on peculiarities in their biometrics, physiological, as well as behavioral data. They are so distinct and in many cases hard to imitate, that they even have been suggested as factors for strong authentication tasks [339–341]. A prototypical example that is very easy to observe and detect, even in low-quality video or gyroscopic data is the gait when people run or walk.

Attribute disclosure in observational data can inform the provider, for instance, about personal characteristics, such as disabilities, diseases, and mental conditions [342].

Furthermore, recognizing the identity of a trained subject and thus being able to link progress over several training sessions, the providers could potentially also be able to infer the predisposition of individuals to learn certain skills, and consequently potential capabilities and extended personality traits (intelligence, grits, discipline), Attributes that clearly are highly personal and sensitive, yet so far hard to protect [343].

It is therefore critical to invest effort in ensuring users' data anonymity [344]. Removing information that reveal people's identity from collected data will come with several challenges. Traditional privacy technologies follow the strategies of either generalizing, or *coarsening* the data, which consist of reducing or perturbing the information details, via superposing random noise on the data. Both approaches are in direct conflict with the requirement of TaHiL technologies to be extremely precise in collecting data. It is unclear how to strike a convincing balance. Approaches achieving differential privacy represent the only provably secure protection. There exists a large body of work on protecting datasets following these approaches, however, most results are restricted to data in which all entries are statistically pairwise independent [345]. This is clearly not the case for sets of data sequences pertaining to the same monitored individual. Proven anonymity of statistically dependent data currently does not exist and for the idea of the Internet of Skills to become a successful application, there is dire need to develop, or use newly developed approaches to protect the privacy of its users.

4.3.1 Requirements and challenges from a learning perspective

As previously mentioned, skill learning is a complex multifactorial process, and different strategies can be implemented to promote it, including instructions, augmented feedback, and task constraints manipulation. The main challenges in effectively implementing those strategies can be summarized into the following processes: characterize the main variables that influence performance and learning of a skill, and design a suitable intervention to effectively manipulate the detected variables. In simple terms, what and how practitioners can do to promote the learning process.

Characterize the skill of interest

A skill can be characterized examining the information that guides movement and the body-segments coordination underpinning movement (perception and action). Keeping with the previous example in volleyball, a deep analysis of the receiving skill can reveal the key variables underlying successful performance and learning: information about the server's distance and position, ball's trajectory and speed, teammate's and opponent's position, and one's own position in the court guides movement, and the whole body is involved in the movement, which can take different forms depending on how the above information emerges in a specific context and an individual's capabilities.

In characterizing a skill, typically experts are compared to novices in the performance of a skill or successful trials are compared to unsuccessful trials, and mechanisms underpinning the (expected) performance difference are assessed to characterize the key features of the skill. For example, this approach has shown how information from an opponent's movement and racquet kinematics guides successful striking skill in badminton [346], and how rapidly changing information from player's interrelational speed and distance supports skills in futsal [347]. This implies that a learning intervention aimed at improving badminton or futsal skill should consider that information, which can be manipulated to enhance an individual's perception, or should be sampled to recreate a training context similar to performance.

Furthermore, a skill can be categorized based on the size of musculature involved, specificity of where actions begin and end, and stability of the environmental context [296]. Accordingly, a skill can be (*i*) gross (e.g., jumping) or fine (e.g., writing) if the movement involves large or small muscles respectively, (*ii*) continuous when action beginning and end are arbitrary (e.g., cycling), discrete when beginning and end are specified (e.g., kicking a ball), serial—which involves a series of discrete movements—e.g., a dance sequence, (*iii*) open (e.g., driving) or closed (e.g., golf putt) if the environment is changing or stationary, respectively. This categorization can implement skill characterization and provide information on which skill elements should be targeted in a learning scenario. For instance, a training design should consider manipulating environment properties in an open skill, but not necessarily in a closed skill; augmented feedback can be provided concurrently (i.e., real-time) in a continuous skill and, unless it is performed slowly, not in a discrete skill; instructions should address the main muscles involved in a gross skill.

Design an appropriate learning intervention

Once the information and action variables are "characterized", a plan to manipulate those variables has to be designed. The main potential strategies include a manipulation of content, focus of attention, and delivery modality of instructions; presentation modality, timing, and content of feedback; and task elements, such as rules and equipment. Considering the range and diversity of possible solutions, selecting an effective strategy is quite a challenge. Keeping with the volleyball example, an opponent's distance can be increased to extend the time available to perceive information and organize an appropriate action, or distance can be reduced to force a player to quickly perceive and act upon a demanding context. The key question is: How much should such distance be increased or decreased? And, how should the other information from the previous step (e.g., other player's positions) be implemented in a training task?

Principles developed in the field of expertise can guide this procedure. A recently developed framework (SPORT [348]) features three main principles: progression, overload, and specificity. Progression and overload refer to a progressive increase of training load to "stress" a learner's movement system and promote skill improvement. A training task should challenge an individual's current skill ability and improve their perception and action capacity. This means progressively increase the difficulty of a task, and task difficulty can be individualized by assessing skill performance. When skill accuracy is too high (e.g., above 90%) or too low (e.g., below 50%), the task difficulty is too high or too low, respectively, thus task difficulty should be adjusted [349]. Furthermore, specificity contends that training should be specific to the condition and context in which the skill is normally performed, to ensure a positive skill transfer from practice to the intended environment. This means that a training task should be representative of the information-movement coupling that characterize a skill in its own environment. Back to the previous example in volleyball, the following steps can be followed to design a training task for the receiving skill: (i) include a server (not a bowling machine), and (at least) a teammate to pass the ball to make the task specific, (ii) decide a variable to manipulate: server's distance, (iii) decide a strategy for increasing the load: the distance can be progressively decreased, (iv) assess receiver's accuracy to evaluate whether task difficulty is appropriate to the individual, and (v) adjust distance accordingly if needed. When deciding a strategy for manipulating a variable, the aspects discussed in (see Section 4.2.3) should be taken into account to maximize the benefit of a training scenario. For example, instruction can draw a player's attention to the ball trajectory or to the location the ball can be directed to, as opposed to one's own movement, to encourage an external focus of attention and implicit learning.

In summary, the design of a training intervention requires a thorough analysis of the skill of interest and an accurate selection of manipulation strategy to ensure a positive effect on skill learning. These requirements pose challenges from a design perspective (as discussed above), but also from a practical perspective, especially in the Internet of Skills, where TaHiL technology is exploited to promote learning. In this sense, technology will be used to manipulate a training variable, and technical requirements (opportunities and limitations) should be accounted for when a training intervention is designed. The next section discusses requirements and challenges from a technical perspective.

4.3.2 Requirements and challenges from a technical perspective

To carry out the skill learning scenarios described above as effectively as possible, the associated technical systems have to meet a variety of requirements. The first step is to identify the type of physical parameters that essentially describe the action of the object, here primarily the human being. Considering that a human being represents, from a mechanical point of view, a multielement system with integrated joints, various trajectories have to be measured in most skill learning scenarios. This leads to questions related to suitable locations for and amount of required sensory and actuator feedback systems. For example, with reference to the volleyball scenario described above, this means to translate the required information about the player's body posture into parameters and corresponding locations that can be measured or influenced by technical transducers. Whereas the volleyball scenario is quite complex, a simpler scenario is presented in Fig. 4.1 to exemplify how sensors can be located on and assess parameters of a hand. The main idea here is, if the orientation and position of the base component of a multielement system is known, the tracking of the relative movement is achievable.

Typical measurable physical parameters include acceleration, rotation, elongation, bending, as well as pressure and loudness. A variety of fitting sensor and actuator principles are available to measure those parameters. Tables 4.1 and 4.2 show exemplary technical relevant sensory and actuatory feedback systems. Chapter 7 covers possibilities of different sensor and actuator technologies in more detail.

For each of those sensory and actuatory feedback mechanisms, the main technical challenge and related technical implementation arise from these specific key issues: (*i*) Sensitivity: How to accurately measure/manipulate, and what is the corresponding measuring/setting range? (*ii*) Stability: How stable is a system for a long operation period and under changing loads and ambient influences? (*iii*) Selectivity: Are the measured/manipulated parameters influenced by other parameters? (*iv*) Reliability: How to deal with sources of errors to reduce measurement errors? Additionally, electrical questions (e.g., powering, data transmission and processing, and case decision) and mechanical questions (e.g., fixation, size, weight, and stiffness of the transducer systems) have to be considered with respect to the skill learning scenario requirements. From the user perspective, an optimal transducer system would have a wireless powering and data transmission system, and large lifetime and processing capabili-

Table 4.1 Examples of sensor systems for different measured variables.

Parameter	Implementation example	Benefits	Drawbacks
Acceleration	capacitive (MEMS)	miniaturization well possible	
	gyrometer	accurate	relatively complex technology, only limited possibilities for miniaturization
	piezo-electrical	miniaturization well possible	only changes in the measured parameter can be detected
Rotation	gyrometer	accurate	relatively complex technology, only limited possibilities for miniaturization
	piezo-electrical	miniaturization well possible	only changes in the measured parameter can be detected
Elongation	laser	very accurate	very complex and expensive technology
	resistive	easy to measure	standard strain gauges limited in applicability
Bending	resistive	easy to measure	standard strain gauges limited in applicability
Distance/ proximity	capacitive	relatively easy to build	limited reliability, cause of many possible parasitic drag
	inductive	partially complex in construction and measurement	relatively stable sensor signal
	laser	very accurate	very complex and expensive technology
Pressure	piezo-electrical	miniaturization well possible	only changes in the measured parameter can be detected
	resistive	easy to measure	standard strain gauges limited in applicability
	capacitive	relative easy to build	limited reliability, cause of many possible parasitic drag
	inductive	partially complex in construction and measurement	relatively stable sensor signal

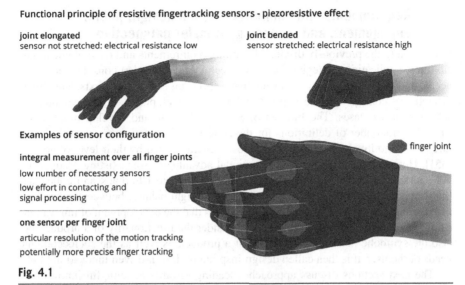

Functional principle of resistive fingertracking sensors - piezoresistive effect

joint elongated
sensor not stretched: electrical resistance low

joint bended
sensor stretched: electrical resistance high

Examples of sensor configuration

integral measurement over all finger joints

low number of necessary sensors

low effort in contacting and
signal processing

finger joint

one sensor per finger joint

articular resolution of the motion tracking

potentially more precise finger tracking

Fig. 4.1

Position detection of a joint, based on resistive sensors (principle *CeTI glove*); (top left)
Example of a glove with fully integrally manufactured textile sensors; (top right) Functional
principle of resistive sensors for finger tracking; (bottom) Examples of possible sensor
configurations.

Table 4.2 Examples of feedback systems.

Addressed sense	Type of feedback	Implementation example	Remarks
Vision	video	beamer/monitor	stationary
		VR/AR glasses	mobile
	light signals	status-LED	meaning of the signals needs to be known
Audio	text	speaker	stationary
		headphone	mobile
	sounds or volume	speaker	stationary, meaning needs to be known
		headphone	mobile, meaning needs to be known
Haptic	tactile	vibration motors	meaning of the signals needs to be known
	kinesthetic	counterforces through cable pull	large actuator superstructures needed at present
	pain	electrodes	ethically controversial

ties with little latency. In addition, it would be very small, light weight, flexible and
perfectly fitting tight. These targets, however, are often contradictory, as detailed in
Chapter 7.

4.3.3 Requirements and challenges from a design, public engagement, and technology transfer perspective

In addition to the previously discussed learning requirements and technological possibilities, application strategies for TaHiL technology will also consider the user's needs, and put all these aspects in relation to relevant societal contexts. This interdisciplinary concept will be applied to both the development of devices and the selection of use cases. The Internet of Skills will create and develop innovations. There are a number of definitions for the concept of innovation in scientific literature [350]. Intuitively, innovations are mostly characterized by their level of novelty [351]. However, technological and functional advantages over existing products do not guarantee success in innovation [352]. According to Utterback and colleagues [353], successful innovation can be defined as the right balance between technology, market, and significance for the user. This means that the development of innovations using TaHiL technology should not only consider the functionality of a product, but also the symbolic and emotional value of a product, as well as the socioeconomic needs of the user. It is then called design-inspired or design-driven innovation [354].

The next sections discuss approaches leading towards meaningful innovations: relevant aspects of the human-centered design process, strategies to utilize use cases, how to engage the public, and how to transfer TaHiL technologies into the market.

4.3.3.1 Human- and experience-centered design

To develop meaningful TaHiL applications, it is vital to involve the user perspective throughout the product development process. The human-centered design approach enables TaHiL technology to "enhance effectiveness and efficiency, improve human well-being, user satisfaction, accessibility and sustainability; and counteract possible adverse effects of use on human health, safety and performance" (ISO 9241-210:2010(E)).

Over the last decade, interactive products became not only more useful and usable, but also more fascinating and desirable. User experience explicitly describes the even more holistic perspective and goes beyond the concepts of pure functionality or usability. It describes technology, which "fulfils more than just instrumental needs in a way that acknowledges its use as a subjective, situated, complex and dynamic encounter." [355]. Our approach enriches human-centered design with user experience. Three levels of user experience are considered when developing TaHiL technology (each level is illustrated by two significant questions) [356,357]:

Function and reliability What do humans want to achieve while interacting with technology? Are they able to complete their task?

Usability and convenience How do humans interact with technology? Do they feel in control?

Pleasure and significance Why do humans interact with technology? What motivates them?

There are numerous design process models from academia, professional associations and companies, which differ in focus, content, structure, and graphical notation.

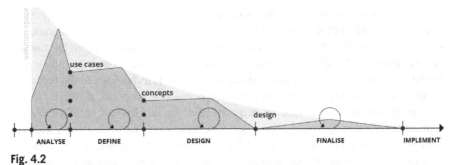

Fig. 4.2

Design process model [358].

In spite of various different formal representations, the majority of models follow the basic design process framework, which includes four plus one stages: *analyze, define, design, develop, and implement*. Fig. 4.2 illustrates the adaptive multilayer design process visualization [358]. The design process complements the product development process and product lifecycle [359]. The design process will be presented shortly with focus on the first process phase *analyze*.

The first phase *analyze* gathers information regarding the design challenges. Whereas *define* is the stage of synthesis on an abstract level (concepts), *design* results in a concrete synthesis. The last and longest phase *develop* elaborates in detail the technical implementation of the design. The end of the design process marks the beginning of production or implementation of the design, i.e., the *implement* phase. Evaluations and iterations inside and in between these phases are a crucial part of the design process [359].

As a way of exemplifying the focus on the user, we will briefly zoom into the first phase. *Analyze* aims to empathize primarily with the end user (main stakeholder), understanding their needs, expectations, and behavior. To achieve this, several methods have been adapted from neighboring disciplines, such as social science (guided interviews and surveys) and business science (market analysis). Two methods have been valuable for understanding task-related and nontask-related requirements from a user perspective: (a) the persona, and (b) the user journey map. Personas are fictional characters, created to represent different target groups, their peculiarities or even different stakeholders. They are based on actual data and bundle information, and help to understand what the user feels and thinks, how they act and what motivates them [360]. Parts of the persona are demographic data, quotes (e.g., *With this device, I am part of a community*), motives, biography, and attitudes.

The user journey map describes the full user's lifecycle and analyzes, from the user perspective, all touch-points the user has with the product (or service), from *heard about* until waste disposal. From these condensed narrative scenarios, product requirements are derived. Furthermore, the user journey map serves as a guide and evaluation tool throughout the process. The persona and user journey map should be combined and implemented in the first process phase when designing technological devices. There is one more benefit associated with these methods: they foster inter-disciplinary collaboration within and in-between disciplines involved in the process. Representing complex research data in a condensed format increases understanding between different disciplines in the research teams themselves as well as understanding for different stakeholders, who will encounter TaHiL technology in the future.

In conclusion, the human-centered design process proposes different stages to identify essential user specific requirements for the Internet of Skills. This application-oriented approach allows the communication and transfer of TaHiL research to the general public and eventually to the economy. The next sections discuss how use cases should include aspects for the public engagement and technology transfer.

4.3.3.2 Public engagement

Public engagement is a core challenge for technological information and also for science in general. The Internet of Skills has great potential in engaging the public as people with common knowledge and any experience levels can easily relate to use cases, such as training sport and learning musical instruments. TaHiL demonstrators are designed to enable learning experiences and thus predestined to inspire people in live events or exhibitions.

4.3.3.3 Technology transfer

In addition to science communication, transferring a developed technology into the economy is a key application issue for the Internet of Skills. Technology transfer includes creating patentable innovations that can be transferred to start-ups or existing companies. In this process, science commercialization represents a key factor for economic growth [361] and an important indicator for society and politics on the success of a research project. Thus choosing use cases relevant to important societal and economic needs is crucial to prepare successful spin-offs. The St. Gallen Business Innovation Model assumes that consumers are predominantly concerned with whether and to what extent an innovation can be integrated into their daily routine [362]. Furthermore, other points of interest could be the potential to provide users with new experiences or benefits [363]. Therefore to develop successful technologies, TaHiL applications need to be developed considering how they can be applied in people's everyday life.

These considerations regarding public engagement and technology transfer add further elements to the development of a use case and learning scenario: (*i*) Define a set of environments, which have a connection to people's everyday life! (*ii*) Select an audience in the form of Personas, such as the end-users and other stakeholders like future investors! (*iii*) Specify a set of actions, in which the developed TaHiL application plays an essential role!

In summary, implementing the human- and experience-centered design approach, public engagement and technology transfer will delineate how use cases can be designed and how their effectiveness can be assessed.

4.3.4 **Application strategies and previously developed demonstrators**

The Internet of Skills is classified into three strategies: machine to human, human to machine, and human–machine unsupervised interaction. Given the early stage of TaHiL technology development, the machine-to-human strategy is discussed in detail, and demonstrators previously developed within this strategy are presented.

4.3.4.1 Machine to human: TaHiL devices promote learning and skilled behavior

This strategy aims at using technology to facilitate skill learning or assist skill performing. For example, individuals can be guided towards biomechanically appropriate and safe movements. This strategy can be applied to a variety of contexts, including physiotherapy and rehabilitation, music and sports.

To show how human-centered and experience-centered design, public engagement and technology transfer can be integrated into the machine-to-human strategy, demonstrators have been developed to be presented at public events. The following issues should be considered when developing such a strategy: (*i*) Human-centered and experience-centered design: What contributes to society? Why and how? How can physiotherapists, carers or patients use TaHiL technologies in their everyday life? (*ii*) Public engagement: What future applications can be used to reach the broad public—from athletes to seniors? What kind of demonstrator can be presented at public events? (*iii*) Technology transfer: How can TaHiL technology support the healthcare system? Which strengths and weaknesses can be addressed? How do experts assess the benefits?

In the following sections, three demonstrators are presented: a vibrotactile and visual feedback system for learning a dance choreography, a tracking system for adjusting volleyball technique, and a real-time force feedback system for supporting and monitoring training in rowing (for further details on this demonstrator see [364]).

Vibrotactile and visual feedback for learning dance choreography

Concept: In the case of a sudden absence in a dance team, a professional replacement dancer is required to learn the team choreography within a short period of time and often detached from the team. A demonstrator was designed to guide a replacement dancer learning a choreography on their own.

Demonstrator: The demonstrator includes a tablet, a vibratory belt, projectors and a monitor connected to a laptop (Fig. 4.3). All components are connected via internet. The belt has eight integrated vibration motors that can independently give a time-variable vibration feedback. A dance sequence with the position a dancer should be in during the choreography is uploaded into the tablet and laptop. The monitor displays the dance sequence, the projector indicates the corresponding formation of the other dancers on a two by four meters area, and the vibratory belt provides the dancer with haptic feedback that indicates the dance directions.

Application: Three professional dancers tried the demonstrators and provided positive feedback. They learned a default sequence within two iteration loops and

Fig. 4.3

Assembly of a dance choreography learning demonstrator: Beamer projection and vibratory belt.

stated that they perceived an enhanced sense of space thanks to their virtual dance partners and the directional impulses of the belt.

Tracking system for adjusting volleyball technique

Concept: Volleyball is a sport that requires precise and skilled movement of the arm/hand to project the ball to an intended target. It is challenging for coaches to spot their player's movement errors, and they can benefit from a system that captures player's movement in real-time. A demonstrator (sensor-suit, "Sensing in Motion") was designed to track and project one's own motion sequence into a screen.

Demonstrator: The demonstrator comprises functional sleeves with nine integrated sensors that send motion data to a computer, and a monitor displays the data in the form of a virtual avatar (Fig. 4.4).

Application: The demonstrator was presented at the 5G Summit 2018 in Dresden. It attracted people's attention, and many visitors tried it out providing positive feedback, as they could see how their movement looked like, and thus refine their technique.

Real-time force feedback system for supporting and monitoring training in rowing

Concept: Rowing is a complex movement that requires a combination of spatial and temporal accuracy. It is a challenging movement for both novices and experts (to a different extent though), and they can benefit from a feedback system that provides information on and adjustments to their movement.

Demonstrator: The demonstrator includes a soft exosuit with six integrated sensorimotors with cable control that can be worn like a jacket (Fig. 4.5). The sensors

Fig. 4.4

Assembly volleyball demonstrator: Sensor bandages and avatar on screen.

Fig. 4.5

Rowing demonstrator: A soft exosuit, which amplifies or facilitates a stroke.

can track an athlete's movement, and the data is displayed on a monitor. Actuators can create local forces with the aid of pulling ropes attached on upper- and under-arm sleeves. A battery and processor pack, placed on the athlete's back, control the motors. Furthermore, the system can provide acoustic and haptic signals.

Application: The demonstrator was presented at the 5G Summit 2018 in Dresden. The highly representative setup (i.e., a rowing simulator) attracted many visitors and encouraged them to try it out. Whereas they liked the concept, they were concerned about the suit's relatively high weight, and impossibility to tailor and customize the system to individual needs.

Implications for further research and development

While the effectiveness of these demonstrators in promoting skilled behavior and learning has not been assessed, they showed how the Internet of Skills can enable the design of new devices using a user-centered approach and can effectively engage the public. These pilot projects were the result of and can lead the way to interdisciplinary collaboration (e.g., mechanical engineering, design, and computer science).

These preliminary studies provided new insights into developing demonstrators for public engagement and technology transfer. From a user-centered design perspective, participatory methods should be included to analyze and identify meaningful skill learning scenarios, data-based personas should be generated, and user experience evaluation tools for interactive demonstrators need to be developed. From a public engagement and technology transfer viewpoint, tangible user-centered demonstrators proved to be suitable means to convey complex and futuristic technologies; interest for new technologies can be raised considerably through examples from popular fields of application, e.g., sport; performance of the demonstrators, e.g., intensity of haptic feedback has been assessed differently, depending on the user's expectations (exoskeletons are more critical than wearables); noninstrumental qualities, such as social influence and hedonic motivation facilitate accessibility to TaHiL applications for the laypersons as well as experts.

4.4 Beyond the state-of-the-art approach: Interdisciplinary collaboration

4.4.1 Developing a learning scenario with TaHiL

Though the volleyball example was used to exemplify the learning, technical, and design challenges, a relatively simpler skill has been chosen as starting point to implement and assess the processes listed above. A whole-body lifting task (e.g., lifting a box or squatting) was chosen, and the next section discusses why it was chosen and how a learning scenario will be developed for this skill. Different user-centered design approaches (e.g., Industrial and Instructional Design Approaches) will be combined not only to design a TaHiL demonstrator, but also to investigate the benefits and constraints of using such demonstrator in pursuing the aims of the Internet of Skills.

4.4.2 Current idea: Whole-body lifting task

A whole-body movement to lift an object represents an important skill, given that it is performed frequently in our daily life and is the core skill of many jobs, such as construction, parcel delivery, and nursing. An incorrect lifting movement is associated with an increased risk of back injury and pain, a condition that affects millions of people worldwide and is one of the leading causes of work absence, which has been estimated to exceed $100 billion per year in the US alone [365]. It becomes apparent that strategies are needed to improve the learning and safe performance of lifting movement.

Whole-body lifting is a gross and discrete (relatively slow in speed) skill, and can be classified as closed or open depending on the context. Muscles (e.g., back muscles and quadriceps) and joints (e.g., hips, knees and spine) of the lower body and trunk are primarily involved in the movement. Lifting can be considered a closed skill if it involves lifting a stationary object and placing it on a stationary surface. A critical feature for a safe and functional lifting movement is how the lifted load is distributed among the muscles and joints involved. Injury on the low back occurs from a lifting movement that overly stress the structures of the lower part of the spine, via excessive flexion, torsion or acceleration of the area. Injuries can also occur at the knees or ankles due to excessive stress.

Augmented feedback can modify a lifting technique and lower the load on certain joints, primarily on the low back. Feedback can be provided on a wide range of parameters related to the movement, such as displacement of joints, joint-to-joint angles, and joint spatiotemporal characteristics. For example, flexion of the spine during lifting is linked to an increased load in the lower part of the spine, and previous research has shown how providing tactile [366] or auditory feedback [367,368] on spine flexion reduced spine bending and (likely) load. Other research has targeted knee alignment, and attaching a laser beam to the knee was effective in augmenting participant's perception of knee alignment and improving performance [369]. Considering that lifting is a complex movement that involves different joints and muscles, ideally (from a learning and performance perspective) all joints and muscles are concurrently tracked and then feedback is provided on the most critical aspects. This, however, will pose technical challenges.

The main challenge in the attempt to assess the body lifting movement is the determination of the skeletal position and the relation between body parts. Direct body-scanning methods, such as x-raying, would be very effective for this task, as they allow direct imaging of the bones and thus their position relative to each other, but they are technically very complex, hardly suitable for recording movement sequences and can be harmful to health. It is therefore necessary to determine the joint or bone positions by means of measurements on the body surface.

In most cases, the movement/posture of a tested person is recorded and digitized with the help of camera-based motion-capturing systems. This method is initially pursued in the Internet of Skills, and user's performance of the body lifting movement will be assessed using a motion-capture system. In addition, textile sensors can be used to record movements and postures independently of complex optical tracking systems. This approach has been shown in a previously developed demonstrator— Rock Paper Scissors Demonstrator (Chapter 10 see the *Rock Paper Scissors Demonstrator*). Another promising approach is a strain measurement at joints or other parts of the body, such as the back. The main challenges are the design of suitable textile sensors, their electrical insulation from the conductive human skin, the nonslip fixation of the sensors on the body parts to be measured, and a signal processing/power supply that has the least possible influence on the user's mobility.

Wearable technology will likely be used in further stages to assess the lifting movement and, from a holistic human-centered design perspective, it is crucial that the wearable technology is comfortable for the user. User acceptance for this body-worn technology is higher when appearing like a lifestyle gadget that supports a positive self-awareness, rather than a medical device, which makes the user feel externally controlled [370]. Altogether, the aim should be to create a user experience of wellbeing, enhanced self-consciousness, and increased motivation when using the lifting device.

Public engagement and technology transfer are important aspects to consider when designing a technology device that facilitates learning and performance of body lifting. As already mentioned, whole-body lifting is a physical movement performed by a large number of people in their profession (e.g., construction workers), but also in the fitness sector (squat and dead-lift are widely performed in the gym) and everyday life (e.g., lifting bags or boxes when shopping or moving house). Due to this broad applicability, the general public can likely relate to and recognize the relevance of developing a device and learning strategy to promote safe lifting movement.

Interactive demonstrators for exhibits of the body lifting device can be developed to enhance public engagement. For this, explicitly communicating the demonstrator's relevance to visitors (by relating it to common knowledge and experience) and allowing the demonstrator to be experienced by everyone should be carefully considered. For example, robust and reliable hardware is needed for quick and hygienic changes between users. The experience should be designed in a way that technology phenomena can be manipulated (e.g., switching feedback modality) and, whenever applicable, social interaction should be encouraged (e.g., two player game). This approach enables informal learning experiences while interacting with the box-lifting demonstrator [371]. For an effective technology transfer, exhibited demonstrators should explicitly emphasize how people can readily use the lifting device in their daily life (e.g., clean casing and small hardware) and how they differ from existing applications in the market (e.g., learning character, real time data transfer). Furthermore, to enhance technology transfer, the device design should highlight the proposed novel solution, e.g., performance and advantages of newly-developed sensors, novel communication technology, and appropriate learning approaches.

4.5 Conclusion and outlook

In conclusion, our vision is to bring together expertise from multiple disciplines, design interventions that implement recent advancements from different perspectives (e.g., motor learning, textile, and design engineering), and exploit TaHiL technology. As a result, new strategies will be developed to promote learning and skilled performance of skills, with implications for a large portion of the population (e.g., specific working populations, athletes, and musicians). As a first step, this approach will guide the development of a suitable strategy to deliver augmented feedback for improving the learning of body lifting movement. For example, currently in body lifting research, spine bending is widely used as an approximation of load on the spine for its relatively easy application; however, load depends on other factors (e.g., load moment and acceleration), and this approximation can be misleading. Ideally, load should be calculated and augmented in real-time. Calculating the load on the spine is challenging though, as multiple information from multiple sources need to be detected and processed, and the TaHiL can help designing assessment strategies that capture the load on the spine in real-time, and use this data for delivering augmented feedback on the load on spine in real-time. The benefits and boundary conditions of promoting safe lifting with TaHiL will be investigated in experimental evaluation studies.

Key technology breakthroughs

Outline

The second part of the book introduces the four interdisciplinary research themes that target key technological breakthroughs, namely haptic codecs, intelligent networks, augmented P/I, and coadaptation. These key technologies support the aforementioned application domains and build on the outcomes of the target primary research fields that will be described in subsequent chapters in Part 3.

Haptic codecs for the Tactile Internet

5

Eckehard Steinbach[a], **Shu-Chen Li**[b], **Başak Güleçyüz**[a], **Rania Hassen**[a], **Thomas Hulin**[c], **Lars Johannsmeier**[a], **Evelyn Muschter**[b], **Andreas Noll**[a], **Michael Panzirsch**[c], **Harsimran Singh**[c], **and Xiao Xu**[a]

[a]*Technical University of Munich, Munich, Germany*
[b]*Technische Universität Dresden, Dresden, Germany*
[c]*German Aerospace Center (DLR), Oberpfaffenhofen, Germany*

Any sufficiently advanced technology is equivalent to magic...
– Sir Arthur C. Clarke

5.1 Scope of haptic codecs

The Tactile Internet with Human-in-the-Loop will enable users and machines to exchange skills across distances with the help of realistic touch experience. Haptic communication is a key technology for the Tactile Internet, supporting remote human-human and human–machine physical interactions. For a transparent and realistic experience, haptic information should be processed and presented with a preserved perceptual quality and minimum latency. Haptic codecs play an important role in this context, as they provide compact representations of haptic information. The goal of haptic codecs is to reduce the data rate significantly, while maintaining high signal quality.

The use cases of haptic codecs cover a wide range of human-to-human or human-to-machine interactions. Examples for this are remote robot control applications or enhancing virtual environments with convincing touch feedback. In this chapter, we will mainly focus on haptic codecs that reduce the rate of transmitted data while using human perceptual models. Applications, such as remote teaching, have received only little attention so far.

5.2 State-of-the-art research and technology

Haptic information is categorized into two different submodalities, namely kinesthetic and tactile. These modalities differ both in terms of the perceived information and their processing chain for efficient transmission. Kinesthetic information refers to the perception of limb position/movement, and also the applied force/torque

Tactile Internet. https://doi.org/10.1016/B978-0-12-821343-8.00016-2

[372,373], whereas tactile information relates to the perception of surface properties via touch, such as roughness, friction, warmth, and hardness [374]. Depending on the target application scenario, either one or both haptic submodalities are considered [9]. However, the encoding and processing of kinesthetic and tactile signals come with very different requirements. For example, a classical use case for the transmission of kinesthetic information is bilateral teleoperation with haptic feedback, where the human operator (leader) and the teleoperator (follower) exchange position/velocity and force/torque data. One important requirement for the stability and transparency of teleoperation systems is the high sampling and packet rate, typically at least 1 kHz [264]. Such a high update rate causes a significant load on the communication network; correspondingly, an important task for kinesthetic codecs is the reduction of the packet rate. Efficient haptic packet rate reduction, though maintaining high perceptual quality, requires the integration of human kinesthetic perception models in the design of kinesthetic codecs. Another important aspect in bilateral teleoperation is the communication delay. Even for small delay in the teleoperation loop, additional stabilizing control mechanisms are required. The state-of-the-art research in this area includes approaches that consider these control methods and the perceptual kinesthetic data reduction approaches jointly. For the tactile submodality, the considered communication scenario is unidirectional and less delay critical. Taking this into account, the state-of-the-art research in tactile data compression has concentrated on waveform-based approaches; however, so far, only for a single point of interaction. Although the data rate for a single point of interaction scenario might seem quite low in comparison to audiovisual data, the necessity to compress tactile data becomes obvious for multipoint of interaction with potentially thousands of parallel information channels [375]. For tactile data compression, the perceptual models play an important role as well, such that the codec ensures not only low bitrates, but also preserves the perceived signal quality. The tactile codecs developed so far mainly focus on vibrotactile information, which is only one aspect of the tactile modality corresponding to roughness and friction perception.

In what follows, we provide an overview of the state of the art in haptic codec design. We begin with perceptual models for the kinesthetic and tactile submodalities. We then continue with a description of existing haptic codecs covering kinesthetic codecs, joint kinesthetic coding and control for time-delayed teleoperation, and tactile codecs.

5.2.1 Perceptual models for somatosensory processing

The human somatosensory perception system processes sensory information through touch (tactile) or movement (kinesthetic) cues. Somatosensory perception relies on the somatotopic mapping of tactile information of the skin onto brain regions in the primary sensory cortex as well as motor and proprioceptive signals indicating where body parts are, and what their actions are when objects are encountered. Generally speaking, tactile cues, also termed cutaneous cues, relate to the perception of the pressure on the skin, whereas kinesthetic cues relate to the perception of force on, and movements of, the limbs or fingers [376]. Mechanoreceptors in the skin capture tactile information, such as frequency, sustained skin stretch and texture, which are first processed in the brain in the postcentral gyrus, also known as the primary somatosensory area. Furthermore, mechanoreceptors in the skin, muscles, tendons, and joints also capture kinesthetic sensory information, such as force and torques acting on the human body and encountered objects, as well as stimulus properties, such as velocity, acceleration, and force changes (see Chapter 9). Moreover, the vestibular system also directly contributes to kinesthetic perception. Through the kinesthetic sense, humans are able to perceive physical stimulus properties, such as mass, inertia, stiffness, compliance and viscosity of currently encountered, touched objects. Thus models of human sensory perception stipulate the fundamental requirements for technical systems that aim at enabling real-time interactions.

5.2.1.1 Perceptual models for kinesthetic processing

Theories and empirical findings about human somatosensory perception are at the foundation for advancing the development of haptic codecs. Detection Threshold (DT) and Just Noticeable Difference (JND) are two important concepts in psychophysics that are particularly relevant for codec development. The DT is usually defined as the weakest sensory signal (absolute threshold) that an individual can consciously perceive; whereas the JND denotes the smallest difference (i.e., difference threshold) between two sensory signals that is required for a person to differentiate between the signals. They are the basis of perceptual models (see Chapter 9).

Weber's law of JND

It is known from psychophysics that the human haptic perception system is limited, and it is often modeled by Weber's law [377,378]. Specifically, this law states that the JND, the size of the difference threshold, is proportional to the amplitude (or intensity) of the initial stimulus itself. Weber's law of JND is represented by the following equation:

$$D_I = k \cdot I, \tag{5.1}$$

where k is a constant and I and D_I denote the initial stimulus and the JND, respectively. The constant k, also called the Weber fraction, depends on the investigated stimulus property, e.g., force, stiffness or velocity and the body part, i.e., the limb or joint where it is applied. A brief summary of the Weber fraction k of human perceptual discrimination for selected haptic stimuli is shown in Table 5.1.

Force perception has been intensively studied. Of note is that the relative perception thresholds of force feedback are not completely independent of the intensity of the forces being applied. Specifically, smaller Weber fractions (in the range of 7% to 10%) were observed for larger forces from 0.5 N to 200 N; whereas for forces with lower intensity (below 0.5 N), the Weber fraction was found to be within the range of 15% to 27% [373]. The JND of Weber's law has been used as a packet rate reduction technique in haptic codec development, the so-called perceptual deadband-based kinesthetic data reduction approach [380,381]. The main idea being that the current

Table 5.1 Weber fraction k of human perceptual discrimination for haptic stimuli [379].

Physical property	k	Experimental conditions
Force	approx. 10%	arm/forearm
Movement	8% ± 4.0%	arm/forearm
Stiffness	23% ± 3.0%	arm/forearm
Viscosity	34% ± 5.0%	arm/forearm
Inertia	21% ± 3.5%	pinch-fingers, at 12 kg

stimuli need to be only transmitted when they are above the JND, meaning the change with respect to the previously applied stimulus is consciously perceivable by humans.

Multiple degrees of freedom in kinesthetic perceptual processing

The aforementioned deadband-based kinesthetic data reduction approach has been extended for multiple degrees of freedom (multi-DoF) interaction scenarios [382]. This approach has also been experimentally validated with human observers in a haptic interaction task, utilizing a remote virtual environment and a commercially available haptic device. The authors asked human subjects to evaluate the effect of different deadband values (presented in randomized order) for force, velocity, and the combination of both on the perceived quality of the VR experience. The internal model of motor control argues that during action performance feed-forward control is used by the central nervous system, which predicts the sensory consequences of an action by using an efference copy of the motor command [383]. A prediction error results as the difference between the predicted and the actual sensory feedback, caused—for example—by temporal delays. Prediction errors are perceptually relevant as they can contribute to a degradation of the link between an action and its consequences and influence the perception of the external environment (see Chapter 9). It can be hypothesized that an increase in deadband values can be equated with an increase in prediction error. That is, with an increase in deadband values, the VR environment becomes less compliant, as reality and prediction do not match, and a decrease of immersiveness can be perceptually noticed. The behavioral data in [382] showed such a common pattern: a progressive increase of deadband values had deleterious effects on the quality of the haptic experience. In other words, negative correlations were observed between the deadband values for force, velocity, and the combination of both and the respective quality ratings. Interestingly, this effect was the least pronounced for velocity ratings, and no additive effect was found for the combination of force and velocity. Hence, it could be taken as evidence that the influence of the force deadband on system quality is higher than that of the velocity deadband [382]. Thus with their work, Hinterseer and Steinbach were able to determine different deadband values in three dimensions that are perceptually relevant. However, the experiment was conducted in a simplified VR environment, and therefore other stimulus properties and their potential effects on visual-somatosensory integration were not experimentally controlled for.

Another approach of extending the perceptual model for haptic codecs is to include the perceptual dimension of force direction. Previous work has taken different approaches investigating the JND for force direction and a range of results, depending on body location, exist. For example, psychophysical experiments by [384] found the threshold to be 33° and suggested that the human discrimination ability of changes in force direction is independent of the direction of the reference force. This effect was shown to be valid for force vectors with varying force intensity [384,385]. Additionally, in [385], the authors manipulated sensory feedback and found congruency effects of haptic feedback in combination with vision. In [386], the authors further investigated the relationship between force magnitude and force direction psychophysically. Their findings indicate that the multi-DoF equivalent of the single-DoF perceptual threshold is not isotropic. Specifically, they investigated the discrimination thresholds of force feedback stimuli of changing directions. Thereby the test stimulus and a reference stimulus were presented pairwise to blindfolded participants. Their results indicate that force perception thresholds are a function of stimulus direction. Thus in the context of haptic codecs, human kinesthetic perception depends on the direction of the pending stimulus and the direction of the stimulus changes. The perceptual thresholds in this case are not directional isotropic, and individual discrimination thresholds to force direction and force intensity should be applied. In summary, multi-DoF kinesthetic perceptual models are more realistic and closer to human perception.

Other relevant properties for kinesthetic perceptual processing

As described in the section above and evident in Table 5.1, a wide range of stimulus properties could influence human perception of kinesthetic signals. Therefore it is important to consider implementing multilayer perceptual models that incorporate parameters reflecting different facets of physical stimulus properties. Furthermore, the models need to be verified and refined by conducting human perceptual quality assessments. For example, there is a considerable amount of research on perceptual processing of texture, and more recently on force compliance [376], as well as stiffness [387,388] and viscosity [389].

5.2.1.2 Perceptual models for human tactile processing

Texture perception is mediated by at least two different mechanoreceptors of the human skin. On the one hand, Pacinian corpuscles are responsible for sensing finer textural surface elements, while on the other hand, coarser surfaces are processed by SA I, i.e., the slowly adapting type I mechanoreceptive afferents [390]. Most empirical research in vibrotactile perception has focused on Pacinian corpuscles, which are sensitive to vibrations in the frequency range from 65 Hz to 400 Hz with the highest sensitivity being around 250 Hz [391]. In a series of experiments, Verrillo and colleagues [392–396] determined the absolute vibrotactile threshold of detectability of sinusoidal stimuli on different locations of the hand. Thereby the minimum stimulus intensity that is consciously perceptible by a human observer serves as the detection threshold. In other words, vibrotactile frequencies below this threshold lack reportability, as they cannot be felt or detected by human observers.

Another important feature of human perception that needs to be considered when developing tactile codecs is the overlay of perceptual effects from other stimuli. Past studies have shown empirical evidence for masking in the perception of vibrotactile signals. The masking effect has been widely studied in sensory perception and describes the phenomenon when the perception of a target stimulus is reduced by the presence of another stimulus, referred to as a mask. This phenomenon occurs in visual, auditory, as well as tactile perception. In vibrotactile perception, on the one hand, masking can occur in the time domain and is referred to as temporal masking. On the other hand, masking effects can also occur in the frequency domain, and is referred to as spectral masking. Furthermore, stimulus competition and stimulus intensity can result in masking that determines the perceptibility of competing tactile stimuli [397–399].

In the context of vibrotactile codec development, stimulus signals consist of various texture patterns, therefore in the following, we will only focus on spectral masking. The authors in [400] observed such masking phenomena for simplified stimuli, where the target consisted of vibrotactile outputs of pure tonal sinusoids in the frequency range of interest ranging from 80 Hz to 380 Hz, whereas narrowband noise (120 Hz, 200 Hz, and 280 Hz) served as the mask. Hence, it can be assumed that masking will also occur for more complex signals, resulting in an increase of perception thresholds around dominant peaks [401]. Thus perceptual thresholds and masking effects can be employed for the design of a vibrotactile codec, as it permits emphasizing perceptible frequency components of a given vibrotactile signal, while penalizing less perceptible ones, or completely filtering out imperceptible information [402]. Rather than spectral stimulus properties, which play a determining role in auditory perception and thus in audio codec design, perceptual processing of tactile information is largely attributable to temporal cues [397,403]. Consequently, temporal cues, such as temporal duration, temporal stimulus delays, and temporal masking effects should be given consideration during the design, development, and evaluation of tactile codecs. The authors in [404] found that Pacinian-mediated texture perception can be predicted by the intensity-based spectral power model, which includes temporal and intensity information, rather than the frequency theory [390]; whereas coarser textural features are coded by spatial variation elicited in SA I afferent firing [390]. Recently, the authors of [405] introduced a time segmented intensity-based model that accounts for relatively slow time-variant vibration patterns. In summary, in the design of vibrotactile codecs (and the subsequent evaluation through quality assessment procedures), these above discussed perceptually relevant features of vibrotactile signal processing, especially temporal and spectral information, are determining factors. Moreover, other factors that affect vibrotactile perception that are relevant especially for codec evaluation with human observers include individual differences between human observers [406,407], body locations [408], cognitive states [397,409], and task situations [400,410].

In the future, we will build such multilayered perceptual models based on psychophysical and neurocognitive experiments in simplified scenes. Additionally, human sensory perception is complex and multimodal. That is, utilizing solely somatosensory perceptual models, as put forth here, might have limitations in comprising effective codecs that represent human perception. We believe multisensory interaction effects, specifically in the context of how another sensory modality, such as vision or audition might influence kinesthetic or tactile perception, merit further investigation. This is of particular importance to provide a truly immersive human experience with promising results for virtual object interaction and teleoperation. Therefore it will be important to investigate and analyze the relationships between physical perceptual stimulus properties and subjective ratings in perceptual quality assessments. Our work will contribute to the design, development, and improvement of the Tactile Internet with Human-in-the-Loop, including both kinesthetic and tactile subsystems.

5.2.2 Existing kinesthetic codecs

The main goal of a kinesthetic codec is to reduce either the packet size or packet rate, or both. Communication of kinesthetic data in teleoperation systems prefers a high data rate of 1 kHz, or even higher. This leads to a packet rate of at least 1000 p/s over the communication channel. Such a high packet rate can quickly overwhelm local networks, as well as the long-distance transmission link. Therefore reducing the haptic packet rate is more important than reducing the packet size.

5.2.2.1 Packet size reduction

Early attempts aimed at compressing the packet size by reducing the redundancy in the kinesthetic data. To this end, statistical properties and predictive models of the kinesthetic signals are exploited. Examples are compression schemes that rely on adaptive differential pulse code modulation [411], a 32-bit IEEE floating-point representation [412], Discrete Cosine Transform (DCT) [413], and wavelet packet transform (WPT)-based compression [414].

5.2.2.2 Packet rate reduction

Perceptual deadband-based kinesthetic codec

The first approach aiming at a packet rate reduction for networked control systems has been proposed in [415]. A new packet transmission is triggered if the difference between the most recently sent data and the current input signal exceeds a fixed threshold. The receiver interpolates the missing samples by holding the value of the most recently received sample Zero-Order Hold (ZOH). This method, however, ignores the fact that human operators have strong limitations in terms of perceivable signal changes.

Later, exploiting the limitations of human haptic perception towards achieving better packet rate reduction performance has been studied. The perceptual Deadband (DB)-based kinesthetic data reduction approach (or in short DB approach) employs

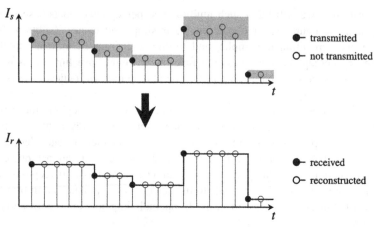

Fig. 5.1

Illustration of the 1-DoF perceptual deadband approach. The sensor readings at the sender (top) and the reconstructed signal at the receiver (bottom). Adapted from [381].

Weber's law to reduce the packet rate, while keeping the signal distortion below human perception thresholds [380,381,416–418]. In general, kinesthetic signals are exchanged only when there are significant perceptual changes. Fig. 5.1 illustrates the DB approach. The input signals are predicted based on the most recently transmitted ones. The current signal needs to be sent if the prediction error is larger than the selected deadband. The deadband parameter (DBP) p controls the size of the deadband. The received signals are extrapolated back to the original sampling rate (e.g., 1 kHz) based on the same prediction algorithm used at the sender. The period in which no update is received is defined as communication interruption (CI). In Fig. 5.1, a ZOH predictor is applied. The DB approach is able to reduce the average kinesthetic packet rate by approximately 80–90%.

Higher-order predictors can achieve better performance and less distortion at the cost of higher computational complexity. Examples of higher-order prediction approaches can be found in [419] for a linear first-order predictor, [420] for a third-order autoregressive model, and [421] for a quadratic curve-based prediction.

Multidimensional perceptual DB-based kinesthetic coding

In multi-DoF scenarios with multidimensional haptic sample vectors, the 1-DoF perceptual deadband becomes a multi-DoF deadzone, which considers both the changes of amplitude and direction. In [382], under the assumption of isotropic behavior, the perceptual deadzone becomes a circle for the two-DoF case, and a sphere for the three-DoF case.

Psychophysical experiments, however, have shown that the assumption of an isotropic multi-DoF perceptual deadzone is only an approximation and rather should be direction-dependent [422]. The author of [422] investigated the force JND as a

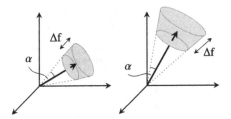

Fig. 5.2

The concept of a direction-adaptive 3D deadzone. Discrimination thresholds for changes in force direction are defined by the isotropically uniform angle α. Perception thresholds for changes in magnitude are a function of the reference force intensity $||\mathbf{f_{i-m}}||$ and described by Δf. Adapted from [423].

function of the force feedback direction and directional changes. Considering the human discrimination thresholds for changes in both force intensity and force direction, a direction-dependent multi-DoF deadzone for haptic packet rate reduction has been proposed [423]. The perception thresholds for changes in multi-DoF kinesthetic signals are defined as a function of the pending direction and the JND of force intensity [384,385,422]. The resulting 3D deadzone in [423] is illustrated as a frustum cone in Fig. 5.2.

5.2.3 **Time-delayed teleoperation**

Haptic devices are mechatronic designs, which deliver to the users force feedback, while interacting with a Virtual Environment (VE) or a Remote Real Environment (RRE). The underlying idea being that the haptic information received should be indistinguishable from a real interaction. This, however, is not always true, the reason being that there is an inherent trade-off between closed loop force-feedback stability and transparency. Reducing all unnecessary parasitic forces and displaying the actual desired physical properties of an object, namely stiffness, viscosity, hardness, texture, etc., is termed as haptic transparency. Thus maintaining stability, while enlarging the impedance range and increasing the rate-hardness, has been a classical control issue. Rate-hardness is the initial rate of change of force versus velocity upon making contact with the environment, and was described as a tool for humans to perceive the actual hardness of the VE [424].

Substantial amount of research has gone towards addressing the issue of stability of haptic and teleoperation interaction, while ignoring the transparency aspect of it. The majority of work is based on the passivity criterion. Scattering and wave variable methods developed in [425] and in [257], reduce the generated energy, which is induced by time delay in the communication channel, to zero. In [426] the authors proposed the Time Domain Passivity Approach (TDPA) to observe and dissipate only the necessary amount of energy. This was also extended as a two-port approach for teleoperation setups to cope with time-varying communication delays [427]. If sufficient communication bandwidth is available, then delayed teleoperation without data

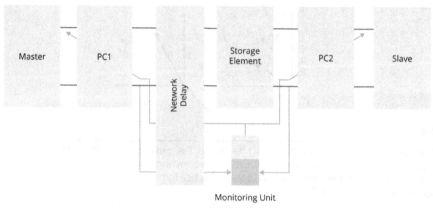

Fig. 5.3

The TDPA-ER architecture proposed in [429]. PC: Passivity Controller, ESE: Energy Storage Element. The follower controller is considered as an ESE.

compression methods is feasible. This was, for example, the case for the space experiment ROKVISS in 2005, in which a follower robot mounted on the outside of the International Space Station was teleoperated from ground [428]. An extension of the TDPA considering energy reflection by the coupling controller, called Energy Storage Element (ESE), was introduced in [429]. This approach, named Time Domain Passivity Approach Energy Reflection (TDPA-ER), is less conservative than the conventional TDPA since it preserves the physical coupling behavior between the leader and follower and avoids position drift. As depicted in Fig. 5.3, the ESE stores the input energy from leader and follower side. The energy exiting to the left and right side of the ESE is limited to the energy stored in the ESE by the passivity controllers PC1 and PC2. Therefore the actual output energy is not larger than the input energy at each sample. This guarantees system passivity.

In a typical teleoperation system, haptic information is exchanged between the operator and teleoperator over a communication network. The used networks are embracing a transition from a private/local infrastructure to a public/wide-area counterpart (e.g., 5G), which inevitably emphasizes the network-related challenges to the design of teleoperation control schemes, including the ability to deal with time delay, delay jitter, packet loss as well as cross-traffic data streams [430]. It is known that communication unreliabilities jeopardize the system stability and strongly affect the usability of teleoperation systems [425]. This is unavoidable for teleoperation systems with geographically distributed leader and follower systems. Therefore high fidelity teleoperation requires joint design and tight integration of haptic data processing, communication, and control. As a result, teleoperation over real communication networks requires a joint consideration of stability-ensuring control schemes and an efficient haptic data communication method [9].

The first joint solution combined the wave-variable (WV) control scheme with the DB approach [416,417] (WV+DB). The WV+DB approach applied the deadband on

Table 5.2 Comparison of methods introduced in this section.

Method	Constant delay	Time-varying delay	Packet loss
WV+DB	[416,417,431]	–	–
TDPA+DB	[434,435]	[434]	[434]
MMT+DB	[437,438]	–	–
ISS+DB	[442]	–	–

the wave variables depending on the absolute or relative changes of the wave-domain signals. This method requires a human perceptual model in the wave variable domain for data reduction. In the literature, however, studies on this topic are very limited. Since human perception has been intensively studied for time-domain signals (e.g., velocity, force), an extended WV+DB approach was proposed in [431] to directly apply the DB scheme on time-domain signals, while using the WV control scheme for stability assurance. Compared to the method presented in [416,417], the extended method [431] achieved higher system transparency and more efficient kinesthetic data reduction.

The above mentioned joint solutions are applicable only for constant delays. In fact, many stability-ensuring control schemes are able to deal with time-varying delay, e.g., the extended WV approach [432]. However, the WV scheme is a relatively conservative control approach [433]. Recently, a combination of the DB approach with the TDPA [427] has been proposed to deal with nonconstant delay TDPA+DB [434,435]. Compared to the existing WV+DB approaches, the TDPA+DB method achieves reduced conservatism of system control, and thus improves teleoperation quality. It has also comparable data reduction performance as the WV+DB approach, and is able to robustly deal with time-varying delays and packet loss without explicitly knowing the network characteristics.

Both the WV+DB and TDPA+DB methods are based on passivity-based control schemes. To combine the DB approach with nonpassive control schemes, [436–438] incorporated a perception-based update scheme into a Model Mediated Teleoperation (MMT) control architecture (MMT+DB). The stability of the MMT architecture requires a stable and precise parameter estimation method to model the environment on the follower side. To this end, combining the DB approach with the MMT for online environment modeling of static objects [439], deformable objects [440], and movable objects [441] have been investigated. In other attempts, the DB approach has been combined with the Input-to-State Stability (ISS) control scheme to achieve the same goal [442]. Experiments showed that the MMT+DB approach performs better for large delay (>50 ms), while the TDPA+DB approach does so for medium delay, and the ISS+DB approach does the same for small delay (<25 ms) [442].

A comparison overview of the above-mentioned joint solutions is shown in Table 5.2 in terms of their ability to deal with different types of communication unreliabilities. In Fig. 5.4, the system structures of these joint solutions are illustrated.

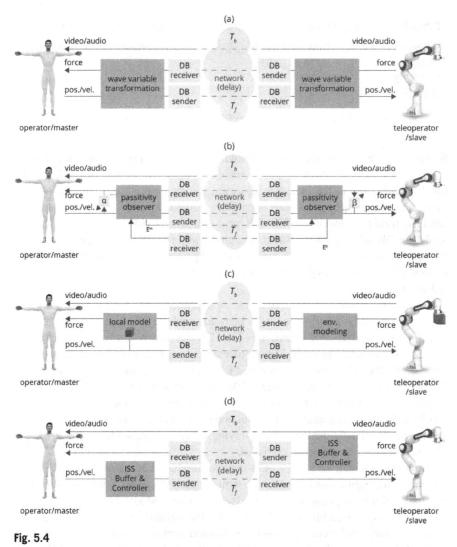

Fig. 5.4

Overview of the existing solutions of haptic data reduction for time-delayed teleoperation. (a) Combination of the WV control and DB approach. (b) Combination of the TDPA control and DB approach. (c) Combination of the MMT control and DB approach. (d) Combination of the ISS control and DB approach.

5.2.4 Existing tactile codecs

Several vibrotactile compression schemes have been introduced in the past. In [443], a compression technique based on the DCT was introduced that is able to reduce the data rate by 75%, while maintaining perceptual signal quality. In [444], vibrotactile signals were compressed again using a DCT in combination with the exploitation of perceptual aspects.

In [400,445], the authors show the adaptation of a speech codec for vibrotactile signals. The used Code-Excited Linear Prediction (CELP) codec is based on linear predictive coding. The input signal is analyzed with a predictive model and the residual coefficients are then quantized and stored or transmitted. In CELP, the quantized residual coefficients are taken to reconstruct the signal in the encoder based on which the error with respect to the original signal is computed. This error signal is processed through a psychophysical model, which takes human perceptual properties into account. The output parameters are then used in the synthesis of the signal from the residual coefficients. Thus the codec forms a loop. By tuning the parameters, in [400], it was shown that good compression performance can be achieved. The codec takes the absolute threshold of perception as well as masking phenomena into account. To accomplish this, a mathematical model of masking thresholds was developed through psychophysical experiments.

5.3 **Key challenges**

To support efficient and transparent human-human and human–machine remote physical interaction, important challenges need to be addressed.

Perceptual model quality Human haptic perceptual phenomena have been studied to some extent as explained in Section 5.2.1. However, the found principles and models are far from being complete. Data compression performance directly depends on the quality, accuracy and completeness of the underlying perceptual models. Thus it is of high importance to enhance the perceptual models by conducting further investigations into human somatosensory processing.

Quality metrics To assess perceptual transparency, human user studies need to be conducted. This is due to the lack of quality metrics that mimic subjective quality evaluation. Such user studies are costly, inefficient, and time-consuming and should therefore be minimized. To accelerate progress in the field, user studies need to be replaced by objective (mathematical) metrics. These allow for fast, efficient, and automatic quality assessment.

Communication delay It has been shown that in the presence of communication delay, teleoperation setups become unstable. This phenomenon is amplified when the transmitted data is compressed. Therefore it is crucial to develop and implement stabilizing control schemes that work seamlessly with the haptic codecs.

Setup variability Reproducing sensor or actuator setups on the human body with exact precision is close to impossible. The difference in sensor or actuator placement can degrade the user experience for an ill-designed coding scheme. To this end, the developed codecs need to account for this setup variability.

Multidimensionality To create a truly immersive human experience, haptic stimuli need to be displayed on a large portion of the human body. For kinesthetic applications, this implies recording and display a large number of force and torque signals simultaneously. For tactile scenarios the data rate increases immensely due to the large number of points of interaction on the human skin, and the abundance of tactile modalities to be processed.

Learning performance Remote transfer of skills from humans to robots is one of the most important aims in the Tactile Internet. The development of haptic codecs that are not only perceptually transparent, but also learning-oriented is of high importance to ensure minimal learning performance degradation introduced by lossy compression and communication impairments.

5.4 Approaches addressing challenges and beyond the state of the art

Addressing the key challenges above requires interdisciplinary effort. The development of codecs heavily relies on research on human haptic perception. These findings need to be transferred into computational models of the human sensorimotor system and perception, which drive coding of haptic information. Combining these perceptual codecs with innovative sensors and actuators and beyond the state-of-the-art communication and control mechanisms, a significant data reduction can be achieved, while maintaining pristine perceptual transparency and stability. It is also of vital importance to assess the influence of haptic communication on the performance of learning schemes targeting remote skill transfer.

5.4.1 Standardization of haptic codecs within IEEE

In IEEE, the standardization group P1918.1.1 is working on a first generation of haptic codecs for the Tactile Internet (TI). Specifically targeted TI application scenarios are Human-in-the-Loop scenarios, such as teleoperation or remote touch applications. P1918.1.1 standardized codecs address both closed-loop (kinesthetic) and open-loop (tactile) communication. The codecs are required to work with stabilizing control and local communication architectures for time-delayed teleoperation. Handshaking protocols, i.e., the exchange of information on capabilities, are also addressed.

In the following two subsections, we describe the vibrotactile codecs currently under investigation within P1918.1.1. They both use a model of human vibrotactile perception to compress vibrotactile signals with the goal of minimizing the bitrate and maximizing the perceptual quality. Then in the subsequent subsection, we describe the current proposal for kinesthetic codec for time-delayed teleoperation.

5.4.1.1 Perceptual wavelet-based vibrotactile codec

The first perceptual vibrotactile codec is described in [375]. It uses the encoding structure depicted in Fig. 5.5. In general, the codec is inspired by the MP3 audio

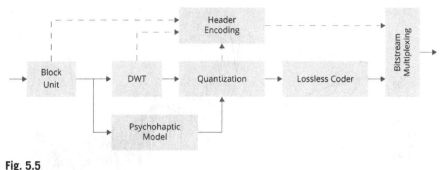

Fig. 5.5

Encoder structure of the wavelet-based vibrotactile codec.

codec [446]. However, the components differ significantly and are adapted specifically for vibrotactile signals. In the following, we describe the function of each individual component in more detail:

1. Block unit: The vibrotactile input signal is split into consecutive blocks. The block length can be chosen as 32, 64, 128, 256, or 512 samples.

2. Discrete Wavelet Transform (DWT): The DWT analyzes the signal blocks with CDF 9/7 filters. We choose the CDF 9/7 filters because they have a symmetric impulse response, which implies linear phase. This allows us to obtain the same number of wavelet coefficients as input signal values. Most importantly, we found that these filters perform best on vibrotactile signals in terms of decorrelation and energy compaction. The number of levels of DWT l_{DWT} depends on the block length L_B as

$$l_{DWT} = \begin{cases} 4, & L_B = 32, \\ \log_2(L_B) - 2, & L_B \in \{64, 128, 256, 512\}. \end{cases} \quad (5.2)$$

3. Psychohaptic model: The psychohaptic model adapts the quantizer such that the introduced distortions remain mostly below perception thresholds. These thresholds are determined by the so-called absolute threshold of perception and additional masking thresholds [400,447]. Masking means a strong stimulus at any particular frequency will render nearby stimuli imperceivable. This phenomenon can be leveraged for a codec by taking the magnitude spectrum of the current block and determining peaks that have a certain prominence and level. From these peaks, respective masking thresholds are computed, which are added together with the absolute threshold to obtain the so-called global masking threshold. Then, the Signal-to-Mask Ratio (SMR) for each wavelet band is computed from this by dividing the energy of the signal through the energy of the global masking threshold. The computation of the global masking threshold is illustrated in Fig. 5.6.

4. Quantization: In the quantization unit, each wavelet band is allocated a certain bit budget according to the psychohaptic model. In general, where the SMR is high, there should be more bits, since this implies significant perceivable signal infor-

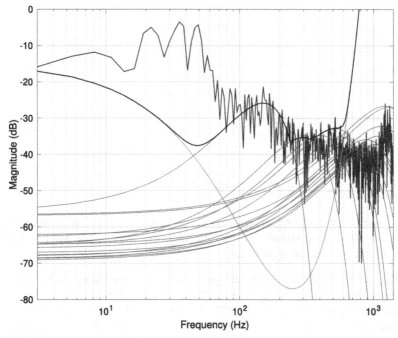

Fig. 5.6

Working principle of the psychohaptic model with an exemplary signal spectrum (blue, dark gray in print version), the absolute threshold of perception (green, light gray in print version), the masking thresholds (red, mid gray in print version), and the global masking threshold (black).

mation. Conversely, for low SMR, fewer bits should be spent. The bit allocation procedure follows the same approach as in the MP3 audio codec [446].

5. Lossless coder: A lossless coding method is used to remove remaining redundancy present in the signal. In this codec, Set Partitioning In Hierarchical Trees (SPIHT) is employed for this purpose. It is presented and described in [448].

The codec is evaluated using the objective metrics Peak Signal to Noise Ratio (PSNR) and Signal-to-Noise Ratio (SNR) over the Compression Ratio (CR). The corresponding scatter plots are given in Figs. 5.7 and 5.8. It is clearly visible that signal quality decreases as we compress more heavily. At a CR of 10, we have on average a PSNR of 53 dB and a SNR of 10.5 dB. The test signals are 280 recorded vibrotactile signals.

5.4.1.2 Perceptual vibrotactile codec based on sparse linear prediction

We next describe the second vibrotactile codec, named PVC-SLP, under investigation in IEEE P1918.1.1. PVC-SLP has been designed and built upon the concept of human sensitivity to vibrations. The two main objectives of PVC-SLP are: (*i*) to augment the

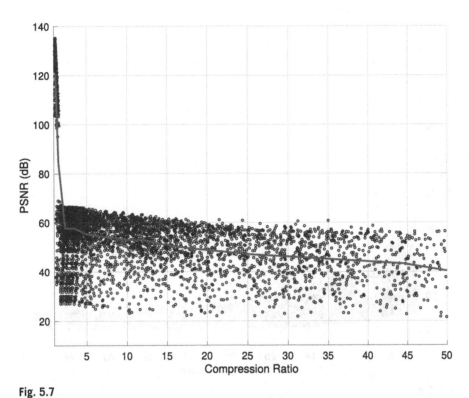

Fig. 5.7

PSNR over CR for all 280 test signals encoded at different rates (blue, dark gray in print version) and mean PSNR over CR (red, mid gray in print version).

classical coding paradigm of redundancy removal with local joint optimization of a sparse solution, and (*ii*) to hide imperceptible coding distortions by applying a locally perceptual quantization scheme. Unlike traditional linear prediction encoding methods, PVC-SLP uses both prediction residual and coefficients into the coding approach to maintain both the temporal and spectral properties of vibrotactile signals [449].

Sparse Linear Prediction (SLP)

SLP coefficient estimation is used to remove the short-time correlation in the vibrotactile signal by modeling the signal $y(n)$ as an AR process of order K,

$$y(n) = \sum_{k=1}^{K} a_k\, y(n-k) + e(n) = \hat{y}(n) + e(n), \qquad (5.3)$$

where $\hat{y}(n)$ is a linear prediction of $y(n)$ and $e(n)$ is the corresponding prediction residual. Then, a sparsity constraint is imposed on the minimization problem of the

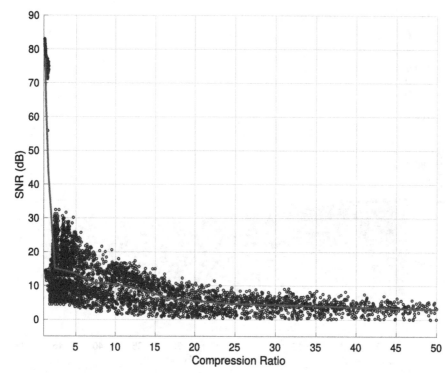

Fig. 5.8

SNR over CR for all **280** test signals encoded at different rates (blue, dark gray in print version) and mean SNR over CR (red, mid gray in print version).

l_1 norm of the prediction residual and the prediction coefficients vector $\mathbf{a}$,

$$\hat{\mathbf{a}} = \arg\min \|\mathbf{y} - \mathbf{Y}\mathbf{a}\|_1 + \gamma \|\mathbf{a}\|_1, \tag{5.4}$$

where γ is the regularization parameter to control the trade-off between the sparsity of the solution vector $\hat{\mathbf{a}}$ and the sparsity of the prediction residual [449]. γ is determined as the maximum curvature of the L-curve $\|\mathbf{y} - \mathbf{Y}\mathbf{a}_\gamma\|_1$ against $\|\mathbf{a}_\gamma\|_1$.

Acceleration Sensitivity Function (ASF)

Neurophysiological studies of vibrotactile perception led to the idea of a multi-channel sensory system, more precisely four channels of information processing in the glabrous skin [395,396,450]. The detection of vibrations requires four separate mechanoreceptive nerve fibers, which are often labeled as Pacinian (P) and non-Pacinian (NP). These four classes of nerve fibers have threshold curves with overlapping frequency ranges. The overall vibrotactile thresholds are often thought to be determined by the nerve fibers which have the highest probability of detecting the stimulus being applied [396].

Detection thresholds are often reported in terms of peak amplitude displacement in μm units. Various studies have shown that the displacement threshold curve exhibits flat to U-shaped frequency dependency regions with minimum displacement thresholds at frequencies between 150 Hz and 350 Hz. The slope of the displacement threshold curve controlled by the Pacinian receptors drops by approximately −12 dB per octave [393,451].

To express the detection threshold curve in terms of acceleration m/s^2 units, we adopt a simple mathematical approach. Assume the applied displacement varies sinusoidal with time as $x(t) = a \sin(\omega t)$, then the resulting acceleration is $d^2x/dt^2 = -a\omega^2 \sin(\omega t) = -a\omega^2 x(t)$. Put in words, the acceleration is proportional to the displacement by a factor of the squared frequency ω^2. Since the sensitivity is inversely proportional to the detection threshold, the reciprocal of the acceleration detection threshold is normalized and labeled ASF [449], as depicted in Fig. 5.9.

Encoder

The encoder takes as input a frame of the vibrotactile signal, which corresponds to 71 ms (200 samples) at a sampling rate of 2800 samples per second. For each frame, the SLP analysis is performed. The prediction coefficients are first estimated and scalar quantized. The prediction residual is then computed using the quantized/dequantized coefficients to reflect the quantization distortion in the residual computation branch. The Reflection Coefficient (RC) representation is used instead of the SLP filter parameters to alleviate possible filter instability problems. A Huffman dictionary is created for the RC representation according to each quantization scale. The prediction residual is transformed using DCT. The ASF serves as a local perceptual quantizer, which is used to quantize the DCT coefficients of the prediction residual. The perceptually quantized DCT coefficients are then Zero Run-Length and Huffman encoded. Finally, the encoded RCs and the quantized DCT coefficients are packed into the compressed bitstream.

Decoder

The received bitstream is first unpacked, followed by a series of operations that reverse the encoding process: Zero Run-Length and Huffman decoding, perceptual dequantization, inverse DCT for both the prediction residual and coefficients branches separately. The RC representation is reverted back to SLP filter parameters. Finally, the reconstructed vibrotactile signal is obtained by the SLP synthesis process. (See Fig. 5.10.)

Performance evaluation

For the performance evaluation of the PVC-SLP, PSNR, and SNR are used as quality measures. Scatter plots between PSNR/SNR against CR for 280 vibrotactile signals data traces are shown in Fig. 5.11.

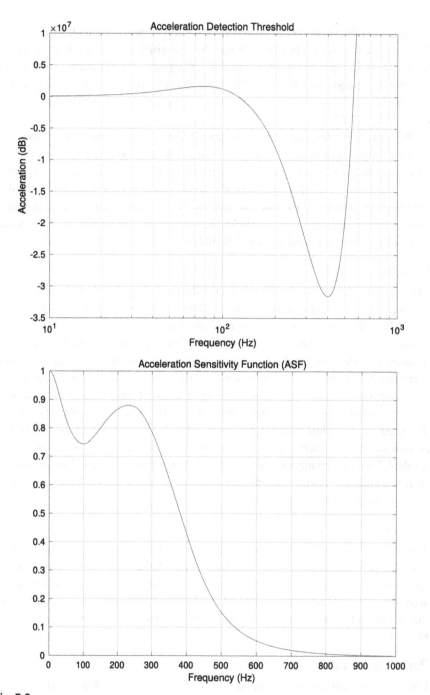

Fig. 5.9

(top) Acceleration threshold curve (in dB) obtained by multiplying the displacement threshold by a scale factor of ω^2. (bottom) ASF.

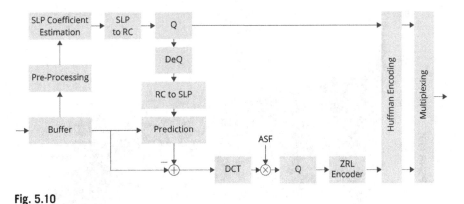

Fig. 5.10

PVC-SLP encoder under consideration in the IEEE 1918.1.1 Haptic Codecs for the Tactile Internet task group.

5.4.1.3 The IEEE proposal of kinesthetic codec for time-delayed teleoperation

System overview

The aim of the kinesthetic codec for time-delayed teleoperation under investigation within IEEE P1918.1.1 is to ensure stable teleoperation over a communication network with reduced packet rate, while preserving high fidelity of haptic interaction quality. The proposed approach, depicted in Fig. 5.12, combines the state-of-the-art time-domain passivity approach considering energy reflection (TDPA-ER [429], see Section 5.2.3) with the ZOH predictor of the DB approach (see Section 5.2.2.2). From the control perspective, this approach is able to preserve physical coupling behavior between the leader and follower and transmits higher impedance. It also avoids position drift and is less conservative than the previous solution [434]. From the communication perspective, the proposed approach dramatically reduces the packet rate and achieves high fidelity teleoperation.

As illustrated in Fig. 5.12, the packet update is triggered by the change of velocity or force signals and the power information is exchanged in combination with the update of velocity or force signals. According to [452], the integration of the DB senders and reconstructors into the TDPA-ER architecture is able to provide the highest degree of transparency at very low packet rates.

Extensions to improve haptic feedback quality

In addition, two extensions were developed that reduce system conservatism and improve the force tracking capability between the leader and follower:

Energy-based TDPA-ER

In the TDPA-ER approach of [429], power is exchanged over the communication network, which leads to a conservative behavior in case of packet loss. The approach for IEEE P1918.1.1 considers the exchange of energy information (energy-based

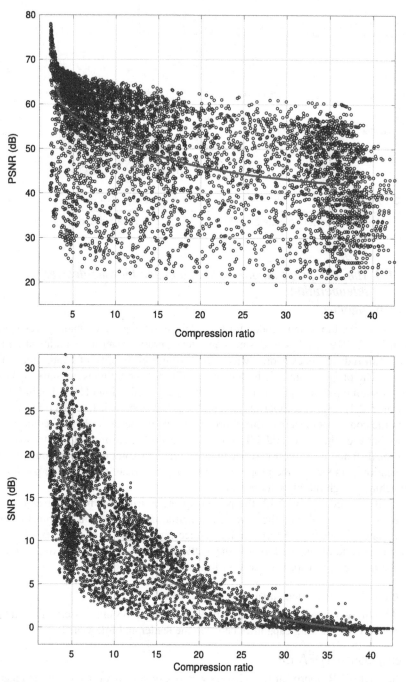

Fig. 5.11

PSNR/SNR vs. CR scatter plot using 280 vibrotactile signals. The red line (mid gray in print version) represents the average performance per compression ratio value.

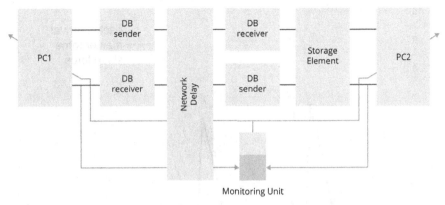

Fig. 5.12

The proposed joint solution: A combination of the TDPA-ER and the DB approach [452].

TDPA-ER) instead of power information to avoid conservative loss of power during Communication Interruptions (CIs). Then, the power is integrated to the transmitted energy value on the follower side. Although no energy change is received on the leader side during the CIs, once the communication recovers, the latest energy information with the accumulated power information is received, which renders the approach less conservative.

Time-based update trigger

During the CIs, the leader side passivity controller (PC) acts in a conservative manner since it has to dissipate according to the upper energy bound of the last received energy value. Therefore the IEEE standard proposal considers, besides the normal DB trigger, a time-based update trigger (T-trigger), which enforces packet transmission, if the CI duration exceeds a predefined threshold. Thus the duration of CIs is shortened and unneeded energy dissipation is reduced.

Evaluation

The experiments presented in Fig. 5.13 were performed with leader and follower devices (Novint Falcon and Geomagic Touch) connected to the campus Eduroam WiFi network with time-varying network delay. Additionally, a round-trip time of 200 ms constant delay was simulated. The energy-based TDPA-ER with the T-trigger update scheme is compared with the former state of the art, which is the original TDPA with DB approach [434]. In Fig. 5.13 (top), the mean packet rate is 74 p/s (the packet arrival plot signal is dense, because of regular transmission and resolution of the x-axis). In Fig. 5.13 (bottom), the mean packet rate is 92 p/s. The position drift caused by the original TDPA with DB approach [434] leads to smaller contact force, even for similar leader motion/penetration. This disturbs the contact perception and leads to low effective impedance/stiffness, which is the major drawback of the original TDPA with DB approach.

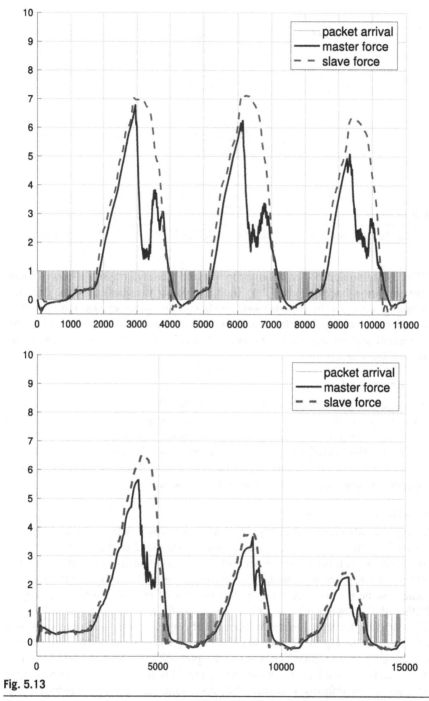

Fig. 5.13

Comparison between (top) energy-based TDPA-ER with the T-trigger [452] and (bottom) original TDPA-ER with the DB approach [434]. The deadband parameter is 0.2 for both.

5.4.2 **Novel approaches for efficient teleoperation under time-delay**

Above paragraphs showed that substantial amount of research has gone towards addressing the issue of stability of haptic and teleoperation interaction, while ignoring the overall system performance of a teleoperation system. This system performance, that is often considered as transparency, may lead to more intuitive teleoperation with increased ease of use, less user fatigue, and improved task fulfillment. The conventional control approaches mentioned in Section 5.2.3 guarantee stability at the cost of system performance. The TDPA gained high attention in the community for countering instability in a teleoperation system caused by communication delay, due to being the least conservative of all passivity-based control approaches, robust to variable delay, jitter and packet loss [427]. However, it comes along with undesirable position drift (for admittance-type Passivity Controller (PC)) and force jittering (for impedance-type PC). In a position-force teleoperation architecture affected by high time-delay, the admittance-type PC (implemented on the follower side) dissipates a significant amount of energy, which leads to a large position offset between the commanded leader and follower position, and the impedance-type PC (implemented on the leader side) causes the feedback force to jitter. Novel control approaches that are covered in this section, focus on the open research question of how to increase performance and enhance the transparency for teleoperation systems with varying time-delay, while maintaining stability, namely (*i*) improving transparency for TDPA [453], (*ii*) Observer-Based Gradient method (OBG) [454], and (*iii*) Successive Force Augmentation (SFA) approach [455]. A lot of research has been conducted to remove the positional drift between the leader commanded and actual follower position [456–458]. The effectiveness of the drift compensators depends on time-delay. The authors in [453] propose using the measured force feedback for the TDPA energy observations. The advantage of this controller is that the operator can freely move the follower robot when no contact is made, thereby considerably reducing the position drift. Upon contact, the admittance-type PC dissipates most of the energy, such that the follower does not apply a large force, unless the operator haptically perceives and approves it. Such a solution not only removes drift, but also makes the system safer to work with. Experimental validation with high time-delay of three seconds for tasks, such as slide-and-plug-in, and pick-and-place were carried out, where the kinesthetic force feedback remained of sufficient quality.

Compared to removing position offset, relatively little work has been done in eliminating high-frequency force vibrations that may lead to misinformation and, in the worst case, jeopardize the task. In [427], a virtual mass and spring (VMS) was used locally at the leader side to filter out these vibrations. The downside of which was the need for proper tuning, which is application based, and it also makes the system sluggish in some cases. Another disadvantage is that though the operator tends to move out of contact with the remote environment, the cued force still keeps increasing, due to the effect of time-delayed feedback. Such an effect might cause misinterpretation regarding the environment's property and state of motion. The OBG [454] method rectifies the feedback force by removing the unnecessary increase in force, thereby reducing the triggering of impedance-type PC. It introduces an observer on the local leader side, which detects the intent of the operator, and thus the force is modified such that the gradient between position force error and force is always positive. The OBG is used as an add-on controller for TDPA, therefore stability is assured. The OBG can be added to any teleoperation system without having any prior system information, irrespective of its dynamics. The shortcoming of the controller is that it further enlarges the position drift.

In bilateral teleoperation, human perception of the remote environment is of the utmost importance. Conventional stability-based controllers either bound the feedback force or introduce physical or virtual damping to stabilize the system. Although such methods guarantee stability, they also distort the stiffness perceived by the operator [459, pp. 81–101]. Haptic interfaces also encounter similar disadvantage. To overcome the aforementioned issues for haptic interaction, the SFA [455,460] approach takes a very different route at circumventing the trade-off between stability and transparency. Unlike TDPA, which introduces an adaptive element to dissipate the excess generated energy, the SFA introduces a feed-forward force offset that adds allowable positive energy into the one-port haptic interface to increase its impedance range. This is achieved by using a low enough stiffness force feedback, which maintains stability since the generated energy is dissipated by the inherent damping of the haptic device, while taking advantage of the feed-forward force that is independent of stability. The feed-forward force is increased in small steps, which is unrecognizable by the operator. The SFA is able to display higher perceived stiffness of the VE than any other rendering method, even for devices having low inertia, such as Phantom Premium 1.5 (as shown in [460]). As this control approach currently works only for VE, it needs to be extended to teleoperation systems for remote skill transfer scenarios, enhancing the rate-hardness during contacts, and increasing the impedance range of the leader haptic device. This would enable the operator to distinguish between different remote obstacles, including virtual-fixtures in a similar fashion as was done for VE (as shown in [455]).

The amalgamation of the above described controllers should be investigated in future work. It could potentially remove any position drift, provide a safe interaction with the environment, and enhance the intuitiveness by satisfying the operator's expectation of the force profile based on his motions, even for high time-delays.

5.5 Synergistic links

The key concepts for haptic data compression heavily rely on perceptual models developed by the research group for human perception and action (TP1). Sensing and display technologies developed by the research group TP2 are also used. The communication of haptic data relies on findings from the research group TP3 and the computing infrastructure from TP5. Close collaboration with the research group for intelligent networks (K2) is required for developing network structures intertwined with compression methods. The key breakthroughs discussed in this chapter will enable use cases in medical applications as well as industry.

5.6 **Conclusion and outlook**

In this chapter, we have presented the state-of-the-art research for haptic codecs, including the human somatosensory perception. We have discussed the key challenges to achieve transparent human-human and human–machine remote physical interactions and presented approaches beyond the state of the art, including the standardization activities for haptic codecs within IEEE. In the development of kinesthetic codecs, the main objective is to reduce the number of packets exchanged bidirectionally in a networked communication scenario. In the presence of time delays, the stability ensuring control schemes and kinesthetic data reduction approaches are considered jointly. For the tactile codecs, the target is the minimization of the data transmission rate. Although the requirements for efficient transmission of kinesthetic (closed-loop) and tactile (open-loop) signals differ, the development of perceptual models is of particular importance for both to enhance transparency. Thus future research will be directed towards developing perceptual models based on psychophysical and neurocognitive experiments. Furthermore, a truly immersive human experience can be achieved through displaying the haptic stimuli on a large portion of the human body, which requires the multidimensional extensions of the haptic codecs. The target application scenario and its requirements need also to be considered for further optimization of the codecs for Tactile Internet with Human-in-the-Loop (TaHiL).

Intelligent networks

6

Juan A. Cabrera G.[a], Frank H.P. Fitzek[a], Simon Hanisch[a], Sebastian A.W. Itting[a], Jiajing Zhang[a], Sandra Zimmermann[a], Thorsten Strufe[b], Meryem Simsek[c], and Christof W. Fetzer[a]

[a]*Technische Universität Dresden, Dresden, Germany*
[b]*Karlsruhe Institute of Technology, Karlsruhe, Germany*
[c]*International Computer Science Institute, Berkeley, CA, United States*

1N73LL1G3NC3 15 7H3 4B1L17Y 70 4D4P7 70 CH4NG3.
— 573PH3N H4WK1NG

6.1 Introduction and motivation

From the use cases of the Tactile Internet, described in the chapters beforehand, different types of communication networks are needed to enable efficient and secure exchange of information between humans and machines. First, information is created or consumed by humans or machines with a massive amount of sensors and actuators. Later, this information is conveyed over a local or cellular wireless access network that serves multiple humans and machines competing for communications resources in the same coverage area. Access networks are then connected by wide area networks as shown in Fig. 6.1.

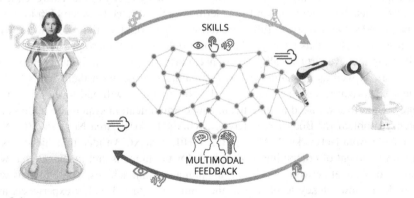

Fig. 6.1

Tactile Internet with skill transfer and multimodal feedback.

Tactile Internet. https://doi.org/10.1016/B978-0-12-821343-8.00017-4

Fig. 6.2

Generic Tactile Internet with Human-in-the-Loop (TaHiL) architecture.

The Tactile Internet is a combination of Ultra-Reliable Low-Latency Communication (URLLC) to enable realistic multimodal feedbacks, Enhanced Mobile Broadband (eMBB) to enable virtual and augmented reality, and Massive Machine Type Communication (mMTC) to handle the massive amount of sensors in the given scenario. It is of utmost importance to underline that URLLC alone is not enough to form the Tactile Internet. URLLC needs to be combined with the developments in softwarized networks to support multimodal sensory data transport and feedback for different types of communication architectures. The adequate consideration of the ultra-low-latency requirements necessitates the inclusion of computation from the edge, close to the users, and all along the network, as well as new approaches based on machine learning and artificial intelligence.

In the following, we use the example given in Fig. 6.2 to illustrate the challenges and the solution space for the research field of intelligent networks. For this scenario, consider that an expert desires to teach a robot certain skills for Industry 4.0, such as described in Chapter 3. For this example, we assume that the expert is equipped with several sensors and actuators in its smart clothing. The sensor information has to be read out and conveyed to the robot, and feedback has to be provided back to the expert to form a closed control loop. Here, we have to deal with two main latency impacts, namely the time to read out the massive number of sensors and the propagation delay within the network. On the way back, the multimodal feedback needs to be fed into the smart clothing of the expert with similar latency problems. If the robot and the expert are in close proximity, the propagation delay is negligible. However, if we wanted to teach a robot in Japan from Germany, the distance between both locations is nearly 9000 km. Even with full speed of light, that would result in a one-way 30 ms propagation delay. This latency value alone would be too high to provide the expert with an immersive virtual reality experience. Furthermore, the individual sensors and actuators as well as the simultaneous communication between them will result in latency. Moreover, the management of the computations will add to the overall latency. Therefore we need to investigate new communication systems to address the latency problem for Body Area Networks (BANs), Local Area Networks (LANs), and Wide Area Networks (WANs) in the TaHiL context. An idea to achieve this is the development of computing models both for the humans and machines that will run as software at network nodes near the humans and machines. These models will provide the low-latency feedback for the immersive virtual reality experience, and they will use the long distance communication with their physical pair (the actual human and the actual machine) to keep themselves accurate and up-to-date.

Our research will benefit from the cooperation with the other research fields. From research on multisensory perception and neurocognitive development as presented in Chapter 9, we learn about the requirements for the multimodal information stream, namely audio, video, and haptic information. With respect to latency and data rates, the requirements differ from 1 ms to 15 ms, and range from several kilobits to megabits per second for haptic and video feedbacks, accordingly. From Chapter 10, we derive the amount of sensors that will be implemented for different TaHiL applications. From Chapter 11, we learn about potential compression ratios for massive number of sensors and latency penalties as well as information theoretical approaches that need to be placed intelligently in the network. Chapter 12 contributes to the architecture for BAN for TaHiL use cases and measurement values for energy consumption and latency for different sensor based use cases. From Chapter 13, novel approaches for security are introduced that have to be placed in the network. All TaHiL use cases will have special requirements for their application domains, and all have to be addressed in our research.

6.2 Evolution of communication networks

Communication networks have tremendously evolved over time. As illustrated in Fig. 6.3, the first global communication network was circuit-switched. The communication systems in early days were initially designed to support voice service. In the 60s, packet-switched networks were introduced for general data communications. Whereas circuit-switched networks can be described as intelligent networks with a hierarchical structure, a packet-switched network is designed to be resilient—making the network less intelligent. In packet-switched networks, the intelligence is pushed to the border elements, such as end devices, servers, and clouds.

In general, communication networks have changed with the needs of its users. Therefore communication networks evolve into computing-centric networks to enable new services, which are mainly demanding low-latency communication. The latter is needed for machine-type communication as well as for new services, such as virtual and augmented reality. To achieve low latency, but also resilience and security, computing is amalgamated into communication networks, as given in Fig. 6.3. The embedded computing enables new concepts, such as mobile edge clouds and network slicing, as described later in this chapter.

To support the aforementioned example, we briefly describe the available technologies and ideas that are currently deployed, starting with 5G networks. The 5G radio networks are the first cellular networks addressing the deterministic latency requirements of the Tactile Internet. Furthermore, new ideas, such as the 5G campus, will enable industry and even rural areas to roll out their own 5G networks, which is a significant deviation from prior cellular network deployments. These networks also require an initial break with the current approach of hardware equipment that is highly specialized in, e.g., routing and switching packets. The multipurpose hardware is required to introduce the required computing in the communication networks,

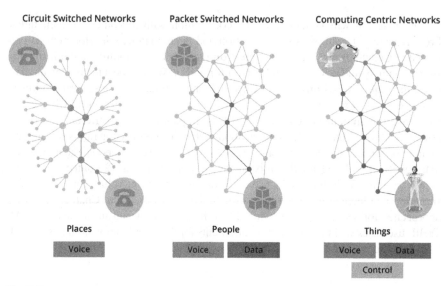

Fig. 6.3

Evolution of communication networks from circuit switched (left), to packet switched (middle), to computing centric networks (right).

which we discuss briefly as softwarization concept, but refer the interested reader to [461] for more details.

6.2.1 5G radio access networks

As a classical approach for larger communication networks (i.e., those consisting of more than a few access-points and more than a small number of users), there are well-known technologies mainly developed for public telecommunication (2G/3G/4G). The newest technology is called 5G, and not only designed for large-scale public networks, but also for small networks and self-managed operations. These latter types of networks are called campus or private mobile network solutions. The overarching standards are under continuous development by the 3rd-Generation Partnership Project (3GPP), an industrial association that has been driving standardization of mobile cellular networks for several decades. Starting with Release 15 (R15) of 3GPP, Fifth Generation (5G) is available in the nonstandalone (4G is needed) and standalone modes. The main focus of this release was on growing the amount of data transferred over mobile networks, enabling the so-called massive broadband. The current release supports up to 1000 MBit/s with 100 MHz bandwidth in real scenarios and 4x4 Multiple-Input Multiple-Output (MIMO) in downlink. With further releases, the main focus will be on Internet of Things (IoT), Time Sensitive Network (TSN), and URLLC. The combination of the latter enhancements is one of the enabling features for the Tactile Internet in the foreseeable future.

For example, with Release 16 (R16) finished and feature-frozen late 2019 to early 2020, one of the main inclusions for the Tactile Internet is the consideration of ultra-reliable low latency. Introduced in release 16, the air-interfaces' latencies decreased due to the introduction of mini-slots. While, normally, a full slot is required for the transmission of a packet, from R16 onwards, a subunit of a whole slot can be utilized. Additionally, the subcarrier-spacing for Orthogonal Frequency Division Multiplexing (OFDM) symbols (the difference between two symbols) further decreases the overall latency. For example, Long Term Evolution (LTE) statically uses 15 kHz. In 5G, it is possible to use multiples of 15 kHz (e.g., 30 kHz, 60 kHz, 120 kHz, 240 kHz), but not in all frequency ranges. The two specifically named frequency ranges are F_1 ($f <$ 6 GHz) and F_2 ($f >$ 6 GHz). In F_1, only 15 kHz and 30 kHz are common (60 kHz is a study item for R16). 30 kHz subcarrier-spacing results in a theoretical air-time of 0.5 ms. In F_2, the theoretical airtime decreases further with a subcarrier-spacing of 240 kHz, resulting in values below 0.125 ms.

New features currently targeted for Release 17 (R17) include support for a large number of devices (mainly for massive IoT) employing multicast and broadcast. A good example is a simplified delivery of firmware to many devices. If there is only a unicast available to deliver the data to N devices, N-times the overall bandwidth for a single device is required. The new reliable multicast/broadcast is developed to transfer the same data to a group of devices or all devices, effectively reducing the required number of transmissions and the amount of data over the air. An additional development is the relaying of data. This is targeted towards scenarios where, e.g., devices are in the shade of the coverage of the network. Here, another device can relay the packets and provide coverage (though at the trade-off of added power consumption due to forwarding).

6.2.2 Softwarized networks

In this section, we briefly describe the main representatives of softwarized networks, namely Software Defined Network (SDN) and Network Function Virtualization (NFV). In addition, we will also describe the principle of Service Function Chaining (SFC), which will not add extra functionality as such, but it will optimize the approaches introduced by SDN and NFV, and Network Slicing (NS). We additionally present the Mobile Edge Cloud (MEC) concept as a direct implementation example of the possibilities enabled through the prior mechanisms. Interested readers are referred to [461] for more details, references, and example implementations. Softwarized networks bring the flexibility required to quickly deploy and manage computing functionalities as well as the communication requirements between these functionalities and the users of the Tactile Internet [462].

6.2.2.1 SDN

SDN advocates replacing distributed static network protocols with centralized and flexible software applications. Legacy networks depend on hardware that implements the standardized protocols directly in the specialized microprocessors used. This means that a new release, or the deployment of a new protocol is a slow and expensive process for the network operators, because the hardware has to be changed. The main idea behind SDN is to deploy the network protocols as programmable software on the devices constituting the network infrastructure. SDN provides an Application Programmer Interface (API) for the developers to easily program the behavior of the router, switches, and other network nodes when, e.g., routing, modifying, and dropping network packets. This property enables the fast experimentation of new ideas in the deployed networks and integration and migration of security measures to arbitrary places across the network [463]. If at some point, for example, the network operator wants to test a new protocol in its infrastructure, then it does not have to go through the long-lasting processes of standardization and hardware exchange. Instead, the new protocol can be developed using the public API of the equipment, and it can be deployed as a software patch. Once the test is finished, the network operator can go back to the previous protocol by simply switching parameters in the configuration files. In the SDN architecture, instead of having smart equipment that knows how to handle the packets, there is dumb equipment that needs to ask a logically central entity called the SDN controller what to do with the packets. This controller can be implemented in a distributed fashion for resilience, but logically it is a central entity. It can be easily programmed to give instructions to the network devices on how to handle the packets and communication flows. Since the controller is a central entity, it has a overview of the whole network. Consequently, it can globally deploy optimal algorithms and protocols. SDN basically allows new functionalities to be deployed in nearly no time, relocated, and upgraded depending on the instantaneous needs of the networks. SDN offers flexibility, but due to its important role, it has become a valuable target for attackers, which requires protection [464].

6.2.2.2 NFV

NFV advocates using general-purpose hardware running software solutions of different network functionalities. In its origin, NFV was conceived to help network operators scale, on-demand, network functions, such as firewalls, among many others. Instead of deploying these functionalities in specialized hardware, they could be deployed, virtually, in general-purpose hardware. Therefore the computing resources of the hardware could be employed for the more demanding applications at a specific time, and if the need for more computing resources arises, then the operators could simply expand their infrastructure by adding extra general-purpose hardware. This equipment is generally cheaper and more flexible than the specialized hardware capable of performing only a few specific tasks. This general-purpose equipment can be deployed at any node of the network. Of course with the introduction of general-purpose equipment, the potential of NFV increased, and it can include many different types of applications. For example, in-network controllers of cyber-physical systems can be deployed as a virtual network function (VNF) to steer and control a driverless vehicle when needed. Since this controller can be deployed in the network close to the controlled entity, the delays due to propagation can be drastically reduced. Similarly, as it is done nowadays in data-centers, network applications are deployed virtually (by means of virtual machines or containerization) in the general-purpose computing equipment. This virtual deployment enables the easy migration and scalability of the applications to optimize their performance in terms of latency, capacity, etc. Furthermore, this enables new functionalities to be deployed in nearly no time as virtual applications in the network. The applications can subsequently be relocated, scaled, and upgraded depending on the instantaneous needs. A conceptual aid to imagine the future of communication networks with NFV is to imagine devices interacting with a data center that is not at the other end of the network, but it is, instead, spanning the entirety of the network. This will allow network operators to rent computing resources to third parties, in the same way as Google and Amazon do nowadays, to run their computing applications, but in hardware that is closer to the users.

6.2.2.3 ICN

The goal of Information Centric Networks (ICN) is to address the mismatch between today's host-centric internet design and its content-centric usage. Instead of enabling the connection of hosts on the basis of addresses, the focus of ICN is on the efficient distribution of data based on named content. This is achieved by modifying the internet core architecture. Besides the focus on named content, another important feature of ICN is in-network caching, which means data can be stored at every network node. This enables a more efficient distribution of data time and location independent from the original creator. Besides naming data, it is also possible to address named functions. Instead of certain content a client expresses interest in the result of a function, which it may not be powerful enough to execute by itself. Examples include statistical values of big data or calculated properties of video files. The network is responsible for finding the function and the function parameters, initiating the execution and transmitting the result. The execution can take place in the network itself or at an end point. Caching of content also allows the caching of function results, which accelerates distribution. The implementations of ICN can be easily deployed with the softwarizations techniques discussed, such as SDN and NFV.

6.2.2.4 SFC

Service Function Chaining provides the flexibility needed for services and applications that are amalgamated from other functions, services, or applications. The virtualized individual functions discussed as NFV concepts can be combined in a logical order (the function chain) through which the packets traverse. An intuitive example for such chaining can be found in modern Linux firewalls based on iptables and Netfilter, where the packets can be manipulated based on their location in the overall delivery path [465]. This provides a powerful tool to combine well-known and well-optimized individual services into more complex and powerful services and applications.

6.2.2.5 MEC

The Mobile Edge Cloud concept enables the placement of computing within the communication network, rather than at the endpoints. To reduce the propagation delay, the computing can be placed in close proximity of the needed execution. For example, the control algorithms for cyber-physical systems can be placed at the nearest base station. MEC is implemented on the network by means of NFV, SDN, and containerization technologies to deploy the services. The cloud functionalities are deployed, scaled, and migrated in the network by means of NFV functionalities, while the seamless communication, e.g., routing of packets, modification of packet headers, and coding is managed on the fly with SDN protocols. As a result, a user communicating with a cloud application running in the network does not have to worry where the actual software is running (this is managed with NFV), or to which address it should communicate (SDN will guarantee that the data packets are routed to the correct server hosting the virtual application).

6.2.2.6 NS

The concept of Network Slicing allows the deployment of several Virtual Network Slices with different characteristics in terms of latency, throughput, massiveness, and resilience in one physical network. With the limitation in the available spectrum, the support of low latency, high throughput, and high resilience at the same time is very challenging and would limit the number of users that could be served by such networks. NS allows us to see individual networks as an aggregated pool of resources. These resources can be virtually sliced to be used by multiple and diverse applications. Most services do not need to be so stringent, simultaneously, in terms of, e.g., throughput, latency, and confidentiality. Some services need low latency, but at the same time, they require only a small throughput. For each combination of requirements, one NS can be established. Whereas in the current Internet packets of different services and service classes are handled in the same democratic way, network slicing allows for prioritization and isolation of certain services.

6.2.2.7 Network security

MEC represents highly valuable targets for attacks that are deployed to devices which are less endowed with hardware resources. Their security hence requires increased attention, and comes with entirely new assumptions and challenges [466]. Denial-of-service attacks, for instance, are a common threat to the availability of services on the Internet. For TaHiL, this threat is amplified by the requirement of low latency and the placement of services in the mobile edge cloud. The low latency makes even short service interruptions more damaging for TaHiL services than for regular services. Additionally, deploying services to the MEC has the drawback that at the edge fewer resources are available to mitigate Denial-of-service attacks. TaHiL will therefore require new research in how Denial-of-service mitigation can be designed with minimal service interruption and resources. NS will be a key technology for TaHiL and 5G, as it allows separating traffic according to its Quality-of-Service requirements and communication requirements. Cloud providers use network tunnels to facilitate the separation of tenants in their environments by creating virtual networks. However, when moving these techniques from secure data centers to the network edge to facilitate NS, it is important to add security measures to protect the confidentiality and integrity of the traffic. Whereas these measures are readily available, it will be important to reduce their latency overhead to a minimum for TaHiL.

6.2.2.8 TLS

Transport Layer Security (TLS) is a layer 5 protocol, which provides authentication and encryption mechanisms. It is the evolution of Secure socket Layer (SSL), which was renamed after its last Version v3.0. TLS handles several cipher suits, where the discrete method is elected during the initialization phase. A cipher suite includes the security algorithms used in the further communication; this includes key exchange, authentication, encryption and hashing. The protocol handshake in the beginning of an session, handles the authentication of server and client (optional) using asymmetric certificates, e.g., X509. It creates a session key, which is used to secure the rest of the session using symmetric encryption (optional). The actual version TLS v1.3 is defined in RFC8446. The TLS definition includes several subprotocols, e.g., Handshake Protocol, Change Cipher Specification Protocol, Alert Protocol, Application Data Protocol. These are handled on top of the TLS Record Protocol, where handshake and change cipher are used mostly at session creation, and the Application Data Protocol manages the transmission of the payload data. The usage of TLS for securing connections is often expressed by adding an S to the used protocol on top of TLS (e.g., HTTPS, SMTPS, IMAPS, and FTPS). TLS assumes a reliable, connection-oriented transport protocol. In the context of the Tactile Internet, retransmitting dropped packets might result in these retransmitted packets being dropped by the application instead: These packets might be too old and they should instead be replaced by fresher information. Therefore we support Datagram Transport Layer Security (DTLS)—a variant of TLS—which uses a datagram-oriented, unreliable transport protocol. DTLS is the correct choice for securely transmitting real-time traffic in the Tactile Internet.

6.3 TaHiL communication concept

The TaHiL communication concept has two main requirements to fulfill. The nature of TaHiL clearly necessitates ultra-low-latency communication as a primary requirement to enable real-time feedback for the multimodality of communication (see Section 6.3.1). The second requirement is the support of different communication streams with respect to latency, massiveness, resilience, and throughput to realize the required multimodal feedback. To fulfill the requirements of the streams, we map them to the technologies introduced in the sections beforehand in Section 6.3.2.

6.3.1 Delay and latency components

Let us first have a glance at the delay and its resulting compounds. Obviously, there is not a single but multiple factors that impact the overall delay. The latency for end-to-end communication is constituted by several components, namely the system delay t_s,

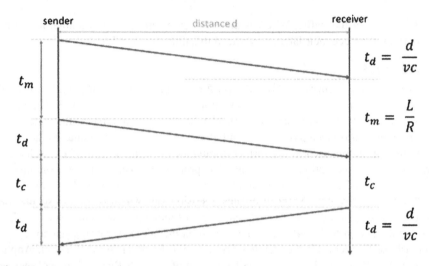

Fig. 6.4

Latency components for TaHiL communication.

transmission delay t_m, the propagation delay t_d, and the computing delay t_c. In what follows, we briefly discuss their characteristics and interplays for TaHiL communications, following a common control loop scenario that is illustrated in Fig. 6.4.

6.3.1.1 System delay

The system delay t_s is responsible for all delays caused by embedded computing, as well as hardware for sensors and actuators. In our current testbeds and demonstrators, we face delays caused by cameras, monitors, or multimedia platforms used by iOS and Android. There is currently no general overview of what delay is actually incurred through these initial data gathering and processing efforts. However, the availability of specialized sensors paired with equally specialized embedded hardware will likely be enabled to reduce the overall delay to around 0.1 ms. To reduce the system delay, we need cooperation with other research groups, such as mentioned beforehand.

6.3.1.2 Transmission delay

The transmission delay t_m depends on the transmission rate R, the message length M, and the channel access delay t_{ca}. Wireless Fidelity (WiFi)6 offers high transmission rates, but the channel access delay depends on "listen before talk" with randomized access delay characteristics. The 5G new radio concept has also high data rates as explained beforehand, but in contrast to WiFi6, it is using reservation-based medium access, which will have a positive impact on the delay. Reservation-based medium access can guarantee that the users will have a specific time slot for the transmissions. It contrasts with the random allocation mechanisms in WiFi6, where multiple collisions due to the randomization process might lead to large delays. Therefore if in certain scenarios a nonstochastic, deterministic delay boundary is required, WiFi6 will not be suitable. For a massive number of sensors, the channel access delay will increase for WiFi6 dramatically. If sensor information is sparse, i.e., they belong to the same hand or body, compressed sensing can be used to read them out in a fast way. The concept of compressed sensing is described in Chapter 11 and can be implemented within the network by using the NFV concept [467,468]. For meshed sensor networks as well as resilient and low-latency coding communication, network coding can be used. This technology is described in Chapter 11 and can also be implemented by means of the NFV concept [469]. For more details regarding low-latency network coding, refer to [470–474].

6.3.1.3 Propagation delay

The propagation delay t_d depends only on the distance d of the communication peers and the employed medium in relation to the speed of light. The propagation delay is the quotient of the distance and the speed of light c multiplied by an effective factor v (e.g., different values apply when using copper or fiber). It seems that all three factors are given, and we would not have any impact on the resulting delay. But in the TaHiL concept, we could use technologies introduced beforehand to reduce the latency. If we come back to our initial example, if a human wants to teach a robot from Germany to Japan, the delay will be around 30 ms one way. To reduce this latency, a model of the robot could be placed virtually with VR or AR in close proximity to the user as shown in Fig. 6.5. Now the operator is teaching the virtual robot with a virtual distance that is significantly reduced. As the virtual robot has to be realized on a computing entity, we use the MEC that we introduced beforehand. The placement of the MEC will take place within the communication network. To minimize the propagation delay, the closest computing platform seems to be the intuitively right choice. However, we will show later that the optimal placement has also to take into account considerations of security questions or latency impacts that result from the computing delay. The implementation of MEC by means of SDN and NFV was shown in [475,476].

6.3.1.4 Computing delay

The computing delay t_c is the last factor along the single direction of a typical control loop. As multiple MEC instances can exist in a given network, the ideal placement of the MEC additionally depends on the computation power and the load situation of a given virtualized platform. In our prior example, we would have placed the MEC as close to the operator as possible to decrease the propagation delay. However, if the closest possible hosting platform is overloaded, there might be a better choice for the placement by additionally taking the computation delay into account. New functions, such as network coding, must be implemented in such a fashion that computing delay is kept to a minimum. For example, first research results take advantage of the implementation of network coding with parallel cores [477–479]. But even the virtualization itself represents a potentially time-intensive overhead, as we have shown in [8]. As the virtualization itself has been introduced to realize MEC to reduce the propagation delay, the approach will increase the computing delay. Here, the usage of SCF can improve the situation as we have shown in [8].

6.3.2 Mapping the multimodel feedback

As our multimodal feedback consists of multiple streams with different technical requirements in terms of latency, data rate, and resilience, the concept of Network Slicing will be used. The individual slices can be mapped to the different modalities' required for network characteristics, e.g., ultra-low latency, but only small amounts of data, or larger amounts of data, but somewhat less tight latency requirements. Tactile and video are two examples for these types of data. Traditional networks would not be able to simultaneously offer both. Furthermore, the logical combination of the different streams for the computational exploitation of redundancy on a metalevel will greatly be simplified via the recombination of the slices at respective nodes in the network. Similarly, consideration of security and resiliency can be facilitated in the logical evaluation of different slices, e.g., through validation sampling of the transported data. An immediate example could be the cross-evaluation of the transmitted kinesthetic information with the (typically later received) video data for a given application scenario. For example, this allows us to periodically provide a validating MEC instance to ensure that, e.g., the positional data sent is in line with visual confirmation.

In Fig. 6.5 we depict (*i*) the *technical requirements*, such as latency, heterogeneity, throughput, massiveness, resilience, security, and energy as explained in Chapter 13 in more detail; (*ii*) the *concepts*, such as MEC, NS, and a new air interface as given in Section 6.2.2.5, 6.2.2.6, and 6.2.1, respectively; (*iii*) the *softwarization aspects*, such as SDN, NFV, and SFC, as given in Section 6.2.2.1, 6.2.2.2, and 6.2.2.4, respectively, and (*iv*) the *innovations*, such a network coding, compressed sensing, and machine learning as given in Chapter 13.

6.4 Architecture discussion

In the TaHiL concept, we have different network architectures ranging from BAN, LAN, and WAN, as given Fig. 6.6 and described beforehand. The BAN is characterized by a massive amount of sensors and actuators. Here the latency can be optimized by compression and filtering to reduce the channel access times. However, the latency minimization activities have all to be seen in interdependency with the goal of ultimately delivering the right level of Quality-of-Service (QoS) and Quality-of-Experience (QoE) to the user (human or machine). Depending on the use case, the network needs to restructure the routing, compression, and coding within an holistic approach. The BAN in the TaHiL concept will require computing for the envisioned coding and compression to reduce the latency and to prepare for multimodal data transmission with network slicing. The LAN has to serve several BANs. Given the limited wireless and maybe computing resources, the LAN has to coordinate the data

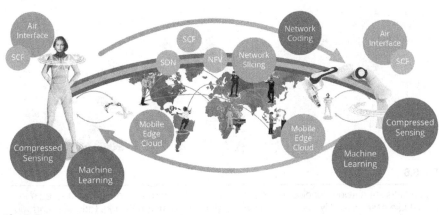

Fig. 6.5

Tactile Internet: Main concepts and introduction of novel technologies.

flow from the BAN, understanding the currently supported use case. LAN can comprise several access points, several computing entities to host MECs, and at least one gateway to the WAN. The WAN is dominating the overall delay when considering the propagation delays. Furthermore, WAN will incorporate softwarized communication elements to embed concepts, such as MEC or NS, in addition to the extended physical distance and incurred propagation delays, the computational delays, and potential transmission delays for interconnected resources at this level. For example, resources might be required on a higher tier or connectivity requirements necessitate utilizing resources locally unavailable. Alternatively, cross-connection from one provider's network core to another provider's core could fall into this category as well. This tiered or hierarchical approach is illustrated in Fig. 6.6 for the WAN scenario. In Table 6.1, we show the impact on the overall delay by its components for the different network architectures. We represent *high impact*, *low impact*, and *medium impact* as +, −, and o correspondingly.

Table 6.1 Impact of different sources of delay on the overall network latency for different network architectures.

	Propagation delay t_d	Transmission delay t_m	Computing delay t_c	System delay t_s
BAN	−	o	−	−
LAN	o	+	+	o
WAN	+	+	o	+

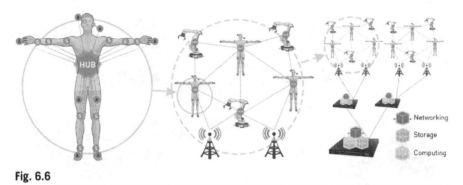

Fig. 6.6

Three different network topologies for Centre for Tactile Internet with Human-in-the-Loop (CeTI) use cases, namely body area networks (left), local area networks (middle), and wide area networks (right).

6.5 TaHiL testbeds

In what follows, we describe three main communication testbeds used by the CeTI teams to perform the research. These testbeds consist of a 1 ms testbed used to show the impact of optical fiber (propagation delay) on communication. A 5G campus testbed to deploy close to the humans and machines the Radio Access Network (RAN) and servers to run the core components of the RAN as well as Virtual Network Functions related to the TaHiL applications. And a WAN testbed to test the impact of computation delays over long distance networks, such as those available in the applications, where an expert would like to train a machine that is located overseas.

6.5.1 Artificial testbed

It is possible to achieve the 1 ms-delayed communications if all the parts of the system operate in a low-latency fashion. As shown in Fig. 6.7, the delay is not only related to the MEC computations and the propagation delay of the communication. If we want to understand how much time of the 1 ms budget we can use for the MEC computations and how far we can physically place the MEC, we need to assign a fraction of the 1 ms to each component of the system. For instance, in Fig. 6.7, acquiring data from the sensors and transmitting it over the 5G air interface might take up to 0.2 ms in total. We consider that we send this data to the MEC for processing and back. We consider that the transmission over the 5G air interface to the actuators and the time it takes for the embedded computing to convert the received message into signals for the actuators adds up to 0.2 ms in total. This means that sensors, actuators, and 5G air interface communication takes up to 0.4 ms of the 1 ms budget. If we assume that the MEC can take up to 0.35 ms to process the information from the actuators, this means that the network has 0.25 ms to transport the information to the computing entities and back. If we consider the propagation delay of optical fiber, then we can

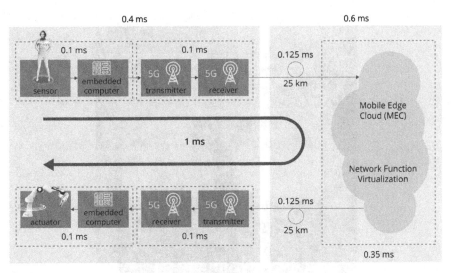

Fig. 6.7

1 ms testbed.

estimate that the maximum physical distance at which we can place the servers of the MEC is 25 km from the sensors and actuators.

For testing the hardware-based latency on fiber, we built a testbed at an optical distance of 36 km. Our complete testbed is on a cable drum with multiple fibers shown in Fig. 6.8, each one with a one-way length of 6 km. We can test different lengths up to 36 km by connecting together as many 6 km sections as wanted. We have two different approaches to measure the round-trip time. Both measurement setups are synchronized via GPS for accurate results. The first approach is only plain cable-length and delay measurement. The testbed consists of two small PCs synchronized over GPS and directly over Precision Time Protocol (PTP). The second connection is made over the fiber-cable drum. With this, we can measure the difference between a direct connection and the fiber. Our setup has a time resolution of 0.1 ns. The second testbed is based on the first. It is expanded by a WiFi link between the computers and the fiber-optic drum. So additional latency can be measured and evaluated. The WiFi link can also be replaced by any other wireless communication technique, to see the difference in different approaches for over-air communication. Fig. 6.8 shows the fiber drum and the switch to which the small computers are connected. In Fig. 6.8, we show the current maximum of distance (36 km), observable at the interconnection of the blue cables.

6.5.2 5G campus testbed

For a minimal set of a campus solution, a core network and a RAN are required. The RAN is the transmitting and time-sensitive part of the mobile network, consisting of passive antennas, power amplifiers, and a baseband for decoding and transforming

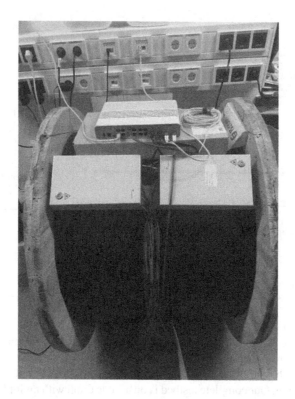

Fig. 6.8

Fiber drum with 36 km length to produce realistic propagation delay.

the data received from the air. The core is the authentication and session management function for the whole network. In the core, all User Equipment (UE) are authenticated against a database and attached to the network. We can implement extra functionalities in the 5G core as a virtual function. The 5G core is based on extra functions, all available in software, to deploy it anywhere on any hardware. This multifunction approach based on software with commodity hardware also drives the flexibility to create localized self-contained 5G network deployments, i.e., the campus networks. A large advantage of campus networks is that the data-paths are very short, and so the latency of the network and all data is covered at a singular, local point. However, a significant hurdle for the implementation of campus networks is the current limitation of air interfaces to only a few specified interfaces usable to connect vendor-independent hardware. An evolving approach is the development of a completely open standard for the radio access part of the network. This project is called OpenRAN and aims at making the antenna and baseband vendor-independent, creating a solution that everybody can adopt with commodity hardware for individual application scenarios. In Fig. 6.9, a campus solution is deployed into a physical container. The container has on board everything needed for a campus network. Inside the container is the core (two servers) with the baseband and power-supply, on top are the antennas with the pico-Remote-Radio-Head (pRRH). It is transportable on a trailer by car. This testbed is used to carry out research for the given applications of CeTI. Furthermore, the testbed is used as demonstrator to convey the TaHiL concept to the people.

6.5.3 National-wide testbed

We will use the infrastructure of the High Performance Demonstrator (HPD), a collaborative project orchestrated by Deutsche Telekom (DTAG), which has built a flexible, SDN/NFV programmable network infrastructure for showcasing new services. It is based on the SASER network, a Germany-wide high-speed research network given in Fig. 6.10. Several partners are involved: DTAG provides the transmission infrastructure with Terabit connections; NEC, ADVA, RedHat, and CheckPoint provide solutions for high-performance and SDN-enabled switches, optical communication, operating system, and security. In addition, the project involves academic partners, such as the University of Stuttgart and KIT. The data/computing centers, such as HLRS, Jülich, and ZIH at the TU Dresden, provide the computation facilities to deploy MEC services. Currently, HPD connections between Dresden, Jülich, and Stuttgart are established, with plans to extend to more locations in Germany. We have been building a Kubernetes-based infrastructure that supports confidential computing and low-latency computing and automatic migration of services. The use of Kubernetes will ensure the support of a large variety of tactile applications. The testbed will include AI classification framework to compute haptic responses in milliseconds latency.

This testbed will be used to carry out research on long-distance learning between machines and humans connecting the research facilities of Dresden and Munich. MEC services can be deployed within the network to reduce the propagation delay, but the computing delay will become an important factor.

6.6 Synergy and collaboration

The intelligent networks developed for the TaHiL provide the framework to communicate the sensors and actuators connected to the humans and machines for the applications described in Chapters 2–4. The different technologies discussed in this chapter will enable the network slicing that supports the multimodal communication required by the different applications. The slicing will be optimized according to the individual requirements of the humans involved, e.g., age and physical condition. This information is available to the networks by the research described in Chapter 9. The data transmission is reduced and the resilience and security are improved in the intelligent networks by exploiting the research described in Chapters 11 and 13. The networks might also share side information with the haptic codes described in Chapter 5.

Fig. 6.9

5G campus solution at TUD for the TaHiL concept.

Fig. 6.10

National SDN/NFV wide area network by Deutsche Telekom.

6.7 Conclusion and outlook

In this chapter, we discussed the importance of the integration of computing into communication networks. These networks are more than a simple set of pipes for bitstreams. Only a communication network that can process and operate over the data will be able to satisfy the communication requirements of the TaHiL. To facilitate the exploitation of the benefits of a computing-capable network, one can make use of softwarization techniques. By means of softwarization, it is possible to create dynamic and adaptive networks that can adapt to the different requirements over time of the different applications of the TaHiL. Furthermore, to develop our research, demonstrate the benefits of intelligent networks, and provide a communication infrastructure to other research groups of CeTI, we have described in this chapter the current testbeds that we have built.

Augmented perception and interaction

7

Annika Dix, Anna Schwendicke, Sebastian Pannasch, Ercan Altinsoy, and Jens R. Helmert

Technische Universität Dresden, Dresden, Germany

Hearing is a form of touch. Something that's so hard to describe, something that comes, sound that comes to you . . . You feel it through your body, and, sometimes, it almost hits your face.
— Evelyn Glennie

7.1 Milestones for building Human-in-the-Loop systems

Perception allows humans to build a stable representation of their environment by identifying, organizing, and interpreting multisensory signals, which stimulate their sensory systems [480]. Accordingly, perception is one of the key components guiding human behavior [481,482]. There exist different approaches to study human perception and determine the relationship between the sensory qualities one perceives and the information available in the physical world. In general, it has been shown that human perception is highly dependent on the person's learning history, expectations, and other cognitive processes, such as attention [480]. Furthermore, sensory information is typically incomplete and rapidly changing, yet people can form an understanding of the outside world, which allows interacting with the environment in a goal-driven manner. The quote at the beginning of this chapter is a description of how Evelyn Glennie, a deaf musician, who switched from playing the piano to percussion after she lost her hearing, describes her perception of sound. Glennie's story nicely illustrates how differently humans can access the information available in their surrounding. Besides this, it points out the flexibility in humans perception and action, and also highlights the limits in human behavior that might origin from reduced sensory information.

If we transfer this example to the context of human–machine interaction, the information available in such interactions similarly influences the potential success or failure of the human's actions. More precisely, today's technology provides information that can differ tremendously from that information typically available in humans' environments. For instance, virtual reality technology mimics the real world more or less exactly, and augmented reality even expands the amount of information avail-

Tactile Internet. https://doi.org/10.1016/B978-0-12-821343-8.00018-6

able in the environment. Still, the information has to be processed by the human user in a way that allows maintaining goal-driven behavior during the human–machine interaction. In some circumstances, reduced information can impede human performance, similarly to a physiologically induced loss of sense. Too much information can be overwhelming and might have contraproductive effects on human behavior as well. Accordingly, the information provided to the human using assistive technology has to be optimized to the human needs regarding perceptual capabilities and goal-driven behavior.

The endeavor of building Human-in-the-Loop systems that account for the needs of the human user gave rise to new interdisciplinary research fields, such as the Tactile Internet with Human-in-the-Loop (TaHiL). One main goal within this field is the development and optimization of integrative multimodal interface solutions for immersive augmented perception and interaction between human and Cyber-Physical System (CPS). For such applications, we first need to understand fundamental mechanisms governing the human–CPS interaction in different task settings, especially in situations involving shifts in task goals and in scenarios when "things go wrong", such as long delays between control command and action execution or high signal noise in sensors, actuators, or control processes. Second, new forms of adaptive interface solutions are required that can integrate massive multimodal inputs from wearable sensors and provide optimal feedback through the actuators to humans on fast time scales.

In this chapter, we start with an overview of past developments and the state-of-the-art research and technology in the field of haptic perception to identify corresponding demands in the context of human–machine interaction. Based on this overview, current and future research questions in this field are highlighted. Following these questions, we outline the key challenges for building Human-in-the-Loop systems as the origin of immersive augmented perception and interaction. In the last part of the chapter, the approach of TaHiL to develop and optimize integrative multimodal interface solutions is presented. We mainly focus on augmented perception and interaction based on the haptic modality, but extend these ideas on the processing of multisensory information during the human–CPS interaction in the latter sections. This is to direct the reader's attention to the transdisciplinary nature of TaHiL with corresponding links to other disciplines and their research depicted in other chapters of this book. The foundation of the present chapter is the following central idea of TaHiL: Combining human factors, such as psychophysical and neurocognitive principles of multisensory perception, goal-directed action as well as age- and expertise-related differences with new sensor, actuator, and electronic technologies and human-inspired reasoning techniques, allows inventing adaptive systems that enable humans and CPS to collaborate, and set up mutual learning during human–machine interaction (see Chapter 9 for discussions about an integrative framework).

7.2 State-of-the-art research and technology

In the following, we present a short overview of past developments and state-of-the-art research and technology relevant to approaches to design Human-in-the-Loop systems with different augmenting devices that introduce haptics for immersive augmented perception and interaction between humans and CPS. By definition, haptic interaction involves touching and manipulating objects, which can happen in real, augmented or virtual environments. Contrary to the functions of the human visual and auditory system, which are just sensory systems and called input systems here (but see [483] for on overview of gaze-based interaction), the haptic system is composed of information input via perception as well as output via action, and critically involves the somatosensory and somatomotor system. Haptic perception of the somatosensory system combines tactile and kinesthetic information, which is also referred to as exteroception and interoception, respectively. Tactile perception describes cutaneous sensations, i.e., spatial and temporal patterns of pressure and stretch (mechanic) or temperature (thermal) on the skin. Kinesthetic perception describes the sense of relative movement and position of body parts as well as muscular effort (proprioception; see Chapters 5 and 9 for related discussions) [484,485]. An illustration of haptic interaction between human and CPS based on these subdivisions of the human haptic sense is provided in Fig. 7.1 and described in more detail below.

The haptic modality allows for active and bidirectional interaction between human and CPS comprising tactile and kinesthetic input and output at both ends [484]. However, successful interaction is only possible as long as the characteristics of the output of the CPS match the capabilities of the input system of the human and vice versa. Even more, Barfield and colleagues [484] state that technological inventions—constructed for the interaction in augmented or virtual environments—must provide perceivable sensory information, which can be processed with the available cognitive resources (e.g., multimodal information can protect humans against visual overload) and interpreted in a way that is meaningful to perform a goal-driven action. All three aspects are a premise to cause feelings of presence (i.e., the perception of being physically present in a nonphysical world) and thereby an immersive coaugmentation between humans and CPS. To expose technological limitations and point out directions for future research and inventions, the authors provided a summary of the capabilities of the human's visual, auditory, and somatosensory system comparing them with the specifications of virtual environment equipment available at that time, which is 25 years ago [484].

During the last 25 years, many technological advances have been made and corresponding research on the advantages and drawbacks of different virtual reality techniques for various applications has been conducted. Typical fields of applications are within the area of education and training (for reviews, see [486–488]), neuroscience research, and therapy (for a review, see [489]) as well as usability (e.g., visual discomfort; for a review, see [490]), user experience (e.g., emotion and presence; for a review, see [491]), human perception (e.g., egocentric distance; for a review, see [492]), and technical aspects (for a review, see [493]). The huge variety of today's

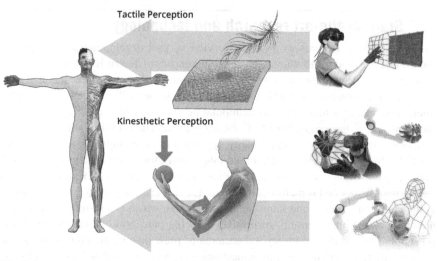

Tactile Perception

Kinesthetic Perception

Fig. 7.1

Haptic interaction with tactile and kinesthetic perceptual input or output for human and CPS.

(commercially available) haptic devices is based on some constantly recurring concepts. But there also exist many research prototypes, which can add new innovative approaches to these technologies. Here, we want to briefly go into some of these concepts of technologies used for the different channels of haptic interaction between humans and CPS. Tactile displays generating tactile output use two basic principles to provide feedback to the skin. They either simulate the characteristics of the surface to be explored or they try to evoke the perception that different surfaces/textures typically elicit by stimulating mechanoreceptors in the skin. While most devices of the first type are stationary, some of the latter can be classed among wearable devices and, accordingly, are mobile. Systems generating kinesthetic output cover a broad variety of force-feedback devices from wearables, including gloves as well as exoskeletons to ground-based solutions. Some of these devices also provide haptic tracking (i.e., tracking of kinesthetic input) or can be used in combination with optical tracking. Haptic devices gathering kinesthetic input do haptic tracking combined with force-feedback (i.e., generation of kinesthetic output) or pure data acquisition without providing additional tactile or kinesthetic feedback. For the design of haptic interfaces, it is essential to consider indications on benchmarks of human perception and influencing factors for the application of the haptic device. The following example will illustrate advantages of this approach, but also reveals some first challenges one might face when bringing together these different research fields (we provide a more detailed discussion of key challenges in the research field of augmented perception and interaction in Section 7.3.2). For more information on how technological approaches use the human somatosensory system and corresponding research on haptic perception for the development of haptic devices, we also refer to [494] for artificial tactile sensing and [495] for an overview how haptic devices induce the perception of different environmental properties.

For a haptic device that imposes movements on the arm or hand of the user (i.e., kinesthetic output of the CPS and input to the human) to provide feedback, this movement has to be recognized by the human. The detection threshold for movements imposed on the forearm, for example, is 0.2° at a speed of 1 °/s. However, at a speed of 0.1 °/s, it is eight times higher [496]. Furthermore, the threshold probably depends on the proportional length change the movement imposes on the muscle spindles and therefore differs between joints. In general, the detection acuity is higher with increasing angular velocity and optimal for the angular velocity of 2 °/s–80 °/s at more proximal joints and 10 °/s–80 °/s at more distal joints [496]. Another factor, that can influence the threshold is muscle activity. Thresholds change, if the muscles have been conditioned beforehand [497,498] and during contraction [498,499] (for a review, see [500]). These different factors have to be considered for the design and application of a haptic device. However, hand controllers, for instance, normally feedback force rather than position information. Accordingly, the information outlined above cannot directly be transferred to the specification of the haptic system. In addition, if a haptic device relies on the principle of force-feedback, the human's perceptual abilities regarding force detection and processing could add instructive aspects to the system design as well. Research on force detection thresholds, in turn, is scarce as experimental methods using very small forces inform about the sense of touch (i.e., tactile perception) rather than the sense of force [501]. Nevertheless, it is assumed that the force detection threshold for the hand is 8.5 g [502]. The force direction identification threshold for hand-held objects is 5 g during arm movement and 10 g for the passive arm [501]. Other factors also influence the perceptual thresholds. For instance, the just noticeable difference for discriminating the weight of hand-held objects is a difference of about 8% to 15% [503]. However, force levels are typically overestimated for low but not higher forces during force matching tasks [504]. This cannot only affect the measurement of the just noticeable difference—information that is used as reference value on the required resolution for the design of a haptic device—but also the percept elicited during the human–CPS interaction. It becomes clear that, similarly to the various technological advancements in the last decades, much knowledge about the capabilities of the human's somatosensory system has been generated that shows potential to add to the optimization of haptic interface solutions. For more detailed information, we refer to some newer work that reviews the literature on human tactile [505] and kinesthetic [500] perception as well as an older piece on the psychophysical principles specifically for electrotactile and vibrotactile stimulation [506]. However, we believe that future research in the field of human perception would benefit from a systematic evaluation of this knowledge. Further, in the following as well as in Section 7.3.2, we also discuss the necessity to verify this evidence on physiological and cognitive principles in human perception, which is for the most part outcome of more basic research, during human–CPS interactions.

Notwithstanding the above-mentioned developments, virtual reality applications are rather new approaches that are not yet well studied. Some of the main research gaps, as pointed out by Muller Queiroz and colleagues [488] in their recent review on immersive virtual environment training in corporate education, is the lack of studies regarding the evaluation of the effectiveness of such training as well as the technology-task fit. Boletsis [493] declares in his review on virtual reality locomotion techniques and their interaction-related characteristics that more user-centric, empirical research approaches, also comparing different settings, are needed. Freina and Ott [486] discuss potentials and risks for different classes of users that need to be verified for the application of virtual environments in education. Also, in different fields, it has been recognized that besides visual augmentation, approaches regarding the auditory, haptic, and olfactory modality should be induced, and the corresponding conditions for immersion as well as the overall effectiveness of these approaches need to be evaluated (e.g., [488]). Already some empirical evidence exists on the benefits of adding haptics to the technology in use, for instance for implementing force feedback in conventional and robot-assisted minimal invasive surgery and virtual reality training [487]. However, especially in the field of human perception and neuroscience, further examination and evaluation are required, for instance with respect to different contexts that will substantially impair the use of virtual reality, but also with the perspective of more real-world settings (e.g., augmented reality using improved sensory displays, including, for instance, touch; see [489,492]). In 1995, Barfield and colleagues already highlighted the relevance of this research line. Their summary of human and CPS capabilities was thought as a first step in developing a conceptual framework to investigate how and why particular interfaces impede the sense of presence or task performance. Their comparison showed that virtual environment technology does not completely match the human perceptual capabilities, which is still the case today (see also Section 7.3 for a discussion on the supposed reasons for this mismatch). Thus as long as this is not solved, the pressing question arises regarding what we can expect from "task performance under less-than-ideal circumstances—in bulky clothing, in space, by teleoperation, with blurred vision, or muffled hearing" (p. 351) [484]. Besides the influences of human factors and the context on human perception, as well as resulting challenges for building Human-in-the-Loop systems, we now discuss reasons, why most haptic devices do not completely match the human perceptual capabilities.

7.3 Identified key challenges of current research and technology

7.3.1 CPS with Human-in-the-Loop: Applications and problems

Challenges of integrating human factors into the implementation of closed-loop human–CPS interactions have only recently begun to attract research attention, for instance in biomedical [3] and automotive [507] applications. During such interactions, at each moment, decisions about what inputs are pertinent and what actions are to be expected have to be taken both from the perspectives of humans and machines, which we describe in more detail below. First, we discuss to what extent the current knowledge on the psychophysical and neurocognitive principles of human haptic perception and goal-directed action can be used to create an immersive coaugmentation between humans and CPS. Future work will need to address this question for multisensory perception, including the interdependencies of the involved modalities as well. Afterwards, consequent design goals for CPS and key challenges in the development of high-fidelity haptic devices are outlined. As the last point, we discuss some methodological issues within the relevant research fields, which have a determining influence on what knowledge is and will be available in the future building the foundation for applications of Human-in-the-Loop systems.

As stated above, the sense of touch plays a critical role during the human–CPS interaction as compared to visual and auditory perception; it allows for active and bidirectional interaction. Although the development of haptic devices serving as an interface between human and CPS has achieved remarkable breakthroughs in the last decades, a systematic investigation in the context of multimodal implementations combining visual, auditory, and haptic interfaces is pending [93,508,509]. Of note, Williams and Carnahan [510] recommended pursuing systematic tests of feedback designs within the haptic modality that promote motor learning before making comparisons to other modalities. The example on the human capabilities regarding kinesthetic perception of movement and force provided in Section 7.2 gives a first impression that already on the psychophysical level many factors need to be considered to identify optimal feedback designs. The empirical evidence so far makes abundantly clear that very different perceptions potentially influencing human behavior during human–CPS interaction might emerge dependent on where a haptic device is positioned, who is using it, and by which external and internal situational factors the interaction is characterized. This, as well as open questions regarding the deduction of benchmarks that will allow the CPS providing appropriate information to the user, is discussed in what follows.

7.3.2 The perspective of humans in human–CPS interaction: Open questions for benchmark setting

Internal and external influences on human haptic perception

It is well-known that perceptual thresholds differ across areas of the body due to differences in skin "thickness, vascularity, density, electrical conductivity, and more derived properties, such as moduli of shear and elasticity" (p. 12-3) [511]. This information, of course, needs to be considered in the design of haptic feedback systems. However, even if we strive to do so, on the practical side, it will be almost impossible to take all these properties of the skin that may crucially impact humans' sensation during human–CPS interaction into account as they, for instance, apply to unapparent individual characteristics, such as the receptor distribution in one's skin. Specifically, threshold values for vibratory stimulation will differ depending on whether stimulation was delivered to Pacinian fibers—typically responding to vibration [512]—or other receptors more sensitive to touch or rapid indentation of the skin [505]. In short, there will be always a certain degree of impreciseness about the expectable perception even if elicited by a haptic feedback design adjusted to the position of the system on the human body. Therefore it will be necessary to define the boundaries in which humans will be still able to maintain goal-driven behavior during human–CPS interaction.

There are other external factors from the technical device or the environment that can alter humans' experiences during human–CPS interaction. These factors are not stable and highly dependent on the characteristics of the situation. We need to find out which one of these influences can be compensated for by a feedback design tolerant towards some variability in the sensation the system generates and for which there is no compensation. For instance, background noise can interfere with the processing of feedback due to a nontarget stimulus that overlaps in time or space with the target causing a so-called masking effect [513]. This masking effect can be reduced by increasing time or space between the two stimuli and thus we might be able to potentially avoid impeding effects on humans' behavior as long as we can specify the situations (i.e., the noise condition), in which a technical device is used. In other cases, it might be more promising to develop flexible CPS, which can adapt to the situational context. As an example, human perception itself adapts to the environment and subsequently also to the feedback provided by a haptic feedback system. Typically, stimulus repetition has been associated with such adaptational processes resulting in a raised threshold for perceiving the succeeding stimulus [514]. Implementing flexible feedback designs into CPS will be necessary to truly augment human perception and action, including the high flexibility of human goal-driven behavior.

Besides these above-mentioned aspects of the used system and environment, the individual characteristics of the user will also form the interaction between human and machine. As the external factors, these characteristics might be more or less stable. The detection threshold for vibrotactile stimulation, for instance, increases as a function of age. Furthermore, women feature a lower threshold than men, but only in the elderly [515]. Besides this, a CPS that adapts to the body shape of the human needs to cover a wide range of differences in body composition, not only depending on age or gender. Available standard norms differ between countries depending on the respective resident population. We need to identify all those human factors that will significantly alter the effectiveness of the feedback system we are using. Among these factors, also intraindividual variability plays a crucial role. For instance, in the field of skill acquisition, such as in sports, it is sufficiently well-known that exercise has an impact on humans' kinesthetic perception: exercise with concentric muscle contractions alters the position sense, but not the movement sense [516]. After eccentric muscle contractions, position errors of the elbow flexors as well as an impaired joint position and force sense [517–519] have been reported (for a review, see [500]). Consequently, particularly technical devices, which provide corresponding knowledge of performance to its user to promote motor learning, have to consider these changes in perception or even integrate feedback methods that help the user to overcome such misleading internal physiological states. However, the development of these methods requires a better understanding of the mechanisms underlying the physiological principles in human perception. Beyond that, a mechanistic understanding could also add to the invention of systems that can predict situations and human states and direct the corresponding human–CPS interactions according to the given goals.

Predicting human behavior during human–CPS interaction

The implementation of models about human unimodal and multisensory perception and action may enable us to deal with the complexity of human–CPS interactions. Therefore we first need to assess the physiological and cognitive principles driving these interactions. By way of illustration, there has been early evidence on concentration differences of Meissner's corpuscles in the human skin dependent on age [520–522] and sex [523,524]. However, only recently Abdouni and colleagues [525] investigated age and gender effects for different biophysical properties of the finger (i.e., adhesive force, Young's modulus, and surface roughness) that play a critical role during tactile perception and corresponding manual interactions with the environment. We do not only need research like this; corresponding evidence also has to become part of the human–machine interaction, which implies that evidence on physiological and cognitive principles in human perception needs to be verified during human–CPS interactions. For instance, the time course of muscle desensitization after maximum voluntary contraction matches the decline in matching errors due to an altered position sense one can notice over time [500]. However, we do not know how it will change the outcome of humans' behavior in the interaction with a CPS when the CPS builds predictions based on the above-mentioned physiological principle, and following this, adapts to the current situational context, at least to some degree. The complexity of such models will also determine the speed and accuracy with which the CPS will be able to interact with the human. Physiological and cognitive principles, as well as situational factors, can influence humans' behavior, but not all of these aspects might be relevant for optimizing the human–CPS interaction. For instance, Proske and Gandevia [500] discuss that the sense of effort, which is altered after exercise, might have "wider perceptual consequences than just signaling muscle fatigue," but also "our current physiological potential to perform intended actions" (p. 1677). Accordingly, we overestimate the steepness of a hill, especially when we are fatigued and wearing a heavy backpack [526]. This evidence could have implications for some future applications of Human-in-the-Loop systems. Still, currently, it would be an open question, which information the CPS needs to consider and what kind of feedback is required, e.g., to help the inexperienced rock climber hanging on the wall to find a suitable route.

Further advancements in creating an immersive coaugmentation between humans and CPS are conceivable when we better understand the interaction effects of factors influencing humans' perception as well as a target-oriented use of perceptual illusions during human–machine interactions. This can also reduce the requirements imposed on the CPS. For instance, the extent to which humans adapt to tactile input depends on the stimulus duration [527]. Accordingly, the need for flexible feedback designs may lapse if we make use of shorter feedback stimuli. Even more, we can use very simple perceptual input delivered by the CPS to elicit rather complex sensations in the human. For instance, sensory saltation is a well-known phenomenon, where a sequence of tapping two regions of the skin, e.g., wrist and elbow, causes a complex sensation, such as sequential taps hopping up the arm [528]. Also, kinesthetic illusions are reported, where different movement sensations occur due to muscle vibrations on different sites with varying timing (i.e., simultaneously vs. sequentially; for a review, see [500]). Only if we know all key factors involved in the emergence of such sensations, we can make use of them for a system's design. This can be nicely illustrated with a movement illusion that occurs when a voluntary movement has been suddenly blocked. Participants overestimate the distance, e.g., of an arm movement that has been stopped by an obstacle. The size of this overestimation depends not only on the impact force of the obstacle but also on the point in time when the impact was encountered [529]. In sum, the main challenge in the research field of augmented perception and interaction from the perspective of humans in the human–CPS seems to be the identification of the key information the CPS needs to consider to predict human behavior and to provide optimal feedback to the human at the right time (see Chapter 8 for related discussions).

7.3.3 The perspective of CPS in human–CPS interaction: Challenges for achieving design goals

In the following, we outline the technical challenges researchers face in the field of haptic technology and some potential restrictions these may cause during the human–CPS interaction. Section 7.2 gives an idea of the scope of haptic interaction in all its forms. A high fidelity haptic system has to integrate multiple aspects to create precise tracking as well as realistic haptic feedback that does not obstruct the user. There are many design goals for haptic technology, including criteria, such as the following: (i) wearabilty (form, weight, impairment, comfort [530]), (ii) precise tracking/calibration/adaptation to the user, (iii) no obstruction (wireless, no bulky constructions), (iv) large/wide working range, (v) duration (long battery life), (vi) strong force feedback, (vii) realistic cutaneous feedback, ($viii$) safety, (ix) low price, (x) computational performance, and (xi) benchmarks. Some of these design goals have yet not been reached, or there is at least still room for improvement. Currently, accurate results often require labor and time-intensive calibration routines. Furthermore, most systems need to be adapted to the specific body shape of the user. To increase performance and user acceptance, quick and simple calibration routines are necessary as well as easy adaptation to the user's physiology. Similarly, haptic devices have to be characterized by high flexibility and a broad spectrum of functions. For instance, high-fidelity force feedback systems have to provide a solution to generate realistic sensations of opposing states of force feedback in the user [531]. This means that in constraint space situations, the force feedback device must allow for a maximum force with high stiffness, stability, and update rate. Contrary, free space should not be perceived by the user, thus the system needs to apply low friction, small mass, and inertia and no backlash [532].

One major challenge in the design of haptic systems is that some of the above-mentioned design goals contradict each other. For instance, strong force feedback currently requires bulky transmission systems and is power-hungry, thus respective systems obstruct the user and reduce battery time. Wearable cutaneous feedback hinders the exploration of real-world objects as do most sensing devices that exploit textile surfaces. Additionally, different device families, such as dataGloves and kinesthetic feedback devices are not easily combined as they often get in each other's way. Further optimization of available techniques, as well as novel inventions, are needed. Miniaturization of actuators and sensors still offers the potential to reduce the overall bulk of haptic devices. New, smart materials that couple sensing, actuation, computation, and communication [533] have an even bigger potential. There will be no device that fits all needs, and the fast advances of new technologies make it almost impossible to compare different prototypes and products across multiple dimensions. Additionally, researchers, as well as companies selling haptic products, rarely reveal full specifications of their systems. Notwithstanding, identifying key user applications and benchmarks for these applications may allow for a better comparison of different technologies and thereby sketch out promising directions for future inventions. On the other hand, technological advancements in the past also shaped the research landscape in the field of human perception and hold the potential to do so in the future as well. This potential, among others, is of special interest regarding the methodological development in this research area, as we illustrate in what follows.

7.3.4 State-of-the-art methodology: Limitations and implications for future research and technology

One critical issue in the research field of human uni- and multisensory perception is the development and application of appropriate methods. For the tactile modality, inconsistent evidence is in part the outcome of a conglomeration of different research methodology and the application of methods, of which each shows its specific deficits. For instance, a typical measure of tactile spatial resolution is the two-point threshold or two-point limen (i.e., the smallest distance at which two stimulations, simultaneously applied to distinct locations on the skin, can be distinguished from one). There are several problems associated with this method, such as the occurrence of up to five perceptual patterns participants report (e.g., point, circle, line, dumbbell, two points), large intra- and interindividual variability, and the evidence of a just noticeable difference (Just Noticeable Difference (JND)) much smaller than the two-point threshold, when participants have to compare the distance of two pairs of stimuli [534]. For kinesthetic measurements, the two-limb matching task is often used to investigate position sense, although the involvement of both limbs affecting participants performance can impede the interpretation severely. By contrast, single-limb tasks, where participants have to reproduce a remembered position, introduce another potentially confounding factor, i.e., memory. Other measurements of the position sense in single-limb tasks are still influenced by the sense of the second limb, which is, e.g., used as a pointer (for a review, see [500]). Thanks to the ongoing development of new measurement techniques, which also rely on more advanced haptic devices, some shortcomings of earlier approaches have been overcome already and further progress may contribute to new advances of psychophysical measurements as well as the understanding of underlying physiological and neurocognitive mechanisms (for reviews, see [534,535]). Reversely, valid psychophysical measurements are required to formulate demands for technical advances in the context of human–CPS interaction.

Further, identifying key user applications and benchmarks for the optimization of technical devices must be based on a thorough inspection and profound knowledge of human sensorimotor function in the corresponding field of application. On that front, methodology plays a critical role as well. For instance, one open problem in clinical research on the recovery of kinematic functions is the lack of standardization regarding kinematic assessment. A recent systematic investigation [536] shows that most clinimetrics used by studies on kinematic assessments of upper-limb sensorimotor function is biased. The authors provide recommendations on the assessment task, measurement system, and performance metrics based on corresponding methodological evidence. Such recommendations are necessary to derive clear implications and also guidelines for assistive technology in the different application areas, such as medicine and rehabilitation. For applications involving multisensory human–CPS interactions, we also need to be aware that the evaluation of multisensory interaction, on which these applications are based, depends on method-specific quantifications [537] potentially restricting its validity in the present context.

Taken together, in this section we provided an overview of current challenges to develop and optimize interface solutions—with a focus on the haptic modality—for immersive augmented perception and interaction between humans and CPS. Fig. 7.2 summarizes these challenges and also how the involved research areas in this field can benefit from each other. Implementing flexible feedback designs into CPS, which take human factors and the context into account, will be necessary to augment flexible goal-driven human perception and action. The invention of such systems requires a mechanistic understanding of physiological and cognitive principles in human perception and action that has been verified during human–CPS interactions. More precisely, a variety of factors can influence humans' behavior, but not all of them might be relevant for optimizing the interaction. Furthermore, the complexity of derived models predicting human behavior will determine the speed and accuracy with which the CPS can interact with the human that will both impact the outcome. Therefore respective boundaries have to be defined in which humans will be still able to maintain goal-driven behavior during human–CPS interaction.

Identifying key user applications and benchmarks for these applications may also allow for a better comparison of different technologies currently available to guide further advances and in the long-run to also approach to date conflicting design goals for high fidelity haptic systems. These technological advances are also required to further improve research methodology, including the validity of psychophysical measurement techniques. Finally, methodological standards have to be defined to derive clear implications and also guidelines for the use of assistive technology in the different application areas of Human-in-the-Loop systems. With that said, we already introduced several implications for future research. In what follows, we present the

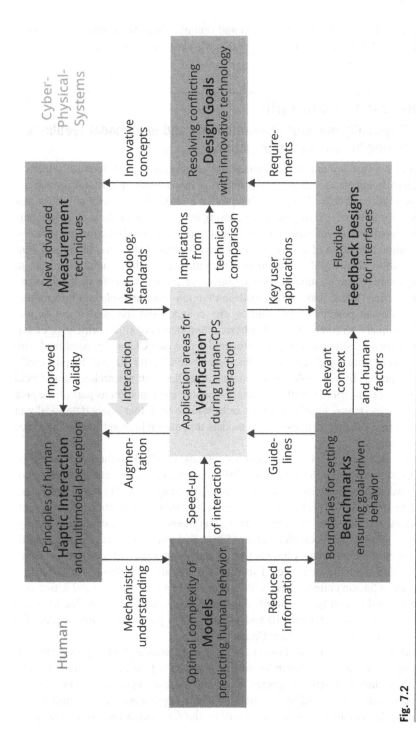

Fig. 7.2

Key challenges for the development and optimization of Human-in-the-Loop systems.

approach of TaHiL that aims to develop and optimize integrative multimodal interface solutions for the human–CPS interaction based on these implications.

7.4 Research within TaHiL

7.4.1 Physically plausible environments and multimodal feedback during human–CPS interaction

TaHiL pools a vast amount of proficiency and knowledge from different disciplines, and thereby holds out the prospect of pioneering innovations, in turn, promoting the progress in each subarea. To develop integrative multimodal interface solutions that will allow for immersive coaugmentation during human–CPS interaction, TaHiL pursues three goals: the development of intelligent sensors, the multiplexing of actuators, and the implementation of individually adaptive technical systems. Intelligent sensors will allow reducing the amount of data and thereby, the computation time to attain quasi-real-time communication required for mutual learning in the human–CPS interaction. These sensors only collect attention- and goal-relevant information based on human perceptual capabilities, the context, and the goal of an action. Transdisciplinary research on psychophysical and neurocognitive principles of human multisensory perception and goal-directed action (see Chapter 9) combined with innovations of sensor and actuator technologies (see Chapter 10) will help to establish these physically plausible environments. Together with researchers in the field of electrical engineering as well as human–machine coadaptation as part of the area of cognitive modeling, body computing hubs will be built that select data locally at the sensor (see Chapter 12), also using models that mimic human filter policies (see Chapter 8).

For the multiplexing of actuator control signals, feedback to the humans will be given based on the task-relevant sensory information. As a systematic comparison of multimodal implementations combining visual, auditory, and also haptic interfaces is pending [93,508,509], TaHiL will systematically introduce haptics into augmenting devices. For different levels of modality congruency (i.e., dominant, conflicting, redundant, additive, and complementary [538]), that will affect the degree of multimodal integration, different solutions (e.g., haptic-visual vs. auditory-haptic-visual) will be contrasted with each other. This research will also rely on new evidence and models about human unimodal and multisensory perception, attention and action (see Chapters 9 and 8) and be combined with new sensor and actuator technology and novel electronics (see Chapters 10 and 12) as well as pioneering haptic codecs for plug-and-play interoperability (see Chapter 5).

Since the capacities of the human operator in Human-in-the-Loop systems will determine the cooperative dynamics of the interaction (see also Sections 7.2 and 7.3.2), the technical system is expected to account for relevant human factors, such as expertise or age. Corresponding constraints on human perception and cognitive control during the human–CPS interaction will be the foundation for building adaptive systems that can also predict the changing cognitive resources of the operator in the control loop (see Chapter 9). With these three lines of research, TaHiL will enable researchers in the field of augmented perception and interaction to provide guidelines, requirements, and solutions for the design of Human-in-the-Loop systems. These guidelines can be used in different fields of application, such as medicine (see Chapter 2); motor learning, such as sports and rehabilitation science or musicology (see Chapter 3); multimedia research, and human–machine coworking industry (see Chapter 2). Below we outline the approach of an ongoing research project that investigates the impact of feedback delays for applications involving complex hand-controlled actions requiring gross and fine motor skills as well as eye-hand-coordination.

7.4.2 Multisensory feedback control and the impact of feedback delays in visually coordinated hand-controlled tasks

Today's CPS are electronic systems that are typically characterized by a high information density, complex computations, and as a result delays between control commands and their execution, including corresponding feedback during the human–CPS interaction. Delays between action and the visual feedback on the outcome have been shown to affect such interactions, as it can alter feelings of control and worsen humans' performance. We conceptualized a project that aims to *i*.) identify processes that are affected by delays in the visual feedback during complex visually coordinated hand-controlled tasks, *ii*.) determine how these processes are related to task performance/learning as well as how they change due to adaptation or training, and *iii*.) develop feedback designs using external augmented feedback to promote performance-enhancing effects of adaptation and training. For this, we conduct studies in which participants play the Wire Loop Game, a complex motor skill task. The game is placed in front of the participant. However, similar to the setting during minimally invasive surgery, visual feedback on participants' hand movements are shown on a display only. In addition, we implement delays in the visual feedback varying in reliability (dynamic vs. systematic) and length (short vs. long). We also manipulate the congruency (temporal overlap) between the visual information and the auditory feedback on performance accuracy (buzzing noise for errors). Besides this, we consider participants' expertise (e.g., surgical training) and provide additional augmented auditory and haptic feedback on relevant movement parameters to facilitate skill learning. We assess participants performance (speed, accuracy) in the game as well as their hand and eye movements using innovative technology (e.g., the eGlove; see Chapters 10 and 12) that can be also used to implement different feedback designs. Results of this project will provide information that is relevant to various applications of Human-in-the-Loop systems. First, we can specify requirements on the speed and reliability of visual feedback for manual motor skill performance during human–CPS interaction. Second, chances and limits regarding human adaptation to feedback delays can be pointed out, and finally, recommendations for the design of multisensory feedback in systems characterized by visual feedback delays can be provided. Accordingly, these studies will help to develop and optimize multimodal interface solutions, such as the eGlove and to derive further guidelines for their design to suit different applications of Human-in-the-Loop systems.

7.5 Conclusion and outlook

In this chapter, we provided insights into the state-of-the-art research and technology, present key challenges as well as new research approaches within the research field of augmented perception and goal-directed action during human–CPS interaction. When building Human-in-the-Loop systems that enable humans and machines to collaborate, the sense of touch allows—other than the visual and auditory modality—for active and bidirectional interaction. In the last decades, the development of haptic devices serving as an interface between human and CPS has achieved remarkable breakthroughs and, for some application areas, there is already first evidence on the benefits of adding haptics to the technology in use (e.g., [487]). Here, we gave a short overview of technological developments, concepts used in recent research prototypes and today's commercially available hardware for haptic interaction as well as research on their application. As a successful interaction requires a matching between the characteristics of the input and output systems of humans and CPS [484], we pointed out the relevance to bring together this research field and psychophysical and neurocognitive research on humans' tactile and kinesthetic capabilities.

Beyond this, we discussed human and environmental factors that may crucially modulate the human–CPS interaction as they limit the resources for processing sensory information and constitute the context in which the information needs to be interpreted in a meaningful way. One key challenge to develop and optimize integrative multimodal interface solutions for immersive augmented perception and interaction between humans and CPS is the implementation of flexible feedback designs taking these human and contextual factors into account. For that account, knowledge of physiological and cognitive principles in human perception and action needs to be verified during human–CPS interactions. For instance, consequences of a substantially impaired use of virtual reality and more real-world settings using haptic and multimodal interfaces need to be examined and systematically evaluated (see [93,489,492,508,509]). In this way, we can satisfy the need to define boundaries, in which humans will be still able to maintain goal-driven behavior and derive benchmarks for future technological inventions to approach to date conflicting design goals for high-fidelity haptic systems. Furthermore, to develop integrative interfaces, solutions of different multimodal combinations need to be compared as the degree of multimodal integration will depend on the level of congruency between the multimodal stimuli.

Generating this knowledge, the approach of TaHiL is to set up physically plausible environments by building intelligent augmenting devices surrounding the body. Moreover, models, which are based on a mechanistic understanding of physiological and cognitive principles in human perception and action and can predict human behavior in different contexts and for different user groups (e.g., experts vs. novices or young vs. old adults), will be implemented. These models will reduce the overall data rate and thereby speed up the computation time and codynamics in the interaction. In an ongoing research project, we aim to specify requirements on the speed and reliability of visual feedback for manual motor skill performance during human–CPS interaction. In addition, we want to derive recommendations for the design of multisensory feedback in systems characterized by visual feedback delays. The interdisciplinary approach of TaHiL holds the promise that corresponding research on augmented perception and interaction will promote the development of guidelines, requirements, and solutions for the design of Human-in-the-Loop systems, allowing fast and accurate goal-driven communication between humans and machines in application areas, such as medicine, rehabilitation, sports, and human–machine coworking industry.

Human-inspired models for tactile computing

Christel Baier, Darío Cuevas Rivera, Clemens Dubslaff, and Stefan J. Kiebel

Technische Universität Dresden, Dresden, Germany

> *Rigid, the skeleton of habit alone upholds the human frame.*
> – Virginia Woolf

8.1 Motivation and aims

For a successful human–machine coadaptation, a portfolio of models, methods, and tools is required that meet the constraints imposed by tactile computing. Since humans and machines inherently differ in their abilities and style to act and reason, a key requirement for a seamless human–machine interaction is the mutual understanding of each others behaviors, goals, and their reasons. To this end, both humans and machines should be enabled to predict future behaviors of each other. Whereas machines can use various means of communication by audio, visual, or haptic effects and information, and their appropriate usage is a main research field in Human–Computer Interaction (HCI), in this chapter we focus on the second direction of interaction, i.e., how the machine can understand and predict behaviors of humans. In this regard, the main aim of this chapter is to present a new decision-making model based on recent neuroscientific insights, as well as model-based approaches based on state-of-the-art learning methods. We present initial steps towards an automated human-style reasoning model that mimics the human ability to filter information for rapidly assessing the current situation and to adapt behavior accordingly. Equipped with a human-inspired model of decision-making, a Cyber-Physical System (CPS) would be able to better understand the current goals of a human with which the system is interacting, essentially endowing the system with the ability to predict not only the short-term future actions of the human, but also predict which forms of feedback would benefit the human–machine interaction and facilitate goal-reaching. Furthermore, such a model could not only provide explanations of human behaviors, but could be also used by machines exploiting the superior ability of human reasoning to filter information, and to timely provide context-dependent decisions. The latter is in the line of requirements for Tactile Internet (TI) applications, where low-latency and context-adaptivity are especially important.

In this chapter, we first focus on one key aspect of human decision-making: the ability to compartmentalize the environment into distinct contexts. This is best exemplified in experiments of task switching, where it is believed that the human brain switches from performing one task to another one by retrieving from memory only the necessary components to perform one task or the other [539]. With this strategy, only the task-relevant parts of the environment and potentially-useful actions need to be taken into account when making a decision in a particular task. The ability to quickly recognize a new context (e.g., a new task to perform) and retrieve the relevant components enables the human brain to quickly make decisions, even in the face of complex tasks. As we discuss in the following sections, this compartmentalization reduces the problem of decision-making to three steps: (*i*) identify the current context, (*ii*) retrieve context-specific information, necessary for making decisions quickly, from memory, and (*iii*) evaluate possible strategies for behavior in this task. Crucially, in this account, context-specific experience helps reduce the number of possible strategies to evaluate to only a handful, previously learned strategies, per context, greatly speeding up decision making without sacrificing flexibility. Using these concepts, we discuss concepts, such as *habitual control* and *goal-directed control* and propose a unifying *prior-based control* scheme.

After reviewing the aforementioned neuroscientific insights and establishing the scheme of prior-based control, we illustrate how an instance of human-inspired model-based learning could be structured, enhancing Reinforcement Learning (RL) methods [540] by contextualization and habitual control. The goal of this approach is to create computing methods that enable the prediction of human behavior, and at the same time improve on classical techniques [540,541] in measures of energy efficiency, latency, and robustness. To accomplish this, we propose the use of stochastic operational models that represent the dynamics and context switches of the environment and incorporate models for human decision-making. Being close to computational models, operational models are likely to have good machine-processing properties and enable the application of many existing techniques for their development and analysis. For instance, a model-based approach enables the use of formal analysis methods to provide guarantees on the achieved level of quality-of-service and robustness, key properties in TI applications.

To this end, our methods mainly tackle the key challenges of Tactile Internet with Human-in-the-Loop (TaHiL) systems to obtain models for predicting human behavior and to verify and analyze combined behaviors of the CPS and the human, indicated, e.g., in Fig. 7.2 of Chapter 7.

8.2 Neuroscientific insights into human decision-making

The interaction between machines and humans, much like the interaction between two humans, requires that one is able to predict what the other might do in the near future. For example, the every-day negotiation of multiple people walking in a confined space without colliding with one another depends strongly on all parties being able to predict each others' movements.

These predictions of human behavior are made at multiple time scales, ranging from fast actions to long sequences of actions, which is often referred to as *theory of mind* [542], i.e., being able to predict what someone else would do in a specific situation. In this chapter, we pursue the hypothesis that such a predictive mechanism is necessary to implement efficient and robust human–machine interactions and discuss how one might build decision-making algorithms that mimic the way humans make decisions, thus endowing machines with the ability to predict human behavior. Our overall aim is to derive a novel conceptual framework of how a decision-making algorithm should be built when based on recent neuroscientific insights into human forward planning and decision-making.

To do this, we first introduce the concept of forward planning, and then briefly review, selectively, parts of the human-decision making literature and focus mostly on what is known about how humans plan ahead to make their decisions.

8.2.1 Forward planning

In real life, the human brain must often make decisions based on which future sequences of actions may follow after a decision. For example, when going from home to work, turning left or right after leaving the building is a decision that cannot be made in isolation. Instead, it must be made by taking into account future turns that may or may not take us to the office. Making this kind of decisions is termed "forward planning" in neuroscience, in general referring to selecting an action based on some knowledge about future consequences of actions.

In our dynamic and uncertain environment, forward planning is in principle a daunting task to undertake, and one that the brain has evolved to perform with apparent (subjectively experienced) ease. For example, while going to work, there are countless possible actions that one could consider, most of which do not aid in achieving the goal of reaching the office. It is easy to see that a naive sampling approach of computing the consequences of some Action Sequences (ASQs), or the brute-force approach of considering all, to select an appropriate ASQ, is doomed to fail because the number of potential future ASQs typically increases massively with the number of time steps to plan ahead.

Although one may assume that the brain uses its abundant computing resources to compute the probabilistic consequences of a large number of possible action sequences, it seems unlikely that the brain wastes its computing power in a brute-force search-tree approach. There is overwhelming evidence in both psychology and cognitive neuroscience that the brain uses specific ways to reduce the number of computations that are required to plan ahead in an online fashion. In this chapter, we first review standard formulations of behavioral control based on forward planning, and sketch a speculative account of what actual computational devices the brain may be using to plan ahead and make decisions in an online fashion. This is followed by first steps to translate these ideas into mathematical formulations of forward planning. The overall aim of this chapter is to work towards a novel framework of human-inspired computing that is able not only to select actions just like a human would do, but also to use this ability to predict what an observed human agent will do in a specific situation.

In neuroscience, behavioral control is usually described by referring to two extremes: goal-directed control and habitual control. These two types of control are usually considered two ends of a control spectrum. Both control regimes are well-studied experimentally and theoretically, and will be briefly reviewed in the next two sections. In the third section, we will sketch a framework that also covers the control regimes between these two extremes.

8.2.2 Goal-directed control

The first extreme is what is usually referred to as goal-directed control. Typically, in experiments, participants have to reach a specific goal over a sequence of trials, e.g., to obtain a minimum number of points [543]. On each trial, participants are given two or more options that they can choose from. To make such tasks interesting and challenging, these options usually have probabilistic outcomes, e.g., the first option gives three points with a probability of 35% (zero points otherwise), whereas a second option gives one point with a probability of 80%. Such experiments, which are often formulated explicitly as multiarmed bandit tasks [544], have been found to be ideal to test hypotheses on how agents resolve the exploration-exploitation dilemma, use heuristics, assess risks, and in general compute goal-directed control by forward planning in a dynamic and uncertain environment. These and similar experiments aim at emulating specific situations in real life, where one cannot rely on well-learned strategies for reaching a goal over time, e.g., when one plays a new, complex board game. In experiments, such as the two-step navigation task [545,546], the limited offer task [547] or the betting task presented by [543], the general research question is: how does the brain achieve the instructed task goals? Goal-directed control in these experiments is often formalized as model-based RL [548] and, more recently, as active inference [549]. With these models, the decision-making agent takes into consideration the state of the environment, as well as the possible transitions between different states, the consequences of actions, and the reward structure of the environment. These computational models have been used successfully to describe goal-directed behavior in human participants [547,550–552] in tasks where the number of states and the number of actions is small enough that a full exploration of the ASQ space (all possible sequences of actions) is computationally feasible. However, the computational requirements to make decision in these models grows massively with the complexity of the task and the planning depth. Without modifications to the way ASQs are selected and evaluated, these models currently used in cognitive neuroscience are computationally too expensive for most real-life tasks. Although there exists a wide range of tools to reduce the number of computations required in RL algorithms [540], we will focus here on the key device the brain may be using to reduce its computations. In particular, a highly efficient way for the brain to reduce its computations required for forward planning is to learn and use habits, which we review next.

8.2.3 Habitual control

The second extreme of how humans control their behavior is typically referred to as habitual control [553]. Most habitual-control experiments are run using animals, training them by making them repeat a specific action hundreds to thousands of times, after which a habit has been developed and can be tested experimentally. In real life, humans rely heavily on habitual control; for example, our morning routine after waking up, the way we make a coffee in our kitchen, the way we enter our office, the way we tie our shoelaces. Clearly, habits are useful as they seem to run on minimal computational resources, thereby freeing us to think about something else when performing the habit. Habits seem to be used by the brain in a context-specific fashion, not only by animals in experiments, but also by us humans in real life [554]. This indicates that habitual control can be described, in contrast to goal-directed control, as a decision to evaluate and select a single ASQ that has been used repeatedly in the same context before. We will come back to this key difference in the next section.

Habits were originally expressed in terms of reinforcements [555]. In operand reinforcements [556], Stimulus Response (SR) associations are learned to avoid punishment or gain reward. Such associations are born from previous experiences with the same decision, by reinforcing actions that have led to reward or avoided punishment. The acquisition of SR associations has been formalized in terms of model-free RL [540,557–559], in which each time that an action leads to reward, the association between the action and the stimulus in which it was taken is strengthened by an amount determined by a learning rate. Due to the simplicity of the calculations necessary for learning and using these associations, SR associations are typically seen as the simplest response-learning mechanism in cognitive neuroscience. There are also different implementations of model-free RL, which have additional parameters, besides learning rate, and ways of calculating SR association strengths. For example, Q-learning [540,559] calculates the strength of the SR association by doing a weighted average of all the possible rewards to be obtained by a given action, an iterative process, which is improved with experience. Similar to habit learning in animal experiments, RL techniques [540] require a lot of training time to learn good decisions (i.e., reward-maximizing decisions).

8.2.4 A novel framework: Prior-based control

One of the key features of habitual control, briefly reviewed in the previous section, is its fast computation and reaction times. This fast response is presumably the consequence of the decision of the brain to not use goal-directed control in a specific context, but rely on a single ASQ, i.e., a habit that has been used before in that or an apparently similar context. Even when not using the concept of habits, an astonishing feature of human decision-making is the speed with which action sequences, which may have significant consequences on a range of different time scales, are selected and executed.

Here, we argue that the brain uses a control scheme that encompasses both goal-directed and habitual control by harnessing the context-specificity of most situations we encounter in real life. In this scheme, for each situation the brain encounters, it retrieves a prelearned prior distribution over action sequences, which reflects which ones have proven useful in past exposures to that specific situation, and prioritizes the ones with high prior probability.

In this view, the brain is like an enormous library of priors, where each book stands for a specific context in which one has been repeatedly in the past. For each context (book), the brain has stored the ASQs that have been successful in the past, and the probability with which it believes *a priori* it will be using this ASQ in a specific future instance of that context. Since for any specific instance of a given context typically only a small number of ASQs has high prior values, these ASQs can be activated and evaluated in a split second. With such a control system, the required online computations reduce to (*i*) inferring the currently active context (In what situation am I?), (*ii*) the evaluation of the ASQs guided by the priors, and (*iii*) if there is a new context, and if under time-pressure, adjusting the most promising ASQs from a similar already-learned context.

The underlying idea here is that a ranking of ASQs based on their prior probabilities enables the brain to evaluate only a few ASQs to quickly decide which one to select. If there is time pressure due to online demands, the brain may just take the ASQ with the highest prior probability and select it, without evaluation and comparison to other ASQs. If there is more time, more ASQs with high prior probability can be evaluated and compared. In this view, habitual control is an extreme point, where there is only a single ASQ to be evaluated and selected.

In Fig. 8.1 (A), we show an overview of the different types of control regimes. In well-practiced contexts, habitual control is used to quickly deploy responses that have been, in the past, successful with very high probability. In most other cases, the brain may use *prior-based control*. Here, previous experience helps to preselect a set of ASQs with high prior probabilities to evaluate and compare to each other. In totally novel tasks, such as many laboratory tasks, goal-directed control is used, i.e., brute-force forward planning with many possible ASQs to be evaluated. Note that all three control regimes are just an expression of prior-based control.

With this approach, the key question changes from how to quickly evaluate many different ASQs, across multiple time scales, to the question of how one can learn and infer contexts. Some contexts can be inferred with minimal uncertainty, e.g., *I am on my way to work*, whereas some other contexts are more difficult to infer, e.g., *What is the intention of someone else?* We will discuss this type of questions further in the following sections, after we provide a basic example of the proposed approach.

Note that several accounts exist for how the brain might arbitrate between using habitual or goal-based control to operate in the full range of decision-making [545,560,561]. Arbitration accounts naturally lead to the idea of a competition between habitual and goal-directed control [560]. Our suggested framework avoids any arbitration by simply using the previously learned prior over ASQs. If there is only a single ASQ with high prior value, the brain will tend to employ habitual control (without the need for online arbitration). If there are many ASQs with high prior value, the brain will use goal-directed control. In summary, the proposed approach

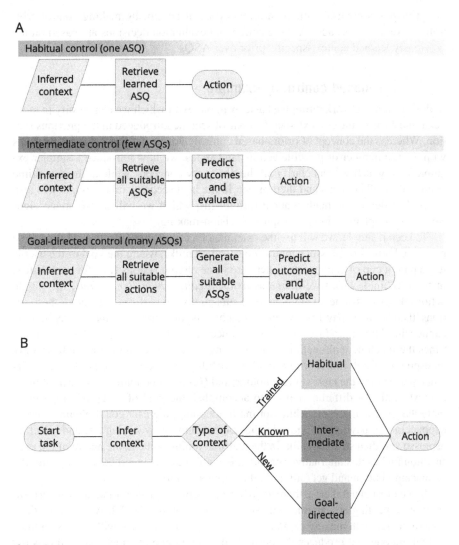

Fig. 8.1

Three different control regimes. (A) Steps for action selection for the three possible scenarios of context inference: (Top) Known and trained context, where experience has reduced ASQs to a single one, which does not need to be evaluated. (Middle) A context that has been encountered before, for which priors over ASQs exist. These priors are retrieved from memory and, using them, only a small number of ASQs need to be evaluated. (Bottom) Totally unknown context, for which no previous experience exists. ASQs must be generated (possibly all combinations of available actions) and evaluated, rendering decision-making very slow. (B) Flow chart of behavior in a task. The first step is inferring context. Depending on the type of context, one of the control regimes in (A) will be used to select an action.

of "prior-based control" unifies different types of control by making control arbitration redundant, because ASQ selection and evaluation decisions are based on the previously learned context-specific prior over ASQs.

8.2.5 Prior-based control by example

In this section, we work through a basic example, with which we can specify in more detail the parts of the context-specific control scheme introduced in the previous section. Whereas the concept of priors-based control finds its best use cases in situations with a large number of possible actions and states, we now introduce a simpler example with which we can clarify all the necessary concepts, while at the same time enumerating all contexts and their priors. In the next sections, this example will also be used to define the mathematical formalisms, which we will use to create algorithms that implement human-inspired decision-making.

To keep it simple, we will use the example of a cappuccino machine. The machine has three buttons (espresso, cappuccino, steamed milk), where the goal of the user is to prepare a cup of cappuccino. There are three atomic actions (one for each button in the machine), which we define as *axiomatically irreducible actions*, i.e., single actions that we assume do not consist of subactions. To perform any one of these actions, the user must pay 1 €. When the machine is functioning normally, pressing the cappuccino button produces a cup of cappuccino; however, it is possible that sometimes the machine malfunctions, and pressing the cappuccino button produces only an espresso or only milk, with a small probability. With this, there are two contexts when performing the task: (*i*) functioning and (*ii*) malfunctioning, and each of these contexts calls for different actions to accomplish the goal of a cup of cappuccino. Note that in this machine, while not malfunctioning, the user could obtain a cup of cappuccino by paying 1 € and selecting cappuccino, or by paying 2 € and selecting espresso and then steamed milk. In the malfunctioning context, the nature of the malfunction makes this alternative (paying 2 € viable) as one could be unlucky with the cappuccino button and get only milk three times in a row.

In the most general case, there are twelve possible sequences of actions to perform (without repeating any atomic action), as can be seen in Fig. 8.2. A person that has no experience with making cappuccino might try to experiment with all possibilities to find the one that produces the desired outcome (a cup of cappuccino); this is the equivalent of brute-force forward planning, and we refer to this as having *no context* (see Fig. 8.2A). Someone familiar with the (properly functioning) machine would already know that a cappuccino cup can be produced either by the single atomic action of pressing the cappuccino button, or by the sequence of pressing espresso and then steamed milk; because the cappuccino button costs much less money, there is no reason for a person to entertain the other option, therefore most ASQs have prior probabilities of zero. Only the cappuccino button has a probability different from zero, as can be seen in Fig. 8.2B.

In the context in which the machine is fully functional, the very precise priors (i.e., with most ASQs having zero prior probability, and one having close to one;

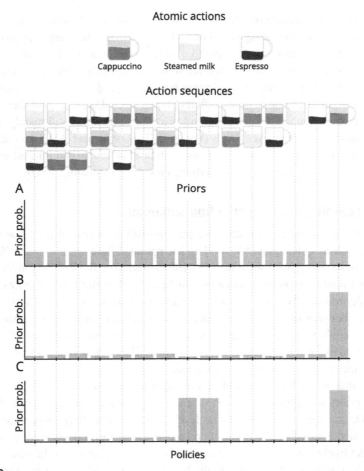

Fig. 8.2

Diagram of the coffee machine example. There are three atomic actions, depicted on top. All possible combinations of these actions (i.e., all ASQs) with no repetition are shown. (A) A person encountering this context for the first time might consider all ASQs equally good *a priori*, which translates into a flat prior over ASQs. (B) Knowing the machine is fully functioning, the user could learn to only consider pressing the cappuccino machine, with all other ASQs having a near-zero prior. (C) In a malfunctioning machine scenario, the two options of espresso + steamed milk should also be considered, although they carry a higher monetary cost.

see Fig. 8.2C) are similar to the traditional concept of a habit. In this case, the task becomes a simple stimulus-response association, where *stimulus* refers to standing in front of the machine with an empty mug and wanting a cappuccino, and the response is pressing the cappuccino button. In the malfunctioning context, a different set of priors would be more beneficial (Fig. 8.2C), as it allows for the consideration of three known ways of getting a cappuccino. Furthermore, a person could use the three ASQs with nonzero prior probability as possibilities for forward planning, predicting the outcome (including costs) of each one of them and using that prediction to evaluate them. As this is done with three sequences instead of all twelve, it greatly simplifies calculations and helps make decisions more quickly. Although a simple example, the main point is that, to implement human-inspired behavioral control, there is no need to differentiate between different types of control, e.g., goal-directed vs. habitual control. Rather, the actual control behavior of an agent will mainly depend on the shape of the prior over ASQs (Fig. 8.2). This prior over ASQs can dramatically change the apparent control behavior from fast habitual to careful goal-directed control, but without changing the underlying process of evaluating ASQs.

8.2.6 Learning priors over action sequences

How does the human brain learn these prior probabilities over ASQs? For habits, models, such as Q-learning [540,559] and temporal difference learning [540,562], provide some answers. If an action has led to reward multiple times in the past, a strong stimulus-response association can be created in the form of priors, to be used in future trials. Whereas in classical learning experiments these SR associations have been between the stimulus and an atomic action, it need not be so. Instead, associations could be created between a context and any action (or sequence of actions) that has lead to reward in the past. This means that SR associations are not necessarily with atomic actions nor need they be unique.

A formal account of a possible mechanism through which repetition leads to learned priors was presented by [563]. The authors presented an account of habit formation and use at the mechanistic level. In their work, they show a decision-making model based on Bayesian inference, in which forward planning is seen as the model likelihood, which is then learned with task repetitions to become a prior over actions to be used in future trials. In other words, habits are stored in terms of a prior distribution over actions, which biases a person's decisions toward a particular set of actions. These prior distributions are created with repetitions of the forward-planning strategy. Note that, in this account, habits need not be a single stimulus-response associations, but instead can consist of a subset of the available actions (or sequences of actions), which are considered for making a decision in a given context.

In the account presented by [563], a system for preselecting ASQs to evaluate could be implemented. For example, only ASQs with a prior probability higher than a threshold could be further evaluated. Determining which mechanism for pruning the ASQ space is in place should be subject of further research.

To complete the picture, in the following section, we discuss how the brain might infer what the current context is and how it uses this information to select the best priors for the current task, thereby greatly simplifying decision-making for most repeatedly-experienced contexts.

8.2.7 **Context**

There is a myriad of possible actions (atomic or otherwise) that a person could perform at any moment. Arguably, the first and most important step the brain must take to make a decision in any given task is reducing the action space (i.e., the set of all actions available) to a more manageable size. Though the psychology literature is rich in studies demonstrating this ability and the computational costs of so-called task switching [564], in this section we focus on one particular mechanism that must precede all others: context inference [565].

Intuitively, we define a context as the situation in which a decision must be made, as was done by [546]. In the simplest animal experiments, the context consists of, for example, a box with a lever and a loudspeaker. The context defines the actions that are possible (e.g., pulling the lever) or impossible (e.g., getting out), the goals (e.g., eat because the animal is hungry) and the rules of the environment (e.g., what happens when the lever is pressed), which may or may not be known to the animal. In RL terms, the context defines the state space, the action-specific transitions between states, and the goal structure.

To identify a context, the brain can make use of at least two sources of evidence. Firstly, observations might help determine which context the person is currently in. In the coffee machine example, a sign hanging in front of the machine may say that the cappuccino button is broken. This would immediately prompt the person not to use the priors in Fig. 8.2C, but use Fig. 8.2B instead. Secondly, the consequences of actions, i.e., observed outcomes, can give clues as to the current context. In our example, pressing the cappuccino button and receiving a cold beverage with no foam will indicate that the context is that of a broken machine, even if there is no sign indicated that the machine is not working properly.

Which mechanisms might the brain use to identify a context is currently a matter of active research [561,566]. Markov Decision Processes (MDPs) provide the necessary mathematical foundation to address such questions. We will discuss them further in the following sections.

In active inference, the best action to take at any moment is treated as another hidden variable which is inferred using Bayesian inference; this is referred to as planning by inference [567–570]. Since context, actions, and all other hidden states are treated equally, they all inform each other's inference; in doing so, the consequences of past actions can be directly used for the inference of the current context. In our coffee machine example, having pressed the cappuccino button and received lukewarm coffee, with no foam, can be used to infer that the machine is broken. The advantage of this account is that it offers a single mechanism, through which the current context and the best action are inferred.

In addition to inferring known contexts, the human brain has the ability to generalize from one context to another in at least two different ways [571,572]: First, the new context might be practically identical to a previously encountered context, such that the same ASQs and priors can be used. An example of this would be encountering a cappuccino machine of a different brand that has the same three buttons. Second, in the case that the new context is not identical, it might still be a good enough approximation to use the same strategies as a first attempt, and only if those fail, resort to more computationally expensive (e.g., full forward planning) or less reliable (e.g., sampling actions randomly) strategies. For example, there may be a more complex coffee machine that has many other buttons (e.g., hot chocolate), but does not have a cappuccino button, though having an espresso and a steamed milk button; with such a machine, the priors in Fig. 8.2B can still be used. If these two approaches fail, the brain might recognize that the current context is too different from anything encountered before, and may try to learn new ASQs for this new context (i.e., the priors in Fig. 8.2A).

It is currently unclear how such context generalization might occur in the brain, and little research has been performed on the subject [571–573]. We believe that understanding this mechanism would complete the decision-making perspective we have outlined so far in this chapter and is of paramount importance for the development of human-like reasoning and the human–machine interaction we strive to develop.

In the next sections, we discuss how we incorporate the concepts discussed so far into an automated approach of decision-making. Moreover, we show how we can use the developed model to make decisions and discuss how to use it in the future to predict a person's choices from current observations to aid in human–machine interactions.

8.3 Human-inspired learning

Based on the neuroscientific findings of goal-directed and habitual control of humans presented in Section 8.2, we now describe concepts for human-inspired learning facilities in a more formal way and present first steps towards an algorithmic implementation. Whereas the previous sections assumed already learned models and policies, mainly focusing on the neuroscientific aspects of how the human brain processes the information towards goal-directed behaviors and habits, in this section we describe how to derive an actual learning method. For this, we rely on stochastic operational models, such as MDPs [557]. Operational models that mimic human behavior have the advantage of being close to the stepwise and state-based behavioral description of computer programs and protocols. Stochastics provide a mathematical concept for uncertainty, e.g., to model environmental effects or partial knowledge. This makes them eligible for using human-inspired policies to solve computational tasks, and to monitor human behavior for providing hints and predictions, supporting the human in a Human-in-the-Loop setting. Humans are strong in contextual inference and filtering information, while still providing an acceptable degree of suboptimal policies. The application of human-inspired learning techniques towards controllers used by machines might thus yield massive state-space reductions and slim implementations, trading mathematical optimality for efficiency. The latter aspect is very important in the embedded system's domain and also contributes to the low-latency requirement of TaHiL applications.

Towards showing the basic concepts algorithmically, i.e., exemplifying facets our novel human-inspired learning method on MDPs, we first introduce MDPs more formally and revisit basic principles of RL. Then, we turn to the simple introductory example of a faulty coffee vending machine, such as the one discussed in Section 8.2.5, and illustrate how cost-nearly-optimal policies for obtaining a cup of cappuccino can be learned. Though our example is very basic in its nature, it illustrates the challenges of human-inspired learning that have to be faced when applied in a Tactile Internet application.

8.3.1 Stochastic operational models

In this section, we recall notions from stochastic operational models, used to introduce our concept of human-inspired learning. Readers familiar with Markov decision processes and their analysis, or those readers not requiring formal insights of the subject, may skip this section and return to it in case there is a need of clarification.

For a set X, we denote by $\wp(X)$ its power set, i.e., the set of all subsets of X. Given a finite set S, a distribution on S is a function $\mu \colon S \to [0, 1]$ with $\sum_{s \in S} \mu(s) = 1$. We write $\mathrm{Distr}(S)$ for the set of distributions μ on S, where $\mu(s) \in \mathbb{Q}$ for all $s \in S$.

Markov decision processes MDPs (e.g., [557,574]) provide a prominent stochastic model that was introduced in the 1950s and widely used for various types of optimization problems with applications, e.g., in operations research, RL, and robotics. Since the 1990s, MDPs have been used as an operational model for distributed probabilistic systems, and various verification algorithms for MDPs and temporal logic have been developed [557]. In this setting, MDPs are state-based stochastic models, where—in each state—actions can be performed that lead to a probabilistic distribution over successor states. During the stepwise behavior of MDPs, the nondeterministic choices between the actions are typically resolved by a *policy* (also known as scheduler, strategy, or adversary) that also could, e.g., stand for decisions made by a human.

Formalization of MDPs An MDP is a tuple $\mathcal{M} = (S, \mathrm{Act}, P, AP, L)$, where S is a finite set of states; Act is a finite set of actions; $P \colon S \times \mathrm{Act} \rightharpoonup \mathrm{Distr}(S)$ is a partial probability function; AP is a finite set of atomic propositions, and $L \colon S \to \wp(AP)$ is a function assigning atomic propositions to states.

A *finite path in* $\mathcal{M}$ is a sequence $\hat{\pi} = s_0 \alpha_0 s_1 \alpha_1 \ldots \alpha_{n-1} s_n$, where $s_0, s_1, \ldots, s_n \in S$ are states, and the α_i's are actions with $\alpha_i \in \mathrm{Act}(s_i)$ and $P(s_i, \alpha_i, s_{i+1}) > 0$ for all $i < n$. We write $last(\hat{\pi})$ for the last state of $\hat{\pi}$. Infinite paths are defined accordingly. A path is said to be maximal if it is either infinite or finite and ends in a state s, for which $P(s, \alpha)$ is undefined for all actions $\alpha \in \mathrm{Act}$.

Policies for MDPs A *policy* for $\mathcal{M}$ is a function that assigns to each nonmaximal finite path $\hat{\pi}$ a distribution $\mathfrak{S}(\hat{\pi}) \in \mathrm{Distr}(\mathrm{Act}(last(\hat{\pi})))$. Given a policy $\mathfrak{S}$ and a state $s \in S$, we write $\mathrm{Pr}^{\mathfrak{S}}_{\mathcal{M},s}$ or briefly $\mathrm{Pr}^{\mathfrak{S}}_s$ for the probability measure on (measurable) sets of maximal paths starting in s induced by unfolding $\mathcal{M}$ from s following $\mathfrak{S}$'s decisions. If φ is a measurable path property over atomic propositions AP, we simply write $\mathrm{Pr}^{\mathfrak{S}}_{\mathcal{M},s}(\varphi)$ for the probability measure of all $\mathfrak{S}$-paths from s satisfying φ. Measurable path properties are, for instance, sets of paths that satisfy the temporal logic formula $\Diamond goal$, i.e., where some state $s \in S$ is reached that is labeled by an atomic proposition $goal \in L(s)$.

For a worst- or best-case analysis, one ranges over all policies (i.e., all possible resolutions of the nondeterminism) and considers the extremal probabilities for satisfying φ:

$$\mathrm{Pr}^{\max}_{\mathcal{M},s}(\varphi) = \sup_{\mathfrak{S}} \mathrm{Pr}^{\mathfrak{S}}_{\mathcal{M},s}(\varphi) \quad \text{and} \quad \mathrm{Pr}^{\min}_{\mathcal{M},s}(\varphi) = \inf_{\mathfrak{S}} \mathrm{Pr}^{\mathfrak{S}}_{\mathcal{M},s}(\varphi).$$

Optimal policies in MDPs can be obtained by linear-programming techniques or policy iteration (e.g., [557]).

Weighted MDPs Plain MDPs can be extended by several additional features, such as a declaration of the initial states, state labels for temporal-logic specifications, or weight functions for reasoning about cost and reward constraints. The latter are important, e.g., in the setting of RL, where policies are learned that either minimize costs or maximize rewards. A *weighted MDP* is an MDP with a weight function $wgt\colon S \times Act \to \mathbb{Q}$ that assigns rational weights to all state-action pairs. The *accumulated weight* of a finite path $\hat{\pi}$ is denoted by $wgt(\hat{\pi})$ and is defined as the sum of weights of its state-action pairs, i.e.,

$$wgt(s_0\alpha_0 s_1\alpha_1 \ldots \alpha_{n-1} s_n) = wgt(s_i, \alpha_0) + wgt(s_1, \alpha_1) + \ldots + wgt(s_{n-1}, \alpha_{n-1}).$$

For a measurable path property φ and a state $s \in S$, we define the (optimal) expected cumulative weight in case $\mathrm{Pr}^{\mathfrak{S}}_{\mathcal{M},s}(\varphi) = 1$ for all policies $\mathfrak{S}$ in $\mathcal{M}$ as

$$\mathrm{Exp}^{\max}_{\mathcal{M},s}(\varphi) = \sup_{\mathfrak{S}} \mathrm{Exp}^{\mathfrak{S}}_{\mathcal{M},s}(\varphi) \quad \text{and} \quad \mathrm{Exp}^{\min}_{\mathcal{M},s}(\varphi) = \inf_{\mathfrak{S}} \mathrm{Exp}^{\mathfrak{S}}_{\mathcal{M},s}(\varphi).$$

Example 8.1. As a running example, we consider a faulty coffee vending machine, like the example described in Section 8.2.5. In this machine, the milk steam facilities interfere with making espresso such that there is a 30% chance to get either only steamed milk or espresso, but not cappuccino, even when selecting cappuccino. The behavior of the faulty coffee vending machine can be modeled as an MDP, as depicted in Fig. 8.3. For this example, a natural goal would be to minimize the expected investment of money to obtain cappuccino. Expressed using the formalism of weights, we could assign weight 1 to the state-action pair $\langle \ell_0, pay \rangle$ (and 0 to all other state-action pairs), formalizing the costs of a drink of 1 €. To describe properties of the states, we label the states with atomic propositions $AP = \{esp, cap, mil, end\}$, where $L(\ell_2) = \{esp\}$, $L(\ell_3) = \{cap\}$, $L(\ell_4) = \{mil\}$, and $L(\ell_5) = \{end\}$. Then, we seek a policy $\mathfrak{S}$ that minimizes the expected costs to obtain a cappuccino, i.e.,

$$\mathrm{Exp}^{\mathfrak{S}}_{\mathcal{M},\ell_0}\left(\Diamond(cap \wedge \Diamond end)\right) = \mathrm{Exp}^{\min}_{\mathcal{M},\ell_0}\left(\Diamond(cap \wedge \Diamond end)\right).$$

An optimal policy without any additional knowledge would always order cappuccino using the *cappuccino* action. Following the sum equation for the geometric power

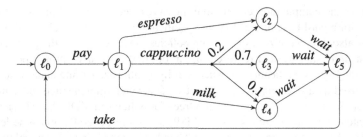

Fig. 8.3

Simple MDP of the faulty coffee vending machine. Arrows indicate the transition probability function, where branchings after a bullet show the probability distribution over the states. When the transition is deterministic, i.e., has probability one, we omitted bullet branchings. We also omitted self-loops of actions that do not have an effect on changing the state, e.g., an action *cappuccino* in ℓ_0 does not have any effect and results in remaining in state ℓ_0. We consider probability measures with respect to paths starting in ℓ_0, shown as the initial state.

series, this leads to an expected investment of $0.7 \cdot \sum_{i=1}^{\infty} i \cdot 0.3^{i-1} = \frac{1}{(1-0.3)^2} \approx 2.04\,€$ until obtaining cappuccino and ending up in a final state.

8.3.2 State of the art

In this section, we revisit learning methods that are relevant in the context of formal methods and are related to the human-inspired learning concept we pursue.

8.3.2.1 Learning techniques and formal methods

Formal methods are understood as a collection of techniques and tools, where systems are specified based on mathematics and formal logic to rigorously reason about their properties [575]. There are several branches of *learning approaches* in the field of formal methods: model-based learning [575,576], learning counterexamples [577,578], using formal methods to verify policies learned by machine learning tasks [579], or using learning techniques to speed up verification tasks [580]. Besides these application areas, one usually distinguishes between *active* learning, which we mainly pursue in this paper and where learning is achieved by actively performing experiments on systems, and *passive* learning, where learning is performed from a set of positive and negative examples given prior to the learning process.

Pioneering work on learning operational models [581,582] by presenting algorithms that generate finite-state systems already showed the potential of using model-based learning techniques in the formal-methods community. Here, learning is understood as a supervised process, where the models are obtained from example sets of behaviors matching or not matching the specification. Based on the models learned, analysis techniques, such as model checking [583,584] can be used to reason about system properties. For an overview on this classical view on model learning, we refer to [585,586].

Besides the computation of extremal measures, such as probabilities and expectations with regard to temporal logical specifications in MDPs, a classical question concerns also the *synthesis* of optimizing policies [587,588]. Here, the goal is to synthesize a finite-state system representing an optimal or nearly optimal policy from a temporal logic specification. A successfully synthesized finite-state system can then be used as a decision-making agent or controller implemented, e.g., in computer systems. Learning (small) decision trees for policies in MDPs have been widely investigated, e.g., in the context of RL [589] and more generally, machine learning [590]. Also here, a training set of good and bad behaviors is given, from which classifiers are generated and concisely represented as decision trees. In [591], an approach to learn compact representations of optimizing policies in body sensor networks was presented. The approach by [592] learns decision trees for optimizing policies in $2\frac{1}{2}$-player games, i.e., MDPs, where an additional source of nondeterminism is present. Learning MDP subsystems that maintain specific properties of the original MDP model was presented in [593]. *Anytime algorithms* to learn approximative policies in $2\frac{1}{2}$-player games, i.e., a synthesis algorithms that provide a policy that is closer to optimality the longer it runs, was presented by [594]. A framework for learning decision trees of memory-less policies was developed and evaluated with various case studies in [595].

In [596], neuropsychological data was used to synthesize controllers for vehicles in a mixed machine-human setting. For this, human behaviors in different contexts learned were integrated in a hybrid system to solve a $2\frac{1}{2}$-player game. Whereas the synthesized controller includes knowledge about human behavior, the underlying model for the human is not operational, but relies on Bayesian networks.

Though the related work mentioned up to now surely cannot claim to be exhaustive, it shows that our aim of learning models and policies that mimic human behaviors has merely been addressed yet by the formal-methods community. Also, the use of neuroscientific insights for novel algorithms to learn operational models is still in its infancy.

8.3.2.2 Reinforcement learning

One prominent instance of an active learning method to synthesizes policies is RL [540], typically applied on MDPs that are either too large for applying exact methods to synthesize an optimal policy or when an unknown environment hinders the construction of the MDP model. RL has a broad area of applications in artificial intelligence, neuroscience, operations research, and control theory [597]. In general, RL is based on the *reward hypothesis*, i.e., that the goal of the policy to be learned can be described by the maximization of expected cumulative rewards [540].

k-armed bandit problem Due to historical reasons, RL is the most well-understood in the context of the so-called *k-armed bandit problem* [598,599]: Imagine a gambler that stands in front of a row of k slot machines (sometimes called *one-armed bandits*, hence the name of the problem) aiming to win as much as possible within a given number of trials or in the long-run. Each machine has a fixed probability distribution over rewards provided by the machine when played, a priori unknown to the gambler.

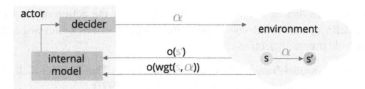

Fig. 8.4

General schema of reinforcement learning.

For every trial, the gambler has to decide which machine to play. Simple policies the gambler could follow would be, e.g., to randomly select a machine and stick on this machine or for each trial switch to another randomly selected machine. However, such policies are likely to perform badly, as they do not include the knowledge about the probabilistic distributions of the machines the gambler could deduce while playing. In fact, following the law of large numbers, the reward distributions of a machine played very often could be estimated quite accurately. Already with this simple description of the problem, the well-known exploration-exploitation dilemma [540] becomes apparent: *exploring* helps to gather more information about the machines reward distributions, possibly leading to better decisions in future, while *exploiting* uses the gained knowledge to select the most promising machine.

Note that when choosing the most promising machine, information about the reward distribution is collected and might lead to revising the knowledge or changing the policy towards some with better outcome.

Basic schema of reinforcement learning Fig. 8.4 depicts the basic schema for RL. An unknown environment reacts on actions α performed by the actor, who receives as feedback the observable part $o(s')$ of the entered state s' and an immediate weight $o(wgt(s, \alpha))$. Based on knowledge about the environment and past trials, the actor builds an internal model based on which a decider component chooses the next action α the environment is confronted with. Within RL, the internal model is rather simple formalized, e.g., as the history of all trials or a learned MDP from the observed states.

Decision-making heuristics To decide which action to choose during RL, several heuristics have been proposed. Classical attempts to solve the exploration-exploitation dilemma in RL are semiuniform policies, such as the following:

ε-**greedy** With probability $1 - \varepsilon$, the currently best action is chosen; whereas with probability ε/N, another action from N admissible actions is chosen.

ε-**decreasing** Just as ε-greedy, but with stepwise decreasing ε to mimic the need for great exploration at the beginning, where less knowledge is present and less exploration in later trials.

ε-**first** Just as ε-greedy within a fixed amount of trials before the policy switches to a purely exploitative behavior.

In the k-armed bandit example, an ε-greedy policy would choose the action that yielded the highest average weight in the past with probability $1 - \varepsilon$, whereas with probability ε, another action would be chosen. When the action of choice is not uniquely defined, one would select one of the selected actions by tossing a fair coin (uniformly distributed selection).

Model-free vs. model-based reinforcement learning   There are basically two approaches to learn optimal policies with RL (e.g., [540,597]):

Model-free   A policy is learned without learning a model of the environment

Model-based   Learn a model of the environment and use it to derive a policy

Both methods have their advantages and drawbacks and various algorithms have been presented in the literature following this classification. For instance, the *adaptive heuristic critic* (AHC) algorithm [600] is a model-free method that uses an adaptive version of *policy iteration*. A very prominent model-free algorithm is the Q-learning algorithm [559], which is a simpler version of the AHC that still guarantees converging to the optimal policy in the limit. However, Q-learning is exploration-intensive and may converge slowly to a good policy. Whereas changes in the environment are not as problematic, switching to a different optimization goal is difficult, as the learned policy in model-free approaches is specifically tailored for a fixed optimization goal. Model-based approaches have the advantage that existing and well-established methods for the model chosen can be exploited to learn a suitable policy. For instance, when an MDP model of the environment is learned, standard value-iteration or policy-iteration algorithms can easily be applied with the full tool support for MDP models. Contrary to model-free approaches, changing optimization goals is not problematic, as one simply has to switch the synthesis task on the learned model, whereas changes in the environment may lead to ill-formed models, and hence unreliable policies.

8.3.3 Towards a new learning approach

In general, we understand human-inspired learning as a learning method that combines the methodology of RL with the concepts of context-based control on priors over policies. To this end, human-inspired learning is a form of active, non-supervised learning that performs a mixture between *exploration* in a goal-directed fashion and *exploitation* using habitual control to obtain a nearly optimizing policy.

8.3.3.1 Goal of the learning approach

The last sections focused on neuroscientific discussions to establish prior-based control (see Section 8.2.4) as a method that can be understood as a mixture between goal-directed and habitual control. There, a portfolio of learned policies was assumed to be given, describing how policies are selected by the human brain under an inferred context. In this section, we mainly describe how these policies can be learned algorithmically and how to represent them in a computational framework. The learning process we describe is, like in RL, unsupervised and is performed during taking action, stepwise refining policies. The concepts of the last sections are applied thereafter also during this learning process to select the policies to be performed and refined. However, an actual implementation of this selection is assumed to be given in this section, focusing on the structure and generation of operational models for learning policies.

8.3.3.2 Components of human-inspired learning

Toward an algorithmic implementation of a human-inspired learning framework, we identify four basic components called *library*, *predictor*, *decider*, and *restructor*, which we describe in more detail in the following:

Library The library component arranges the storage of information and data management for the human-inspired reasoning process, including many facets about the contextual environment, goals, the learned policies, and event knowledge. For the representation of knowledge, a key property of the data management we pursue is an appropriate degree of abstraction and filtering. As already discussed in Section 8.2, the human brain is very efficient in abstracting and filtering information, e.g., not keeping exact information about how many times a repetitive event happened, but an abstracted version that still provides a sufficient level of precision to allow for reasoning. Furthermore, the library component includes the "library of priors", where for each context a portfolio of policies is included (see Section 8.2.4). Policies can be represented by the operational behavior of the environment, similar to model-based approaches of RL, including annotations to store information about the past (e.g., how the outcome of a behavior has been in the last applications), predictions, and other assessments noteworthy for the reasoning process.

Predictor Besides the predictions saved in the library component as manifestations of past behaviors and outcomes, predictions can also be generated by involving a more elaborate reasoning process about knowledge. Predictions could, e.g., follow the model-free approach of RL, but also include more involved reasoning processes. For instance, when the predicted behavior or reward in the learned policies differs significantly from the real one, the system is likely to explore, generating novel knowledge about the environment or policies.

Decider Humans decide based on the observation and library on the state of the environment, in particular the context (see Section 8.2.7). Hence, the decider component follows policies learned, depending on the predicted situations, selecting policies with the best properties regarding the goal, and possibly adapting them when the global goal changes (see *habits* described in Section 8.2.3). Besides deciding for next actions to be performed as in the case of RL, the decision mechanism also switches contexts and explicitly provides switching between exploitation, exploration, and restructuring. In addition to this, the selection of actions performed by the actor in human-inspired learning depends on more facets than in RL. The decider may also switch goals, e.g., to either exploit existing policies following a first-class optimization goal, or explore new policies by switching to a subgoal that aims to increase knowledge and for providing better predictions about the context.

Restructor   During learning, information is aggregated and incorporated into the library. To reduce model sizes and increase efficiency, there are pruning phases that perform abstractions on the learned library. To provide an example, consider that a very suitable policy has been learned for a specific goal. In this case, other policies that are more sophisticated and yield suboptimal results could be abstracted in such a way that they require less space in the library. To this end, the library is dynamically adjusted through filter mechanisms, e.g., updating the library by abstracting away information that turns out to be not as important to achieve the similar predictions and decisions. For instance, when an action did not yield the expected result for several times, it will not be performed any further, similar to the concept of habits, where successful actions are taken without performing predictions and involved reasoning steps to decide for the action to take. Note that, in turn, if predictions do not match the desired outcome several times or the overall goal is not achieved anymore for learned policies, an inverse operation is invoked, restarting exploration for learning new policies (see *priors* over policies detailed in Section 8.2.6).

8.3.3.3 Implementing human-inspired learning

In principle, human-inspired learning is implemented in a similar fashion as RL, but with an enhanced reasoning on the library and predictions. Furthermore, though the basic principle is similar to model-based RL, we pursue here a model-free approach for predictions of the expected costs for reaching a goal, and similarly within value iteration. Specifically, much like RL, human-inspired learning is performed in *trials* that end either in achieving the overall goal, or are restarted when either the goal cannot be achieved anymore or is so unlikely, that a reset and start of a new trial is considered to be beneficial. Each trial comprises several learning *rounds* as depicted in Fig. 8.5. Note that the right part of Fig. 8.5 coincides with the schema of RL depicted in Fig. 8.4. Each round's purpose is to refine models and policies to optimize achieving the goal towards habits (see Section 8.2.3). In each round, predictions are adapted and annotated to the model. We assume to have structured any implementation of the approach into the components described in the last section. The components involved in the proposed learning steps of Fig. 8.5 are library (1)–(4), predictor (2), decider, and restructor (4).

8.3.4 Human-inspired learning by example

We describe our general concept of human-inspired learning applied on the faulty coffee machine from Example 8.1. We only illustrate parts of the library mimicking human knowledge and the learning process that are relevant for the goal of getting a cappuccino while minimizing costs, i.e., minimize expected costs for $\varphi = \Diamond(cap \wedge \Diamond end)$. In this example, there will be only one context and one policy to be learned such that we can incrementally illustrate this one through annotated MDPs.

Initial situation   For the initial situation of our learning method applied on the faulty coffee machine example, we assume our agent does not know this particular coffee machine. In the real world, a human would already have a portfolio of habits and

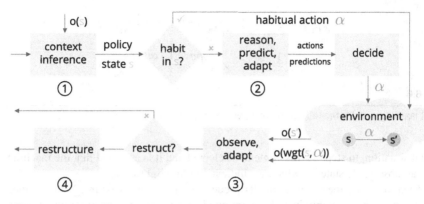

Fig. 8.5

Basic steps of one round in human-inspired learning.

① Reasoning about the context of the system (e.g., as described in Section 8.2.7), the state of the observed system and a goal-directed policy model to be refined or exploited is selected (see Section 8.2.6, selecting priors).

② Based on the currently observed state, reasoning about possible actions and their outcomes with regard to system behavior (state, transition, and cost predictions) and outcome (goal prediction) takes place. According to this reasoning, the operational model for the selected prior policy is updated.

③ After performing an action, observe the system, and update the model by adjusting priors and prediction annotations.

④ On the global level of goal-directed policy models, possibly restructure models in a filtering fashion. Most importantly, possibly turning actions into *habits* (see Section 8.2.3).

reasoning possibilities to establish a basic exploration policy. For instance, an adult would know that one either has to pay first and selects the desired product thereafter, or the other way around, and thus, explore which of the two contexts applies. On the other hand, a young child would simply press buttons to select drinks, observing what happens and later on establish the knowledge an adult has.

In our example, we assume an initial situation of human-inspired learning, where no prior knowledge is present except that the machine exposes its situation that can be perceived for learning, e.g., the machine exposes information about the next step to perform. Hence, there is initially no model learned for the machine, i.e., the initial model of the only context of the faulty coffee machine comprises only an initial state s_0, and no actions.

First learning round The first learning round consists of one run of the steps indicated in Fig. 8.5, applied on the initial situation described above. Fig. 8.6 depicts the MDP models for the policy to obtain a cappuccino, while minimizing costs that are adapted through the learning process. As we have only one goal defined and only one

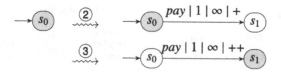

Fig. 8.6

First learning step for the unknown coffee vending machine.

initial situation, in step ① this context is selected and it is assumed that the machine is in an observable state s_0, which we indicate by highlighting s_0.

Then, as in s_0, there are no habits predefined in this context; in ② the actions considered to be available in s_0 are computed. Assuming that the machine exposes possible next steps, this first leads to a single action *pay* available in s_0 with costs 1 €. Besides these standard weight annotations for MDPs formalizing costs, in step ② we further annotate predictions on the optimization goal, i.e., the expected costs of reaching the goal, and an assurance measure that formalizes the trust into the predictions in terms of transition probability distribution, costs, expected costs, etc. This gives rise to a transition from s_0 in the model to a fresh state, which we call s_1, that is annotated with "$pay \mid 1 \mid \infty \mid +$", where *pay* denotes the action, 1 the costs in terms of an annotated weight, ∞ the expected costs until reaching the goal (i.e., the measure to be minimized initially set to ∞, as it is not sure yet to reach the goal), and the assurance $+$. The latter measure heavily depends on the formalization and can differ from application to application. Here, for illustrative purposes, we chose a scale from "$--$" (very unsure that the predictions hold) over 0 (expect a random choice to hold or not) to "$+++$" (very sure about the predictions to hold as annotated). In ③, the only action that is promising to reach the goal is chosen and performed, i.e., the learning algorithm performs action *pay*. As the observed reached state is in fact one that is different from s_0, and it costed 1 € taking the *pay* action, trust in the predicted behavior is increased from "$+$" to "$++$" and the current state is switched to s_1 (highlighted). The last step ④ of a learning round is done by possibly restructuring the models, which does not apply yet here in the first learning round.

8.3.4.1 Learning scenarios

Here, we describe some scenarios of the learning process of the cost-optimizing policy to obtain a cappuccino that arise from different outcomes in different rounds and trials due to the randomized behavior of the faulty coffee machine. As a basis, we assume the first learning step to be done as indicated in Fig. 8.6, i.e., ① deduces the context in a current state s_1.

Success scenario Let us assume that in s_1, the vending machine indicates "Choose a drink". This leads to possible actions inferred in step ② of *espresso*, *cappuccino*, and *milk*. Keeping in mind that the overall goal in the selected context is to obtain a cappuccino, the prior on actions as a distribution over these actions might be as depicted in Fig. 8.7, i.e., there is a high probability of selecting *cappuccino*. Assume

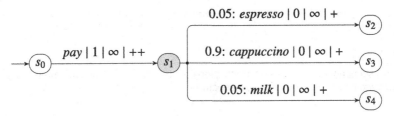

Fig. 8.7

Success scenario after first ② step.

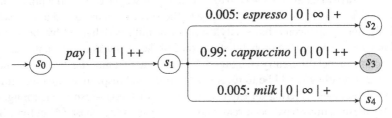

Fig. 8.8

Success scenario after steps ②–④ for selecting drinks and ②–③ after waiting.

that in ③, this action is chosen and leads to the observable state s_3. Here, it cannot yet be determined whether one actually obtains a cappuccino, such that only the trust of this action is increased as a different state than s_0 and s_1 has been observed. However, when continuing the learning steps and selecting a *wait* action, the outcome becomes clear. In this success scenario, we assume that the probabilistic distribution of the operational behavior of the faulty vending machine (see Fig. 8.3) is resolved, eventually leading to a cappuccino. Then, the context is in a situation as depicted in Fig. 8.8, where the priors of the *cappuccino* action have been increased and the predicted costs until reaching the overall goal are adapted to 1 (for the *pay* action) and to 0 (in the *cappuccino* action). Reaching the overall goal finalizes the first trial of the learning process.

In further trials, the policy in the context can be further adapted and optimized. Let us assume that the next two trials also have the same successful outcome of a cappuccino with costs 1 invested. Then it is likely that the *pay* and *cappuccino* actions with the assigned predictions turn into a habit by step ③, i.e., they are immediately performed in subsequent trials when observing states s_0 or s_1, respectively. This situation is indicated in Fig. 8.9. Note that when turning actions into habits, priors turn into Dirac distributions and other actions could be removed in a restructuring phase of step ④. Whereas in the case of the *pay* habit, this does not change the prior distribution, but only the fact of skipping the reasoning step ②, we assumed for the *cappuccino* habit that the possibilities of selecting other drink actions are removed in ④. The resulting policy in the limit is a pure ASQ, as discussed in Section 8.2.1.

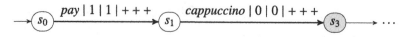

Fig. 8.9

Success scenario after further successful steps of obtaining a cappuccino, turning *pay* and *cappuccino* into habits in s_0 and s_1, respectively.

In the general case, we allow for habitual transitions in the operational model rather than ASQs.

Sudden failure after success scenario Assume that we have learned a policy in the context of the vending machine according to the success scenario and a trial, where suddenly not a cappuccino but solely espresso or milk is delivered by the vending machine, i.e., resolving the probabilistic choice of the *cappuccino* action in the vending machine model differently (see Fig. 8.3). Then, the first modification in the model of the learned policy would be to reconsider actions and respective outcomes for the different outcomes, entering an exploration phase of human-inspired learning. Recall that these actions have been removed in a restructuring phase ④ before. Still, the *cappuccino* action might be left as a habit when there is no strong indication to completely change the behavior of the policy. Here, the states of obtaining a different drink are added and predictions are annotated to the actions.

Faulty scenario Assume now a scenario where directly in the first trial, no cappuccino is delivered by the vending machine. As a result, trust in the vending machine is reduced and still all three actions in s_1 are maintained. After several trials, *pay* is established as habit, since it clearly performs the predicted behavior. Our proposed approach would still prefer a cappuccino action to be chosen, but tries espresso and milk occasionally. In case of many failures, i.e., not obtaining cappuccino, one might learn a policy that is not memoryless as in the examples discussed above and in Section 8.2.5, obtaining cappuccino by sequentially ordering espresso and milk. The knowledge on which this policy is based, i.e., that cappuccino comprises espresso and milk, could either be deduced in a more involved reasoning task on the library in phase ① or by surprise in an exploration phase. Note that the MDP models for policies then contain states that encode information about the history of the choices, e.g., that a cup of espresso is already in the cup and could be turned into cappuccino by adding milk.

Eager scenario A two-step policy that first orders milk and espresso as discussed in the faulty scenario could also be obtained when the predicted expected costs seem to be too high, and reasoning on the library would deduce that further exploration might yield better expected costs. This could be also the case in later learning phases when almost the exact probabilities of the faulty coffee machine have been learned, and thus it is known that the expected costs for obtaining a cappuccino is greater than 2 €. Another situation where this case arises is, e.g., when only a small amount of money is available and risking several failure trials due to variance is not desirable.

Note that the two-step policy above has expected costs of 2 € and has a variance of 0, i.e., obtaining a cappuccino with costs of 2 € is an almost sure event.

8.3.4.2 Further aspects of human-inspired learning

The example we drew here already implemented several aspects of human-inspired learning. Further extensions could be, e.g., to enhance the formalization of knowledge for contextual inference and reasoning about contexts or switching to exploration. Also an appropriate formalism to represent knowledge has to be further elaborated, e.g., to formalize different contexts to also include a *broken machine* sign, increasing priors of actions other than the cappuccino action and reduce its trust. Another aspect which has to be formally investigated is measures of similarity between contexts. Then, contextual inference could first lead to a new context that takes an abstracted version of a similar one with weakening trust, enabling more exploration towards either noticing that the context has been already learned or finally establishing a new one.

Context models might also include *negative habits*, i.e., habits that are actions that reasoning might consider for exploration, but are indicated as known to be not useful for achieving the goal. Similar as the notion of habits, where the habitual action is directly taken without any reasoning in phase ②, negative habits are directly dismissed and not considered to be taken, again without any reasoning in phase ②.

8.4 Synergetic links

The behavioral models that we present in this chapter comprise stochastic operational models that mimic human decision-making to predict human behavior and fulfill the quantitative requirements, such as low-latency and energy consumption, a TI application has to meet. Such a model is of utmost importance in many TaHiL use cases: For robot-assisted medical and industrial training issued in the research groups for medical applications (U1, see Chapter 2), industrial applications (U2, see Chapter 3), as well as in the Internet of Skills (U3, see Chapter 4).

In robot-assisted environments (U1 and U2), the key requirements that the model needs to fulfill are efficiency and suitable explanations to support the human with key information about which tasks will be achieved by the robotic assistant and what is left to the surgeon or industrial worker. To this end, our models might provide the basis to decide when to handover control from machines to humans, and vice versa. Also in critical situations where coordination of all participants (human and robotic) is important, our models could provide predictions on how humans could react and how complications could be avoided.

The human-to-machine transfer via modeling illustrated in Chapter 2 could be also supported by learning human-inspired models. Here, context-aware assistance is in the line of selecting contexts (see Section 8.2.7) and policies of the learned library component (see Section 8.3.3.2).

The Internet of Skills (U3, see Chapter 4) further relies on CPSs to be developed. Following the context-adaptive software concepts (TP5, see Section 13.5), the development of such systems will rely on a model-based approach such that also operational models are at hand. These models could be included into our learning framework to analyze policies depending on the context of the CPS. For using machines to support learning in the Internet of Skills, it is important that machines adapt themselves to the predicted behavior and demands of an individual user. In these applications, our human-inspired learning approach could continuously learn the behavior of the humans and reason about their selected policies and habits.

Besides the general human-inspired learning approach, additional knowledge and refined versions of the models could be included to instantiate general human-inspired models to specific individuals. This addresses also the instantiation of models for the whole lifespan of humans, as illustrated in Chapter 9. For instance, elderly people might stick more to their previously learned strategies, whereas young people may tend to experiment more. This is important not only in rehabilitation use cases (U1, see Chapter 2), but also for teaching scenarios in the Internet of Skills (U3, see Chapter 4). To instantiate learned models for an individual Human-in-the-Loop, classical learning techniques [540,541] can be employed to extract key features of the individual. The long-term goal is to enable automatic system adaptations as well as providing learning and explanatory facilities that fit individual demands and objectives of the user. This goal requires joint forces, e.g., relying on essential technologies for tactile computing issued in Chapter 13. Furthermore, results from TP1 and TP2 (see Chapters 9 and 10) could help to provide input for training our human-inspired reasoning models.

8.5 Conclusion and outlook

In this chapter, we argued that to have cohabitation between humans and CPSs, a CPS must be able to understand human behavior and, when possible, predict it. This is important for TaHiL applications, such as machine-assisted medical and industrial training, in which a clear understanding of human actions and motivations would enable the machine to provide the appropriate feedback and adapt its own actions to the expected actions of the human counterpart.

To achieve this, we proposed a novel family of computational models called *prior-based control*, where previous experience can help make quick decisions in novel tasks. This control scheme mimics human decision-making by reducing behavior to habitual control (i.e., habits) when enough experience with the task has been acquired, and to goal-oriented control (i.e., forward planning) when there is no experience. Moreover, the model also covers situations where some experience exists but cannot be applied directly to form habits. In these situations, much like humans, the model focuses its attention on the most relevant elements of the task only to enable quick and flexible decisions.

We formalized these concepts in a general learning scheme that refines model-based RL with concepts of human reasoning. For this, we used annotated versions of stochastic operational models that formalize decision-making and could be interpreted and analyzed by machines using standard methods. To this end, these models could be also used to understand and predict the future behavior of humans. We believe that the full specification and application of these models in CPSs will greatly widen their applicability in the future.

Fundamental
challenges

Outline

The third part introduces the five target primary research fields, namely human perception and action, sensors and actuators, communication, electronics, and tactile computing. These fields span multiple disciplines, including psychology, cognitive and computational neuroscience, medicine, telecommunication, electrical engineering, electronics, textile technology, and computer science.

Human perception and neurocognitive development across the lifespan

9

Shu-Chen Li, Evelyn Muschter, Jakub Limanowski, and Adamantini Hatzipanayioti

Technische Universität Dresden, Dresden, Germany

> *As you are aware, no perceptions obtained by the senses are merely sensations impressed on our nervous systems. A peculiar intellectual ability is required to pass from a nervous sensation to the conception of an external object, which the sensation has aroused. The sensations of our nerves of senses are mere symbols indicating certain external objects, and it is usually only after considerable practice that we acquire the power of drawing correct conclusions from our sensations respecting the corresponding objects...*
> — **Hermann von Helmholtz**[☆]

9.1 Introduction: Multisensory perception is the gateway for interactions

One of the Talent Pool (TP) for developing Tactile Internet with Human-in-the-Loop (TaHiL) technologies is the basic research on neurocognitive and psychophysical foundations of human multisensory perceptual informational processing and action (TP1 Human Perception and Action; see Fig. 1.15). We humans interact with multiple facets of the environment (physical, virtual, social, and their combinations) through the biological organs (e.g., eye, ear, skin, and muscle) of our senses (e.g., vision, audition, and touch). The transformation of sensation into perception takes place through information processing across a hierarchy of neural networks that turns external physical or chemical signals into sensory and perceptual representations in the brain [601–608]. Furthermore, the sensory-perceptual processes and their underlying brain mechanisms interact with other higher-order cognitive functions (e.g., memory, attention, executive control, valuation, and decision-making) to guide behavior, while also receiving afferent sensory feedback during sensorimotor actions.

[☆] (1857). On the physiological causes of harmony in music. In R. M. Warren and R. P. Warren *Helmholtz on Perception: Its Physiology and Development*, p. 49, John Wiley & Sons, 1968.

Tactile Internet. https://doi.org/10.1016/B978-0-12-821343-8.00021-6

In this way, perception and action are closely linked and resonate with each other [609–611]. Thus perception is inherently predictive and active, with perceptual processing being modulated by the individual's prior expectations that are based on the current task goals and past experiences as well as by the sensory consequences of one's own actions.

9.1.1 The *Go-Senses* framework for closed-loop human–machine interactions

Technological progresses are advancing rapidly in developing adaptive wearable sensors as well as actuators (see Chapter 10), intelligent telecommunication networks (see Chapter 11), ultracompact bendable wireless transceiver chips (see Chapter 12), and secure computing infrastructures (see Chapter 13). These technological developments along with progresses in understanding psychological, neurocognitive, and computational principles of human behavior have together paved the way for the new field of TaHiL research. This emerging field of transdisciplinary research aims to promote next-generation digitalized human–machine interactions in perceived real time. To achieve this aim, mechanisms and principles of human goal-directed multisensory perception and action need to be integrated with technological developments en route to designing new devices and technologies for breakthroughs. Focusing on the human operators in this chapter, we propose a conceptual framework (see Fig. 9.1) to guide experimental investigations on fundamental mechanisms of human *goal-directed multisensory perceptions* (*Go-Senses*) in TaHiL research (see Fig. 1.13) and to highlight the necessary synergies between the human and engineering sciences. The term *goal-directed* is chosen here to encompass the influences of expectations that are derived from prior experiences as well as current intended actions on perception.

The *Go-Senses* framework builds on Hermann von Helmholtz's classical concept [612,613], which postulates that prior experiences (expectations gained through learning and memory) actively guide perception and action. In modern psychophysical research, it is well-established that humans integrate different modalities of sensory signals to form robust percepts [614]. Of particular interest is the concept that behavioral and brain mechanisms of multisensory integration are context-dependent. That is, the weightings of different sensory signals can be flexibly adapted, depending on various factors, such as the reliability and timing of the sensory inputs, task demands, action goals, and the human operator's prior experiences or skill levels [615,616]. For instance, it is known that temporal delays between visual-haptic inputs [617] and sensory noise [618] can bias the perception of softness/stiffness. Furthermore, anticipations based on prior experiences bias audiovisual perception [619,620]. Even the weighting of different bodily inputs can, in some cases, be deliberately adjusted in this way. As an example, recently it has been shown that the instructed task-relevance of visual vs. proprioceptive movement feedback from one's own hand can change the weighting of visual vs. proprioceptive signals in early sensory brain regions [621].

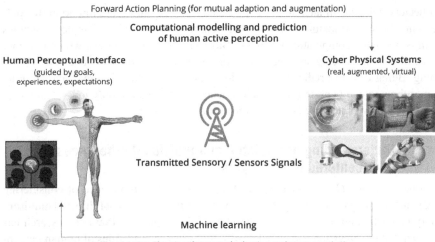

Fig. 9.1

The *Go-Senses* framework highlighting goal-directed multisensory perception in digitally transmitted closed-loop human–machine interactions.

9.1.2 Modeling human goal-directed multisensory perception

To ensure smooth closed-loop human–machine interactions (see Figs. 9.1, 1.9, and 1.10), the *actors* at both ends of the loop need to be able to anticipate the other's actions and action planning. Thus the modeling of human reasoning and behavior is indispensable for developing mutual adaptation algorithms for machine learning. The *active inference* approach [622,623], which applies Bayesian statistical inference and the principle of minimizing prediction errors between sensory sampling and prior expectations, could be a promising means for computationally modeling goal-directed human perception and action. Recently, it has even been proposed that the subjective perception of time arises from minimizing prediction errors and adaptive recalibration of multisensory temporal integration [624].

When humans or machines perceive and act, sensory (or sensor) noise along with certain degrees of processing unreliability and delays are inevitable. The potential amplifications of multiple sources of unreliability and delays are big challenges for achieving real-time remote interactions between humans and machines that are connected via the Tactile Internet. The fusion of multisensory processing with goal-directed action planning is a crucial step ahead for current research on the perception-action dynamics in humans and machines, because (*i*) goals and goal hierarchies predict which actions volitional humans will select at longer time-scales, and (*ii*) goal-directedness enable predictions of how humans dynamically sample and weight different sensory channels depending on changes in task contexts. Take visual and haptic sensing for example, during the simple action of reaching for a ball, humans initially heavily rely on visual inputs to initiate and control the arm's trajectory, whereas in later stages haptic feedback becomes more important for perceiving the surface texture, softness, and weight of the ball. Such complex dynamics can be modeled computationally using Bayesian inference models, which consider goal-directedness being an integral part of multisensory processing and action planning. Being able to predict human action planning would allow machine learning algorithms to compensate for operation delays to achieve not only low-latency processing, but even negative latency at the system's level.

9.1.3 Understanding age-related and individual differences for user-centered technologies

Furthermore, the *Go-Senses* framework emphasizes the importance of considering age- and learning/experience-related individual differences in goal-directed multisensory perception en route to new technological developments. Previous research has shown that the precision and speed of different sensory modalities of human perception as well as cognitive control mechanisms vary across the lifespan and dependent on the individuals' skill levels. These differences can, in part, be attributed to age-related changes in neurocognitive mechanisms. Of note, a computational theory of brain aging postulates a sequence of effects that link attenuated neuronal gain control with increased information processing noise, which consequently reduces processing efficiency and computational capacity [625]. Such age-related and individual differences are important from the perspective of usability and user-centered engineering design (see also Chapter 4), which need to be systematically investigated in the emerging field of TaHiL to ensure that technological breakthroughs can benefit wide ranges of user populations.

The sections below start with detailed overviews of neurocognitive and psychophysical foundations of perception and multisensory processing, which are then followed by outstanding issues that still need to be addressed to advance the TaHiL research. In particular, challenges associated with the temporal properties of the different senses and impacts of brain development and aging, as well as individual differences on perceptual processing are discussed. To go beyond the state of the art, examples of synergistic efforts across the fields of TaHiL research are highlighted.

9.2 State-of-the-art research on multisensory perception

During the awake state, the human brain consistently samples physical or chemical signals in the environment through different sensory channels (e.g., visual, auditory, haptic, olfactory, or gustatory). The information flow from environmental stimuli to the perception of integral representations of objects for the senses of vision, hearing, and touch is schematically shown below in the overview figure and table (see Fig. 9.2 and Table 9.1).

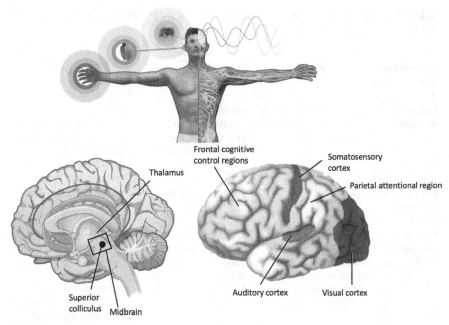

Fig. 9.2

From sensory organs to brain regions underlying subjective sensory-perceptual experiences. Graphs of relevant brain regions shown here are adapted from CNX OpenStax http://cnx.org/contents/GFy_h8cu@10.53:rZudN6XP@2/Introduction (reuse of the images is permitted by CNX OpenStax).

9.2.1 From sensation to perception

Sensory signals originating in the environment are transmitted through the sensory organs and registered first by the respective receptors of the senses, specifically the central photoreceptors (rods and cones) in the retina of the eye for vision, the central mechanoreceptors (hair cells) in the cochlea of the ear for hearing, the peripheral mechanoreceptors and thermoreceptors in the skin for the tactile aspect as well as the proprioceptors in muscles and tendons for the kinesthetic aspect of the haptic sense (see [602,626] for reviews). From these receptors, the sensory signals are then transmitted to the respective brain regions for further processing. In humans, the functional specializations of sensory processing for the senses in the cortex are the occipital (visual) cortex for vision, areas around the superior temporal gyrus (the auditory cortex) for hearing, and regions in the post-central gyrus (the somatosensory cortex) for the haptic sense that encompasses tactile and kinesthetic sensations (see Fig. 9.2).

The initial processings of registered information from the sensory receptors take place in regions in the thalamus and in the midbrain. The thalamus is a small structure within the Central Nervous System (CNS) that is located just above the brain-stem, between the midbrain and the cerebral cortex. Sensory signals from different modal-

Table 9.1 Information flow from sensory receptors to perception: Examples of vision, hearing, and touch in the case of tennis playing.

Information flow ⇓	Modality		
	Visual	**Auditory**	**Haptic**
Sensory receptors	central photoreceptors (rods and cones) in the eye on the retina	central mechanoreceptors (hair cells) in the cochlea in the inner ear	peripheral mechanoreceptors and thermoreceptors in the skin as well as proprioceptors in the muscles and tendons
Impulse in sensory fibers	optic nerve fibers	auditory nerve fibers	somatosensory nerve fibers
Impulse reaches CNS	lateral geniculate nucleus (LGN) of the thalamus, midbrain, layers in visual cortex and other regions in the cerebral cortex	medial geniculate complex (MGC) of the thalamus, midbrain, auditory cortex and other regions in the cerebral cortex	ventral posterolateral (VPL) and ventral posteromedial (VPM) nucleus of the thalamus, midbrain, sensorimotor cortex and other regions in the cerebral cortex
Sensation	a small, round stimulus	a fast and short sound	a small force and pressure
(memory, prior knowledge, goal)	seeing the shape, size, and color of a tennis ball	hearing the sound of the racket hitting the ball	feeling the force and hardness of the ball, muscle tension, and hand motion
Perception			

ities are processed in different subregions within the thalamus, specifically, the optic nerve fibers project to the Lateral Geniculate Nucleus (LGN), the auditory nerve fibers project to the Medial Geniculate Complex (MGC), whereas the somatosensory nerve fibers project to the Ventral Posterolateral Nucleus (VPL) and the Ventral Posteromedia Nucleus (VPM). Another brain area that is also involved in early sensory processing is the superior colliculus, which is a small structure within the midbrain that sits right below the thalamus. The initial sensory representations formed within the different regions of the thalamus and in the superior colliculus are then relayed to other brain areas, which include regions in the midbrain (e.g., the basal ganglia) and in the cerebral cortex (e.g., visual, auditory and the somatosensory cortices), as well as regions in the parietal and frontal lobes that are important for attentional and cognitive control processes (see [627] for review). These later regions are engaged in downstream processes that (*i*) compare and (*ii*) bind sensory representations acquired from the different modalities, as well as (*iii*) interpret the current subjective sensory experiences as guided by attention and memories of past experiences. These processes together eventually give rise to the perception of an integral whole of an object, such as a tiny ball that is flying across the court as in the case of tennis playing. Most sensory signals are attained or, at least, modulated by motor sampling processes; thus perception inherently also involves sensorimotor processes.

9.2.2 Principles of multisensory integration

In navigating through the manifold of sensory signals in the environment, the brain's sensory and perceptual processes need to distinguish between signals that belong to the same event from those that do not. When sensory information of an object or event is provided by more than one input modality, three general processing principles are involved in multisensory integration. The first and second principles account, respectively, for the effects that signals which are closer in time and in space are more likely to originate from the same event. At the neuronal level, the spike timing-dependent plasticity captured by the simple Hebbian rule [628], which states that neurons that fire together wire together, is a general learning principle that operates in various neural circuits over a broad spectrum of species, from insects to humans (see [629] for review). Thus multisensory signals (visual-auditory, visual-haptic, auditory-haptic, or visual-auditory-haptic) that reach the respective sensory registers closer in time (temporal contiguity) and in space (spatial proximity) enhance the underlying neural responses. In contrast, temporal or spatial disparities would result in weaker or failed sensory integration. A third principle, known as *inverse effectiveness*, describes the effect that the benefit of multisensory integration is proportionally greater for those unimodal sensory cues that are weaker (see [607,630] for reviews).

Classical perceptual illusions are special examples of multisensory integration. For instance, the ventriloquist effect involves audiovisual integration and arises from the perceptual biasing of sound location towards the location of a dominant visual stimulus. The perceived location of the human voice from the ventriloquist drifts towards the location of the puppet, which is a more salient visual stimulus in the context of a puppet theater [631]. The McGurk effect also entails audiovisual integration, in which the visual cue of lip movements affects how a concurrent speech-sound is perceived [632]. Regarding visual-tactile integration, the rubber-hand illusion demonstrates that watching a rubber hand being stroked, while one's own unseen hand is synchronously stroked, could cause the rubber hand to be perceived as one's own body part [633,634].

9.2.3 Neurocognitive bases for multisensory perception

Most stimuli in the environment encompass multiple types of sensory information to convey different facets of the external world. Considering again the example of tennis playing: one hears the sound of the opponent's racket hitting the ball, sees the ball approaching, prepares to catch the ball and subsequently feels the force and impact of the ball while catching it with the racket in one's hand. The majority of early sensory processes that take place in the thalamus and in the superior colliculus are modality specific, with information from different sensory receptors projecting to distinct subregions in these structures. However, to achieve a complete perceptual experience that includes different facets of a moving tennis ball, the brain needs to compare and combine sensory information from the different modalities as well as to track and predict variabilities in the sensory signals to perceive the ball and to adaptively monitor changes in its trajectories for planning actions to catch it.

Multisensory neurons

Based on evidence from animal research, other than the thalamic modality-specific sensory pathways, there are also multisensory neurons that respond to auditory, visual, and somatosensory signals in the superior colliculus. Modality-specific neurons and pathways allow parallel unimodal sensory processes for comparing features of a stimulus across different modalities to detect cross-modal correspondence or incongruence, whereas multisensory neurons support computations that integrate different sensory signals into a common representation. Relative to modality-specific neurons, the proportion of multisensory neurons in the superior colliculus increases from about 1/6 during early infancy to about 1/3 during adulthood. This age-related process requires experience-dependent development involving cross-modal experiences that closely parallel the maturation of unisensory processes during the early postnatal life [635]. Multisensory integration is a key aspect of the functional development of the superior colliculus, with sensory representations for the visual, auditory, and haptic senses developing in a manner that multisensory neurons constitute overlapping topographical maps of these three senses. This feature of overlapping sensory maps seems to be general across many species (see [630] for review). Moreover, the sensory maps also overlap with a common motor map for initiating movement orientations of the sensory organs and body limbs.

Interactive cortical sensory brain regions

Moving further downstream of the sensory-perceptual information processing stages, apart from transferring the registered uni- and multisensory representations to the sensory-specific regions in the cortex, the superior colliculus also has rich reciprocal connections with a range of other subcortical and cortical regions (Fig. 9.2). Of note, these include the basal ganglia [636], the parietal association cortex [637,638], and the frontal cortex [639]. Neural information exchanges between these cortical regions allow the intricate interplay between external stimulus-driven and internal volition-guided multisensory perception and action. In humans, various areas in the association cortex are key regions for multisensory integration at the cortical level (Fig. 9.3; see [640] for review). Using Functional Magnetic Resonance Imaging (fMRI) to measure brain activity while exposing human participants to visual, auditory or tactile stimuli, researchers can observe whether a given brain region responds to multiple modalities. Furthermore, certain neurons in these areas respond supra-additively to multisensory stimuli; i.e., show stronger responses in such cases than the sum of responses to both unisensory stimuli together [641,642]. Such supra-additive

(a)   (b)
Unimodal areas   Candidate multimodal areas

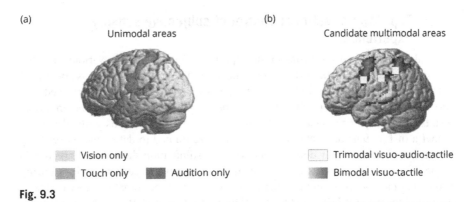

| | Vision only | | | | Trimodal visuo-audio-tactile |
| | Touch only | | Audition only | | Bimodal visuo-tactile |

Fig. 9.3

Schematic representations of (a) unimodal and (b) multimodal areas in the human brain. Images adapted with permission from [640] © 2005 by Elsevier.

response is in line with principles of multisensory integration [643]. Empirically, cross-modal responses to both visual and tactile stimuli have been shown in the parietal, temporal, and premotor brain regions (e.g., [644]), i.e., colored areas in panel (b) of Fig. 9.3 (see [640] for review). Of note, such integration between sensory modalities, involving fusing sensory information from multiple modalities, is important for higher-level abstractions of sensory experiences for perception. Take the example of perceiving the shapes of physical objects. Although for most humans in most common situations, sensory signals about object shapes in the environment are primarily processed visually, the objects can also be explored through the haptic sense. Thus bimodal visual-haptic sensory integration also plays a role in shape perception. In this regard, a small region in the occipitotemporal lobe, known as the lateral occipital complex (LOC), has been identified as a key player in multisensory shape perception. This region is situated in the middle level of the visual pathway and processes higher-level, more abstract sensory information. In particular, the LOC is activated both when humans see or touch the physical objects and may be involved in underlying processes for configuring the geometrical shape of an object through visual and somatosensory inputs [645,646]. Regarding tri-modal sensory integration, thus far the research findings have identified the Superior Temporal Sulcus (STS) as well as other subareas in association cortex, i.e., the Intraparietal Sulcus (IPS), Inferior Parietal Lobe (IPL), and Ventral Premotor Cortex (VPMC), as trimodal regions (white boxes in the b panel of Fig. 9.3) that respond to visual, auditory, and haptic stimuli [647,648]. Of note, neurophysiological and brain imaging studies have shown that these multisensory areas play a very specific role in the generation of a sense of bodily selfhood, by integrating sensory inputs generated by one's own body with the brain's internal "body model" [649–652]. Taken together, multisensory neurons in the superior colliculus at the cell level and the multisensory regions in the cortex are the neural substrates for multisensory perception.

9.2.4   Top-down attentional control of subjective sensory experiences

Other than multisensory neurons in the superior colliculus and the various uni- and multisensory regions in the cortex, high-order cognitive processes, such as attentional control, are also involved in modulating the subjective perception of experienced sensory signals. Focusing here on vibrotactile perception as an example, the empirical results of a study [653] that recorded neuronal activity in various cortical regions (panel a of Fig. 9.4) of awake monkeys that were trained to detect the presence or absence of a mechanically delivered vibrotactile stimulation show that the representation of a sensory stimulus unfolds in time across several cortical regions during perceptual judgments: from the order of below 10 ms in the somatosensory cortex to about 150 ms in the VPMC and the Dorsal Premotor Cortex (DPMC). At the neuronal level (panel b of Fig. 9.4), activities of neurons in most of the brain regions are very sensitive to changes in basic, physical sensory parameters (amplitude of vibrotactile stimulation in this case). Albeit not having a sensitivity function as steep as that of neurons in the somatosensory areas, neurons in brain regions implicating top-down attention, particularly the VPMC and the DPMC, also respond differently to different stimulus amplitudes. At the behavioral level, as the amplitude of the vibrotactile stimuli increases, the percentage of yes (stimulus present) responses of the monkeys also increases in a manner that follows the typical psychophysical signal-detection function (panel c of Fig. 9.4). Of particular interest, under situations with substantial sensory uncertainty, such as when processing near-threshold vibrotactile stimuli, the subjective experience of signal detection was found to be correlated with activities in the frontal attentional control regions (e.g., the premotor cortex), but not with activities in the sensory regions [653]. This suggests that attention-guided cortical gain control [654–656] plays an important role in reducing sensory uncertainty for the subjective experience of perception.

9.2.5   Neuromodulation of subjective sensory experiences

Other than anatomical structures, functional processes in the brain are modulated by several neurotransmitters, which play important roles in regulating signal transmissions between neurons. Through widespread projections in many brain regions, neurotransmitters have pervasive effects in regulating brain dynamics in different networks. Here we focus on the dopamine system, which originates from dopaminergic neurons in the midbrain, particularly in the Substantial Nigra Parc Compacta (SNc) and the Ventral Tegmental Area (VTA). The dopamine neurons widely innervate other brain regions through three main pathways: (*i*) the nigrostriatal pathway with fibers of dopamine neurons, projecting from the SNc to the caudate and putamen in the dorsal striatum; (*ii*) the mesolimbic pathway, projecting from the VTA primarily to the Nucleus Accumbens (NAcc) in the ventral striatum, but also to the hippocampus and amygdala; and (*iii*) the mesocortical pathway, projecting from the VTA to the frontal, cingulate, and perirhinal cortex (see [657] for overview). It is established that dopamine enhances synchronized neural firing as well as the sensi-

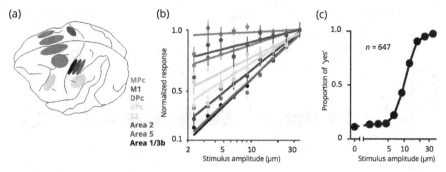

Fig. 9.4

Neural activities in various brain regions and subjective perceptual responses are sensitive to changes in basic physical sensory parameters (shown here is the amplitude of mechanical vibrotactile stimuli). (a) Regions from which neuronal activities were recorded in awake monkeys. (b) Normalized neuronal responses as a function of the amplitude of vibrotactile stimulation recorded from brain regions shown in panel (a). (c) Psychophysical function depicting the subjective, perceived presence of vibrotactile stimulation. Reprinted and adapted with permission from [653] © 2006 by the National Academy of Sciences, U.S.A.

tivity and specificity of neural responses to auditory signals [658]. As for modulating subjective sensory experiences of touch, the midbrain dopamine system plays crucial roles in interfacing between the registered sensory parameters (e.g., amplitude, frequency, noise) in the somatosensory thalamus (i.e., the VPL and the VPM) and in the primary somatosensory cortex during sensory processing and the interpretations of sensory representations during perceptual processing [608,659]. Apart from coding reward prediction (see Fig. 9.5), which is a well-established general function of dopamine [660], it has been found that dopamine neurons in the midbrain of monkeys trained to detect vibrotactile stimuli also increased their firing rates as the stimulus amplitude increased [659]. Of note, this sensitivity to the amplitude of vibrotactile stimulation was only observed when the animals correctly perceived the presence (hit) of the stimuli (see middle part of Fig. 9.5). The response time course of midbrain dopamine neurons during vibrotactile perception is longer than the response latency of neurons in the somatosensory cortex, and more closely matches the onset of perceptual processes in the frontal cortex. Altogether, these findings suggest that midbrain dopamine activities modulate the subjective experience of perceived intensity of tactile stimuli [608,653,659,661].

9.2.6 Formal models of multisensory integration

To quantify the relationship between basic parameters of physical stimuli (e.g., loudness of sound, frequency of light, pressure of touch) and the subjective sensory and perceptual experiences they elicit, basic laws of psychophysics have been derived for unimodal sensation and perception. Two classical models in this regard are the Weber's law and the related Fechner's law (see also Chapter 5 for related discus-

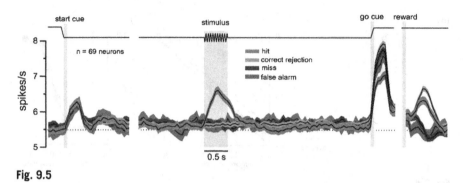

Fig. 9.5

Dopamine modulation of subjective vibrotactile sensory experience. The average of recorded activities from sampled midbrain dopamine neurons is plotted as a function of the time course for different response types during vibrotactile sensory signal detection. Reprinted with permission from [659] © 2011 by the National Academy of Sciences, U.S.A.

sions about these models). The Weber's law focuses on the smallest perceivable change in a stimulus or difference between stimuli, known as Just Noticeable Difference (JND). Specifically, the relation between JND and the amplitude of the initial stimulus is mathematically defined as $dS = K * S$, where dS is the measured JND, S is the reference stimulus, and K is a scaling constant. A related measurement version of this law was later expanded by Fechner, which formulates that the perceived stimulus intensity (p) is proportional to the logarithm of the actual stimulus intensity (S) at a given threshold (S_o) measured by physical instruments: $p = k \ln S/S_o$. Taken together, both laws capture the perceptual principle of subjectively experienced signal differences between physical stimuli, that is, simple differential sensitivity to differences of the stimuli is proportional to the size (amplitude, intensity) of the physical signals, whereas relative differential sensitivity remains the same regardless of size [662].

Uncertainty reduction and sensory integration

In the context of multisensory integration, empirical findings and theories suggest that, under most situations, human sensory-perceptual systems combine sensory information from multiple modalities in an approximately Bayes-optimal manner. Central to several models of multisensory integration that capture processes either at the behavioral or neuronal levels are two key features: (i) estimations of variance (uncertainty) in sensory inputs from different modalities, and (ii) the principle of uncertainty reduction [663–667]. To this end, Bayesian statistical methods are commonly applied to estimate the prior distributions associated with the different sensory inputs. The estimate $(\hat{S}_i)$ of a sensory signal derived from a physical stimulus (S) of a given modality is a function (f) of the brain's sensory-perceptual processes. On the one hand, the sensory signals are usually contaminated by environmental noises and, on the other hand, neural information processing in the brain is highly time-dependent and varying [668]. Thus both external and internal sources of variability render the estimations of sensory signals to be associated with certain degrees of uncertainty. Take haptic perception as an example, it is known that sensory noise affects the perception of force and stiffness [618]. Thus uncertainty reduction plays a crucial role when multiple sources of sensory inputs of a given physical event are integrated. Assuming independent Gaussian noise distributions and uniformly distributed Bayesian priors, as proposed by Ernst and Banks [664], the principle of maximum-likelihood allows optimal estimations that reduce uncertainties by computing the weighted sum ($\hat{S}$) of sensory estimates ($\hat{S}_i$) of the different modalities, with the weightings (w_i) being determined by their normalized reciprocal variances defined as

$$\hat{S} = \sum_i w_i \hat{S}_i, \text{ where } w_i = \frac{\frac{1}{\sigma_i^2}}{\sum_j \frac{1}{\sigma_j^2}}. \tag{9.1}$$

For example, by applying maximum likelihood estimation to combine estimates of visual (V) and haptic (H) sensory signals when picking up a ball, the integrated multisensory visual-haptic (VH) estimate can be stated as

$$\sigma_{VH}^2 = \frac{\sigma_V^2 \sigma_H^2}{\sigma_V^2 + \sigma_H^2}. \tag{9.2}$$

Considering another situation when estimates of visual cues are more precise (smaller variance in the distribution of visual signals) than the estimates of the auditory cues (larger variance in the distribution), then integrating both visual and auditory information would shift the combined distribution towards the subjective perception that is based on vision, the more reliable modality (see panel a of Fig. 9.6, [666]). If sensory estimates of cues from both modalities are about equally reliable, then the combined distribution will result in multisensory perception that is not biased by either of the two modalities (see panel b of Fig. 9.6).

Empirical findings from studies investigating different types of multisensory integration can be well captured by these models. For example, by adding noises into the visual displays of a visual-haptic size discrimination task to increase the uncertainty in visual signals, the perceived size—as indicated by perceived subjective equality—becomes increasingly dependent on haptic signals (see Fig. 9.7, [614,664]). Taken together, both formal models and empirical data of multisensory integration suggest that the sensory-perceptual processes and the underlying neural information processing combine multimodal sensory information in a Bayes-optimal manner, by taking into account the precision of the unimodal sensory signals to arrive at robust multisensory percepts (see [614] for review).

Prior expectation and sensory integration

Besides considering the precision of sensory estimates of physical stimuli in the environment, in order to interpret the integrated multisensory information, the perceptual system needs to infer the most likely state the sensory estimates may represent. In

(a)

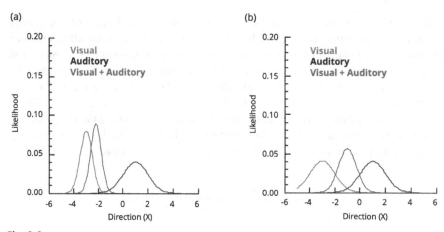

(b)

Fig. 9.6

Schematic depictions of principles of multisensory integration based on the maximum-likelihood integrator model. (a) An example showing that estimates of multisensory perception drift towards the more reliable visual modality. (b) An example showing unbiased multisensory perception with equal precision in visual and auditory modalities. Figure adapted and reprinted with permission from [666] © 2004 by Elsevier.

line with Helmholtz's [612,613] classical concept of perception as a process of inference, factors such as expectations built upon past experiences as well as current task contexts and anticipated actions also feature important roles in perceptual processing [669,670]. To address the effects of these factors, other models, such as the *active inference* theory (e.g., [622,623]), further emphasize that perceptual inference entails minimizing prediction errors between the estimates of currently sampled sensory signals and prior expectations, which can be informed by past experiences or guided by action goals. A similar Bayesian formalism has also been used to devise computational frameworks for developing artificial agents, such as hierarchical Bayesian modeling of multimodal active perception in robotics (e.g., [665]).

9.3 Outstanding challenges in current research

As reviewed in the sections above, researches on psychophysical principles and neurocognitive mechanisms of sensation and perception have come quite a long way. However, so far technologies for Human–Computer Interaction (HCI) have mainly been dominated by vision- and audition-based approaches and the research on multisensory HCI is still in an early stage [671]. The currently available empirical data and theories on human perception and multisensory integration can inform technological developments that take the Human-in-the-Loop approach [3]. To integrate haptic perception into closed-loop systems that support human–machine interactions

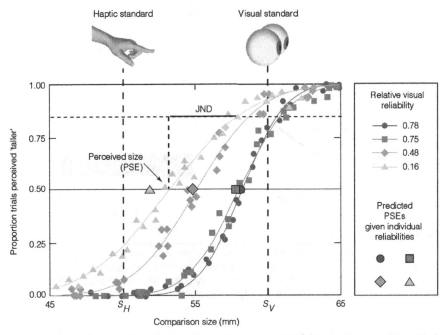

Fig. 9.7

Empirical findings showing impacts of visual noise in affecting the weighting of visual and haptic cues during visual-haptic integration. These results support the maximum-likelihood integrator model of multisensory integration. Figure reprinted with permission from [614] © 2004 by Elsevier.

in perceived real-time, the following challenges and outstanding issues still need to be tackled.

9.3.1 Temporal requirements of the senses

For developing quasi-real-time human–machine interactions that are connected through digital communication networks, the latencies of current technologies at multiple levels are relatively slow. Even ignoring network communication delays, which are at least on the order of 100 ms, most current technologies have device-to-device delays that are slower than the neurocognitive information processing speed of humans. For instance, recent developments in ultralow delay videos [672] are currently on the order of approaching 15 ms, which is getting close to the temporal requirement of human visual sensory processing. However, the haptic sense, though still less used in HCI, operates much faster than vision and audition. The cutaneous mechanoreceptors in the skin respond to sensory stimulation in the submillisecond range. Moreover, such early sensory nerve responses already carry considerable information about the physical properties (e.g., force direction and curvature) of the touched surface [673].

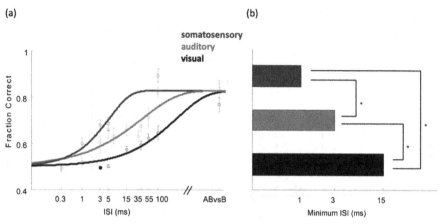

Fig. 9.8

Different temporal requirements of the senses. (a) Mean best performance as a function of ISI for rats stimulated in the somatosensory (barrel), auditory, and visual cortex.
(b) Minimum detectable ISI for the different sensory brain regions. Figures reprinted with permission from [4] © 2012 by the Society for Neuroscience.

At the cortical level, the somatosensory cortex also exhibits higher temporal precision than the visual and auditory cortex. Using the method of direct cortical stimulation to control for between-modality differences in the time required for basic sensory transduction and subcortical processing, it was found that the minimum detectable temporal difference between two stimuli, known as Inter-Stimulus Interval (ISI), is around 1 ms in the somatosensory cortex, but is considerably longer, i.e., about 3 ms and 15 ms in the auditory and visual cortex, respectively (see Fig. 9.8, [4]).

Such differences in temporal requirements of different sensory modalities and the precision-dependent weightings of sensory signals as discussed in the previous section pose challenges for designing sensors, actuators, as well as digital data coding and compression schemes that could together provide the coherent, naturalistic multimodal feedback for humans in different application use cases (see Chapters 2, 3, and 4) for which technological developments in the new field of TaHiL aim to enable. Basic science research on mechanisms, through which sensory experiences as well as action goals can shape the human operator's expectations and perceptions of the environment, can provide insights for tackling technical challenges in engineering naturalistic multimodal human–machine interfaces. In this regard, models of human perception that apply Bayesian principles to sensory adaptation (e.g., [674]) or causal inference (e.g., [622,623,669,670]) as well as models that integrate sensory filters on multiple time scales (e.g., [675]) could be helpful. Furthermore, the Bayesian statistical principles behind these models may allow smoother methodological synergies between models of human perception and a class of data compression methods recently applied in engineering sciences that also relies on Bayesian inference, such as the case of probabilistic programming [676].

9.3.2 Age-related and individual differences require user-centered engineering design

There is a further challenge for developing multimodal human–machine interfaces that could serve broad populations of users, who are of different ages or different skill levels. Accumulating empirical findings from lifespan psychology and cognitive neuroscience show that brain development and brain aging greatly impact perceptual and cognitive processes. Such age-related differences in perception and cognition challenge the one-size-fits-all assumption in technical designs and will likely break usability for users at the two ends of the lifespan.

Protracted maturation of higher-order perceptual processes

The receptors in the sensory organs already undergo considerable development before birth and mature rapidly until they reach the full functioning level during early postnatal life [630,677]. In contrast, neurochemical processes and brain networks for top-down attentional modulation of subjective sensory experiences and perception develop rather gradually during childhood and adolescence [678,679]. Concerning auditory perception, although children's hearing acuity is better than adults, in situations where the direction from which auditory signals reach the ear conflicts with the direction one needs to attend to, children perform much worse than adults in detecting and identifying auditory information [680]. Regarding vision, the protracted maturation of sensory cue integration in the visual cortex until late childhood (around 11 years of age) contributes to children's lower ability relative to young adults in recognizing visual objects with multiple features [681,682]. Similarly, it is only until late childhood that the sensitivity for size discrimination by touch becomes mature [683]. As for multisensory integration, empirical evidence also indicates that the ability to combine information about visual disparity and relative motion for depth perception [681], as well as the ability to integrate visual and haptic information for size and orientation discrimination [683] do not mature until late childhood. Taken together, higher-level perceptual processing that requires top-down attention as well as processes for optimally weighting sensory cues from different modalities are still not mature during early and middle childhood. Therefore multisensory integration in children relies more on the characteristics of physical stimuli in the environment, whereas adults are more flexible in weighting the reliability of sensory signals and use prior expectations gained from past experiences to inform multisensory perception [677].

Aging and gain control of information processing

Regarding effects of aging during the adult lifespan, it is well established that the acuity for auditory and visual [684,685] as well as tactile processing [406,686] all decline substantially in old age. Furthermore, brain aging entails substantial changes at the anatomical, neurochemical, and neurofunctional levels (see [625,679,687,688] for reviews). Focusing on dopamine, which also plays an important role in modulating subjective tactile sensory experiences, as discussed above (see Fig. 9.5, [653,659]), the functions of its receptors start to show age-related decline already in early adult-

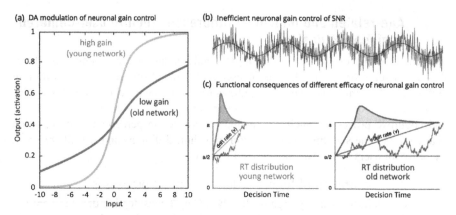

Fig. 9.9

Simulated effects of declined dopamine neuronal gain control on perceptual processing noise with consequences on uncertainty of sensory estimates and processing efficiency [625,690,691]. (a) Aging-related deficiency of dopamine modulation simulated by attenuating the gain parameter of the sigmoidal function that transforms presynaptic inputs into postsynaptic outputs. Attenuating the gain parameter reduces the slope of the neuronal response function. (b) Attenuated gain control increases random processing fluctuations, which functionally reduces the Signal-to-Noise Ratio (SNR) and hence increases the uncertainty of information processing. (c) Depicted here is a simple decision process between criterion 0 (signal absence) or a (detection threshold). Decreased SNR of information processing limits the rate of sensory evidence accumulation (drift rate, v) for simple perceptual decision and attenuates the precision of information processing, as revealed by the broader Reaction Time (RT) distribution (i.e., comparing simulated *old* with simulated *young* networks). Figure reprinted with permission from [692] © 2017 by the authors.

hood, with an estimate of about 10% loss per decade in many brain regions (see [625] for review). Dopamine's role in regulating the signal-to-noise ratio of neural information transmission between neurons (i.e., neuronal gain control) has been computationally modeled as the gain parameter of the sigmoidal activation function in feed-forward multilayer neural networks [689]. The impacts of aging-related decline in dopamine-regulated neuronal gain control of uncertainty (noise) during neural information processing have also been established in computational simulations [625,690,691], which show clear negative consequences of increased processing noise and reduced processing efficiency in simulated *old* neural networks (Fig. 9.9).

Noisy and less efficient processing in children and older adults

Accumulated empirical findings support the simulated computational effects of noisy and less efficient neural information processing with negative effects on perceptual and cognitive performance in individuals at both ends of the lifespan, whose neurotransmitter functions are either not yet mature or have already declined. When making perceptual decisions, children and old adults show lower levels of processing speed and process robustness (i.e., higher degrees of random processing fluctuations) than

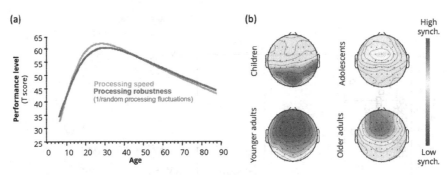

Fig. 9.10

(a) Lifespan age differences in cognitive processing speed and processing robustness. Figure adapted and reprinted with permission from [7] © 2004 by the American Psychological Society; (b) Lifespan age differences in temporal synchrony of brain psychophysiological responses. Figure adapted and reprinted with permission from [693] © 2013 by Elsevier.

young adults (Fig. 9.10, [7]). Relatedly, at the brain level, the temporal precision of neuronal signals (as indicated by synchronized brain electrophysiological activities) is weaker in children and in old adults, relative to young adults (Fig. 9.10, [693]). There are also other prominent aging-related declines in frontal brain processes of attention (e.g., [694,695], executive control (e.g., [696–698]), and valuation (e.g., [699,700]) that could influence perceptual processes in the elderly. Taken together, these age-related constraints on perceptual and cognitive functions either in children or older adults need to be scrutinized and compensated when designing multimodal interfaces for human–machine interactions for these user groups in real, virtual or mixed reality environments.

Individual differences in perceptual processing

Besides age-related effects, individual differences in neurobiological factors (e.g., genetic predispositions affecting neurotransmitter functions), past learning experiences, or skill levels, can all contribute to considerable between-person differences in sensory and perceptual functions, even among people within the same age groups. Take attention-guided auditory processing as an example, empirical evidence [701] shows that, even in a given age range, there are substantial differences between the listeners' ability to selectively attend to speech streams coming from different directions under conditions with no background echo (anechoic) or with a middle or high level of background reverberation (see panel a in Fig. 9.11). In general, adding background reverberation reduced performance accuracy in detecting targeted speech-sound; furthermore, large individual differences in performance were observed in all conditions.

In investigating potential neurobiological factors that might contribute to such individual differences, findings from another study [702] showed that individual differences in genetic predispositions affecting neurotransmitter functions are associated with brain psychophysiological signals that underlie attentional control of auditory

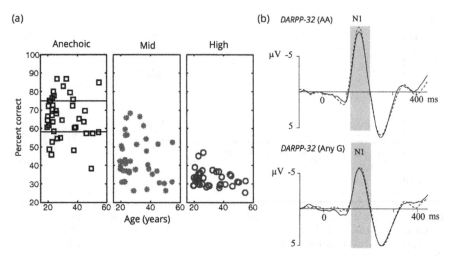

Fig. 9.11

Attentional control of perceptual processing. (a) Individual differences in attentional control of auditory processing under conditions with different levels of background reverberation. Each symbol in the figure depicts the individual performance level (% accuracy) in detecting targeted speech stream (see text for details). Figure reprinted with permission from [701] © 2011 by the National Academy of Sciences, U.S.A. (b) Effects of the Dopamine- and cAMP-Regulated Neuronal Phosphoprotein (DARPP-32) gene, which is relevant for dopamine-regulated neurotransmission, on brain-evoked potential (the N1 component) when individuals selectively attending to auditory signals that are in line (solid line) or in conflict (dashed line) with the direction of attention (see text for details). Figure adapted and reprinted with permission from [702] © 2013 by Elsevier.

perception. The DARPP-32 gene is associated with the functioning of dopamine and other neurotransmitters in the human striatum, with carriers of the AA genotype showing higher neurotransmitter function. Correspondingly, AA carriers also exhibited larger brain psychophysiological responses during attention-guided auditory processing. Specifically, the AA carriers of the DARPP-32 gene showed a larger amplitude of the N1 component during early auditory processing (see panel b in Fig. 9.11) and a larger amplitude of the N450 component during attentional regulation [702]. Neurotransmitters (e.g., dopamine) regulating neuronal gain control (Fig. 9.9) can affect the perceived uncertainty of subjective sensory experiences (see Fig. 9.5, [653,659,661]), which, in turn, could contribute to between-person differences in sensory and perceptual processing.

Individual differences in depth perception is another example. In the extreme, some individuals could not perceive 3D information in stereoscopic scenes [703]. Even within the regular functioning range, different individuals may rely on different cues for depth perception [704]. In an ongoing study of depth perception and learning, we found clear individual differences in the sensitivity to disparity cues that is reflected in the psychophysical signal detection function. Furthermore, we

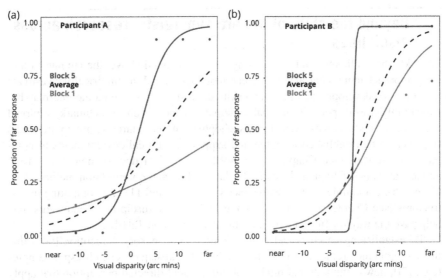

Fig. 9.12

Effects of individual differences and learning on depth perception (see text for details). The slopes of the psychophysical signal detection function depict the sensitivity to visual disparity cues. Shown here are performance data of two participants from the first and last block of the experiment as well as the average across five learning blocks (data from an ongoing study in the authors' lab).

also observed an effect of learning, which sharpens the sensitivity to depth cues. Nevertheless, the effect of learning, which is reflected in the increased slope of the psychophysical function in the last task block, also differs substantially between individuals (see Fig. 9.12).

Taken together, development and aging affect brain functions in several regions relevant for uni- and multisensory perception and attention. Furthermore, the efficacy of neurotransmitter functions, such as dopaminergic modulation highlighted in this chapter, not only undergo clear maturation- and aging-related changes across the lifespan, but also differ between individuals. Neurotransmitter systems can affect the SNR of neural information processing in several brain networks with consequences on the precision, efficiency, and capacity of human perception and cognition. Automatic signal gain control is an important principle for information processing and dynamic process control in neurobiological sensory-perceptual systems as well as in electronic signal processing systems and digital communication networks. Advances in developing sensors and actuators, signal filtering, and compression algorithms as well as communication networks need to closely consider age-related and individual differences in neuronal gain control and their impacts on multisensory perception to develop next-generation technologies for human–machine interactions that can serve broad populations of users.

9.4 Beyond the state of the art: synergistic research across disciplines

To tackle the challenges and outstanding issues discussed above, the Human-in-the-Loop approach that is central to the research of TaHiL needs to empirically investigate the effects of development, aging, and skill acquisition on the dynamic interplay between multisensory perception, goal anticipation, and action. Benchmark empirical data on age- and expertise-sensitive psychophysical and neurocognitive parameters still need to be established to guide and support theoretical developments of computational models (see Chapter 8), as well as technical developments of sensors and actuators (see Chapters 10 and 12), data coding and compression methods (see Chapter 5), communication networks (see Chapters 6 and 11) and computing infrastructures (see Chapter 13). The following lines of interdisciplinary endeavors may help pave the way for further synergies in the research on TaHiL.

By establishing an empirical foundation for multisensory perception [607] that involves goal hierarchies and flexible switches of goal sequences in lifespan samples (e.g., [7]), new computational models of human goal-directed perception that apply principles of hierarchical Bayesian inference (e.g., [620]) can be developed in collaboration with the key technology field of human-inspired computing (see Chapter 8). Once developed, such models can be utilized to determine age-related and individual differences in key processing parameters to inform the development of sensor and actuator technologies (see Chapter 10). Furthermore, collaborations between basic research on human perception and the development of tactile and haptic codecs (see Chapter 5) could help adapt test procedures for subjective quality assessments that are frequently used by engineers to compare different codecs. Specifically, procedures for quality assessments can be adapted by choosing parameters and psychophysical experimental setups that are in line with age-related differences in perceptual thresholds and sensitivity. Such psychophysical perceptual quality assessment could pave the way for identifying more realistic and comprehensive human operator models, with the ultimate aim of developing age-adjusted objective quality assessments of compressed vibrotactile signals in the future.

Basic research on age- and learning-related differences in multisensory perception would also be helpful for refining the designs of virtual reality and mixed reality environments that use multimodal sensor and actuator devices (see Chapters 7 and 10) for applications in, for example, music and movement training (see Chapter 4) as well as E-learning in schools. In particular, age-related differences in the capability of the human perceptual system to optimally weight multisensory signals as a function of delay or signal precision may provide insights for developing multimodal feedbacks that can compensate for limitations of multisensory perception in children and in the elderly. As for designing multimodal feedbacks for surgical training in medical applications (see Chapter 2), assessing individual differences in psychophysical parameters of depth and haptic perception can help designing assistive devices that can flexibly adapt to the trainee's perceptual sensitivity and learning progress. In terms of technologies for protecting the privacy of digital data that may reveal personal characteristics (see Chapter 13), investigations of psychophysical and neurocognitive mechanisms that underlie person perception, including the perception of human bodily motions, may be an innovative route to take. In this case, perceptual properties that are important for identifying a person through movement, facial, and voice characteristics may shed lights on a set of features that computational algorithms can perturb for privacy protection.

Another important outstanding question that needs to be tackled in future research is how existing body models in the human mind and brain maybe affected by interactions in virtual, augmented, or remote environments. Often, such interactions with Cyber-Physical Systems (CPS) rely on a contextual readjustment of internal body representations in the brain, such as a temporary recalibration to a novel sensorimotor mapping. For example, the user may have to adapt to slight delays or distortions when controlling a virtual avatar [621,705–710]. Thinking further, this could also mean adapting to the novel movement range of a whole new (e.g., virtual or robotic) body. This may be the case when a surgeon carries out a surgery through manipulating a robot via telepresence (see Chapter 2), but also when a user plays an avatar in a video game using virtual reality headsets and sensors at home (see Chapter 4). Since these processes rely on the very basic multisensory mechanisms discussed in this chapter, it will be important to examine them experimentally and to look at the corresponding behavioral and brain data. The insights from these experiments can be used to refine computational models of human goal-directed action and perception [711–715]. Moreover, they can reveal individual and age-related differences in the underlying neural processes and the potentially resulting behavioral consequences. Thus this line of research can contribute to the user-centered design of hardware, i.e., sensors and actuators (see Chapter 10), interfaces (see Chapter 7), and software for cyber-physical interactions (see Chapter 8).

9.5 Conclusion and outlook

In biological or technical systems, sensory or sensor noise as well as certain degrees of process unreliability and delays are inevitable. During closed-loop human–machine interactions, the potential amplifications of multiple sources of unreliability and delays are big challenges for current technologies to enable quasi-real-time remote interactions between humans and machines that are connected via the Tactile Internet. Focusing on the human operators, the *Go-Senses* framework, as proposed in this chapter, can serve as the basis for guiding transdisciplinary research across the fields of TaHiL to advance towards solutions for these challenges to enable applications in the use cases highlighted in this volume. Such synergistic research can, on the one hand, shed new lights on how humans dynamically sample and combine multiple sensory signals that are transmitted to them through other humans or machines in real, virtual, or remote environments; and, on other hand, help to advance technological developments that are well grounded in psychophysical and neurocognitive processes of human perception and action.

Sensors and actuators

10

Ercan Altinsoy[a], **Thomas Hulin**[b], **Uwe Vogel**[c], **Tina Bobbe**[a], **Raimund Dachselt**[a],
Konstantin Klamka[a], **Jens Krzywinski**[a], **Simone Lenk**[a,c], **Lisa-Marie Lüneburg**[a],
Sebastian Merchel[a], **Andreas Nocke**[a], **Harsimran Singh**[b], **Anna Schwendicke**[a],
and **Hans Winger**[a]

[a]*Technische Universität Dresden, Dresden, Germany*
[b]*German Aerospace Center (DLR), Oberpfaffenhofen, Germany*
[c]*Fraunhofer-Gesellschaft, Dresden, Germany*

An artist does not fake reality – he *stylizes* it.
– Ayn Rand

10.1 Sensors and actuators of the future

In our modern world, the interface between user and technology is still mostly uni-modal, relying on flat displays. Haptic feedback and immersive audio reproduction is insufficient in most of the cases. Natural touch, coherent optical feedback, and realistic sound can greatly enhance digital experience and human–robot interaction and control as shown in Fig. 10.1. To achieve this, the development and integration of intelligent wearable sensors and actuators for humans (and robots) has to use knowledge from interaction design and human sciences. A highly customizable, modular approach allows easily adapting for different individual needs. Hardware will soon become soft and bendable to tightly fit the human body. Low latency, low weight, low cost, easy integration, and high quality of user experience are the main goals of future human–machine-interfaces.

The main objectives of this chapter have to do with wearable (bendable and stretchable) sensors and actuators with multimodal feedback for human perception, action, and cognition. Sensors and actuators for three main sensory modalities are addressed: hearing, vision, and touch. Haptic feedback improves presence, object, and self-perception, while interacting with the digital world. Wearable integration has to be enabled for fast haptic peripherals. State of the art from soft robots to tactile robots has to be revolutionized, meaning that the design of the next generation of robots has to be equipped with a full tactile sensory and soft motor system. Visual sensing and display technology have to be improved by two orders of magnitude in display latency and power consumption. For example, very low-latency near-to-eye Organic Light-Emitting Diode (OLED)-on-silicon micro-display for instant visual feedback should have µs response time at ultralow power consumption ($< 10\,$mW per module)

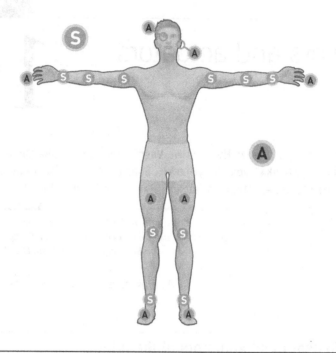

Fig. 10.1

TP2 develops intelligent sensor and actuator solutions.

for at least all-day battery life. Furthermore, coherent sound when interacting with virtual items is essential. Wearable audio reproduction solutions have to be developed that can be integrated locally at the main haptic interaction points. Additionally, projecting speaker systems for ambient sound will be perceptually optimized. All the rich interaction possibilities facilitated by new sensors and actuators require novel design tools of multimodal interaction and feedback. Therefore the perceptual characteristics of each modality, their suitability for particular scenarios and tasks, and the interaction between modalities need to be understood. This allows carefully designing the interaction experience and interplay between human and Cyber-Physical System (CPS). Perceptual knowledge and multimodal interface approaches feedback to all levels of sensor and actuator development.

10.2 State of the art

Whereas technology in recent years was often restrained to flat displays, e.g., tablets or smartwatches, the focus of this chapter is on wearable technologies of the post-smartphone era, allowing rich sensory input and output modalities. Several solutions exist for individual modalities, e.g., vibro- and electrotactile feedback [716], or dielectric elastomer actuators [717], and partly even for their bimodal combination, but currently, there are no approaches for the close interplay and adaptation of auditory, visual, and haptic modalities. Interaction design needs to expand beyond traditional desktop computing approaches and current mobile devices as we know them. It is largely unclear how adaptive feedback [718] has to be designed for humans interacting with a CPS. A rich, multisenosory feedback allows focusing the attention at important parts of tasks, as Ayn Rand put it in [719]: *An artist does not fake reality – he stylizes it. He selects those aspects of existence which he regards as metaphysically significant – and by isolating and stressing them, by omitting the insignificant and accidental, he presents his view of existence.*

10.2.1 Textile smart wearables with haptic feedback and robotics

Various prototypes and early products exist for haptic user interfaces, including glove and stylus-based devices. The authors in [720] provided vibrational friction, pressure, and vibration, at the same time, using the principle of electrovibration and electrical stimuli. Pin-arrays are collection of skin contactors to exert pressure onto the human skin, preferably the fingerpads, to provide tactile information, such as braille dot patterns or a virtual surface. They can incorporate electromagnets [721], piezoelectric crystals [722], shape memory alloys [723], and pneumatic systems [724] to cue useful feedback, which mostly require a separate actuation system, making these devices nonmobile. To stimulate the weight sensation of a virtual object, the authors in [725] designed a wearable device that applies normal and shear stress to deform the fingerpad. Another wearable tactile device that deforms the fingerpad to convey grasping, squeezing, pressing, lifting, and stroking information was presented by Schorr and Okamura [726]. The authors in [727] and in [728] developed a portable device that could apply 3-Degrees of Freedom (DoF) force onto the fingerpad by actuating a rigid platform. The latter also installed a voice coil to deliver vibrational feedback along with orientation information [729]. Of all the actuation principles, vibrational actuation is the most common form of tactile feedback, because of being lightweight, compact, and portable. Some of the common actuators to generate nondirectional feedback are voice coil motors, piezoelectric materials, shape memory alloys [730], and electroactive polymers [731,732]. They can be used to generate vibrational patterns for video games, Virtual Reality (VR) [733], wearable belt [734], vest [735] or wristband [736], and also for communicating emotions [737].

Several limitations—which are addressed within Tactile Internet with Human-in-the-Loop (TaHiL)—restrict broad application to date, e.g., insufficient wearability, lack of precision, and unrealisitc presentation. One of the unsolved challenges is the combination of different haptic dimensions (force-feedback, vibro-tactile feedback, temperature, friction, etc.) in one device. Therefore to date no interface solutions with breakthrough success exist. State-of-the-art soft robots can make use of highly sensitive joint-torque sensors to interpret haptic interaction, such as gestures invoked by humans. Next-generation soft robots utilize collision detection and low-level reflexes [738] for safe physical interaction with humans and their environment [739]. Such feedback can also be used in virtual-reality applications by connecting it to wearables.

Fiber-based sensors and actuators for e-textiles and complex 3D-fiber structures [740] have been subject of recent research. Amongst others, research has been performed on strain sensors, pressure sensors, piezoelectric fibers, shape-memory alloy-based actuators and stretchable polymer tubes filled with a liquid conductor.

10.2.2 Smart vision

Compact 3D-imaging sensors, such as low-cost range imagers and solid-state laser-scanning devices, have improved recently. Photogrammetry, with its focus on optimizing precision and reliability of 3D-object information, has the potential to turn these imaging sensors into powerful measurement devices. All display technologies on the market, including OLED microdisplays [741], have been addressed by frame patterns, which are transmitted sequentially. The continuous refresh across millions of lines and pixels consumes significant electrical power, limiting near-to-eye display battery life to less than one day. Today's high display resolution results in high latency above 10 ms. We will innovate OLED *eyeables* and significantly improve reaction time and power consumption.

10.2.3 Immersive 3D audio integration

Various 3D audio rendering approaches can be found today. A principle architecture of such an audio renderer as applied in virtual-reality applications is shown in Fig. 10.2 [742]. Often the position of the user's head is required as input information for the auditory virtual environments. Sound source and sound fields are two architecture modules. Sound-source signals can be obtained through recording or synthesis. A detailed overview of sound-source synthesis has previously been documented [538,743]. The behavior of the sound field in which the user and sounds stay should be either physics- or perception-based. A complex model might result in a more authentic reproduction, but result in longer computer processing times (i.e., reproduction delay). Perception-based models are derived from psychoacoustic investigations (for an overview, see [744]). Another important part of auditory rendering is the reproduction-based renderer (e.g., Wave-field synthesis, Ambisonics, etc.). Depending on the reproduction technique, the user's head-related transfer functions may be necessary.

The sound-source signals, reflection and directivity filters, filter parameters of the Head-Related Transfer Functions (HRTFs), and algorithms of the reproduction renderer are the input parameters of the signal-processing module. An example of a high-quality, state-of-the-art 3D audio rendering system is the Wave Field Synthesis (WFS) setup in the multimodal lab of the Chair of Acoustic and Haptic at TUD. It was decided to build such a system to (*i*) create a virtual auditory scene over a large listening area, (*ii*) produce plane waves that are localized in the same direction throughout the entire listening area, (*iii*) enhance the localization of virtual sources, the sense of presence and envelopment through a realistic reproduction of the amplitude distribution of a virtual source, and (*iv*) create focused sources in

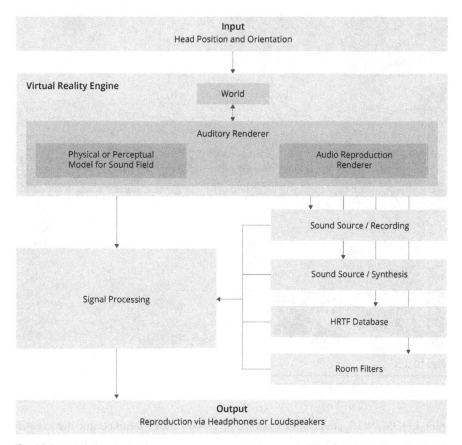

Fig. 10.2

General architecture of an auditory rendering approach.

the field between the listener and the loudspeakers [745]. A WFS reproduction system was installed to the laboratory. This system consists of 464 loudspeakers and four subwoofers. Loudspeaker panels, which consist of eight channels (six tweeters, two bass/midrange drivers), are shown in Fig. 10.3. The theory of WFS assumes a spatially continuous distribution of secondary sources, which is not practically implementable. Therefore some spatial aliasing artifacts occur in the reproduced wave field [746]. To reduce the aliasing artifacts, the tweeter loudspeakers were installed with a very little spacing of 6 cm, which is the smallest spacing in current installed WFS systems. Each individual loudspeaker (468 separate channel) is driven by wave field synthesis signals generated from the incoming audio signals and spatial parameters assigned to each sound source.

Simultaneous rendering of 32 virtual sound sources (focused source, point source, or plane wave) is possible. Maximum delay is about 30 ms. Object-based authoring

Fig. 10.3

Loudspeaker panels for WFS system.

plug-in IOSONO Spatial Audio Workstation (SAW) allows the experimenter to easily create and edit complex sound scenes (Fig. 10.4). Using SAW, the motion paths can be created, moved, rotated, scaled, and grouped.

The control unit manages all central functions: audio server functionality, signal-processing, and a routing system that handles internal and external connections. Using eight rendering PCs, the signal processor computes the wave field synthesis algorithm in real-time. To avoid the noise of PCs, they are placed in a technique room. A room simulation module is implemented either to import and use the measured room impulse responses or to simulate different soundfields. The simulation is based on the mixture of perception and physical based models. To enable 3D analysis and processing of impulse responses in a room, a prototype spherical array measurement system was developed for our lab by Fraunhofer Institute, TU Ilmenau, and TU Delft. A microphone mounted on a mechanical arm moves along a Lebedev distribution of discrete positions on a spherical surface [747]. In some cases, where wave field synthesis, but also higher-order ambisonic, may face some limitations, the combination of both sound spatialization technologies is very promising. Therefore the loudspeaker setup can also be used for higher-order ambisonics reproduction. Combination of the stereophony and WFS can also provide solutions for the WFS limitations [748].

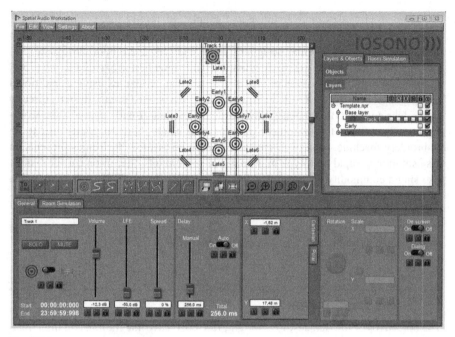

Fig. 10.4

Screen shot of the IOSONO Spatial Audio Workstation.

10.2.4 Rapid prototyping of interactive textile and body-worn user interfaces

Current research on body-worn and wearable technologies as well as material science investigate how novel functional yarns, smart fabrics, and printed electronics could be fabricated and seamlessly integrated into textiles. Such approaches provide rich digital functionalities and enable new promising use cases in medical applications, smart fashion and accessories, active wear and furniture, as well as for ubiquitous everyday life interfaces. Projects like *Jacquard* [749] already demonstrated the technical feasibility of interactive digital textiles at large scale by introducing a comprehensive production process that shows how smart fabrics and garments can be manufactured and distributed commercially in the domain of textile fashion.

Conventional prototyping used in a development process, typically focuses on sequential steps. For instance, textile artists create designs, technical engineers develop the sensor and actuator integration, and textile machinery experts fabricate iterations of physical prototypes. Unfortunately, these approaches are often time-consuming and expensive since the designer's input and the resulting physical prototype are separated from each other in time and involved workflows. As an example Buechley et al. [750] pioneered the idea of sewable electronics by introducing the *LilyPad Arduino* and crafting techniques [751] as an easy and accessible approach. To make development processes more intuitive and simple, *interactive fabrication* [752] and *personal fabrication* [753] approaches propose new interfaces for digital fabrication that aim to closely couple the designer's input and physical output during the prototyping process. Another research example in this direction is *Sketch&Stitch* [754], an interactive embroidery system that is able to create e-textiles using a traditional crafting approach. Sketch&Stitch allows drawing envisioned circuit designs directly on the fabric using colored pens. Afterwards the system visually captures the working area, converts it to embroidery patterns, and hands these programs over to an embroidery machine. Moreover, current research on printed electronics presents an easy solution to rapid prototype multimaterial soft circuits on iron-on carrier materials using a commodity inkjet printer. Such *soft inkjet circuits* [755] make it possible to design, print, and iron-on functional sensor and actuator components in a simple and fast way. Furthermore, iron-on sensor patches, like *ZPatches* [756], aim to enable easy prototyping or ad hoc modifications of existing garments by providing prefabricated hybrid resistive and capacitive designs.

10.3 Key challenges

The objectives of low-latency, low-weight, low-cost, easy integration, and high quality of user experience can be separated in four research fields: (*i*) Textile smart wearables with haptic feedback and robotics, (*ii*) Smart glasses and attributed 3D-data provision, (*iii*) Immersive 3D audio integration, and (*iv*) Smart adaptive interaction and multimodal feedback design. Whereas the first three fields focus on different modalities (haptics, visual, and acoustic), the fourth integrates—in close co-operation with TP1—the developed technologies to provide novel means of interaction.

10.3.1 Textile smart wearables with haptic feedback and robotics

Haptic devices are a special kind of human–machine interfaces that let users perceive information by their sense of touch. Usually haptic devices are classified into two groups. Tactile devices display information to human skin, whereas kinesthetic haptic devices affect the position of human joints by applying forces (impedance-type haptic devices) or limiting human movements (admittance-type haptic devices) [757]. Tactile devices, compared to kinesthetic haptic devices, are typically more lightweight, have lower power consumption, and are less obstructive [758,759]. Also, they enable easily implementing distributed feedback on limbs or the human body through systematically arranged actuators that are controlled simultaneously. In comparison, kinesthetic haptic devices could provide more realistic feedback, and hence enable more intuitive interaction with the environment. Moreover, force-feedback allows the user to intuitively control the applied forces and torques in distant or virtual environments, in case of teleoperation or virtual-reality simulations, respectively. Grounded kinesthetic devices can typically provide high forces and do not require external tracking system, as the position of the human hand can be easily determined by the device position sensors. However, such devices have limited workspace and introduce additional inertial effects on the human hand during interaction.

10.3.1.1 Key challenges in the application of haptic feedback

To efficiently employ tactile or kinesthetic haptic devices, they all need to provide clear and convincing haptic feedback and to obstruct the user as little as possible. However, on real haptic systems, it turns out that improving these two design targets is challenging as they often require contrary modifications. Improving a system with regard to one design target causes a deterioration of the other. Hence, haptic system design is always a trade-off. The great challenge is to develop considerably better haptic feedback systems in terms of the five main objectives that were identified in the introduction of this chapter, i.e., low-latency, low-weight, low-cost, easy integration, and high quality of user experience. Instead of a purely iterative process, completely new paradigms could be a key to reach the goal of leveraging haptic technology and to make it accessible to the general public.

10.3.1.2 Textile integration of sensory and actuator functionality

Textile smart wearable systems are highly suited to arrange haptic systems within a stable, comfortable and preferably functional carrier layer. Those textile smart wearables systems, which include—amongst others—eBodies and eGloves, are based on three general concepts: textile compatible functionalization (foil-based bendable or ultrasmall transducers applied to the textile), textile processable transducers that can be integrated into the textile carrier, and fiber-based transducers (see Fig. 10.5). In this order, the degree of integration rises, leading to a higher automation level of fabrication, higher material compatibility, and highest user compliance in terms of wearing comfort and breathability. In principle, the production of textile smart wearable systems is conceivable with all textile processes. As a rule of thumb, weaving, warp knitting, and nonwoven processes are more economical for the production of flat structures. The implementation of textile structures suitable for final contours is only possible to a very limited extent with these processes. In contrast, the weft knitting process has a decisive advantage: it permits the flexible manufacture of products already in batch size 1 and the local introduction of a wide variety of functional materials and structures, such as conductive, sensory and actuatory yarns, directly in the formation process. This enables the production of tailored functional systems within the textile structure, such as sensor and haptic feedback systems with high lateral resolution, electrical data and power transmission systems as well as mechanical load transmission components. Additionally, weft-knitted textile smart wearable systems can be processed as spacer structures, which makes it possible to bring the functionality into the third dimension. The resulting key challenge for the textile smart wearable concept is to generate suitable layouts for the functional and carrier component of the eBodies and eGloves, to adapt the machine technology based on this, and finally, to verify their functionality within the overall TaHiL context.

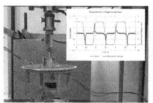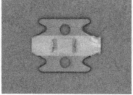

Fig. 10.5

(left) Example of textile compatible functionalization; test of the piezo-electric properties of a textile compatible PVDF-foil. (middle) Example of textile processable transducers; Light-Emitting Diode (LED) to be placed and connected by embroidery. (right) Fiber-based transducers; braided sensor-yarn.

10.3.2 Smart vision

In addition to the development of audio- and haptic-feedback sensors and actuators, visual feedback is important for completing the entire human–robot interaction within TP2. Therefore strong collaboration of the Chair of Acoustic and Haptic Engineering and Fraunhofer FEP is required. In this respect, wearables (such as near-to-eye displays), also referred to as *eyeables* or smart glasses, as given in Fig. 10.6, are developed at Fraunhofer FEP. A typical OLED-on-silicon microdisplay architecture is schematically shown in Fig. 10.7. The microdisplay consists of a silicon wafer with integrated Complementary Metal Oxide Semiconductor (CMOS) electronics and the patterned anode. Different OLED layers are deposited by thermal evaporation in ultrahigh vacuum. Finally, color filters and encapsulation are put on top. Typically, OLED microdisplays consist of a white emission layer and lithographically etched color filters. Within the last years, several micropattering approaches (e.g., fine metal masks, lithography, e-beam direct writing) have been investigated to realize red, green, and blue OLED pixels instead of the white OLED in order to increase efficiency, color gamut, and contrast. However, all these techniques came along with new challenges, such as resolution limitations or yield issues.

Increasing pixel density for high-resolution, reducing latency and power consumption, and the application of multicolor displays are the key challenges for OLED microdisplays nowadays. Therefore new OLED-on-silicon display backplane architectures are required and need to be implemented in a deep-submicron process node. Besides new backplane circuitry concepts (see Section 10.4.2), the OLED structure itself needs to be highly efficient and stable. This requires a careful design of the electronic, optical, and excitonic properties. To meet the criterion of long lifetime, typically fluorescent emitters are used in OLED microdisplays, which allow only for maximum 25% internal quantum efficiency due to spin statistics. Phosphorescent emitters allow for 100% internal quantum efficiency, but they are expensive and the lifetime, especially of blue emitters, do not meet industry requirements.

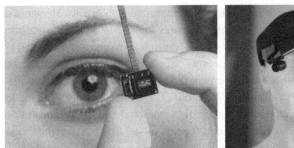

Fig. 10.6

Microdisplay (left) and *eyeable* (right). © Fraunhofer FEP, photographer Anna Schroll.

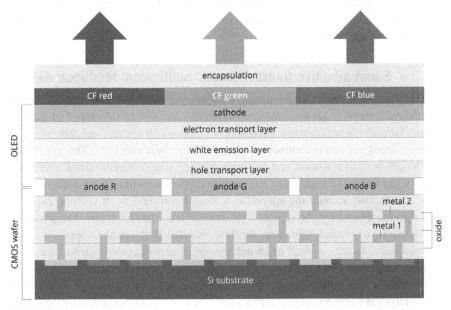

Fig. 10.7

Schematic setup of the OLED micro display. The top metal of the CMOS wafer serves as anode contact for the OLED. Different emission color is achieved using RGB color filters (CF) (not to scale).

10.3.3 Immersive 3D audio integration

All current 3D audio rendering systems have challenges regarding their reproduction quality. The quality elements of auditory virtual environments include its frequency resolution, bandwidth, spatial and temporal resolution, as well as its dynamic behavior [748]. Alternatively, its HRTFs, number of mirror sources, position of mirror sources, reflection filter, and late reverb tail generation are also quality elements of auditory virtual environments [760]. Quality features include its loudness, auditory spaciousness, timbre, localization accuracy, reverberation, dynamic accuracy, and artifacts [760]. Numerous elicitation experiments have been conducted to define the quality attributes of multichannel audio (e.g., reproduction and rendering) that are a part of the auditory virtual environment [761–763]. These attributes also include localization, source width, envelopment, source distance and depth, space perception, and naturalness [764]. Previous research developed a model that contained these attributes to predict multichannel audio quality [765]. One very obvious example of challenges regarding current loudspeaker-based 3D audio reproduction systems is the need for many distributed loudspeakers. Such systems might be possible in a research laboratory, such as the multimodal measurement laboratory at TUD (see above), but they are not suitable, e.g., for standard living rooms. Therefore new approaches for high quality, immersive 3D audio reproduction systems are developed, as described in Section 10.4.3.

10.3.4 Smart adaptive interaction and multimodal feedback design

Within the global research community a number of interesting solutions for smart interactive textiles, novel wearables, and smartglasses will be developed in the near future. However, it still remains unclear how humans will actually use this variety of on-body input and output channels to interact at their best with CPS. This constitutes challenges beyond hardware and software design. Adaptive multimodal interaction approaches for mixed reality and human–robot interaction have to be designed, and have to take into account the specifics of these novel devices. It is to be expected that the smartphone will be replaced by them in a not-too-far future. Supported by appropriate user studies, which are partly conducted in the wild as opposed to traditional usability labs, as well as an iterative development cycle, which of the novel sensory technologies are most suitable for a given task has to be investigated. In close collaboration with psychological and neurological research, interaction design for a cyber-physical world has to take into account the user's goals, experiences, and physical and mental abilities. To augment humans in an unprecedented way also includes risks and cognitive challenges, which have to be carefully investigated and avoided wherever possible. Novel ways of interaction and control have to be provided for the control of CPS, for example by means of gestures, gaze, touch, speech, tactile input, etc. Therefore feedback modalities to optimally support human co-habitation in a cyber-physical world will be designed, blended, and adapted. Further design goals are feedforward techniques to previsualize consequences of human and machine actions. Here, machine-learning approaches, which learn from previous user behavior and interaction, will play a key role. In the line of these sketched research challenges, latest research results and representative examples for smart adaptive interaction as well as multimodal feedback will be presented in Section 10.4.4.

10.3.5 Rapid prototyping of interactive textile and body-worn user interfaces

Designing and prototyping interactive and smart textile and body-worn interfaces is still challenging for smart fashion and interface designers since they need fast and functional iterations during their development processes. Unfortunately, most approaches for e-textile prototyping require expert machinery and knowledge that is typically not available in a small-scale design and prototyping environment. Future interface designers will need further tools to solve the emerging challenge of prototyping with the novel class of next-generation ultra-thin, multilayered as well as printed sensors and actuators. Therefore an important key challenge will be to empower designers and technology-enthusiastic makers with fabrication tools that support them to create, iterate, and evaluate textile-integrated and body-worn input and output components at an early design stage. Current results that take this challenge into account will be presented and discussed in Section 10.4.5.

10.4 Beyond the state-of-art approaches

Although the technologies described in Section 10.3 already offer much scope for improvement and the fusion of the various approaches to multimodal TaHiL systems will create major challenges and need for adaptation, research is already being conducted in all areas on further technological steps. An exemplary outlook on current and future research topics in the field of TaHiL-relevant sensor and actuator technology is given in this section.

A substantial role for the success of further developments in the context of TaHiL will be, independent of the actual research topic, a further enhancement of interdisciplinary thinking and cooperation (see also Section 10.5).

10.4.1 Textile smart wearables with haptic feedback and robotics
10.4.1.1 Kinesthetic feedback

A first step to generate kinesthetic feedback is to develop a lightweight hand-exoskeleton to generate kinesthetic feedback. As an alternative, a kinesthetic feedback device opposed to the hand is a promising solution in terms of high forces. The purpose is to compensate the weight of the finger device, to generate additional force-feedback to the human wrist, and to enable a large workspace. Additionally, various eBody solutions have to be explored (wrist band, jacket, etc.). Textile smart wearables systems are based on three concepts: (*i*) textile compatible fictionalization (foil-based bendable or ultra-small transducers applied to the textile), (*ii*) textile processable transducers that can be integrated into the textile carrier, and (*iii*) fiber-based transducers as given in Fig. 10.8, [740].

According to these three approaches, innovative multifunctional actuators and sensors will be developed and evaluated. On the actuator side, electrotactile, vibrotactile, e.g., piezoelectric fibers, and motion-causing actuators (like shape memory

Fig. 10.8

Sensor/circuit integration and intelligent robot skin.

alloys or multilayered electroactive fiber structures) with a wide range of actuation frequencies (from 0.1 Hz to 10 kHz) will be developed to allow self-movable or stiffening soft peripherals as given in Fig. 10.9, [716]. To improve the capabilities of the actuators, e.g., regarding virtual object recognition, audio-haptic illusions should be used. Tracking information, such as finger and arm position, is required for the interaction. Position and motion sensors, e.g., capacitive stretch sensors, and other sensors monitoring heart rate, finger temperature, or skin conductance, have to be applied. The simulation-based integration of all actuator and sensor solutions into textiles will be an important task. In parallel, TP4 will develop the necessary stretchable electronics. In tight interaction with the wearable design, a novel robot hand–arm system will be designed that serves as the core system of the ideal tactile avatar, pushing significantly further the boundaries of intelligent robotic embodiment, developing new human-like actuator, and artificial nervous tissue and deeply integrated low-level reflexes. The resulting high-performance robot will have human-like range of motion, feasible dynamics/forces, multimodal nervous capabilities, rich and dense sensory feedback for soft controls, and dexterous grasping and manipulation abilities.

10.4.1.2 Smart textiles

Textile smart wearable systems based on conductive knitted fabric structures have already been considered for some time in research and industry comprising, e.g., resistive strain sensors made of conductive nylon yarn and silver-coated conduc-

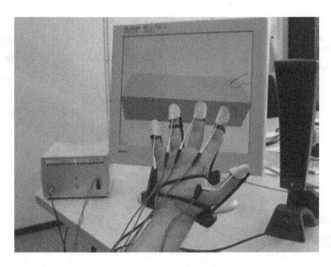

Fig. 10.9

Electrotactile feedback with eGlove.

tive polymer yarn, piezoresistive respiration monitoring sensors, temperature sensors made of braided nickel wire, pressure sensors made of copper-coated acrylic-cotton yarn, and textile heating elements [766–769].

All the structures mentioned have only one function to be implemented. The structure-integrated combination of functions (haptic feedback, position and gesture recognition, energy and data transmission) aimed here requires the simultaneous processing and arrangement of different materials in the knitted structure. To permanently ensure the various functions to be fulfilled, the materials must be knitted into carrier structures in such a way that, among other things, a mutual influence and falsification of the signals is guaranteed by adequate separation (spatial, shielding). The fulfillment of this requirement is a complex interplay of material selection, textile-technological structure development, knitting technology processing and has not yet been addressed or solved in this complexity.

To enable integral textile smart wearable systems with very high functional density innovative multifunctional yarn-based haptic feedback and sensor components is very expedient. Amongst others, multilayered electroactive fiber structures based on dielectric elastomers and piezoelectric polymers (see Fig. 10.10) with a wide range of actuation frequencies (from 0.1 Hz to 10 kHz) as well as shape memory approaches will be developed to allow eBodies and eGloves with intrinsic force-feedback capabilities.

While such textile smart wearables are able to sense information and provide feedback simultaneously at different areas of interaction, in some cases or for some body parts, a different solution for generating haptic feedback could be advantageous. In particular, for stably manipulating objects with the hand, it is crucial for the human to feel contacts and interaction forces by the sense of touch [770]. Textile fabrics

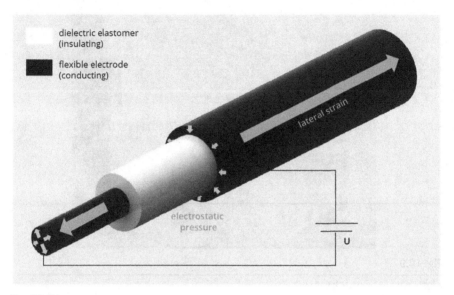

dielectric elastomer (insulating)

flexible electrode (conducting)

lateral strain

electrostatic pressure

U

Fig. 10.10

Concept of multi-layered electro-active fiber structures.

between the skin and the object could impair perception and respectively hinder intuitive manipulation.

10.4.1.3 Tactile feedback

A novel concept for tactile interaction devices aims at circumventing this drawback during direct physical manipulation tasks. It makes use of the limited spatial discrimination capabilities of vibrotactile stimuli at the skin and generates a tactile stimulus at a location that is not covered by or in contact with an interaction device. Instead, the vibrational feedback is induced at two sides of this location of perceived stimulus. As a consequence, a human can perceive real interaction with manipulated objects and tactile feedback from such tactile device at the same time and the same location. This concept is denoted as augmented haptics in analogy to the notation of augmented reality, where the human can simultaneously see real and virtual objects.

Fig. 10.11 shows an illustration of the FingerTac, which is a novel concept for a mobile augmented haptics tactile thimble [771]. This device generates tactile feedback in the form of vibrations at the bottom side of a fingertip. The vibrations are transferred from the actuators to the skin via two vibration transmission elements that are located beside the area at which the vibration is perceived, such that this area is unobstructed. The body structure that holds the microcontroller and the battery is mechanically decoupled through hinges from the vibration transmission elements. With additional sensors, for instance distance sensors that detect the distance to obstacles, the device can provide useful information on the environment to the user.

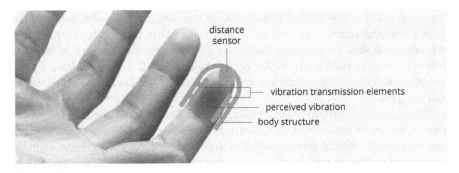

distance
sensor

vibration transmission elements
perceived vibration
body structure

Fig. 10.11

Conceptual sketch of the FingerTac [771].

The FingerTac concept is especially promising for complex manipulation tasks, in which users can benefit from augmented haptic feedback. With such feedback, additional information may be displayed to the user, or it can be used to guide the hand to reach a higher task performance. Hence, potential applications span from industrial applications (U2) to surgical tasks (U1). In future research, we will prepare functional demonstrators and investigate with them the efficiency of the concept for basic manipulation tasks. Compared to pure vibrational feedback, it would be advantageous to have a wearable tactile device that could transmit both orientation and texture information simultaneously. Recently, the authors in [772] fabricated a ferrofluid based tactile device to achieve this purpose. The device uses one miniaturized motor and a solenoid to actuate a neodymium magnet and ferrofluid, which simulates 3-DoF forces on the fingerpad. Future research would focus on optimizing the design and conducting a thorough user study for feasibility analysis.

10.4.2 Smart vision

Typically microdisplays are used in smart glasses to present moving video images. The challenge: Regardless of the screen content, large data volumes are transferred and processed by the system electronics and the microdisplay. This leads to a short battery runtime and a noticeable heat generation, and latency is strongly linked to frame rate. Moreover, all necessary electronics are limiting the miniaturization of the entire system design. In contrast, many wearable applications focus much more on long battery runtime and a slim and lightweight design rather than presenting high-resolution videos. According to these requirements, a special implementation of a fully digital microdisplay can achieve an extremely low power consumption and a small and simplified driving electronics.

The major challenge here is to overcome current display's latency limitations due to frame-based video data transfer. One approach is to replace frame transfer by event and command transfer, which will be enabled by random pixel-addressing scheme. The basic idea for the reduction of both latency and power consumption is the minimization of the necessary data transfer and, at the same time, the elimination of the normally needed refreshing cycles within the display. Therefore the display pixels are equipped with static memory and arranged in a freely addressable matrix. In this way, only the changing parts of the display need to be updated. If nothing is changing, the complete data transfer electronics in and outside of the display can go to idle mode and safe power.

This appears specifically suited to Augmented Reality (AR) applications, where most of the user's field of view is provided by her/his real scene perception, whereas just very locally virtual screen content is superimposed. Moreover, most of those pixels contributing to the virtually overlaid scene often do not change much from one event to another, thus dispensing the need for time-consuming continuous frame transfer, but instead to store the individual randomly-addressed pixel values locally in the display backplane. In combination with much faster deep submicron CMOS backplane Integrated Circuit (IC) technology (instead of slower submicron or even Thin Film Transistor (TFT) backplane processes) new levels of low latency could be reached.

Fraunhofer FEP intends to improve visual display technology by two orders of magnitude in latency and power consumption. We will achieve very low latency near-to-eye OLED-on-silicon microdisplay for instant visual feedback (μs response time) at ultralow power consumption ($< 10\,$mW per module) for at least all-day battery life. A very promising concept to overcome efficiency limitations of OLEDs is the use of the so-called 3rd emitter generation of Thermally Activated Delayed Fluorescence (TADF) emitters [773] that offer the possibility of 100% internal quantum efficiency. Using the concept of hyperfluorescence, e.g., the combination of a TADF emitter and a fluorescent emitter, highly efficient and long-living monochrome OLEDs were already demonstrated on macroscopic scale [774].

All these progressions will be of great significance for the work, namely (i) in K3 to support applications of immersive augmented human–CPS interaction, (ii) in U1 to provide medical assistance and AR/VR-based training, and (iii) in U2 for testing and evaluating human/machine-learning and human-avatar interaction scenarios.

10.4.3 Immersive 3D audio integration

Multichannel 3D spatial sound reproduction is a challenging task, partly because of the need for many distributed loudspeakers. In many scenarios, the placement of speakers at arbitrary positions is not possible, because of functional or aesthetical restrictions. A solution is the invisible projection of sound on walls or other reflective surfaces. The aim is to create auditory events from various directions using this reflected sound. To this end, a sound projector can be applied. Fig. 10.12 shows an exemplary sound projection scenario.

A sound beam is created by a phased speaker array. A schematic directivity pattern of such an array is shown in the background for mid frequencies (gray). The main lobe is steered towards an acoustically hard room boundary, where it is reflected before it reaches the listener. This projected sound (solid line) is used to create

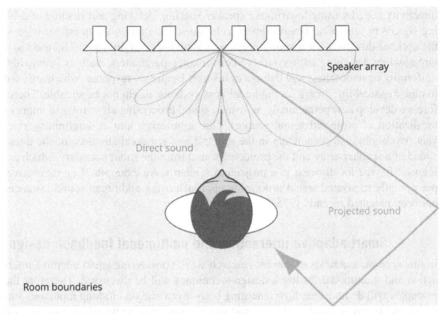

Fig. 10.12

Sound-projection scenario using wall reflections: a speaker array is applied to create an auditory event behind a listener using a projected sound beam (solid line). Direct sound (dashed line) arrives somewhat earlier at the listing position, because of the limited focusing ability of the array. A schematic directivity pattern is shown for mid frequencies (gray).

an auditory event in the direction of the last reflection behind the listener. However, a speaker array radiates additional (usually unwanted) side lobes, inter alia, in the direction of the listener. This direct sound (dashed line) arrives earlier at the ears than the projected sound, because of the shorter propagation path. In real listening rooms the delay is typically below the echo threshold. If the direct sound would be undamped, it would dominate the localization, because of the law of the first wave front, also called precedence effect. For very short delays, summing localization can occur. However, the directivity of the speaker array reduces the amplitude of the direct sound compared to the projected sound. If this damping (between main lobe and side lobes) is strong enough, the projected sound can inhibit the preceding directed sound. The spatial location of the auditory event is then dominated by the projected sound. Unfortunately, the focusing ability of real sound projectors is physically limited and frequency dependent. A schematic damping characteristics in the direction of the direct sound is shown as a gray line in Fig. 10.12. Significant damping can be achieved at midrange frequencies. Damping at low frequencies is limited because of limited array lengths. Damping at high frequencies is restricted by strong grating lobes. These high frequencies lobes can be shifted to some extent in frequency by changing the speaker spacing. There are further methods to increase damping and influence the directivity, e.g., by using logarithmic speaker spacing, delaying and shading of driving signals or numerical beamforming techniques. This raises the question: what is the optimal directivity of a sound projector? Which directivity should be used as a target, when designing such systems? Performance parameters, such as beamwidth uniformity or smoothness and flatness of off-axis frequency response, which are used to rank classical line arrays for public address systems, might not be suitable. Therefore we develop new perceptually motivated signal processing algorithms to improve localization of projected sound sources. Our approaches aim at determining relevant psychophysical parameters in the spectral and temporal structure of the direct sound of a speaker array and the projected sound from the room boundary, which are relevant for the localization in a projection scenario. An example of an innovative, perceptually motivated signal processing approach using additional sound instances has been patented recently [775].

10.4.4 Smart adaptive interaction and multimodal feedback design

In this section, examples of current research work considering smart adaptive interaction and multimodal feedback design techniques will be discussed. Therefore the examples will demonstrate how emerging body-worn and on-clothing input and output channels can be combined in a smart way to support adaptive interaction, as well as multimodal feedback strategies that aim to support users best for different contexts. Smartwatches are very popular, however, their input capabilities and available screen sizes remain limited. To address this issue, *Watch+Strap* [776] explores how commodity smartwatches can be enhanced with interactive watchbands with embedded display and touch technologies to provide smart adaptive interaction techniques by utilizing the watch straps. Therefore such StrapDisplays are able to enable rich visual adaptive strategies, such as adapting content arrangements depending on the arm position, showing context-aware glanceable information or seamlessly extend the watch display (see Fig. 10.13, A). In addition to the enhancement of digital gadgets and smartwatches, also the future role of visual feedback enhancements for unobtrusive e-textile interfaces was explored and possible types, positions, time dimensions, as well as visibility aspects were discussed [777]. Therefore Klamka et al. [778] started to visually augment cords for mobile interaction using smart fabrics and AR glasses with ARcord. *ARcord* aims to extend the interaction and application repertoire of body-worn cords by contributing the concept of visually augmented interactive cords using state-of-the-art AR glasses. This novel combination of simultaneous input and output on a cord has the potential to create rich AR user interfaces that seamlessly support direct interaction and reduce cognitive burden by providing multimodal tangible as well as visual feedback (see Fig. 10.13, B).

Besides the approach of dual-using and enhancing existing textile accessories, it could be also promising to investigate new tiny body-worn devices that are able to provide multimodal feedback and an adaptive set of DoF. Therefore *CHARM* [779] was developed as a combination of a belt-worn interaction device, utilizing a retractable and deflectable cord to enhance AR input capabilities with physical feedback controls and spatial constraints (see Fig. 10.13, C). By wearing such a belt-worn

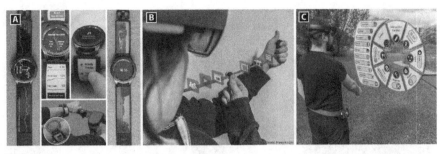

Fig. 10.13

With *Watch+Strap*, [776], explores how commodity (smart) watches can be enhanced with interactive watchbands to investigate smart adaptive interface principles (A). *ARcord* [778] provides a novel combination of simultaneous input and output on a textile hoodie cord, providing visual and graspable feedback (B). Furthermore, *CHARM* [779] introduces a belt-worn controller that combines a retractable cord and a tangible handle to provide a set of interaction techniques to enhance AR input capabilities with physical controls and spatial constraints (C).

smart device, the user can just grab a multipurpose handle and pull, deflect the cord, as well as additionally use further physical controls buttons and thumb-joysticks for complex 3D navigation tasks in a body-relative interaction space. Depending on the users task, different control strategies can be realized. Further goals are to continue the previous work in the field of smart adaptive interaction and multimodal feedback design to develop novel interaction concepts that enable and further simplify the use of CPS. For this purpose, multimodal interaction techniques and feedforward techniques will be designed. As can be seen in the examples shown, novel input and output devices are developed that enable these new interactions. A further approach to achieve this are the rapid prototyping methods described in the next section, which can help to develop prototypes and implement planned concepts more quickly in the future.

10.4.5 Rapid prototyping of interactive textile and body-worn user interfaces

Based on the idea of directly applying sensor and actuators to garments, *Rapid Iron-On User Interfaces* [165] was developed as a new fabrication approach for creating custom textile prototypes that directly transfers adhesive functional tapes and patches onto fabrics by using a handheld ironing tool (see Fig. 10.14).

These functional tapes and patches are made of smart fabrics and printed thin-film technologies, offering a rich variety of electronic functions, while preserving soft and flexible properties. The new handheld iron tool allows the designer to rapidly add functional patches and tapes, similarly to making a sketch with a pen. For instance, it is possible to create conductive traces with desired properties by *sketching* lines, or to add functional patches for specific I/O functionality, circuit design, power supply, etc. By flexibly combining, layering, and juxtaposing, it is finally possible to

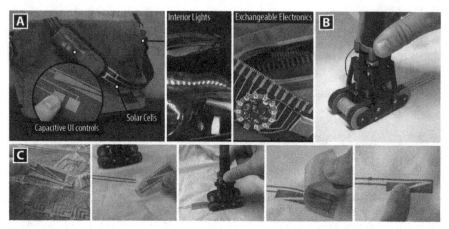

Fig. 10.14

Rapid iron-on user interfaces [165] constitute a new fabrication approach for creating custom textile (A), by using a handheld ironing tool that directly transfers adhesive functional tapes (B) and patches (C) onto fabrics.

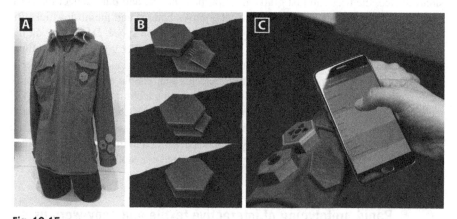

Fig. 10.15

BodyHub [780] introduces a modular wearable approach that allows users to realize their own smart garment applications by adapting and configuring exchangeable functional modules.

create custom interfaces with rich functionality. To explore how highly customizable and modular on-clothing approaches can be achieved for garments, *BodyHub* [780] was developed (see Fig. 10.15, A) as an approach that introduces a reconfigurable wearable system for on-clothing interaction by integrating an interconnected socket system directly onto the garments (see Fig. 10.15, B).

The unique possibility of exchanging a rich set of input and output modules allows reconfiguration adapted to the users' needs and contexts, even at run-time, using a smartphone-based interface (see Fig. 10.15, C). These systems will help to gain a deeper insight into the possibilities of body-worn interfaces by conducting different user studies in the future. The flexibility of these allows making the necessary adjustments and test different concepts quickly and easily. Furthermore, these systems can improve the cooperation with the other research groups by creating early prototypes without significant additional technical effort.

10.5 Synergistic links

Through cooperation with TP1, TP2 contributes to K3 to exchange perceptual knowledge, which is needed for holistic sensor-actuator development. Sensor and Actuator prototypes are evaluated by TP1 members. Another strong collaboration exists with TP4 in terms of developing flexible electronics and circuits for interfacing/processing. Together with TP5, there is collaboration on world model acquisition and augmented reality for interaction and multimodal feedback.

One example of collaboration across different disciplines is the development of a demonstrator for the IEEE 5G Summit 2019. It aims to show the possibilities of Tactile Internet in combination with robots through the game *Rock Paper Scissors*.

Different components, provided by different disciplines, work together to mirror finger, hand, and arm movement by the robot-player, as shown in Fig. 10.16. Finger movement is tracked using a sensor glove with novel integrated textile yarn sensors. An optical Vive tracking system monitors the arm movement. The robot-player consists of an end-effector modeling forearm and hand, mounted on an industrial robot arm, allowing the whole arm to move.

 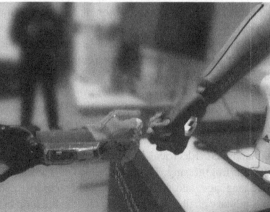

Fig. 10.16

Demonstrator *Rock Paper Scissors*.

The robotic hand and forearm have been developed and adjusted based on the open source robot platform InMoov. The goal was to make it appear more friendly, so conference participants would be attracted to interact with it. Furthermore, we advanced the design regarding integrating the hinges and being able to do maintenance and repair work. The data glove with integrated sensor yarn has been designed to fit to the overall Centre for Tactile Internet with Human-in-the-Loop (CeTI) Corporate Design and communicating the technology.

The tracking of finger movements is realized via structural integrated strain sensors within the sensor gloves. This approach is based on the fact that solids under bending are elongated on their convex side, this extension can be measured using strain sensors. The movement of fingers is, in a broader sense, a bending, which is made possible by the joints. The strain effect is therefore most significant in the joint areas. Strain sensors on the upper (convex) side of the fingers can detect this elongation. In contrast to other sensor gloves, the CeTI demonstrator does not have discrete sensors attached to a glove. Instead, textile sensors were knitted directly into the fingers of the glove during its production, which enables a high precision, optimal fit, and comfortable wear. Based on the actual sensor signals, the three-scenario-specific states, rock, paper, and scissor, are derived by processing within a commercial microcontroller board. A miniaturized more energy-efficient board is already under development and testing (see TP4 presented in Chapter 12). The specific design of the board with space-saving components tailored for the special application has significantly reduced the required installation space. In addition, it was possible to design the geometry of the board in such a way that it can be integrated into a glove system much better than the electronic components used to date, from both ergonomic and aesthetic points of view. A further miniaturization of the electronic components is also planned in the ongoing project. In principle, the whole sensor glove technology can be transferred to completely different layouts and is also to be used in modified form in other CeTI research groups (e.g., U3, box-lifting-scenario in Chapter 4).

Developing the demonstrator aimed at two main objectives: Firstly, being a *boundary object*, it fosters communication and understanding between different disciplines in TP2. Boundary objects are artifacts, which help mediate in the boundary between actors with different perspectives, knowledge, or skills. The authors of [781] describe them as *(...) an analytic concept of those scientific objects which both inhabit several intersecting social worlds (...) and satisfy the informational requirements of each of them. (...) they have different meanings in different social worlds but their structure is common enough to more than one world to make them recognizable, a means of translation.* To develop the demonstrator in TP2, four disciplines collaborated: Electrical Engineering, Robotics, Textile Engineering, and Industrial Design Engineering. It indeed helped to gather understanding for each other, find one language, and will most probably support future collaborative projects. Cross-disciplinary challenges regarding closing the loop in the Tactile Internet, such as how to achieve a stable and fluent movement representation transmitted by the data glove to the robotic hand as well as the combination with a collaborative robot and its sensor system, were uncovered.

The second aim for developing the demonstrator was the purpose of communicating science out of TP2 to the scientific community as well as to the general public. Hence, knowledge transfer, as well as raising interest and motivating people to engage, condensed into several requirements for the demonstrator concept, such as high attracting power, having a low threshold, and being interactive and enjoyable. The demonstrator itself has been exhibited firstly at IEEE 2019 for a rather scientific audience. Since then, it has been used on several events, where pupils, as main audience, wanted to learn about the Tactile Internet.

The *Rock Paper Scissors* Demonstrator has been seen as an ongoing long-time project rather than a completed one. Several improvements will be made during our ongoing research. Firstly, a new sensor chip will be implemented in the gloves hardware, tracking not only the finger movement, but also the hand rotation. Furthermore, the casing will be smaller and located on the back of the hand (instead of the forearm). Secondly, the interaction itself is going to be improved from recognizing three statuses (rock, paper, and scissor) to a continuous mirroring of movements. Additionally, the glove is going to be redesigned to have a higher usability when putting on and taking off.

To sum it up, the first collaboratively developed demonstrator *Rock Paper Scissors* connected several disciplines in TP2, including their research, which forms the basis for further collaborative work. Furthermore, the resulting demonstrator has a great potential for communicating our scientific work to the scientific community and the general public. The playful approach attracts people and raises their interest—interacting with it is fun and fascinating. Experiencing the Tactile Internet first hand has a high potential for the public to engage and discover further.

10.6 Conclusion and outlook

The interaction between human and machines, environment, or virtual objects in real-time will be much more profound and collaborative relationship in the near future, thanks to the Tactile Internet. However, the low-latency, low-weight, low-cost, easy integrable sensor and actuator technologies are necessary for the high quality tactile internet applications, such as VR. Therefore the focus of research is directed towards textile smart wearables, eyeables, smart glasses, gesture recognition and adaptive interaction, and multimodal feedback design. Textile smart wearables include eBodies and eGloves and require fiber-based sensor and actuator technologies. Among other solutions, the development of multilayered electroactive fiber structures based on dielectric elastomers and piezoelectric polymers is very promising scientific task to achieve above mentioned aims. In the future, one important topic will be the intelligent and adaptive usage approaches of sensors and actuators depending to the application and content requirements. Therefore our investigations aim to develop algorithms, which give these decisions automatically and in an effective way. Perceptually motivated signal-processing approaches will allow us to generate plausible multimodal feedback, like immersive 3D audio, smart vision, and haptic feedback.

Communications and control

11

Frank H.P. Fitzek[a], Eckehard Steinbach[b], Juan A. Cabrera G.[a], Vincent Latzko[a], Jiajing Zhang[a], Yun Lu[a], Merve Sefunç[a], Christian Scheunert[a], René Schilling[a], Andreas Traßl[a], Andrés Villamil[a], Norman Franchi[a], and Gerhard P. Fettweis[a]

[a]Technische Universität Dresden, Dresden, Germany
[b]Technical University of Munich, Munich, Germany

The fundamental problem of communication is that of reproducing at one point either exactly or approximately a message selected at another point.
– Claude Shannon

11.1 Motivation

The motivation for research in the field of communications and control is to design resilient communication systems that are robust enough for novel control algorithms, as well as control algorithms that give enough leeway for imperfections in wireless communication, such as packet errors and jitter in latency. This research field aims to provide fundamental tools, directly or indirectly required by the Key Technologies and Methods (K) and Use Cases (U) levels (see Chapter 1) of research to improve performance of communication systems, namely techniques, such as network coding, compressed sensing, functional compression, optimization, and machine learning. Though these techniques are beneficial in specific application fields on their own, their potential is not sufficient for the envisioned Cyber-Physical System (CPS). For Tactile Internet with Human-in-the-Loop (TaHiL), the major challenges lie in combining the aforementioned approaches to bend the multidimensional hyperspace of constraints and objectives, named in Chapter 1 and shown in Fig. 11.1, in CPS as well as complete tactile communication and control networks with Human-in-the-Loop. The key contribution to the overall TaHiL objectives is to provide other research fields with fundamental tools for communications and control techniques, ranging from theoretical results on achievable performance measures to detailed information on algorithms.

For TaHiL, it is central to focus on joint optimization of communication and control approaches for a massive number of interacting CPS, even for different communication architecture, ranging from Body Area Networks (BANs), over Local Area Networks (LANs), to Wide Area Networks (WANs). Thereby, erroneous and delay-prone feedback in the control loop of each CPS needs to be considered along with time variant dependencies of CPS due to mobility, task, and human interaction. The

Tactile Internet. https://doi.org/10.1016/B978-0-12-821343-8.00023-X

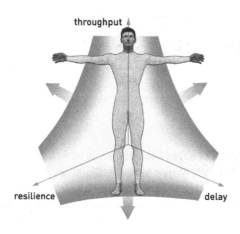

throughput

resilience   delay

Fig. 11.1

Multidimensional hyperspace of constraints and objectives by means of throughput, resilience, and latency.

massive numbers of CPS with Human-in-the-Loop, and hence demands for short latencies, essentially require a data rate reduction that necessitates techniques for compressed sensing and haptic coding to avoid the transmission of redundant data. Furthermore, mobility constraints require transmission paths to be jointly used by different data packets at the same time. Therefore techniques, such as network coding are needed. To fulfill the overall system requirements, to obtain efficient approaches and algorithms, and to achieve a high level of flexibility and adaptability for remote physical interaction scenarios, several of the mentioned techniques will be jointly optimized and provided to other research fields described in this book. In addition, fundamental results, measures, and algorithms on combined techniques to soften the hyperspace trade-off for control and communication in CPS are of major concern. Such systems face a large number of mutually dependent constraints, such as latency, resilience, throughput, security, and energy. In state-of-the-art systems, usually only a single objective out of many is optimized at a time. For TaHiL, however, multiobjective optimization is needed. This will soften the dependency hyperspace. Instead of having only a small fixed set of system operating points, in this chapter functional expressions to describe a surface of available points of operation in a higher dimensional space with respect to given fixed and variable constraints and requirements will be described. Corresponding results are required by the research group for intelligent networks (K2) in Chapter 6. Medium access techniques for wired and wireless communication are another central component of TaHiL. To meet the demand of quasi-real-time interaction of humans with CPS, low latency and determinism are of crucial importance. Traditional definitions of resilience, such as packet-error-rate and recovery time, need to evolve to take into account the models of human perception, cognition, and action. These requirements call for an extremely efficient and flexible Physical Layer (PHY), Medium Access Control (MAC), link adaptation, and innovative Radio Resource Management (RRM) approaches. Moreover, flexibility and scalability of radio resources are needed to support multiple tenants with diverging service requirements. Corresponding results directly support the research towards the research group for haptic and tactile codecs (K1) (see Chapter 5), intelligent networks (K2) (see Chapter 6) and flexible electronics (TP4) (see Chapter 12).

11.2 Research in the field of control

The classical control of dynamical systems is based on a perfect feedback of information, which is crucial to reach a desired state of the system. For TaHiL, the availability of multiple sensors and different types of feedback (e.g., visual, auditory, and haptic) will help the Human-in-the-Loop to achieve a desired Quality-of-Performance (QoP) for different applications. Moreover, the interconnection of CPS allows the systems to be distributed across long distances with multiple computing units that allow the system to implement distributed control mechanisms.

However, the imperfections of the network and wireless link, e.g., packet losses, transmission intervals, queuing delays, among others, imposes additional challenges to the design of the control system affecting the Quality-of-Control (QoC). The theory of Networked Control Systems (NCSs) provides a first insight to analyze the aforementioned problems [782]. Fig. 11.2 depicts a NCS with a human interacting with the control system through a Human–System Interface (HSI) and receiving the respective feedback information depending on the application. It is important to notice that in case the network block of Fig. 11.2 does not fulfill the Quality-of-Service (QoS), both sides (human and the physical system) will operate in an open loop configuration, putting at risk the performance of the system. Therefore the study of control strategies that cope with different imperfections of the communications system and network are necessary.

Basic concept

The QoC of a NCS changes drastically depending on the choice of the communications protocol. An information queue can be harmful to the stability of the system as shown in [783], since this queue would provide the NCS outdated information, producing control actions for an old state rather than the most current one. Furthermore, the wireless communications produces a sampled information of the current status on the CPS, hence, the controller action is limited to discrete information. To ensure the stability of the system, it is necessary to receive an information update within a certain time interval. In the research in [784], this time interval is referred to as the Maximum Allowable Transmission Interval (MATI), which depends on the dynamic behavior of the system. Moreover, the synchronization of the received packets is not perfect and will experience some intrinsic delays. The Maximally Allowable Delay (MAD) was calculated in [785] and depicted that for an increasing MAD, the MATI is reduced. Therefore delays could increase the demand of radio resources to satisfy the latency requirements of the system.

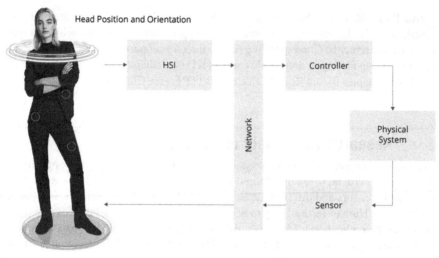

Head Position and Orientation

Fig. 11.2

Networked control system with Human-in-the-Loop.

On the other hand, the imperfections of the PHY and network layer cannot always guarantee the arrival of packets to the NCS. To reduce the dependency to an update and the effects produced by these lost packets, additional control strategies can be implemented to estimate the state of the other CPS [786]. The estimated value should ensure a lower error between the sampled signal in the receiver and the observed signal of the physical system, thus increasing the MATI.

Moreover, it is possible to reduce the transmitted amount of haptic data by utilizing the methodologies for teleoperation described in [419], where estimators feed back information to the human system interface and the teleoperator. If the model error surpasses a threshold, an update has to be sent. Furthermore, a predictive controller can determine an optimal control action if the channel model probabilities are known [787] and can produce the estimation of K future states of the system. The research in [788] showed that $K - 1$ consecutive packet drops can be tolerated without becoming unstable when considering packetized predictive controllers, and the work in [789] depicted that the number of predictions K can be determined by optimizing a function of the spectral transmission power density, bandwidth, and packet duration.

Similarly, the MATI assumes that a NCS will become unstable if it is perturbed and no updates are received. Then, the scheduler has to assign radio resources at most each MATI. However, these conditions can be relaxed if the state of the system is close to an equilibrium point. Thus an event-triggered control scheme (see [790]) can relax the constraints on the scheduler by designing a triggering condition that determines when a packet has to be sent. This, scheme could liberate radio resources for other functionalities, however, the triggering condition should be designed, such that it does not require more resources than transmitting periodically at each MATI.

Relevance for the Tactile Internet

When considering an NCS for a Tactile Internet application, it is important to determine and ensure its latency requirements. Usually, the dynamic behavior of the system will determine how often the communications system has to send an update. As an example, consider Cooperative Adaptive Cruise Control (CACC), where the distance between automated vehicles is regulated. In this example, wireless communication is needed to guarantee safety and short intervehicle distances. The work in [791] depicted that the dynamic response of the system can be used to calculate the MATI of the CACC system.

Furthermore, if teleoperation is considered, not only would it be important to consider the MATI of the teleoperated entity (robotic interface), but also the MATI of the human. Thus depending on the type of feedback [1], the Human-in-the-Loop needs to receive the information of the sensors in a timely and synchronized manner to ensure the desired QoC [430]. Then, the idea for TaHiL is to investigate the dynamic behavior of the human and its interaction with the environment to determine the MATI and give the user the adequate real-time experience.

Moreover, the distance between CPS is limited by this latency requirements due to the speed of light and known delays in the network. Therefore the prediction, estimation, and compression of data is fundamental, as described in the research group for haptic and tactile codecs (K1) (see Chapter 5), to guarantee the controllability of the CPS to compensate for the imperfection of the communications systems and relax the stringent latency requirements of the CPS. Using the example of CACC, the work in [792] showed that the estimation of control signals kept the desired performance when a channel with packet losses was considered. On the other hand, [430] reviewed different methods for the compression and prediction of haptic data to reduce the necessary data rates for teleoperation. Thus a human model is necessary to generate proper estimation and prediction systems to extend the range of TaHiL applications with the required resiliency to channel and network imperfections. The codesign of communications and control systems produces a challenge to satisfy all the requirements on QoS and QoC. Moreover, a design specification of one system can produce detriments on the performance or impose requirements that might not be achievable by the other system. Therefore the joint optimization of the requirements is necessary to achieve the best performance of both systems.

11.3 Research in the field of communications

Flexibility and scalability of radio and network resources are needed to support multiple tenants with diverging service requirements. On the one hand, multiple objectives, including—but not limited to—spectral and energy efficiency as well as latency and delay, comprise the service level agreements. Conversely, sophisticated signal processing at the wireless sensor nodes and fusion centers enables flexibly adapting to the QoS and Quality-of-Experience (QoE) requirements with minimum resource consumption. Using compression techniques at the sensor and intermediate nodes and bringing computational power and distributed computation closer to the edge of the access network are necessary techniques to develop TaHiL from the communications side. Distributed and centralized resource allocation for spectral and energy efficiency maximization in wireless interference networks is well understood [793]; the same applies to the impact of latency constraints and packet loss bursts characterized in [794]. Precoding and nonregenerative relaying for multihop multiantenna systems and its distributed implementation are considered in [795]. The understanding of the structure of the underlying nonconvex fractional programming problems is necessary to develop the distributed efficient algorithms to approach local or global optimal solutions. The quality of the air interface is susceptible to changes in working environments and interference. Current air-interfaces, such as 3rd-Generation Partnership Project (3GPP) Long Term Evolution (LTE), have a latency of 30–50 ms, whereas 5G aims to bring down latency to 1 ms. However, applications, such as virtual reality in high-definition, need throughput in the order of 100 Gbps, within 100 µs of reaction time. These requirements are beyond the capabilities of Service-oriented Architecture (SoA) systems and even the next-generation 5G networks. Furthermore, resilience terms of packet-error-rate and recovery time objective need to take into account the models of human perception, cognition, and action. Therefore extremely efficient and flexible PHY, MAC, link coordination, and innovative RRM approaches are required. We are well positioned to understand the challenges of Tactile Internet as we have extensively worked in the area of 5G networks, investigating massive Multiple-Input Multiple-Output (MIMO), Ultra-Reliable Low-Latency Communication (URLLC), and industrial communications. To support the aforementioned control use cases, communications need to support different communication architectures derived from the research group for intelligent networks (K2) (see Chapter 6), namely BANs, LANs, and WANs. In addition, communications have to be optimized for different technical parameters as there are:

Latency   Any control loop, embedded in machines or humans, is prone to latency. In general, the lower the latency in the feedback of a control loop (see Section 11.2), the more efficient is the application it controls. Here novel physical layer access technologies are needed that currently do not exist. Furthermore, new concepts, such as compression or computing, should help to reduce the latency further, rather than increasing it.

Resilience   Besides latency, the feedback loop also suffers by packet losses. Due to the wireless medium, which is error-prone, new mechanisms to increase the resilience are needed. Normally we have to trade off resilience with throughput and latency. In Section 11.3.2, novel techniques for low-latency and resilient coding are described.

Massiveness   From several applications foreseen from the use case application areas, U-research, we expect a large number of sensors or actuators. In some cases, the number becomes even massive. The pure number will definitely have a negative impact on the delay if we would rely on state-of-the-art approaches to read those entities out. This leads to novel compression algorithms (see Section 11.3.1) that would work in a distributed fashion and decrease the amount of data that has to be conveyed, and in turn decrease the latency.

Heterogeneity Based on the massiveness described beforehand, we will see a large variety of communication devices with quite heterogeneous characteristics, such as air interface, battery capacity, computing power, and others. This requires scalable solutions for those heterogeneous devices in letting them communicate with each other, rather than classifying them to different service classes. Here we will introduce new techniques, such as Fulcrum codes (see Section 11.3.2).

Energy To extend the operational time for battery driven sensors, the energy consumption plays an essential role. Therefore also energy is one of the parameters we have to optimize for. Here the comparison of energy for wireless communication and energy for compression and coding have to be compared.

Security and privacy It would be of utmost importance to build means of security and privacy into the initial design, considering solutions provided by the research group for tactile computing research (TP5) presented in Chapter 13. In Section 11.3.2 security is provided by coding, considering energy consumption.

Here we shortly introduce novel techniques that will be used from a communications perspective to realize the multidimensional hyperspace of aforementioned parameters.

11.3.1 Compressed sensing
Basic concept

Classic sensing/sampling problems rely on the pioneering work of Nyquist, Whittaker, Kotelnikov, and Shannon, where signal recovery is based on the so-called Nyquist rate sampling. Unfortunately, the required theoretical sampling rate is too high/costly/time-intensive in many modern applications and a realization can even be impossible in practice. To address the problem caused by such high-dimensional data, one usually depends on compression. One possibility is known as transform coding, which relies on finding a basis or frame providing a sparse or compressible representation, i.e., a signal of the dimension d can be represented with only $K \ll d$ nonzero coefficients, such as in the JPEG, MPEG, and MP3 compression standards. The Nyquist-Shannon theorem states that a certain number of samples is required to recover an arbitrary band-limited signal perfectly. However, if the signal is sparse in a known basis, we can reduce the number of samples dramatically. This is the basic idea of compressed sensing based on the work by Candès, Romberg, and Tao [796] and Donoho [797]: rather than first sampling at a high rate and then compressing the data, we would like to sense the data directly in a compressed form, i.e., a sparse signal can be recovered from a small set of linear, nonadaptive measurements. Basically, compressed sensing is a subtle mathematical application with connections to the area of information-based complexity, considering the general question of how well a function f from a class $\mathcal{F}$ can be approximated from a number of m sample values. The recovery process relies on the sparse decoder, e.g., l_1-minimization. That is

$$\min \|x\|_1 \quad \text{s.t.} \quad \|\Phi x - y\|_2 \leq \lambda, \qquad (P_{1,\lambda})$$

where $y = \Phi x + \xi$ is a noisy measurement of a vector x, which is sparse or compressible in a redundant dictionary represented by an $m \times d$ matrix Φ with $m \ll d$. Furthermore, ξ is additive noise, which is small in some sense. The smaller the noise ξ is, the more accurate one expects the recovered approximation of x to be. If the noise is bounded in the l_2-norm as $\|\xi\|_2 \leq \lambda$, an upper bound on the approximation error can be given as (11.1). However, if ξ is assumed to be Gaussian noise, such an upper bound can only hold with a certain probability. To overcome this problem, an alternative decoder, such as the Dantzig selector can be applied [798, p. 25], [799].

The first ever sparse decoder was a Reed-Solomon, a similar decoder was developed by Prony [800] around 1795. He tried to identify the frequencies and amplitudes in a nonharmonic trigonometric domain. More information is provided in [801]. However, an explicit appearance of l_1-minimization was Logan's Ph.D. thesis in 1965 [802]. A related method, called total-variation minimization, appeared in the 1992 for image processing, see Rudin, Osher, and Fatemi [803]. Also during this time, an early theoretical work on l_1-decoders was presented by Donoho and Logan [804], which was later popularized by the work of Tibshirani [805] on the least absolute shrinkage and selection operator (LASSO). Independently, the theory of sparse approximation and associated algorithms began in the 1990s [806–808]. An important feature of the l_1-decoder is its projection over a narrowed solution set from all column candidates in Φ. The corresponding upper error bound in compressed sensing by solving $(P_{1,\lambda})$ was given by Candès, Romberg, and Tao in 2006 [809], namely

$$\|\hat{x} - x\|_2 \leq C_0 \frac{\|x - x_K\|_1}{\sqrt{K}} + C_1\lambda, \qquad (11.1)$$

where C_0 and C_1 are constants and x_K is the best K-sparse approximation of x. Obviously, there are two major error sources, one is given by the additive noise and the other by sparse approximation.

Consider (11.1) in the noiseless case, i.e., where $C_1\lambda$ vanishes. If Φ possesses the k-order restricted isometry property (RIP) [796], then it satisfies the null space property (NSP) as well. For $C_0 > 1$, however, it cannot guarantee the uniqueness of the l_1-minimizer anymore. If Φ is very ill-conditioned, then the vectors in the null space of Φ are dominated by a small number of support columns only, i.e., the solution is unstable. In other words, the l_1-decoder is unable to achieve a sufficient sparsity level in such cases. Theoretically, an l_p-nonconvex decoder with $p \in [0, 1)$ [810] can handle this problem better, however, it suffers from the local minima problem. Alternatively, one can find a suitable frame so that x is more compressible or even exactly sparse. For example, a signal representation in a more redundant dictionary, which usually leads to sparse x, was studied extensively [811,812].

To recover a proper approximation of x in problem $(P_{1,\lambda})$, the noise needs to be treated appropriately. For this, however, an optimal regularization parameter λ is required, which is unknown a priori and has to be estimated. More information about typical estimation methods can be found in [813–817]. Unfortunately, these methods are usually rather complex and unstable, in particular for ill-conditioned Φ and, even worse, available for particular scenarios only. In contrast, other algorithms, such as greedy methods [812,818,819] can provide similar results, as in (11.1), without estimating $\|\xi\|_2$. However, they require an estimation of the sparsity level K, which is a priori not available either. Alternatively, l_1-decoders can also lead to similar results without requiring any estimation of the noise or the sparsity level, i.e., $\lambda = 0$ in $(P_{1,\lambda})$, denoted as (P_1), if Φ satisfies the l_1-quotient property [820,821]. This leads to an upper error bound

$$\|\hat{x} - x\|_2 \le C_0 \frac{\|x - x_K\|_1}{\sqrt{K}} + C_2 \|\xi\|_2, \tag{11.2}$$

where C_2 is a constant, depending on the singular values of Φ. It is, however, not the usual case that Φ enjoys the l_1-quotient property, and it is only available for special cases, say, a Gaussian matrix. It is most likely impractical if Φ is very ill-conditioned, i.e., C_2 is then very large.

A natural generalization of sparse vector recovery is the passage to recover low-rank matrices, which gained increasing attention over the last years. Driven by the *Big Data* topic, related questions emerge from matrix completion, factor analysis, vector autoregressive processes, multivariate regression, as well as from compressed sensing and other contexts. In the compressed sensing case, consider a vector valued linear map $\mathcal{A}$ defined on some space of rectangular matrices with the minimization problem

$$\min \operatorname{rank} X \quad \text{s.t.} \quad \|\mathcal{A}(X) - y\|_2 \le \lambda, \tag{11.3}$$

where $y = \mathcal{A}(X) + \xi$ is the received signal of length p. For $\lambda = 0$ follows $\xi = 0$ in the model, corresponding to an exact low-rank recovery. The case $\lambda > 0$ corresponds to nearly low-rank recovery, and it is possible to include additive noise, which will be pathwise bounded by $\|\xi\|_2 \le \lambda$ on a set of high probability. If the matrix variable in (11.3) is enforced to be diagonal, then the rank minimization problem reduces to the classical compressed sensing problem $(P_{1,\lambda})$ of finding a vector with minimal support in an affine subspace. Since l_1-norm minimization is used for vectors, it coincides for diagonal matrices with minimization based on the sum of singular values, i.e., the nuclear or Schatten-1-norm. To be a genuine extension for general rectangular matrices, additional constraints are necessary to guarantee the low-rank recovery. Using the fact that an l_1-heuristic can be seen as a special case of the nuclear norm heuristic, many of the results from the classical compressed sensing literature can be extended to the more general rank minimization task [822].

Consider first the optimization problem in terms of the nuclear norm

$$\min \left\{ \|X\|_* + \beta \|X\|_F^2 : X \in \mathcal{C} \right\} \quad \text{s.t.} \quad \|\mathcal{A}(X) - y\|_2^2 \le \lambda, \tag{11.4}$$

where $\mathcal{C}$ denotes the cone of feasible directions, $\|\cdot\|_F$ and $\|\cdot\|_*$ the Frobenius and nuclear norms, respectively, and β is an optional parameter to smooth the problem (if $\beta > 0$). Known constraints are expressed through conditions for the linear map $\mathcal{A}$ and, in the presence of a noise term, for the additive noise ξ.

Since the previously mentioned NSP is hard to check, other conditions have been formulated instead. First of all, the RIP condition can be transformed to the matrix case [822]. Let $\mathcal{C}_r$ be the space of considered rectangular matrices restricted to rank of at most r, then the smallest value δ_r of a linear operator $\mathcal{A}$ such that

$$(1 - \delta_r)\|X\|_F^2 \leq \|\mathcal{A}(X)\|_2^2 \leq (1 + \delta_r)\|X\|_F^2 \quad \text{for all} \quad X \in \mathcal{C}_r \qquad (11.5)$$

is said to be the RIP constant. Much effort has been spent to find appropriate upper bounds for δ_r much smaller than one. For $\beta = 0$ and $\lambda = 0$ in (11.4), such bounds have been worked out in [822–826]. This was further investigated for $\beta > 0$ and $\lambda = 0$ as well as for $\beta > 0$, $\lambda > 0$ and an arbitrary bounded noise ξ in [825]. In the last case, upper error bounds are given additionally. For real applications, linear operators $\mathcal{A}$ with such a property are generally very hard to find. Instead, one can show this property with very high probability for certain matrices with random entries. It leads to the definition of nearly isometric random matrices. Cases with i.i.d. Gaussian entries are treated in [822,827,828], with sub-Gaussian entries in [829,830] and with i.i.d. symmetric Bernoulli entries in [822,831] using the interplay of the Johnson-Lindenstrauss lemma [832] and the RIP condition [833,834]. Random operators with different structure have also been considered, e.g., subsampled bounded orthonormal systems in [835,836], random convolution systems in [837,838] or block random diagonal matrices in [839]. However, many of these statements have been worked out only for the classical compressed sensing problem, but not for the matrix-based problem of finding low-rank solutions.

A further alternative condition on $\mathcal{A}$ is the spherical section property (SSP). It has the advantage of being invariant under left-multiplication by nonsingular matrices if $\mathcal{A}$ is presented as a matrix [840]. One says that $\mathcal{A}$ enjoys the SSP if there is some $\delta > 0$ such that

$$\frac{\|X\|_*}{\|X\|_F} \geq \sqrt{\frac{p}{\delta}} \quad \text{for all} \quad X \in \ker(\mathcal{A}). \qquad (11.6)$$

Under some additional constraints, the NSP holds for all $X \in \ker(\mathcal{A})$ [825]. Furthermore, for some $\beta > 0$ and $\lambda = 0$, it is sufficient for the optimization problem in (11.4) to recover the low-rank solution. Here the operator $\mathcal{A}$ is regarded as having no random entries.

Another condition is the restricted strong convexity (RSC), which is weaker than RIP since only lower bounds on $\mathcal{A}$ are needed and the constant $\kappa(\mathcal{A})$ can be arbitrarily small [841,842]. One says that $\mathcal{A}$ satisfies the RSC if some $\kappa(\mathcal{A})$ exists such that

$$\frac{1}{2p}\|\mathcal{A}(X)\|_2^2 \geq \kappa(\mathcal{A})\|X\|_F^2 \quad \text{for all} \quad X \in \mathcal{C}. \qquad (11.7)$$

In [841], it was shown that for certain dependent Gaussian entries in $\mathcal{A}$, the RSC condition holds with high probability; thus one can achieve low-rank recovery in the noiseless optimization problem (11.4) with $\beta = 0$ and $\lambda = 0$. For the following, consider additive noise ξ, i.e., $\lambda > 0$, and a Lagrangian version of the optimization problem in terms of the nuclear norm

$$\min \left\{ \nu_p \|\mathcal{A}(X) - y\|_2^2 + \gamma_p \|X\|_* : X \in \mathcal{C} \right\}, \qquad (11.8)$$

where $\nu_p > 0$ and $\gamma_p > 0$ are regularization parameters. Under such circumstances certain assumptions on the random structure of $\mathcal{A}$ and the noise ξ have been made to ensure that $\mathcal{A}$ has the RSC condition with high probability. Moreover, some lower bounds on γ_p and upper error bounds for solution $\hat{X}$ have been established using techniques from the concentration of measure theory [843,844], [845, pp. 145–198]. In [841] this is done for some general dependent Gaussian entries in $\mathcal{A}$ or any (pathwise) bounded noise or sub-Gaussian noise entries, respectively. Gaussian white noise and i.i.d. symmetric Bernoulli entries in $\mathcal{A}$ have been treated in [846]. To capture the dependence structure one can use covariance models for the Gaussian noise entries, which is quite popular in mathematical biology, see [847]. In the very different context of mathematical biology, the covariance models have been used.

Relevance for the Tactile Internet

Compressed sensing enables a drastic data rate reduction in real-world measurements. It requires little amount of resources in hardware and energy for compression. The properties of the signal, such as sparsity or compressibility, are fundamental to strongly undersample the measured signals. The combination of the powerful mathematical techniques of compressed sensing, and kinesthetic and tactile information will contribute to the development of efficient perceptual haptic codes.

One may consider the benefits of compressed sensing by an example of data acquisition, processing, and transmission from sensors to the data network. Fig. 11.3 depicts a smart glove equipped with several sensors for the acquisition of motion data. The glove is also considered to contain a small computing unit that can transmit or receive data to or from the BodyHub on the person's body. The BodyHub itself can exchange data with the network via access points, and thus has access to extensive computing and storage resources. However, access to such resources is associated with a higher latency. A given control or documentation task can be adapted to the resource and latency requirements by exploiting compressed sensing. For this purpose, kinesthetic and tactile information can be utilized to react to the requirement profile in a short time and flexible way by adjusting the sampling rate, by shifting the compression/encoding from or to the BodyHub, and by exploiting the large computational power of the Edge Cloud.

Transaction data can be used, for example, to control graphical user interfaces, to teach robots or to document work steps that have been performed. These modern input methods with or without haptic feedback will revolutionize the human–machine interface by supplementing or replacing the conventional input devices. In terms of

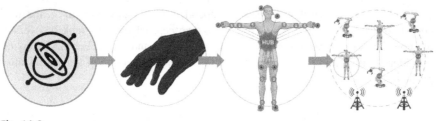

Fig. 11.3

Compressed sensing signal processing chain.

the required resources and latency constraints, the following questions are of interest:

Resources Which resources, in terms of computing power, data rates, and energy, are available at which point in the data transmission and processing chain from the sensor to the network? Regarding the data transmission and data processing, which path has to be chosen to make optimal use of the resources? What is the relationship between the costs for compression/encoding of the data and its transmission? Is it advantageous to process the data at various stages on the transmission path? What are the requirements of perceptional haptic codes, i.e., kinesthetic and tactile codes, compared to the classical approaches, such as compressed sensing and network coding? How can these techniques be optimized together?

Latency What is the maximum permissible time delay from the moment that the measurement data is acquired, processed, and transmitted to the application? Is it already possible to obtain negative values for latency during data acquisition by means of prediction? How and at which point of the transmission and data processing chain should the measurement data be compressed/encoded? What is the benefit of the appropriate choice of kinesthetic and tactile information, i.e., what is the influence of the right choice of the models used on the latency of data processing?

The answers to these questions will lead to the development of powerful computational models of the human sensorimotor systems, and perception and haptic codes. Thus compressed sensing, network coding, and efficient control methods will play a crucial role in the path to the success.

11.3.2 **Network coding**

Basic concept

Network coding and its more disseminated type, Random Linear Network Coding (RLNC), are key enablers for efficient low-latency and massive mesh networks of heterogeneous nodes. It enables intermediate nodes of the network to participate in the coding process adding redundancy when and where it is needed, which increases the reliability and capacity of communication systems, while reducing the latency.

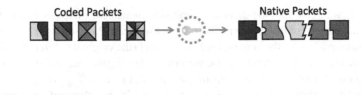

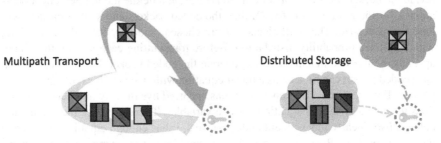

Fig. 11.4

A destination needs to collect enough mixtures. They can come from multiple paths or storage clouds.

In RLNC [848], the data is divided into multiple blocks (also called *generations*) consisting of equally sized packets called *source packets*. Before delving into the details, let us consider the process from a high level point of view. We can think of the source packets as colored pieces of a puzzle. We might be interested in transmitting a message (i.e., a full puzzle) from a sender to a destination. If we do not use coding, and transmit the pieces of the puzzle over a nonreliable network, some of them might get lost. If we require reliable communication, the sender must retransmit an exact copy of the lost pieces, so the destination can collect all the colors. If the sender uses RLNC, it produces instead mixtures of the original pieces. It can produce, practically, as many mixtures as desired. It can combine, in different proportions, all the colors of the source packets. The destination would simply need to collect enough mixtures to recover the original pieces. In principle, this is how traditional block codes operate. The novelty of RLNC is that the proportions used for mixing the colors are not planned, but chosen randomly instead, and they are appended as a header to the mixtures. This brings advantages to RLNC, namely decentralization and the ability for recoding. Any node in the network can recombine a set of mixtures to create a new valid mixture in a process known as recoding. The benefits of recoding will become evident shortly. However, we can see already the advantages in terms of decentralization of the code. If all the destination requires is a certain number of pieces, then, these can be stored in different storage nodes and transmitted through multiple paths without any planning, as shown in Fig. 11.4. Furthermore, if an eavesdropper does not have access to enough mixtures, then it cannot recover the source packets. This adds an inherent level of confidentiality to the transmission. As long as the destination collects enough pieces, it can decode.

If we delve into the details, an encoder creates the mixtures by producing linear combinations of the source packets. For example, if $\mathbf{P}$ is the vector containing the five source packets of the generation $\{p_1, \ldots, p_5\}$ (the original colored pieces), then each coded packet c_i (mixture) is the sum of all the original packets multiplied by a coding coefficient (the proportions of the colors), i.e., $c_i = \sum_{k=0}^{5} \alpha_{i,k} \cdot p_k$. If we further divide the packets into *symbols* of a fixed size 2^q, and define the mathematical operations over the finite field $GF(2^q)$, then the coded packets and source packets will have the same size. The coefficients $\alpha_{i,k}$ are chosen randomly from the finite field, with a uniform probability distribution. Before transmitting each packet, the sender appends the five coefficients used to generate the coded packet. For the destination, each coded packet can be seen as a linear equation with five variables, i.e., the source packets. If we remember our algebra lessons, we need five linearly independent equations to solve a system of equations of five variables. Therefore once the destination collects five linearly independent coded packets, it has enough equations to decode the source packets. Recoding is possible by combining two or more coded packets. For example the recoded packet R_1 can be generated as follows: $R_1 = \beta_1 c_1 + \beta_2 c_2$, where c_1 and c_2 are two coded packets. What coefficients should we then append to the recoded packet? Since all the operations are linear, we can perform the same operations over the coding coefficients of the coded packets, i.e., multiplying them by the corresponding β_i and adding them together. The resulting coefficients are the valid coefficients of the recoded packet. Notice that the size of the header of coefficients do not increase with recoding.

Recoding has the potential to improve the capacity and latency of communication networks with multiple hops. We illustrate this with an example in Fig. 11.5, where a sender needs to transmit four source packets to a destination over a packet erasure channel, i.e., some of the packets might get lost. Since the source and the destination are distant, they need to communicate through a relay. Each communication link in this example has a symmetric erasure probability of 50%, i.e., on average, half of the packets are erased. We compare three strategies implemented at the relay. In Fig. 11.5 (left), the relay only forwards the received packets. In this case, for the given erasure pattern, the sender and relay must transmit in total 23 packets to overcome the erasures of both links, and the transmission takes 15 time slots. When the relay can recode, i.e., Fig. 11.5 (right), it can generate redundancy where it is needed by producing recoded packets. This reduces the number of transmissions to 15 packets, and the transmission delay is also reduced to 8 time slots, i.e., almost half of the time. Fig. 11.6 shows the case when traditional block codes are used. The relay waits until it can decode the information, and then transmits it, once again encoded. This approach is as equally good as RLNC in terms of total transmissions, however, the latency advantage is not present. The relay must wait to decode the data, because it cannot produce valid coded packets from a set of already coded packets, i.e., recoding.

RLNC has many more applications (e.g., low-latency point-to-point codes, reliable massive multicast, signaling-free distributed storage) that improve communication networks' capacities and latencies. However, they are outside of the scope of this book.

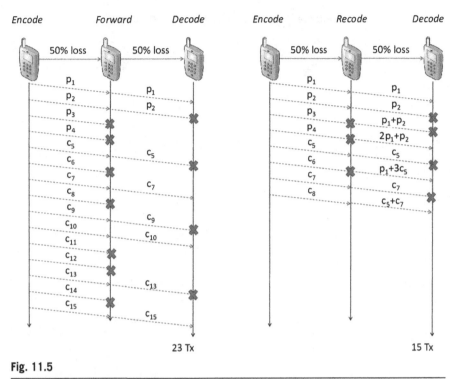

Fig. 11.5

The potential of recoding in multihop networks: (left) forwarding, (right) recoding.

Relevance for the Tactile Internet

The relevance of network coding in the Tactile Internet is many-fold. Its ability to perform in a stateless and decentralized fashion allows it to increase the performance of multipath communication networks, as those present in the Tactile Internet, and described in Chapter 6. Since the receivers are interested only on obtaining a fixed number of pieces, no matter which ones, and no matter where they are coming from, then it is easy to schedule transmissions through multiple paths [849]. Similarly, it is easy to orchestrate in-network caching, which contributes to the low-latency performance of the communication. In-network caching of coded packets allows the devices to obtain data they are interested in from geographically close network caches, thus reducing the latency. Furthermore, the performance of network coding in multihop and multipath environments, make it a perfect code for mesh networks [850,851]. One of the most interesting factors of network coding is its flexibility. As illustrated in Fig. 11.7, current communication protocols based on Automatic Repeat Requests (ARQ), such as Stop and Wait ARQ (SW-ARQ) and Selective Repeat ARQ (SR-ARQ), can operate on only two points. SW-ARQ enables low-latency communication, but at the cost of reduced throughput. On the other hand, SR-ARQ brings high throughput, but at the price of increased latency. There is a fundamental and unavoid-

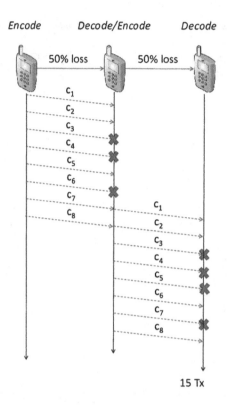

Fig. 11.6

A decode and encode approach does not provide the low-latency benefits of network coding.

able trade-off between resilience, latency, and throughput; however, network coding allows us to operate not only in optimal operation points along this trade-off, but it also allows us to be flexible and choose how much throughput and resilience are we willing to sacrifice for reduced latency [471]. As illustrated by the curves in Fig. 11.7, network coding and traditional block codes enable multiple operation points in the throughput-latency trade-off. However, network coding outperforms block codes and ARQ-based protocols by achieving the same throughput with a lower delay. This flexibility and high performance allows network coding to build the network slices for the multimodal communication of the Tactile Internet. Furthermore, the random nature of network coding allows it to be efficiently coupled with compressed sensing [852].

11.3.3 Functional compression

Basic concept

Network Functional Compression (FUNc) is a novel concept that can be seen as the generalization of one of the problems of traditional information theory, i.e., the dis-

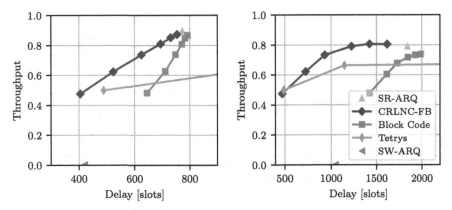

Fig. 11.7

Throughput and 99 percentile delay for an erasure probability of 10%, a mean 64 successive packet losses, and different protocols using ARQ strategies, and network coding strategies.

tributed compression of correlated sources over the network. The original problem addressed by Claude Shannon in his seminal paper of 1948 required a destination to reconstruct the original message transmitted by a source of information. The problem in FUNc consists in the reconstruction of a function of the message. For instance, if the source produces the original message x, and the destination is interested in a function $f(x)$, then the question of FUNc is how to compress the message to reduce the network traffic needed to compute $f(x)$. The initial problems of information theory regarding compression can be seen as a especial case of FUNc, where the function to compute is the identity function, i.e., $f(x) = I(x) = x$. In other words, the destination is interested in the original message.

The problem of FUNc gets more complicated when there are more sources involved. For instance, if two sources transmit x and y, which can be correlated, and a destination is interested in computing $f(x, y)$. The problem of source compression for the computing of the identity function $I(x, y)$ has been addressed by Shannon [853] when there is side information available, i.e., when one of the sources is available at the destination. The same was done by Slepian and Wolf [854] for the case of no side information. Beyond traditional research focused on the reconstruction of the sources at the receivers, i.e., computing the identity function of the sources, FUNc generalizes the problem to the computation of any function. The problem of FUNc with side information has been addressed in [855]. And for the case of distributed FUNc in [856–858].

The researchers in [858,859] have described a clever technique through graph coloring to find the codes to compress the sources for the computing of different functions. Let's illustrate their technique with an example. Let's assume that two sources, namely X and Y want to convey information to a destination to compute the function $f(X, Y) = (X + Y) \mod 2$. The source X can produce the symbols

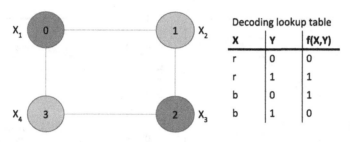

| Decoding lookup table | | |
X	Y	f(X,Y)
r	0	0
r	1	1
b	0	1
b	1	0

Fig. 11.8

Graph coloring of the source X for the computing of $f(X, Y) = (X + Y) \mod 2$, given Y as side information, and the corresponding lookup table for decoding the value of the function.

$0, 1, 2, 3$ with equal probability, while the source Y can produce the symbols $0, 1$ with equal probability. Therefore we need two bits and one bit to encode the sources X and Y correspondingly if the destination is interested in knowing which symbols were produced. Let's further assume that the destination has as a side information with the value of Y. So we only want to transmit through the network the symbol produced at X. To find the code to compress X for the computing of $f(X, Y)$, we can draw a graph, where the nodes are the symbols that the source can take, as illustrated in Fig. 11.8.

We connect two nodes of the graph with a vertex if they have to be distinguished at the destination for computing the function. For our example function, the symbol 0 and the symbol 2 will yield the same result of the function $f(X, Y)$ for all the values of Y. Therefore we do not draw a vertex between them, as illustrated in Fig. 11.8. When we draw all the vertices of the graph, we can color it, i.e., choose a color for each node such that the two nodes connected by a vertex do not have the same color. There are many ways to color a graph, but the coloring that uses the less number of colors yields the code that require the less number of bits to encode. The source X, now only has to transmit these colors to the destination. In our example in Fig. 11.8, the colors blue (light gray in print version) and red (dark gray in print version) require one bit to encode them (the entropy of the source producing colors is 1 bit, given that the colors are likely equal). Now, we can build at the decoder a lookup table for decoding the colors and the values of Y, as shown in Fig. 11.8. In this example, we reduced the bits needed to be transmitted over the network by 50%, i.e., from two bits to represent the four symbols of X, we went down to one bit to represent the coloring of its characteristic graph.

Functional compression algorithms also have the potential to increase the security of the communication. In the previous example, it is shown how compression enables one sender to transmit information to a destination to compute a certain function. One can see in that particular example that the information available at the source is compressed, and only enough bits to compute the function are transmitted over the network. We can see that to reconstruct the information at the source, i.e., the variable X, the destination would need 2 bits (i.e., the entropy of X). However, to compute $f(X, Y)$, only 1 bit is transmitted. This means that a nontrusted receiver will have enough information to compute the function, but not enough information to reconstruct the value of the original variable X.

Relevance for the Tactile Internet

As we have seen in this and the previous chapters, Tactile Internet communication networks will do more than transport of bits. These networks will include storage and computing of information. In that context, functional compression will provide algorithms to reduce the latency due to the transport of information. On one hand, functional compression will allow multiple sources performing secure computations over large amount of data to reduce the network traffic needed to exchange information relevant for the computations. On the other hand, it will enable algorithms that allow trading computing power and network traffic to perform distributed computations among network nodes [860]. Furthermore, its compression will enable potentially secure communications at nontrusted destination nodes. The users may request eavesdropper network nodes to perform computations, while providing the minimum amount of information and preserving the secrecy of the involved arguments of the computations.

11.3.4 **Machine learning**

Basic concept

In the last few years, especially thanks to the recent advancements in the field of deep learning, machine learning has drawn a lot of attention. One of the main driving factors of the *machine learning hype* is that it offers a unified framework for introducing *intelligent decision-making* into many domains. Machine Learning (ML) is a wide umbrella term encompassing a plethora of theories and algorithms (e.g., statistical learning, Bayesian networks, and self-organizing maps) developed over 70 years. The ultimate goal for a parametrized model is always to make a correct prediction based on unseen input. To this end, data is used to guide models towards useful behavior. The distribution of this data is learned and plays a crucial role in model performance, especially at deployment. Algorithms can be grouped into three macrocategories, namely *supervised*, *unsupervised*, and *reinforcement learning*. Supervised learning strives to find a desired output given some input, unsupervised learning algorithms aim at extracting information from unlabeled data (e.g., Clustering), whereas reinforcement learning algorithms address the problem of teaching an agent to interact with an environment based on an observation of its current condition.

Typical uses of ML algorithms are consequently pattern matching and machine vision, pattern recognition and clustering, as well control problems. The prowess of recognition is often exposed to higher level applications, such as anomaly detection, provisioning or configuration of systems. Common to the various fields is the use of data driven models, where the internal model parameters θ are tuned to the specific problems at hand, the *training*. Models are evaluated by problem-adapted loss functions that act as measures for prediction quality. Generally, the prediction problem is cast as an optimization problem over the model parameters via the loss function and the data. Overwhelmingly, ML involves (possibly large) collections of multidimensional *datasets*.

Supervised learning is searching for the approximation of unknown functions when a set of input and output pairs is available. Models are tasked with mapping from input to output. Given a model $\hat{f}$, the goal is to find the optimal parameter vector $\boldsymbol{\theta}^*$ such that a specific distance measure between the true label and the prediction is minimized for all the dataset $\mathcal{D}$. $\mathcal{D}$ is typically split into three disjoint subsets: the *Training set* ($\mathcal{D}_{train}$), to find optimal model parameters; the *Validation set* ($\mathcal{D}_{val}$), to monitor prediction accuracy; and the *Test set* ($\mathcal{D}_{test}$), for a final assessment of the model performance. Fig. 11.9 shows the generic training process that is repeated until convergence: At each step s, a model $\hat{f}$ is fed a set of data points x_i and the corresponding targets y_i, typically in randomly-sampled batches, as large as computationally feasible.

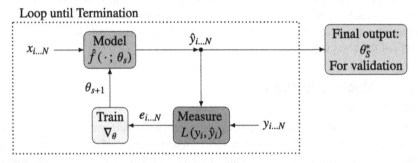

Fig. 11.9

Machine learning algorithm's learning loop: Data pairs $(x_i, y_i), i = 1 \ldots N$ are supplied to the model, which predicts $\hat{y}_i$ and receives updates to its parameters θ via the error measure for some steps $s \leq S$. $\hat{f}$ may, among others, predict probability distributions over labels, clusters, or actions.

The model produces predictions $\hat{f}(x_i; \boldsymbol{\theta}_s)$ for each sample based on its parameters $\boldsymbol{\theta}_s$. They are compared to the actual values according to an error measure (often called *loss* functions) $\mathrm{L} : \mathbb{R}^k \times \mathbb{R}^k \to \mathbb{R}^+$, producing the errors e_i. Common loss functions include *Mean Squared Error* and *Categorical Cross-Entropy*, each averaged over $\mathcal{D}$. Method-specific changes are applied to the parameters, yielding $\boldsymbol{\theta}_{s+1}$, which improves model performance. The training process is, in fact, the minimization of the loss function L with respect to the model's parameters, i.e.,

$$\boldsymbol{\theta}^* = \underset{\boldsymbol{\theta} \in \Theta}{\mathrm{argmin}} \left(L(\hat{f}(y, \boldsymbol{\theta}), y) \right), \tag{11.9}$$

an unconstrained (often nonconvex) optimization problem. Well-known second-order solvers (quasi-Newton, Broyden-Fletcher-Goldfarb-Shanno (BFGS), etc.) or Gradient Descent approaches have proven successful in driving down the errors, often yielding updates in the form

$$\boldsymbol{\theta}_{s+1} = \boldsymbol{\theta}_s - \alpha \nabla_\theta \left[L(\hat{y}, y) \right] = \boldsymbol{\theta}_s - \eta \nabla_\theta \left[L(\hat{f}(y, \boldsymbol{\theta}_s), y) \right], \qquad (11.10)$$

with the gradient operator ∇ and the *learning rate* α, taking small steps towards the optimum. Due to the aforementioned batches, this is called Mini-Batch Stochastic Gradient Descent (MGD). It relies on the assumption that any sampling is representative of the whole data distribution, and that the latter does not change during training, preventing *distribution mismatch*. Model performance is gauged by inspecting its predictions on the unseen $\mathcal{D}_{val}$. $\boldsymbol{\theta}^*$ is flanked by the regimes of underfitting and overfitting: the former refers to a model's incapability to learn the relationships between input features needed to correctly predict the output, and may be solved by using a more expressive model. When a model overfits, it correctly predicts on $\mathcal{D}_{train}$, but fails at $\mathcal{D}_{val}$ or $\mathcal{D}_{test}$. A remedy is lowering the model complexity, or using regularization terms in the loss function.

In the unsupervised learning setting, no *truth* values exist, but the principal approach is very similar. Models need to find patterns in the unlabeled or partially labeled data themselves, which is strongly linked to the clustering and density estimation problems. Notable approaches that lend themselves for exploration are Autoencoders (neural networks with a bottleneck) and mixture models. Though they are not parameter free, modern frameworks search the relevant hyperparameters (e.g., the number of clusters) automatically, if so desired.

Finally, Reinforcement Learning (RL) trains an *agent* to interact with a (possibly dynamic) system by sensing its *state* and, based on this information, taking an *action* that results in a *reward* signal, also triggering a *state transition* in the system. Its goal is to find a *policy* (i.e., a probability distribution over the available actions in the current state) that maximizes the expected cumulative reward encountered. A mathematical framework particularly suited to model the problem is that of finite Markov Decision Processes. Five elements always stand out in RL: a specific system (or environment) and an agent; the set of all the possible system states (i.e., the *state space*) $\mathcal{S}$ with random variables $S_i \in \mathcal{S}$ as the states of the system; the set of all the possible actions $\mathcal{A}_s$; and the instantaneous reward R_i. The ultimate goal of an agent is to maximize the expected cumulative reward, i.e.,

$$G_i = \sum_{j=0}^{\infty} \gamma^j R_{i+j+1} = R_{i+1} + \gamma G_{i+1}, \qquad (11.11)$$

where $\gamma \in [0, 1] \subset \mathbb{R}$ is the so-called *discount factor*, used to emphasize closer rewards. To maximize the return, an agent needs to be able to evaluate the *quality* of a state-action pair via a *value function*. Since the return depends on the future state transitions that depend on the future actions, the probability distribution of the ideal actions given a specific current state needs to be known. Such a probability distribution is called *policy*, and is defined as

$$\pi(a \mid s) = P[A_i = a \mid S_i = i]. \qquad (11.12)$$

Often, the goal in RL is to find the best policy leading to optimal actions, $q^*(s, a) = \max_{\pi} q_{\pi}(s, a)$. With large problem spaces, function approximators, such as artificial neural networks are commonly used.

Relevance for the Tactile Internet

Whereas the concepts above certainly hold some promise in and on themselves, their applications for TaHiL are quite essential. They enable tackling a variety of challenges that have resisted engineered solutions for years, if not decades.

Robot recognition and control Robots tend to have cameras to feed data into their processing, and these data need to be used to accomplish a task. For any grasping task for example, first finding exactly the position and angle of the object being grabbed with respect to the grasping tool is essential, and ML is employed widely.[1] Furthermore, ML has been shown[2] to be able to end-to-end learn their task based on visual input and external guidance, and ML is the crucial element in this success.

Augmented reality Interaction at a distance strongly benefits from contextual additions, and the visual nervous system is the highest bandwidth input to human brains. Augmenting some raw camera footage with information based on the image is therefore beneficial, and ML is the prime method to find, localize, and highlight these additions.

Network optimization Properly orchestrating a potentially transcontinental network in an end-to-end manner is a challenging problem. QoS needs to be maintained or at least predicted for higher layers to cope with deteriorating conditions, and ML is poised to revolutionize both fields. Finding nonlinear, composite patterns in data is the prime use case for unsupervised learning, whereas deploying and scaling functionality is essentially an involved control problem, for which RL has emerged as one of the most powerful approaches known to man.

Sensors Sampling reality with many sensors at high rate will inevitably produce massive amounts of data (see above). If data is processed at receiving or aggregating nodes already before being put in transit towards a silo or database, significant savings may be made. This includes intelligent strategies for compression and aggregation, both spatially and temporally.

11.3.5 Physical layer

Additionally to the previously described software based methods, usually it is worth considering hardware implemented PHY solutions when it comes to wireless connectivity. The wireless channel, due to its random nature, poses a special challenge for

[1] MIT-Princeton at the Amazon Robotics Challenge, http://arc.cs.princeton.edu/.
[2] TossingBot: Learning to Throw Arbitrary Objects with Residual Physics, https://arxiv.org/abs/1903.11239.

the main technical parameters throughput, delay, resilience, and massiveness. Some methods to tackle these challenges found on modern chips and novelties for future developments are briefly introduced in the following:

Basic concepts

In the last decades the PHY was mainly designed for ever-increasing data rates, which according to the theorem formulated by Shannon and Hartley, can be achieved by raising the communications bandwidth. As the bandwidth on the historically used carrier frequencies is already heavily occupied, researchers are developing methods to utilize higher frequencies, where more bandwidth is available, e.g., recently also considering the terahertz band [861]. However, the main challenge for utilizing these frequency bands is the high signal attenuation. The second major development for raising the communications bandwidth relates to the number of antennas used. Truly massive numbers of antennas are considered for future base stations, which make the utilization of the spatial domain additional to the time-frequency domain possible [862]. Simultaneously multiple antennas enable the use of beam forming, which allows focusing the energy of the transmission, and therefore help to overcome the pathloss in higher frequency bands. In all cases, when handling such high data rates, the energy demand of the device becomes a major problem. This requires research for more energy-efficient PHY solutions, i.e., by reducing the amplitude resolution of the analog digital converters in the receivers down to a minimum of one bit [863].

Additionally to throughput limitations, wireless transmissions are limited when it comes to resilience. The wireless channel, due to its random fading, has an unreliable nature, which is problematic for the overall resilience of the system. In particular, radiated waves from the transmitter are reflected by scatterers in the environment. These waves interfere with each other and, depending on the location, can sum up constructively or destructively. Hence, when moving through space, the receiver will observe quickly changing conditions in terms of receive power as depicted in Fig. 11.10.

In the event of strong destructive interference, denoted as an outage, the Signal-to-Noise Ratio (SNR) is too low to recover the transmitted information from the received signal. This inevitably results in transmission errors and is a problem for the reliability of the wireless link. The challenge when designing a reliable wireless communications system on the physical layer is to prepare for such fading-induced outages. One approach is to employ redundancy in the form of physical layer multiconnectivity. In multiconnectivity systems, the device is simultaneously connected over multiple links. Both connections to a single transmitter and connections to multiple spatially separated transmitters can be considered [864]. The scenario of a human wearing special clothing connected over multiple wireless links simultaneously is depicted in Fig. 11.11.

The fundamental idea behind multiconnectivity solutions is that, when connecting over multiple links in parallel, outages only become a problem when they occur on all links simultaneously. The probability of such a critical event is significantly lower compared to the single link case, since every link sees a different fading realization dependent on the correlation between the links. Therefore the correlation becomes

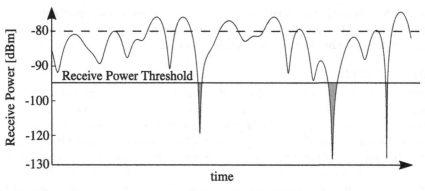

Fig. 11.10

Rapid variations of the receive power (fading) due to interfering radio waves. Transmission errors become highly likely when the receive power is low.

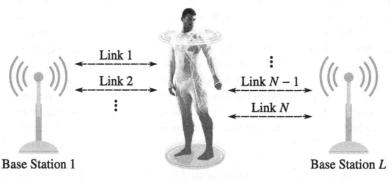

Fig. 11.11

Concept of multiconnectivity for resilient wireless communications.

crucially important for the performance of the approach. The links in physical layer multiconnectivity scenarios can be separated in frequency, the spatial domain or time. In frequency-separated multiconnectivity approaches, redundant data is served on multiple carrier frequencies simultaneously. Spatially separated multiconnectivity approaches use multiple antennas for redundant data transmission. When using time separation, data simply is retransmitted after a first transmission attempt.

The physical layer also adds latency to the loop due to the extensive processing required, e.g., for employing channel coding and decoding or waveform generation. Low-latency communications on the wireless link has to be considered together with the resilience of the transmission, as there is usually only time for a single transmission attempt within the targeted round trip times. Simultaneously, this also means that some methods to increase the reliability fail when it comes to latency, namely time-separated multiconnectivity approaches.

When designing PHY methods for reliable and low latency wireless communications the question for massiveness becomes important for practical systems. For example, the available bandwidth greatly limits the number of links that can be assigned to each user in multiconnectivity approaches with frequency separation when a massive number of devices needs to be connected. TaHiL-optimized RRM approaches are envisioned to allow for high numbers of devices, and due to channel monitoring capabilities, enable the necessary flexibility between the described specialized methods. By knowing the fading state, the number of resources can be accurately adopted to achieve a certain QoS in terms of reliability and throughput. Moreover, resources that are in outage for one user can be operational for another. Thus the number of servable devices can be optimized if the scheduling decisions are dependent on these state information. Such QoS-aware scheduling is a MAC task, however, it is enabled by the monitoring capabilities of PHY. The problem when monitoring the wireless channel is that from the point of observation to the point of data transmission, the channel continues to vary. In the worst case, a channel is monitored to be operational but is in outage during a transmission attempt shortly after, therefore requiring methods to predict these outages [865]. Since monitoring as well as prediction are subject to error, such predictive methods need to be designed carefully to the needs of the Tactile Internet. The basic idea for TaHiL-envisioned prediction based RRM approach is depicted in Fig. 11.12 (left). By introducing a threshold for the detection of outages greater than the outage threshold, the probability for critical prediction errors, which falsely predict an operable fading state can be reduced. Such missed outages lead to an unreliable system when occurring too often, therefore making it unusable for TaHiL operation. The detection threshold can be flexibly adopted for different reliability requirements to suite varying use cases and system states. As depicted in Fig. 11.12 (right), fading prediction combined with the additional threshold for outage detection P_{thr} can overcome the monitoring delay and achieve reliable channel monitoring for TaHiL operation. Dependent on the detection threshold, the possible single link reliability is improved by orders of magnitude. With monitoring available, the challenge is now to design a TaHiL–RRM solution achieving the QoS needed by the use cases described in Chapters 2, 3, and 4.

Relevance for the Tactile Internet

When it comes to wireless connectivity, transmissions are generally unreliable due to the random nature of the fading. Moreover, the PHY is an additional source for latency as outlined above. Since these parameters are highly important for the realization of the Tactile Internet, the development of novel PHY techniques becomes a major research focus for lowering the overall round trip latency down to a minimum. For the improvement of wireless reliability, detailed information about the channel is only available at the PHY. Therefore methods to increase wireless reliability and to simultaneously achieve an acceptable latency are most effective when employed at the PHY. Simultaneously these PHY nonidealities are relevant during the development of Tactile Internet applications as well. The remaining possibility for packet loss needs to be considered when it comes to controller design and intention prediction.

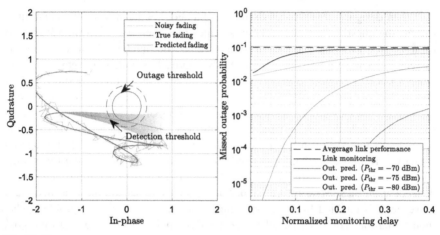

Fig. 11.12

TaHiL-envisioned prediction based RRM: (left) Concept of outage prediction. Predicted fading is prone to prediction errors. An outage detection threshold different from the outage threshold is introduced to adjust detection certainty. (right) Outage prediction performance for different outage detection thresholds P_{thr}. Plotted for Rayleigh fading, 20 dB SNR, −70 dBm mean receive power and −80 dBm outage threshold. Normalized by Doppler frequency.

As portable wireless devices need to be battery powered the energy consumption of the PHY becomes highly relevant for the Tactile Internet as well. For the system as a whole, novel wireless chip design and energy optimized PHY can significantly help to generally reduce power consumption.

11.4 Conclusion and outlook

For the ambiguous requirements of the Tactile Internet, a wide variety of specialized methods were presented in this chapter. One has to understand that these solutions have differing suitability for different network scales. Such common network scales are BANs, which connect sensors around the human; LANs for the connection of devices in a locally limited space; and WANs for everything beyond. For ease of classification the discussed techniques and their suitability within the Tactile Internet are compared in Fig. 11.13.

Compressed sensing is most suitable for BANs, as most of the raw sensor data for the Tactile Internet can be found close to the human. Network coding will have the most impact on WANs, since network latency is dominating over large distance, but can also be a suitable solution for smaller sized LANs. The same holds for functional compression, which is also envisioned for WANs. Machine learning solutions can find application over all network scales due to their wide variety of application areas.

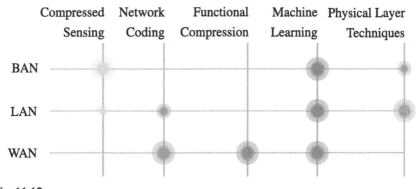

Fig. 11.13

Overview of solutions and suitability for different network types. Bigger circles indicate a greater suitability.

In addition, physical layer solutions can drastically improve when wireless connections are present, i.e., in BANs and LANs.

Although we have presented a wide variety of techniques and research topics that will be the base over which the networks of the Tactile Internet will be built, we have not exhausted the subject. There is still plenty of research in early stages that is promising for future communication networks. The communication community has started to look into post-Shannon communication strategies that could largely increase the efficiency of communication networks. Examples of post-Shannon communications are identification codes [866–869], and exploiting the medium chosen to transmit the message as part of the encoding strategy [870,871]. The research on quantum computing and quantum networks is promising in regards to increasing the reliability, and the efficiency of future communication networks. Quantum networks will enable resources, such as quantum entanglement and quantum teleportation that will increase the communication and computing capabilities of the networks [872–874]. Moreover, there are researchers looking into more efficient and ultralow-latency decoders for physical-layer communications. A good example is the Guessing Random Additive Noise Decoding (GRAND) decoder, which achieves the efficiency of a maximum likelihood decoding, but guessing the noise in the channel instead of the transmitted codewords [875,876]. This allows the decoder to be a *universal* decoder that can decode different forward error correcting codes (e.g., Reed-Solomon, Polar codes, and network codes), with the same low-latency and noise-guessing approach with a better performance in terms of latency compared to other decoders.

Tactile electronics

12

**Jens Wagner, Frank Ellinger, Diana Göhringer, Karlheinz Bock, Christian Mayr,
Ronald Tetzlaff, Gökhan Akgün, Krzysztof Nieweglowski, Johannes Partzsch,
Jens Müller, and Dirk Plettemeier**
Technische Universität Dresden, Dresden, Germany

Any sufficiently advanced technology is indistinguishable from magic.

– Arthur C. Clarke

12.1 Goals

As discussed in Chapter 1 under Objective 2, the development of soft exoskeletons
is essential for the research on Tactile Internet with Human-in-the-Loop (TaHiL). In
this chapter, we focus on the requirements and the development of the electronics that
will enable these. These have to be ultrasmall and bendable to integrate with the garments, and at the same time collect various sensor inputs on different body parts. To
build human–machine interfaces with Human-in-the-Loop, electronic systems need
to fulfill new requirements. They need to be small, fast, and reliable to allow a natural interaction between human and machine. To have a holistic representation of
the user in cyberspace, a large number of sensors and actuators on the body is necessary. To reduce the system size, ultracompact fully integrated transceivers at very
high frequencies have to be developed with integrated on-chip antennas. The power
consumption has to be massively reduced by aggressive duty-cycling techniques to
minimize battery size. These transceivers will be integrated with flexible and stretchable substrates, thus enabling ergonomic and natural interfaces. Furthermore, the user
has to be located within its environment; therefore a heterogeneous positioning system with ultralow latency will be developed. The huge amount of sensor data cannot
be processed in a traditional store and forward manner. Instead, it has to be preprocessed in the Body Computing Hub (BCH), an intelligent, distributed sensor/actuator
processing network (see Section 12.4.3 for details).

Tactile Internet. https://doi.org/10.1016/B978-0-12-821343-8.00024-1

12.2 State of the art

12.2.1 Transceiver

Conventional wireless systems operating at a few GHz require an area of several mm^2 per node. To connect a large number of sensors, e.g., for eGloves, this is too large. The transceiver area is mainly determined by the antenna area. We will exploit the fact that antenna area scales antiproportionally with frequency by the power of two. Hence, the antenna area is decreased by orders of magnitude if we go to very high frequencies of up to 100 GHz. As a consequence, the antennas will be integrated on chip. It has been shown that transceivers at such high frequencies can be fully integrated [877]. We have demonstrated very compact wireless 190 GHz transceivers with up to 50 Gbps at a relatively low dc power of 154 mW and an efficiency of $3.1\,pJ\,bit^{-1}$, which is the best reported so far in silicon [878]. The power consumption of these works is still way too high and will be reduced by orders of magnitude to allow an integration within the Body Area Network (BAN). In the past, several other groups have demonstrated ultralow power transceivers [879–882], however, the power consumption of these designs is still too high for TaHiL application. To our knowledge, the work of [883] is the only one with a power consumption of around 1 mW, however, in this case, a narrow-band approach is used with only a few kbps data-rate possible.

12.2.2 Localization

One key property of Cyber-Physical System (CPS) with Human-in-the-Loop is the continuous awareness of the location and posture of the user and its surroundings with high accuracy and low latency. A precise radar-based indoor 3D-localization system with accuracy better than 10 cm was designed by Ellinger's group [884]. Moreover, data-fusion based, resilient heterogeneous positioning systems allow improved positioning accuracy and universal outdoor and indoor coverage. However, up to now, these heterogeneous positioning systems need around 100 ms per measurement of the position. Other groups have published wireless localization systems until recently ([885], [886], and [887]), however, both accuracy and latency are not adequate for TaHiL demands.

12.2.3 Mechanical integration

Until now, we know electronics as more or less bulky and rigid devices. However, research in the past years has shown that it is possible to develop electronics on flexible and organic substrates. Stretchable electronics have become a crucial demand within the last decade. It is also the template to integrate electronic functions for medical and health applications as well as for smart wearables, enabling humans to seamlessly cooperate with both Internet of Things (IoT) and robotic environments. The majority of the flexible and stretchable interconnects have been based on electrically conductive particles (e.g., carbon or silver) within a polymer matrix [888],

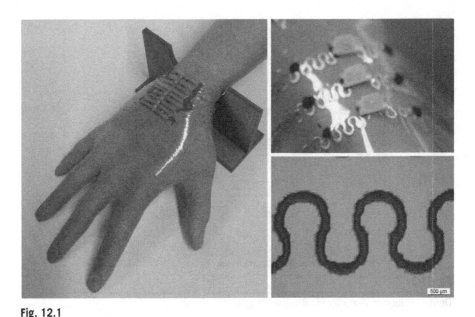

Fig. 12.1

Stretchable polyurethane based ink-jetted horseshoe interconnect.

thin metal tracks in μm- [889,890] or nm-layers [891], with additional mechanical resilient design (e.g., horseshoe Fig. 12.1), conductive liquids (ionic or metal) within microchannels [892,893], or conductive yarns for textiles [894].

The performance limitations of these technologies push the researchers to combine ultrathin and flexible silicon-based high-performance Integrated Circuit (IC) technologies. Ultra-thin sensor [895] (see Fig. 12.2) and thinned-chip packages [896,897] were designed and fabricated on flexible polyimide substrate materials by Bock's group.

These package assemblies are combined with stretchable substrates to support the system integration for the different application scenarios. To fulfill the high demands on miniaturization, multimaterial and very fine feature sizes for fabrication of flexible and stretchable interconnects, alternative electronics manufacturing methods are further applied [898].

12.2.4 Low-power sensor data processing

The development of the BCH is grounded in Mayr's work in the Human Brain Project [899], as well as previous experience on real-time sensory processing. In particular, dynamic, distributed power management techniques and memory/processing integration strategies developed there will be instrumental in meeting the significantly tighter power budget of wearable computing (10 mW/chip). State-of-the-art energy efficiency values for machine learning [900] still pose challenges for implement-

Fig. 12.2

Demonstration of flexibility of ultra-thin magnetic field sensors.

ing sufficiently powerful processing algorithms in wearable computing. A promising way for decreasing computations and memory is exploiting sparsity for decreased memory footprint and processing effort [901]. In particular, weight sparsity can be increased by dedicated learning algorithms [902]. The ultralow latency (< 100 ns) of the hardware adaptation features [899] is a key prerequisite for their application in a latency-constrained processing chain. A highly energy- and performance-efficient runtime system is required to operate the BCH. A base for such a runtime system was presented by Göhringer in [903]. In this work the orchestration of an adaptive computing system was investigated and evaluated with several applications from the domain of image- and signal processing.

12.3 Research challenges

How can the latency and bandwidth requirements be fulfilled? To have a holistic and accurate representation of the human body in cyberspace, the number of sensors will be very high. In a first guess, we can assume at least one sensor per joint of the human body, which results in 15 sensors on one single hand. Not only is the number very high, but also the demands in terms of precision and latency are very challenging. If an update rate of 10 to 100 Hz per joint at a latency of 1 ms is assumed, the processing and forwarding is an obvious challenge. The easiest approach would be to have all these sensors connected by wire to a computing platform with enough computing power to handle these demands. Because the sensors have to be integrated with garments, this is not possible. To have reliable and fast sensor data transmission, a wireless transmission is unavoidable.

What are the implications of the required garment integration on the electronics system design? Furthermore, the system has to be integrated within the garments in a way that feels natural to the user. Above all, this means that the electronics must not be the bulky and rigid devices we know. They have to be ultrasmall and flexible to allow a natural interface and enable a high number of use-cases. With system size being that critical, we must first of all reduce power consumption to an absolute minimum. This is because in a wireless system, the battery size will be one of the major contributors to the form factor.

How can the size of the wireless transceiver be minimized? For wireless transceivers, another aspect becomes apparent. Our above considerations are very challenging in terms of latency and update rate; the accumulated bandwidth, however, is not that high and could be realized with low-frequency (and thus low-power) data transmission. Nonetheless, the size of these is usually determined by the antenna size since these scale with the wavelength. Thus very high frequencies were chosen with the difficulty to fit in the power budget.

How can we minimize the energy consumption of the computing platform? The processing of the sensory data will be performed in the BCH. The most difficult task here will be to process the big amount of data very fast and yet comply with the extremely tight power budget. To achieve this, the platform has to be extremely agile regarding its adaptivity. The computational power has to be adaptively scaled down whenever possible to save power. This requires also an optimization of the DC/DC converters to make full use of the potential. Last but not least, the possibilities of the BCH hardware have to be exploited fully by the operating system. Thus hardware and operating system have to be tightly integrated. Furthermore, the algorithms to classify the motion of the human body need to perform as efficient as possible, making use of the special hardware capabilities as much as possible.

How can the electronics be integrated within the garments and allow natural user interaction, while still delivering reliable sensor data measurements to the system? To mechanically integrate electronics with garments, several challenges have to be faced. The Application-Specific Integrated Circuits (ASICs) have to be integrated with flexible and stretchable substrates. Especially for the stretchable substrates, manufacturing conductors with reliable and reproducible resistances, even while stretching is still a tremendous challenge.

Though there are challenging research questions in each of the above mentioned fields, the most important one will be the collaboration between those. For the goals we have set, it is not sufficient for every area to optimize its respective component, for a tight collaboration is essential.

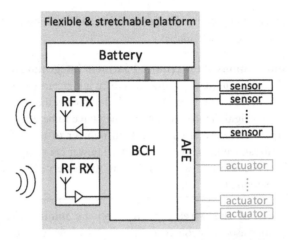

Fig. 12.3

Architecture of wireless sensor processing platform.

12.4 Research approaches

To address all these research questions, the system design has to be optimized for TaHiL requirements. To handle the high amount of sensor data, a hierarchical structure of the sensor network was chosen. A certain number of sensors will be concentrated and preprocessed locally before being sent to a central processing platform, which then generates the overall data stream. In the course of this chapter, this platform is called the eGlove (when referring to the glove) or the eBody (when referring to the bodysuit). The architecture of the eGlove/eBody is shown in Fig. 12.3.

This platform can handle up to 15 sensors in the first step. Later, a number of actuators are interfaced with the platform to provide feedback to the users. The sensor data will be preconditioned by an Analogue Frontend (AFE), digitized, and processed by the BCH and then transmitted wirelessly using the RF transceiver covered in Section 12.4.1. Both the RF transceiver and the BCH are integrated within a flexible and, ultimately, even stretchable substrate to be compatible with the mechanical demands of the garments. The electronics are paired with a rechargeable battery. We aim to operate the wireless sensor processing platform from a commercially low-profile battery for at least 10 hours. The electronics heterointegration is discussed in Section 12.4.2. To meet the demanding power restrictions, the BCH and its operating system are to be tightly integrated. This is discussed in Section 12.4.3 and 12.4.4, respectively. Last but not least, the algorithms to process the sensor data have to be extremely efficient to not exceed the limited computational resources and at the same time allow for power-saving as much as possible. This topic is covered in Section 12.4.5.

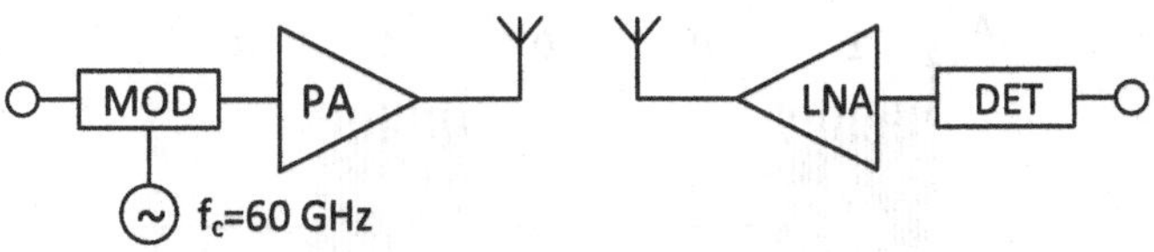

Fig. 12.4

Block diagram of basic OOK transceiver.

12.4.1 **RF transceiver**

Before the actual design of the RF transceiver started, certain specifications had to be fixed. First, we decided to choose the 60 GHz band for transmitting the data. This seemed a reasonable compromise among system size (the antenna can still be integrated on silicon), transmission range and power consumption. To go for even higher frequencies would certainly allow for even smaller antennas, yet the power consumption would be higher and it seemed unlikely that the required range could be met. Furthermore, we decided to use an On-Off Keying (OOK) modulation scheme. The reason for this choice was mainly that such a simple modulation scheme allows for extremely simple receivers without the need for coherent demodulation. Above all, we wanted to eliminate the necessity for a Phase-Locked Loop (PLL) to save both power and the need for synchronization of the nodes, although this definitely means a sacrifice in transmission range compared to robust modulation schemes, such as Frequency Shift Keying (FSK). On the other hand, we do not really need the bandwidth efficiency provided by modulation schemes like, for example, Quadrature Phase-Shift Keying (QPSK) or Quadrature Amplitude Modulation (QAM).

A block diagram of a very basic OOK transceiver is shown in Fig. 12.4, consisting only of a Voltage Controlled Oscillator (VCO), a modulator, and Power Amplifier (PA) on the transmitter and Low Noise Amplifier (LNA) and detector on the receiver side. One of the challenges in this design will be the full integration of the transceiver, including the antenna. Whereas there are several publications on integration of antennas at 60 GHz, we will also investigate alternative possibilities. In the past, we have demonstrated that antennas can be formed using bondwires for very high frequencies, see [878] for a reference. Apart from the antenna integration, the power consumption of the transceiver will be a tremendous challenge. As one example, the excellent work from [882] demonstrates a transceiver with a continuous power consumption of around 67 mW. Since our goal is a power consumption of only 1 mW, we have to reduce this already excellent value by roughly a factor of 100. Apart from block optimization, it has been decided that such a huge step can only by achieved by introducing aggressive duty-cycling or wake-up functionality. In this approach, the receiver is not switched on all the time, but is rather powered with a certain duty-cycle.

The wake-up principle is sketched in Fig. 12.5. Let us assume we have an OOK signal of an arbitrary quantity A_{RF}, which is received by the receivers antenna, where the presence of a 60 GHz signal signifies a digital *1*, whereas its absence signifies a *0*. To receive this signal in a robust way, we have to guarantee that the receiver is

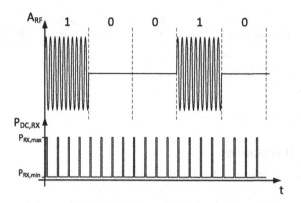

Fig. 12.5

Duty-cycling or: the wake-up principle.

switched on at least once per symbol. However, if we consider drift between the clocks of transmitter and receiver, this might not be enough. Thus we have to introduce what is called oversampling. This means that the receiver will be switched on four times per symbol, thus guaranteeing even for a reasonable amount of clock drift that the receiver is able to receive at least three times per symbol. In this example, a four-times oversampling is employed. If the receiver has an instantaneous power consumption of $P_{RX,max}$, its average power consumption can be calculated by

$$P_{avg} = P_{RX,max} \cdot \frac{T_{on}}{T_{on} + T_{off}}, \tag{12.1}$$

assuming that $P_{RX,min}$ is zero. If we apply this approach theoretically on the receiver published in [882], we could conclude that by using a duty-cycle of 1%, the power consumption of the receiver will be reduced to around 720 µW. A duty-cycle of 1% could be realized with $T_{on} = 10$ ns and $T_{off} = 1$ µs. This results in a reduced data-rate of 250 kbps, assuming four-times oversampling. Whereas all this sounds rather straightforward, this is where the circuit design challenges begin. If the on-time of our receiver is only 10 ns, this means that within this time, the circuit has to be operational for a long enough time to be able to reliably receive the symbol. That is, the required turn-on time of every block in the receiver must not be bigger than only a small fraction of this time. For now, a turn-on time of below 1 ns seems like a reasonable specification for the receiver blocks.

12.4.1.1 Receiver

The most important block in the receiver is the LNA, and thus this was the first block to be designed. A two-stage cascode amplifier with transmission line matching was chosen for this task; a layout picture is provided in Fig. 12.6.

The circuit was submitted for fabrication in a cutting edge 22 nm Fully Depleted Silicon-on-Insulator (FDX) Complementary Metal Oxide Semiconductor (CMOS)

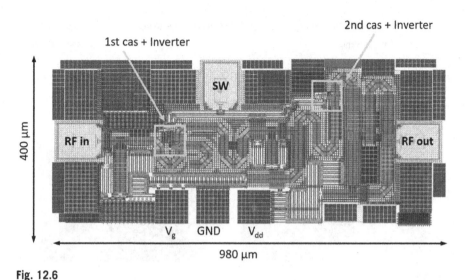

Fig. 12.6

Layout of the low noise amplifier (LNA).

technology from Globalfoundries and covers an area of around $0.4\,\text{mm}^2$. The silicon has been received at the beginning of 2020 and first measurement results were promising. However, the complete characterization is still pending.

12.4.1.2 Transmitter

On the transmitter side, the concept of duty-cycling can be implemented directly, since we always know when we want to send data. We only have to make sure the transmitter does not consume power when it is not transmitting. Since the amplitude of the transmitted signal is quite big, we really have to optimize the transmitter for efficiency. For very efficient PAs, the class-E topology is a promising approach. The authors of [904] show that this architecture allows being directly used for designing an oscillator with high output power, thus eliminating one block from Fig. 12.4. For this project, a possibility for switching off the Direct current (DC) current has to be implemented. Furthermore, the possibility to directly modulate the output in an OOK manner is needed.

12.4.1.3 Positioning

To accurately measure the users position within its environment, a variety of technologies have been reported in the past. To have precise positioning measurements at both low latency and low power is still a challenge. Primary pulse radars promise to accommodate all these requirements, while at the same time allowing to detect the user without the need of additional devices at the body. In our work [905], we show that it is possible to generate these pulses with extremely low-power requirements. With this pulse generator and the specially designed receiver, we have shown that persons can be detected with a latency of only $100\,\mu\text{s}$.

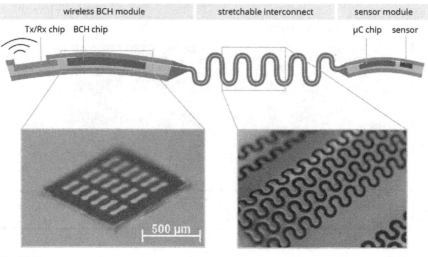

Fig. 12.7

Concept of heterointegration of BCH- and sensor module with embedded chips and stretchable interconnects (upper picture): embedded test-chip in polymeric chip-carrier using 3D-printing process and screen-printed horseshoe interconnects (lower pictures from left).

12.4.2 Electronics hetero-integration

To integrate the developed electronics with the human body, heterointegration of different materials will be researched. To achieve the requirements of TaHiL, common semiconductor electronics have to be integrated with flexible and stretchable materials to be worn on the human body or integrated in robotic systems. This will enable an unimpeded cooperation between human and robots, even in close contact. At the same time, electrical energy has to be provided reliably to the BAN and the BCH. Because of the huge number of existing sensors and actuators and the need for customization, for the integration of such heterogeneous electronics, cost-effective fabrication methods should be considered from the very beginning. Therefore Additive Manufacturing (AM) methods are considered for integration of sensors/actuators with computing and transceiver electronics. We will cooperate on the heterointegration of wearable very thin BCH and sensor modules with stretchable or flexible interconnects, as shown in schematic picture in Fig. 12.7.

The aim is to modify a manufacturing process with suitable AM-steps to combine rigid, flexible, and stretchable assemblies with end-to-end continuous reliable interconnects and contacts. We developed a novel concept for chip embedding consisting of face-up chip mounting in film laminates, using laser patterning of vias and sequential building of metal/dielectric multilayer for signal redistribution. The embedding process will result in miniaturized stiffened fan-out chip package islands of low thickness (below 100 μm), which will be investigated for extension towards the high frequency embedded chip-carriers for wireless transceivers and BCH. The simultaneous components embedding and chip-carrier fabrication using 3D-printing and molding processes and subsequent deposition of metal contacts are investigated for very thin miniaturized assemblies. In Fig. 12.7, a test-chip embedded in polymeric chip-carrier using 3D-printing process is depicted. These miniaturized transceiver and sensor modules are to connect with each other using stretchable interconnects. Printing of flexible and stretchable conductive tracks using inks or pastes based on stretchable polymeric matrix (Polydimethylsiloxane (PDMS), Polyurethane (PU), Thermoplastic Polyurethane (TPU), etc.) filled with conductive particles (carbon, graphene, nanotubes, flakes, etc.) will be developed. In this way, a hundred percent stretchability has been already reported in literature, however, reliability of such solutions is still very limited. For the aimed smart skin heterointegration cases, a more reliable stretchability in the range of 10–25% should be sufficient for safe applications. Another approach to achieve bendable interconnects, which will be exploited in the future, is the usage of the mechanical flexibility of very-thin materials. By dedicated design of conductive paths to, e.g., wave- or horseshoe structures and embedding of metal tracks into polymer matrix, high stretchability will be evaluated. This approach leads to results of inkjet printed silver nanoparticle structures on polyurethane substrate, which has been stretched for 10% and 20% for more than 400 cycles with alternating resistance around an average (to be published in September 2020 conference). Further miniaturized feature sizes will be achieved using unconventional inkjet technology. The electrohydrodynamic printing technology with the printer of the company SuperInkjet Technology allows manufacturing of conductive tracks of only few micrometer and additively via filling [898]. The biggest challenge next to providing stretchable but reliable interconnects, which ensure stable supply voltages for the transceiver, will be connecting reliably rigid and stretchable materials.

12.4.3 Body computing hub

12.4.3.1 Overview

The BCH will be a multiprocessor system-on-chip targeted for real-time, mobile applications in the tight energy budget of wearable computing. Its system architecture concept is shown in Fig. 12.8. The BCH will allow preprocessing and compression of both analog and digital sensor information. Analog sensor inputs, e.g., originating from tactile sensors, will be converted by low-power Analog-to-Digital Converters (ADCs). All sensor information can be fed into a preprocessing module, which allows for filtering, initial classification, and compression. In particular, temporal sparsity of information, e.g., in slowly changing signals, will be exploited for a significant reduction of processing operations. Encoding and compression schemes will be researched that allow for efficient hardware realization, but at the same time effectively compress the sensor information.

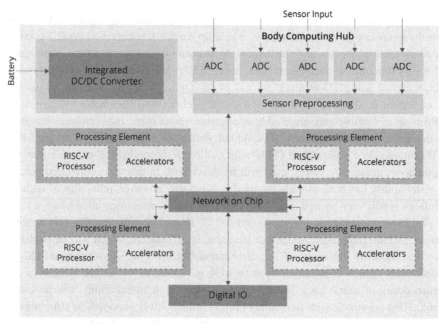

Fig. 12.8

Architecture of the BCH chip.

12.4.3.2 Processing element

Preprocessed and aggregated sensor information is processed by the Processing Elements (PEs) of the BCH. These consist of a RISC-V processor and dedicated hardware accelerators, which are memory-mapped into the processor's address space for easy-to-use software integration. The processor performs all general-purpose programmable tasks, such as system control, task scheduling, and communication with the outside world via a digital interface to the BAN transceiver. All communication inside the system will be done via a light-weight network-on-chip.

For achieving sufficient processing power in the tight power budget of wearable computing, emphasis is put on hardware accelerators for machine learning and sensory processing. Specifically, neural network accelerators will be developed that utilize the temporal and structural sparsity of the preprocessed sensor information. For this, encoding and compression schemes from the sensor preprocessing will be further exploited in neural network layers. Initial studies showed that sparsity greatly varies dependent on the sensor input and the neural network structure. These varied between less than 10% and approx. 90% for activations in different network layers, and almost no sparsity ($< 1\%$), and more than 95% sparsity for weight matrices. This wide range poses a challenge in exploiting sparsity memory-efficiently. As a first step, coding schemes for hardware-efficient storage and retrieval of sparse weight matrices are developed and evaluated regarding their compression potential. Furthermore, they are assessed with respect to efficient hardware implementation. Complementary to this, neural network training is adapted to increase sparsity in weights and compensate for lossy compression schemes for input activations. These investigations will also integrate work in K1 on haptic codecs, which potentially open up very powerful ways of performing sensory processing on compressed data. As a first step, research will be put into hardware-efficient realization of haptic codecs via dedicated hardware blocks.

Memory for storing state variables and weights is a limiting factor in embedded devices, such as the BCH, even at comparatively small sizes of a few 100 kByte dominating silicon area and significantly contributing to leakage power. The latter is especially critical, as it results in a constant power draw, irrespective of the operations performed on the system. All the research on compression and sparsity exploitation described above is instrumental in allowing more powerful sensor processing, e.g., with more sensory inputs and bigger neural network-based classifiers, on the BCH.

12.4.3.3 Implementation considerations

Previous work on ultraefficient processor systems will be combined with new design implementation strategies for leakage and dynamic power minimization in a battery-powered system. The goal here is to bridge the gap between the design implementation, i.e., concrete synthesis and place-and-route runs, and the application scenarios. By combining information of both, the final implementation of the BCH can be steered towards the minimum energy point for the targeted operation, where a close-to-optimum trade-off between leakage power and dynamic power is achieved. For allowing a compact, energy-efficient overall system, DC/DC converters will be integrated on-chip for providing the necessary core voltages from a single supply battery.

It is expected that sparsity exploitation will result in varying work loads of the processors and hardware accelerators. In the first place, this will make the system dynamically fluctuate around its minimum energy point and lower its overall energy efficiency. However, these workload variations can be exploited by dynamic voltage and frequency scaling schemes, which adaptively change supply voltage and clock frequency of processing elements dependent on the current processing requirements. This allows running the system close to its energy-optimal operating point, even at varying work loads. Integrating dynamic control of the DC/DC converters into the overall dynamic power management concept of the BCH will help in further increasing energy efficiency of the overall system.

12.4.4 Energy-aware runtime management

The need for energy-conscious designs arises for battery-powered embedded devices. The development of applications and the management of the available resources have to be adjusted to this key requirement. Further requirements as real-time, performance, and reliability restrict the proper design of applications on processing elements. Therefore power-saving techniques have to be performed to minimize energy consumption at run-time based on their workload and operating conditions. One such technique is the Dynamic Voltage and Frequency Scaling (DVFS).

This technique has to be part of the running software. The software manages the different types of tasks, such as the power management unit on the processor. The main problem is that tasks cannot be executed concurrently, and therefore have to wait for the allocation of CPU resources to launch their execution flow. However, a scheduler coordinates the execution flow and allocates Central Processing Unit (CPU) resources for tasks in the software. Thus a lightweight Real-Time Operating System (RTOS) has to be performed on the processor. It has a different type of scheduling scheme considering priorities that has to allocate CPU resources to the tasks in the correct order with context switching and maintain the real-time requirements for the running application.

When a context switch occurs, the running task pauses and saves its states in the stack. The scheduler selects another task from the list and restores its states. Afterward, the new task resumes its execution. The context switching is part of the scheduler in the RTOS kernel. To have an energy-aware context switching, the power management unit has to be integrated into the RTOS kernel. After selecting the new task from the list, the power management has to adjust the voltage and frequency of the processor and dedicated hardware accelerator. Therefore a different type of energy-aware scheduling algorithm has to be explored and implemented in the RTOS kernel.

The continuous context switching has also an impact on system performance. To enhance the performance and guarantee more reliability, some parts of the RTOS kernel have to be implemented as dedicated hardware accelerators. It alleviates the context switching, while executing the application tasks. The RTOS kernel is able to adjust faster the voltage and frequency pairs for the tasks.

An application has to be evaluated with the defined constraints regarding the feasibility on the aforementioned hardware and software architecture. Due to the complex design of the system architecture, it is valuable to explore the whole design in a simulation environment. This type of design exploration has the advantage that the system architecture can be easily adapted to the changing design specifications and estimate, for instance, the power consumption, based on the recent workload and operating conditions. Thus the overall energy consumption can continuously be kept at a minimum with the adapted RTOS design in the simulation.

12.4.5 Low-power signal processing

The BCH will provide the hardware basis for a high-quality multimodal feedback system with adaptive decision-making capabilities as developed in K4. Its ultralow latency hardware adaptation features will allow the system to dynamically recognize and interpret the inputs from different modalities, and provide multimodal outputs for various human senses, forming a learning "awareness cloud". New multivariate signal processing algorithms will be developed for exploiting the adaptation features of the BCH. Machine learning algorithms will be customized to the mathematical characteristic of the hardware elements of the BCH to minimize the power consumption during a real-time application of the sensor-processor structure.

In this context, different approaches for the assessment and classification of biosignals are implemented and compared. The first method is based on strict separation of feature extraction and a subsequent classification. Typically, linear and nonlinear features are derived in the time and frequency domain, with a strong focus on parameters of multivariate autoregressive processes and features from nonlinear dynamics. It has already been shown, that such a signal processing is real-time capable when implemented on a highly parallel cellular processor array [906].

The second approach relies on the application of deep neural networks—typically convolutional neural networks—for the extraction and identification of features from multivariate biological time series. In intracranial Electroencephalography (EEG) analysis, these topologies already outperform the more traditional methods using "hand-crafted" feature vectors [907,908]. However, an evaluation of their real-time capability when implemented on an integrated low-power devices is still an open task.

12.5 **Collaboration**

From the start, there was an intensive collaboration on the codesign of the textiles and the electronics with the respective target primary research group TP2. As a first prototype, the glove with resistive sensors was chosen. To start with the textile/electronic integration as soon as possible, we decided to design a wireless sensor platform consisting of commercial components. This platform should provide already more or less the functionality of the final platform, however, it is of course a lot bigger. In Fig. 12.9, we can see a first joint prototype, consisting of a glove with resistive sensors developed within TP2 and a battery powered, wireless sensor platform from TP4. This early joint prototype will allow us on the one hand to define and verify the interfaces between electronics and textile sensors, and on the other, allow first user tests and feedback on the user interface. There has been a close collaboration on the use-cases both within the use-case research groups U2 and U3; refer to Chapters 3 and 4 for details. In the future, there will be an intensive collaboration with TP3 and TP5 on wireless communication and algorithms, respectively. However, this will not start until hardware components are available.

12.6 **Conclusion and outlook**

In this chapter, the mission for designing the underlying hardware for eGlove and eBody has been presented. The system architecture is finished and entails an embedded processing platform, a fully integrated transceiver, and the integration on flexible substrates. The objectives for the fully integrated transceiver are ultrahigh energy-efficient electronics, optimized antenna patterns, and wireless links combined with

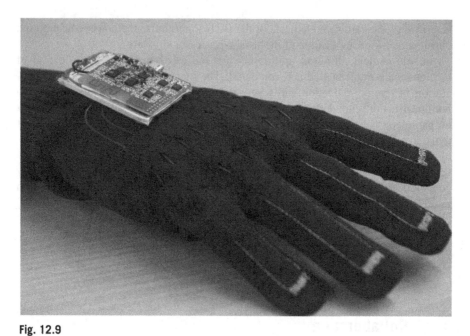

Fig. 12.9

eGlove prototype with wireless sensor platform.

low latency and resilience. The goal here is to design an ultrasmall and highly integrated mm-wave transceiver and antenna system on chip. Both operating system and algorithms have to be codesigned with the hardware to fully support the specialized features and comply with the tight power budget.

A first version of the eGlove has been designed using commercial components. Whereas this is still very far from our goals in terms of size, flexibility, and power consumption, it allows us to already start to integrate this with the textiles. It will also be used as the testbed to characterize the textile sensors to optimize the sensor interface in future versions. Most importantly, a usable hardware platform is already available for cross-discipline research projects to gather feedback from the human trials and optimize the platform.

Tactile computing: Essential building blocks for the Tactile Internet

13

Uwe Aßmann[a], Christel Baier[a], Clemens Dubslaff[a], Dominik Grzelak[a], Simon Hanisch[a], Ardhi Putra Pratama Hartono[a], Stefan Köpsell[a], Tianfang Lin[a], and Thorsten Strufe[b]

[a]*Technische Universität Dresden, Dresden, Germany*
[b]*Karlsruhe Institute of Technology, Karlsruhe, Germany*

The whole is greater than the sum of its parts....
– Aristotle

13.1 Introduction

Tactile Internet with Human-in-the-Loop (TaHiL) will comprise a large variety of tasks at different complexities and a variety of devices and machines for their execution. Implementing TaHiL use cases as detailed in Chapters 2, 3, and 4 therefore requires many different building blocks from various research disciplines. In this section, we focus on the computer-science aspects of Tactile Internet applications and present five building blocks that involve different research areas. Together they build the foundation of the TaHiL computing infrastructure: Cohabitation of humans and machines implies that humans can understand, predict, and inquire about the behavior of the machines, and likewise that the machines have a general understanding of the world in which they are interacting with the humans. Envisioning close integration of humans and machines, TaHiL computing infrastructure has to guarantee safety, security, and privacy of the users and their data. Representing the cognitive world model of the users and the interaction space of the machines, reacting to events, and processing the data at scale and very low latency are key requirements.

Based on these requirements the following five main research objectives will be addressed: (*i*) Safe and secure infrastructures: TaHiL applications and infrastructure must guarantee the safety of the users at all times. Data and system integrity have to be maintained. The close integration into the daily lives of users demands their privacy to be retained. (*ii*) Scalable computation: Considering the heterogeneity of tasks and hardware, the platform needs to recognize and match resources and demands, to allocate machines (considering their architecture and location in face of the characteristics of tasks), and to guarantee responsiveness at very low latency. (*iii*) World capturing and modeling: Both humans and robots act in three-dimensional environments over time, which have to be captured, recognized, and immersively augmented for human–machine interaction, as well as efficiently modeled for prediction and planning. (iv) Context-adaptive software: Humans need appropriate tools and languages to intuitively teach and program machines, such as robots without the need for extensive learning or prior competences relying on the time-spatial context of human commands and interactions. (v) Self-explanation: In a human–machine coadaptive setting, humans and machines have to be continuously well-informed about the current situation, to understand the reasons for automatically triggered events, as well as to capture the causes and potential risks in critical situations. This also requires tools that intuitively explain machine behaviors at different levels of abstraction for humans and machines. This requires a software infrastructure that depends on the context of a scene, i.e., what human actions, such as gestures, Augmented and Virtual Reality interactions, audio commands, or demonstrations mean, and how these commands and interactions are translated to programs.

In the following sections, we will describe how we will approach our research objectives. In this regard, every section has the same internal structure: based on the state of the art, we present the general approach, the goals, and our methodology. We will identify research questions and formulate our hypotheses as well as expected results. We elaborate how we plan to answer our research questions and give a related roadmap.

13.2 Safe and secure infrastructure

In recent years, many cases of safety-critical systems have shown up around us. The Tactile Internet, including the automation of industry and healthcare systems, will become vital as they are relied upon by many people in their daily life. These safety-critical systems must meet two properties: safety and security [909,910]. Safety and security used to be two different topics with two different communities working on them. However, nowadays, both are intermingled with each other. There is no safety without security. This means that if security cannot be guaranteed, then external adversaries can compromise the system to make the system unsafe.

When the Internet was first designed its security was of no concern due to its limited scope and the limited number of participants. As the Internet grew and became the critical infrastructure it is today, this oversight has been manifested in its insecure design, which makes adding more security inherently hard. Now in the design of the Tactile Internet, we must learn from this mistake and consider security from the start. When comparing the Tactile Internet to the Internet, we find that it differs in two main regards, that are relevant for security: First, it will require low latency and second, it will transport large amounts of biometric information.

The low latency will be a challenge for network security methods, such as secure network tunnels and packet filters, because both of them induce latency overheads on network traffic. Furthermore, both measures are mostly optimized for throughput and not for latency. The low latency demands will also require the processing servers to move closer to the users and therefore to the network edge. Moving the processing out of large data centers exposes it to higher risks, because individual nodes become smaller, and are hence less capable of fending off denial of service attacks on their own. In addition, the increase of locations increases the risk of physical tempering, reducing the level of trust of the computing hardware.

The biometric information transported and processed by the Tactile Internet is privacy-sensitive, as it allows the identification of individuals. Additionally, the information can be used to infer personal characteristics, such as gender or age. For the Tactile Internet not to become a privacy nightmare for its users, it is imperative to build the tactile infrastructure in a privacy-preserving manner. The most important question will be if the private aspects of the tactile data can be anonymized, while the utility of the data is preserved.

On the safety spectrum, most of the challenges are coming from bitflips, either in the Central Processing Unit (CPU) or memory region. Bitflips can lead to not only information leakage, unauthorized access, and stolen secrets, but also undetected computation errors called Silent Data Corruption (SDC) that may introduce catastrophic, unpredicted events in a safety-critical system [911]. Therefore protecting against bitflips is the first step toward safe applications.

13.2.1 State of the art

One attempt to guarantee security on untrusted hardware is by enveloping the intended application in the Trusted Execution Environments (TEEs), such as Intel Software Guard eXtensions (SGXs). Intel SGX is in instructions set that add hardware support to differentiate trusted and untrusted code to run in a specialized-secure environment [912].

Although Intel SGX already came to the market in early 2015, utilizing its feature is quite some work for developers since they have to refactor the program's source code. Several frameworks tackle this issue, for example Graphene [913] and our previous work, SCONE [914]. SCONE is a framework to run secure applications inside an enclave, a secure environment based on SGX [914]. SCONE enables an unmodified application to take advantage of the isolation introduced by SGX. SCONE is compatible with many programming languages and only requires the program to be recompiled against the SCONE custom library.

In system safety, the one that produced acceptable performance usually comes from mixing both software and hardware approaches [915–920]. HAFT is a software hardening approach that provides low-cost detection and recovery using Intel Transactional Synchronization eXtensions (TSXs). Mixing software-based fault tolerance with various hardware, such as microcontrollers [917], commercial off-the-shelf processors [921], and commodity hardware [922], has been rising to ensure program safety. All of those researches provides fair detection rate and acceptable performance penalty. Some of them also can recover from faults.

In our previous work SeCoNetBench [923], we measured the performance overhead of secure network tunnels and packet filters on container networks, concluding that packet filter only induces minimal overhead, whereas secure network tunnels add between 100 and 600 nanoseconds of overhead to network traffic. This shows a need for improvements in the state-of-the-art of secure network tunnels regarding latency. Related works, such as [924] *Less is More: Trading a little Bandwidth for Ultra-Low Latency in the Data Center*, suggest that trading some of the bandwidth could decrease the latency of secure network tunnels.

With its unique position in the network, edge computing has spawned new ideas for combating distributed denial of service attacks, such as [925] *Towards Internet of Things (IoT)-Distributed Denial of Service (DDoS) Prevention Using Edge Computing*, in which the authors suggest filtering DDoS attacks at the network edge. However, so far, little work has been done on protecting edge computing itself from DDoS attacks. Traditional DDoS defenses require a large number of servers and enough bandwidth to filter incoming DDoS traffic, both of which are not available at the network edge.

The usage of kinesthetic information for identifying individuals has been a research topic for a while now, and gait recognition [926] is already in widespread use in surveillance systems. Furthermore, the usage of hand motions for identification has been proven in [927]. Whereas it is clear that kinesthetic information can be used to identify persons, the privacy-preserving processing of this information has not yet been studied. Approaches, such as privacy-preserving machine learning using differential privacy [928] look promising for adaptation.

Despite the active attempts of both safety and security on strengthening their fields, to the utmost of our knowledge, there is no research that considers both safety and security aspects in a one-integrated solution. Most, if not all, security providers assume that the hardware beneath their system is reliable. We argue that this issue is incomplete if only seen from one side. Therefore the need of ensuring the safety and security of applications is crucial for not only the Tactile Internet, but also other safety-critical systems.

13.2.2 Key challenges

Achieving security at a low-latency overhead will be one of the key challenges for TaHiL. Security mechanisms, such as secure network tunnels and packet filtering add to the latency of packets as they require additional cycles to complete. It is therefore essential to push down the overhead of such methods to a minimum, while keeping an adequate level of security. Furthermore, TaHiL will require these solutions to be flexible and adapt to new situations on the fly, making automatic reconfiguration and scalability additional challenges.

The hosting infrastructure of TaHiL is faced with a challenge in trust since edge computing will move applications out of trusted data centers to less trusted locations at the edge of the network. This requires the hosting infrastructure to detect and prevent unauthorized changes to applications and data. In addition, the edge computing paradigm will most likely add more to the tangle of infrastructure providers, hosting providers, and service providers. For the user, it is no longer evident who has access to its data and applications. To reestablish trust and accountability, the user requires methods to verify the correct execution of applications and the confidentiality of its data.

The applications running on TaHiL will collect and process biometric information about their users. This information is privacy-sensitive, as it can be used to identify individuals and infer additional personal information, such as age, gender, or sexual preferences. It is therefore necessary to anonymize the biometric data to ensure the privacy of the users. Anonymizing data often has a negative impact on the utility of the data in question. Striking a good trade-off between privacy and utility is hence another challenge for TaHiL.

13.2.3 Beyond state of the art

For the foreseeable future, we want to explore the following research questions: (*i*) How can we design DDoS protections for edge clouds nodes?; (*ii*) Can we use secure enclaves to guarantee the safe and secure execution of applications?; (*iii*) How can we optimize the throughput-latency trade-off of secure network tunnel for edge clouds?; (*iv*) What are good anonymization strategies for biometric information in the Tactile Internet?

The latency requirements of the Tactile Internet make it very challenging to ensure its availability when faced with a DDoS attacker. We seek to develop DDoS protections of edge computing to preserve the latency of connections. Our main design goal is to prevent the generation of DDoS traffic in the first place, making costly DDoS protection unnecessary.

Reflecting on safe and secure properties on applications in the domain of the Tactile Internet makes it evident that finding an approach to ensure the safeness of a program inside an isolated environment is imperative. Enclave, which is one of the isolated environment concepts implemented in SGX hardware, will be our main target since it focuses more on security. On top of that, we are obliged to ensure the safety of both user data and program execution in the edge computing environment.

A safe and secure environment would not come without any trade-off. Investigation of what is the trade-off of such an environment is essential. Exploration on metrics and notions that might be related and have a significant impact on performance, safety, and security must be conducted to fully understand the system. We believe this will be beneficial not only in the short term, but also in an extended period after the project is ending. We also wish to explore the trade-off between latency and throughput for secure network tunnels to decrease their latency. We plan to do so by building a secure network tunnel that is configurable in regards to its throughput latency trade-off.

Tactile and kinesthetic information is biometric information that requires adequate anonymization methods to protect users' privacy. At the same time, we want to keep the utility of the data for applications as high as possible. We want to develop new methods to anonymize tactile information building on top of existing approaches, such as k-anonymity and differential privacy, while ensuring the processing is both safe and trusted.

13.2.4 Synergistic links

Safe and secure infrastructure is a core building block for TaHiL to be built upon, and must be a design goal from the start. Hence, we will be involved in developing the intelligent networks (K2) infrastructure early on. Furthermore, we plan on contributing to the development of the use cases as they handle privacy-sensitive information. The research groups industrial applications (U2) and Internet of Skills (U3) will especially require privacy and security protection since data in those use cases will be shared with third-parties. We expect the applications employed in research group medical applications (U1) to be fault-sensitive so that the safety guarantee is crucial.

13.2.5 Intermediate conclusion

Safety and security undoubtedly touch most of the topics of TaHiL. In our research, we try to provide adequate solutions that offer a safe and secure environment, while keeping the balance between data privacy and utility. Since adding those at the end may introduce vulnerability and performance overhead, they must be designed, modeled, and integrated from the very start, beginning with the design and development of TaHiL. We believe our infrastructure will enable many TaHiL use cases to be trusted, safe, and interactive in a straightforward way.

13.3 World capturing and modeling

In the past several decades, Three-dimensional (3D) scanning technology has been applied in a wide range of fields, such as heritage preservation, healthcare, industrial design, etc., which is the main way to capture and model the real world. Currently, with the popularity of commodity depth sensors, real-time 3D scanning has big potential in the fields of robotics and human–computer interaction. For 3D capturing and modeling in the Tactile Internet with Human-in-the-Loop, combination of 3D scanning and immersive environment would be very promising. Immersive scanning reconstructs the 3D scene by using fully parallel algorithms to guarantee low latency. Humans are presented with a multimodal experience in an immersive Virtual Reality (VR)/Augmented Reality (AR) environment. They can interact with machines and the 3D scanner directly through controllers and sensors. We expect that our immersive scanning technology will significantly improve the immersion for humans and allow for new coworking approaches between machines and humans based on the captured model of the real world. Immersive scanning will serve as the foundation for a variety of Human-in-the-Loop systems in the Tactile Internet.

13.3.1 State of the art

The first 3D scanning technology was developed at the end of the 20th century. The initial 3D scanners in the 1980s used contact probes. With the advance of technology, laser scanners, Computed Tomography (CT) scanning, photogrammetry, and

Red Green Blue and Depth (RGB-D), cameras became more and more efficient and accurate. Laser scanners collect millions of points per second in three dimensions, generating detailed point clouds, which can be used in a variety of applications, ranging from large infrastructure to small objects. In industry, the most commonly used method for precise measurement of 3D surfaces is based on structured light techniques. To overcome the limited field of view, several scans are merged together. On the consumer-level color and depth (RGB-D), cameras are now widely available and affordable to the general public. The main types of RGB-D cameras are based on structured light (e.g., Kinect v1, PrimeSense), Time of Flight (ToF) (e.g., Kinect v2), or Stereo cameras (e.g., Tango).

With the popularity of RGB-D cameras, in 2011 [929], the real-time 3D reconstruction system Kinect Fusion was proposed. It stimulated follow-up research [930–933] aimed at robustifying the tracking and expanding its spatial mapping capabilities to larger environments. Zhang [934] proposed a method to obtain handheld object 3D model by using a Deep Neural Network (DNN) to segment hand and object in real-time. These above-mentioned systems greatly improved the effectiveness and accuracy of 3D reconstruction, but the resulting models cannot directly be used in interaction. Recent work started to study the combination of VR and scanning: Stotko et al. [935] designed a VR system for immersive robot teleoperation and scene exploration within live-captured environments for remote users based on virtual reality and real-time 3D scene capture. The approach in [935] is limited to the exploration and measurement of 3D scenes. For immersive interaction, more advanced functionality, such as annotation and tracking of dynamic objects need to be developed. In summary, to allow for seamless interactions between humans and machines, an immersive scanning system will be the foundation.

13.3.2 Key challenges

For 3D capturing and modeling in TaHiL, there arise two new challenges. Firstly, the question of how to incorporate tactile and kinesthetic information in the interaction between humans and machines. Secondly, low latency for transmitting a large amount of information over a long distance is required.

To conquer the above-mentioned challenges, innovative 3D scanning techniques need to be invented for the Tactile Internet. On the one hand, to present a 3D scene in a plausible and immersive way not only visual and acoustic, but also tactile, kinesthetic sensations need to be provided to the user. On the other hand, low latency demands real-time faithful 3D reconstruction for scenes. Moreover, for seamless interaction with arbitrary objects in the scene, real-time object segmentation, labeling, and classification are also necessary.

In light of the foregoing, an immersive 3D scanning pipeline for world capturing and modeling will be designed in TaHiL. The immersive 3D scanning will support real-time capturing and reconstruction of a real-world model that allows for transmission with low latency. Also, it will provide an immersive environment, in which humans can interact with other humans, machines, and their environment. Besides visual and acoustic feedback, kinesthetic and haptic feedback is also essential for a natural interaction. For these objectives, the basic requirements are described respectively as follows:

Real-time capture Due to the rapid increase of network speed, human and machines need to obtain and control 3D models remotely. Likewise, the scanning area and position should be updated with very low latency. As a result, it is necessary to implement a real-time method for world capturing and modeling.

Portability 3D scanner should be as small as possible such that it can be deployed on smart-phones or wearable computers, etc.

Affordable price As the envisioned applications on the Tactile Internet address a huge number of potential users, 3D scanning technology needs to be cost effective.

Color acquisition Besides increasing the realism in a virtual environment, color information can boost the performance of computer vision techniques, such as segmentation, classification, and labeling of the 3D objects.

Immersive experience 3D world models need to be renderable in stereo with frame rates that never drop below 60–90 Hz when being used in immersive environments. Furthermore, acoustic, kinesthetic, and haptic sensations need to be synthesized in parallel with high demands on synchrony.

13.3.3 Beyond state of the art

To push state of the art further and fulfill above basic requirements, we propose a multilayer immersive 3D scanning framework as shown in the Fig. 13.1. The first layer is the 3D scanner layer that supports capturing raw 3D data from portable RGB-D camera, registering 3D scans into a global coordinate system and robust tracking in real-time. The rendering layer reconstructs and renders the 3D world model in real-time for the immersive environment. On the interaction layer, a variety of interacting operations especially based on acoustic, kinesthetic, and haptic sensations will be provided with immersive experience. Furthermore, a large amount of user study will be conducted to refine the framework to achieve a high performance on handheld, real-time 3D scanning, and real-time rendering techniques.

On the top layer—the application layer—we plan to support object labeling, segmentation, and classification in the immersive environment. For this we can build on existing machine learning-based computer vision approaches [936–938]. A current drawback with these methods is their execution time, which needs to be reduced for the operation in an immersive environment. Moreover, we will track the objects in the scene, aiming to allow recording of skills in an unprepared environment.

Based on the state-of-the-art and current trends in the development of enabling technology, the following open questions will guide our research: (*i*) Low latency is critical to the Tactile Internet. To address the low-latency requirement, we want to explore the relationship between scanning accuracy and latency. Based on the result, parallel computing and compression of 3D models will be taken into consideration.

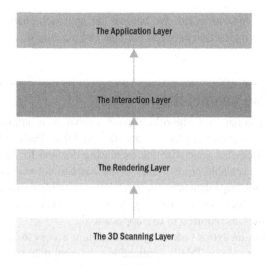

Fig. 13.1

Multilayer immersive 3D scanning framework.

(*ii*) In the Tactile Internet, haptic and kinesthetic information will be added to 3D objects and scenes during the process of immersive scanning. It is currently unclear how to design a corresponding transfer format that fits to the demands of the wide variety of application domains. (*iii*) Through the immersive environment our methods are human-centered and, currently, it is unclear how to design immersive scanning systems that can adapt to the specific needs of different humans.

13.3.4 Synergistic links

World capturing and modeling is a foundation of other research of TaHiL and offers versatile potentials for the same. To let humans interact with other humans, machines, and environments, our immersive scanning framework will provide an interactive immersive environment for U1 (Medical applications), U2 (Industrial applications), and U3 (Internet of Skills). On the other hand, for natural interactions and acoustic, kinesthetic, and haptic sensations, we will build on sensors and actuators from TP2 as well as the communication technology developed in TP3.

13.3.5 Intermediate conclusion

In summary, world capturing and modeling plays an important role in TaHiL, and a new immersive 3D scanning framework for world capturing and modeling is proposed. Immersive 3D scanning reconstructs high-quality immersive environments and provides a variety of interactive operations for related research of TaHiL. It extends the space of possibilities in TaHiL, and provides a way to seamlessly bridge the virtual world and physical world. We believe our approach enables research topics that require remote interaction between humans and machines.

13.4 Scalable computation

While scalable computation algorithms have shown their applicability, e.g., for compute-intense simulation tasks for decades in High Performance Computing (HPC), new requirements arise with the increasing demand for large-scale learning applications. Here, scalable computation is the basis to solve either (*i*) large-scale problems processing huge data sets, for example, in data intensive applications, such as deep learning models, or (*ii*) needing good strong scaling capabilities to reach short response times, as required in IoT applications.

Nowadays, data-analytics tasks in general require access to large, complex, heterogeneous, and, in case of TaHiL applications, external data sources, which might even change over time. Especially for highly parallel applications, this imposes an additional stress in the efficient execution of parallel tasks, due to an increased communication delay between distributed compute elements and storage components, which result in complex and not well predictable communication patterns. On the hardware layer, this is partly caused by large differences in access times to different parts of the compute system, such as distributed nodes, storage elements, or remote sensor systems.

13.4.1 State of the art

To relax this situation and avoid the degradation of parallel performance, concepts, such as the semiautomatic movement of data prior to its computing closer to the relevant compute elements, have been proposed to avoid long waiting times imposed by data access. Therefore previous work focused on data staging in multilevel storage hierarchies within HPC environments [939] with an HPC-driven focus to support data-intensive proven scalable simulation tasks. Here, one promising option is to include nonvolatile memory (NVME-DIMMs) [940] as new memory hierarchy layer in the programming model to reduce access times to remote storage locations.

In general, an important requirement for scientific computing is the incorporation of measurement or observation data in complex and large-scale analysis workflows, or in combination with extensive simulations [941]. Data is no longer generated exclusively in central systems, but also partly remotely in external production facilities or sensor systems. To process such data efficiently and to extract knowledge, e.g., new dependencies or relation between data instances, they often have to be pre-processed, aggregated, or reduced (filtered) prior to the analysis steps. Furthermore, the increasing need for fast processing of data that is not generated in central systems (for instance, edge computing, IoT), but whose analysis requires high computing power, requires new methods. Here, in the HPC domain, besides the prior data movement, no general work schemes have been established, leaving this challenge still to the user.

The aim to reduce or preprocess large amounts of data in a timely manner, in parallel, and to include this data in the analysis process was a driving element in the development of big data frameworks, which arose from the rather simple MapReduce approach [942]. Nonetheless, the handling of variable, and more important not reliable, data-access operations requires the temporal storage of large amounts of data before it can be processed holistically. Distributed computing concepts for batch, and nowadays more interestingly for stream data analysis, have been proposed to provide methods for shaping the data stream as preprocessing capability in data management systems, such as Apache Spark [943], Flink [944], or Kafka [945]. With that, methods are available from the users perspective to define and adjust data stream processing semantics as input for various analytics scenarios. The user can concentrate on the data transformation and analysis tasks leaving the distribution of the workload to the framework routines. From the performance perspective and for very time-critical applications this is a drawback, since the user methods to steer those transformation and the degree of parallelism is not straightforward and not transparently supported by said frameworks.

Finally, the topic of detailed performance monitoring and analysis is a long established one in HPC, because it is imperative to pay close attention to the efficient use of parallel resources and good scalability. There is a set of well-understood methods for instrumentation, run-time data collection, and specific analysis techniques, as well as high quality software tools that implement them for all relevant programming languages and parallelization models in HPC [946]. They have also been extended for Java and languages relying on the Java Virtual Machine (JVM) to be extended onto big data frameworks for better control of the distribution of tasks [947]. They need to be extended towards relevant communication protocols as well as edge computing and IoT programming languages, since the monitoring and investigation of performance bottlenecks of heterogeneous parallel applications become increasingly complex analytics problems themselves [948].

13.4.2 Key challenges

Based on TaHiL application requirements and the current state of the art in technology, a major key challenge is to efficiently map analysis workflows in the IoT area on HPC infrastructures for cases, where large compute capacities are required. Here, the typical HPC processing models needs to be combined with real-time, and on-demand usage patterns and real-time programming models need to communicate with world-modeling applications (dynamic data loads trigger dynamic resource scheduling).

In highly adaptive systems, this includes for distributed computing applications, the mapping of different access pattern in (e.g., scalable learning applications) or the separation of relevant data preprocessing steps prior to analysis, it would be highly beneficial to suggest compute architecture(s) based on requirement analysis for workloads as it relates to workflow steps to instance necessary data movements, compression, or transformation of data (semi) automatically.

Furthermore, programming models need to be expanded to allow a single application to migrate or collaborate on all devices from a low-power mobile device all the way to a highly parallel HPC cluster.

The extension of monitoring and tracing techniques of parallel and distributed applications and workflows, allowing a suitable level of detail when analyzing how well they execute on the heterogeneous hardware landscape is important; analysis of this monitoring data will reveal optimization potentials and account for their effectiveness during evaluation.

13.4.3 Beyond state of the art

One promising approach to realize Tactile Internet applications is to combine highly scalable parallel computing in the data center with distributed and edge computing. The challenge still remains to bridge between the differing computing systems and consequently different technologies used for computing tasks. Furthermore, this requires mapping a complex analysis chain, including data processing across different technical systems and resembles a complex nonstatic workflow, which might impose variable and data-dependent compute demands on the compute system. To cope with different architectures and kinds of analysis steps, a detailed understanding on the different requirements is necessary to ideally map the workflow on the compute and network infrastructure. Here, trace-based workflow and task performance monitoring tools provide methods to map the I/O, communication, and compute demands and hints at performance improvements.

A first step is to closely analyze mixed execution environments from IoT devices, mostly mobile and partly available, continuously available server components (gateways, databases, storages), and HPC processing capabilities, either by addressing batch processing if sufficient, or addressing cloud access patterns providing resources with high availability. To realize a dynamic interaction between systems, combinations of parallel programs in an HPC environment with distributed processing on mobile devices or on the Mobile Edge Cloud (MEC) needs to be investigated. Here, especially the communication patterns and data rates needs to be preprocessed to define a reliable input stream for the central computing component. Depending on the latency requirements, different strategies could be realized, either by direct integration into an HPC cloud with fixed resources, or a more staging-like mechanism, as buffered input stream, e.g., for a machine-learning scenario. Here the challenges still remain on the data side to cope with possible network collisions, varying latency's over time, or interrupts. A trade-off needs to be found for every application between resiliency and low-latency access patterns. In cases where data prestaging is required, separating the data gathering (with few dedicated compute nodes) from the intensive computation (highly parallel) would be beneficial to keep control of the input data stream, while maintaining performance of the compute intense part of the workflow. There are data-staging approaches for HPC, yet not in a flexible way that can be interleaved with traditional HPC processing. To realize the mapping between systems, first of all an overall monitoring scheme alongside the workflow will collect necessary system data for steering the subprocesses, or to take action if some design criteria are not met (e.g., dynamic compute node allocation, rerouting of data streams, etc.).

For complex workflows constructed of central HPC processing and distributed processing or edge computing, the performance analysis needs to look at the combined monitoring data. It must not look at the components in isolation, because this is likely to hide performance effects of one component to another, which also cannot be optimized in isolation. Therefore collecting monitoring data to a central location for combined analysis requires additional data transmissions. This needs to be carefully designed and balanced to allow detailed enough data, while limiting the effects on the primary data transfers by the workflow under observation.

13.5 Context-adaptive software for the Tactile Internet

Owed to the rapid advancements of mobile devices and various gateway solutions located at the network edge, a high number of compute resources equipped with sensors, actuators, and network connectivity, are readily available to gain more value from the underlying increasing connectedness. Prospective mobile devices, wearables, and mobile broadband networks, among others, will lead to a platform shift towards the Tactile Internet, possibly complementary to edge and cloud computing.

Their adaptive context-dependent resource-provisioning nature at run-time mainly characterizes Tactile Internet applications. The development of such large-scale distributed systems that may self-adapt to different contexts come along with several additional challenges, as compared to the traditional desktop application development [182,949,950]. We argue that the context-aware computing paradigm enables software engineering of applications of the Tactile Internet in the first place in terms of autonomously varying context-dependent behavior, where software verification becomes a first-class citizen.

It is widely acknowledged that the variety of emerging sensor technologies and their continued development will play an important role for context-aware systems. In this regard, context sensing is the ability to acquire any information from devices, environments, and humans, compute and embed context in applications for enabling awareness of physical conditions. In essence, the increased number of IoT devices magnifies the sensing density; at the same time, both equally sensor fusion, and machine-learning algorithms enable more precise context processing and detection. Over time, digital information can be translated into actionable knowledge: First, sensors gather raw data; context is then analyzed and transformed into specific knowledge, which is finally translated to concrete actions (see [951]). Actions can even be more personalized and specific to individual environments. Such actions are made directly available to programs and systems to automate Artificial Intelligence (AI) and make applications smarter. Furthermore, such information can be distributed to or requested by other network participants (e.g., applications, mobile devices, or gateways) which *is crucial for reducing the complexity of reasoning in AI applications* [952, p. 563].

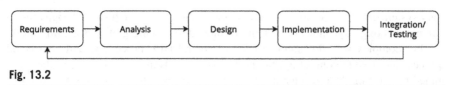

Fig. 13.2

A traditional software development lifecycle.

13.5.1 State of the art

When we look at software development, we might be inclined to associate the developer's daily work processes only with terms, such as programming languages, reusable components, or build tools. However, in the domain of software engineering, the boundaries of development are continuously shifted to cope with the increasing complexity; and, as a result, has driven ahead through innovations by introducing concepts and approaches, such as object-oriented paradigms and modularity. Though, due to the software-related complexity of Tactile Internet applications, most of these established *on-demand* or *ad-hoc* development approaches are a less viable option for safe and context-aware software.

Considering a typical software development lifecycle, as shown in Fig. 13.2, we observe that requirement analysis and design are the earliest phases of development. Instead of having to abide by concrete programming paradigms, we tend to look at context modeling approaches to facilitate engineering of Tactile Internet applications to be independent of the architecture. This is in line with the model-driven approach of context-aware systems in [953], where the authors argue that *context adaptation needs to be handled in the preceding development stages especially when developers face more complicated cases* [953, p. 345].

To perform context-adaptive software engineering at all, the software development process can be supported by two types for realizing the actual implementation (see [211,955]). First of all indirectly, by *architectural frameworks and distributed infrastructures* (e.g., through graphical and interactive interfaces, tool-based, or Application Programmer Interfaces (APIs)) that software engineers can employ to foster the development approaches by providing the necessary frame, easing the implementation process. The other type of assistance comes from a higher level. This direct model usage is achieved by languages or theories [955]. Therein, context models are directly used as the primary artifact for the specification and implementation. In this line, formal context models can be complemented with the model-driven paradigm. Therein, models are the core entities of the specification, implementation, and deployment, mostly incorporating automated processes, such as code generation by model transformation. These two branches are classified by the taxonomy depicted in Fig. 13.3, consolidated and extended after [211,955,956]. Refer to [956] concerning the differences between modeling formalism and modeling language.

Regarding the development process of context-adaptive software, we begin by taking a high-level point of view and consider context models only, thus leaving out the discussion about middleware-related or architectural approaches, which are set-

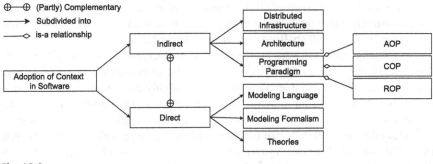

Fig. 13.3

A taxonomy of context adoption approaches within software products, also relating Aspect-Oriented Programming (AOP), Context-Oriented Programming (COP), and Role-Oriented Programming (ROP) (from [954]).

tled at a higher level. The major reason is that the abstract notions of *context* contained in purely middleware-centered approach are not natural, and most context models incorporated in these infrastructure frameworks are informal and lack the necessary features and expressiveness (see [949,957]).

Spectrum of context models

Since Mark Weiser's vision of ubiquitous systems, many researchers have studied the foundations of context modeling (e.g., [182,225,958–964]). This has resulted in a large quantity of different model types being developed. Specifically, context models can be divided into the following eight different categories: object-role based models, spatial models, ontology-based models, key-value models, object-oriented models, markup scheme models, graphical models, and logic-based models (see [959,960,963]). This clearly shows the widespread interest of this research area and the importance of context models for prospective software systems. Note that this is not a comprehensive digression about this particular topic; much has been investigated on this subject, and we refer the reader to the surveys mentioned above. Furthermore, we leave out context-aware architectures for the future. A detailed discussion of the commonalities and differences is out of the scope for this section, and we refer the reader to the surveys above instead.

13.5.2 Key challenges

The exploitation of sensor data and the distribution of numerous available mobile devices has significantly changed the way software engineers view the development, management, and deployment of software. Prospective software systems will continue to evolve to accommodate more sophisticated models of data collection and processing. From a software engineering perspective, the development of such systems and applications of the Tactile Internet become harder in the future because of the distributed nature, mobility, uncertainty, and unstructured environment. System engineers must take account of all local and global aspects of a system to correctly and consistently conceptualize and manage such systems. Undoubtedly, prospective informatics systems will challenge our understanding [223].

Specifically, such software systems inherit an increased complexity (i.e., both functionally and structurally, cf. [176, pp. 21–44]) compared to classical software systems, and as an inevitable consequence, reusability and maintainability is dramatically reduced and lead to higher development costs. The number of parameters within a system increases, and incremental changes in complex systems become increasingly difficult to handle and maintain. Due to the large-scale distributed nature of mobile and global computing systems, data consistency, synchronicity, code complexity, and interface heterogeneity, to mention a few, are well-known challenges, also in software and system engineering communities. This also applies to applications of the Tactile Internet. These problems are further accentuated due to the uncertainties and unpredictability of the physical surroundings in which such software is deployed. Especially, when multiple contexts are involved, ensembles of different entities lead to unpredictable collective behavior (see [964]). However, there is no explicit support in most traditional programming languages, or for procedural and object-oriented programming paradigms, to support context-dependent behavior. Consequently, application and systems without contexts as first-class citizens, programmed in object-oriented style produce mostly tangled, scattered source code that is difficult to manage, incorporating complex logic, patterns and frameworks, which lead mostly to a dependency explosion (see [204]). Software that lacks the necessary modularity is rather rigid and makes reusability difficult (see [204,949,965,966]).

Concerning software verification, such *systems are harder to verify than in earlier days* [967, pp. 701–736], and software developers spent most of their time on testing and verification. Moreover, the reliance on the functioning of such systems is an important point to be considered, which goes hand in hand with the objective of the absence of bugs [584].

13.5.3 Beyond state of the art

We argue that the former model-driven approaches, context-oriented languages and context models either do not take the large-scale distributed nature into account or do not incorporate the notion of context-awareness very sufficiently to be also used as a computational model. We consider computational resources to be rigorously hierarchically organized with respect to the spatial structure. Location-awareness, device heterogeneity, and low-latency are only some characteristics of the Tactile Internet, which makes it a nontrivial extension of the Cloud, or traditional programming paradigms, and architectural patterns. Therefore we propose a two-fold solution strategy to meet these new requirements. First, a generic metamodel is required that is capable of representing the static and dynamic core properties of Tactile Internet applications, which is formally sound and expressive to derive domain-specific languages. The semantics of such a metamodel defines the meaning of an abstract syntax, where concrete syntax representations are derived from. The system itself is then an implementation of one or more domain models, which represent the actual domain under consideration. Second, on the basis of these models, we require a model-driven reactive programming approach for the development of such applications to make the overall synergy between formal mathematical models and software engineering practices possible at all.

To investigate some preparatory steps of context-adaptive software engineering for TaHiL, we wish to explore the following open research questions: (*i*) What are the main contextual and behavioral properties of a generic metamodel for context-adaptive software in the perspective of TaHiL?; (*ii*) How can a formal model be complemented by model-driven techniques to be usable for systematic engineering of large-scale distributed systems for TaHiL?; (*iii*) What is the most suitable programming paradigm for context-adaptive software engineering?

13.5.4 Synergistic links

Context-adaptive software engineering is an essential element in the TaHiL frame of reference, aiming at a scalable, maintainable, verifiable, and systematic software development process. The models and programming paradigms may be directly applied in the targeted primary research field of tactile computing (TP5) and are strongly associated with world modeling in Section 13.3, by providing a mathematical reasoning framework, and furthermore, the model-based feature analysis described in Section 13.6. Moreover, the proposed approaches also have some merit for the use cases in the research group for industrial applications (see Chapter 3), where a model-driven architectural framework for human–robot cohabitation is proposed. The methodical and consistent analysis of safety-critical architectural aspects throughout all software-levels may be treated to enable the efficient development of distributed cobotic cells. This equally applies to the other research groups for medical applications (see Chapter 2) and for the Internet of Skills (see Chapter 4).

13.5.5 Intermediate conclusion

Undoubtedly, to reduce problems mentioned above, prospective applications *must become more versatile, flexible, resilient, dependable, energy-efficient, recoverable, customizable, configurable, and self-optimizing by adapting to changes that may occur in their operational contexts, environments, and system requirements* [968].

A step forward towards an improvement of former software engineering processes is context-aware adaption, which is, in the parlance of Grassi and Sindico [969], *an important feature for pervasive computing applications* [969], directly applies to applications of TaHiL.

Context-adaptive software has the potential to render the capabilities of disruptive technologies, such as robotics, IoT, and 5G, which are currently complex to implement. However, applying the context-awareness to programming, one can transform data into information and finally to knowledge to stimulate the development of intelligent context-adaptive software that may self-adapt itself or parts of it to some changing conditions from its environment (physical and virtual). Context as a resource has the potential to automate software parts, e.g., decrease the use of artificial intelligence in the application itself, and automatically make applications and devices *smarter*. By reducing programming complexity, we may minimize dependencies and, in turn, write less code, make the software more scalable, and maintainable.

13.6 Self-explanation for Tactile Internet applications

Intuitive cohabitation requires that humans and machines both can inquire about their goals, reasons for decisions, and dependencies between behaviors. In particular, humans need to understand why machines have taken specific choices and what they have to expect from machines when influencing them with their own actions. This requires the machines to be capable of explaining themselves, which is also underpinned by the current EU General Data Protection Regulation. There, every automatic decision-making is demanded to be *fair and transparent* and to provide *meaningful information about the logic involved, as well as the significance and the envisaged consequences* [970]. The development of user-understandable explanation facilities is an emerging field, but even basic principles are by far not well understood. In this section, we describe how to establish an explanation framework for Tactile Internet applications, discuss requirements, and present first concepts towards an implementation.

13.6.1 State of the art

Most approaches towards explanation frameworks aim at understandability at design-time. There is, however, only very little work on systems with integrated explanation mechanisms that inform the user at run-time. We briefly revisit existing work on notions of causality, counterexample-based causality, certification, and feature-oriented systems.

Causality

The notion of causality has been extensively studied in philosophy, social sciences, and artificial intelligence (e.g., [971–974]). Most prominent, causality is understood as a binary relation over events that declare the constraints for an event C to be cause of another event E (the effect). These rely on the concept of *counterfactual dependencies* [975], stating that E would not have happened if C had not happened before. Halpern and Pearl formalized counterfactual actual causality using a structural-equation approach [976–978]. Whereas algorithmic reasoning following this approach is computationally hard in the general case [979,980], tractable instances for deciding causes and explanations could be identified in [981]. Probabilistic causation attempts to provide criteria for the stochastic evidence to treat C as a cause for another event E. As illustrated, e.g., in [974,982], various nuances of probabilistic causation have been defined in the literature. Most of them share the idea that causes raise the probabilities for their effects and rely on a formalization using conditional probabilities. Some of these definitions use an explicit notion of time and require that causes occur before their effects, whereas others use a formalization of the *screening-off principle* that attempts to avoid the spurious correlation of events due to common causes. Pearl's probability-raising approach [973] relies on the distinction between observation and intervention, where the latter overrides the original model and zooms into what would have happened if system variables are forced to take a certain value. Other probabilistic extensions of Halpern and Pearl's structural-equation model have been considered recently under philosophical aspects [983] and in a logic-programming approach [984].

To involve behaviors of programs, such as the ones developed in the context-adaptive framework described in Section 13.5, the aim of a self-explanation framework for Tactile Internet applications should rely on operational models, such as Markov Decision Processes (MDPs). MDPs are state-based stochastic models, where actions can be performed in states to switch to a successor states following an action-dependent probabilistic distribution. Usually, the probabilistic distribution models several aspects of the environment, whereas actions correspond to decisions of the program or the user. MDPs that contain only a single action are called *Markov chain*. For more details on stochastic operational models, we refer to standard textbooks such as [557] (check also for Chapter 8.3.1). Research on probabilistic causation in operational models, such as MDPs and Markov chains, is comparably rare. An exception is the work by Kleinberg [985,986], who formalized probabilistic causation in the context of temporal logics for Markov chains to infer complex causal relationships from data and to explain the occurrence of actual events. In [985], a notion of *prima facie causes* that relies on requirements formalized as formulas of Probabilistic Computation Tree Logic (PCTL) has been introduced. To reason about probabilistic causation in Markov chains, [985] combines standard PCTL model checking [584] with statistical hypothesis testing.

Counterexamples and witnesses

During design-time, the cause for a violation of a system property is of utter interest for debugging the system. *Counterexamples* are likely to contain an event that causes the property violation. They are usually provided as error traces or in a tree-like structure to guide through the systems source code or operational behavior. In the nonprobabilistic setting, model-checking approaches are prominent for generating counterexamples as a byproduct of the verification process [584]. The generation of counterexamples in the probabilistic setting is, however, more involved [987–989]. In [990], Leitner-Fischer and Leue describe an approach to generate counterexamples based on event variables in a specialized event-order logic and use them to reason about causality, implemented in the tool SPINCAUSE [991]. Causality-based approaches that analyze counterexamples and extract user-understandable visualizations have been suggested only recently [992,993]. Inspired by the concept of certifying algorithms [994,995], some work has tackled the generation of certificates for positive verification results [996]. A subsystem that violates a reachability property could serve as certifying counterexample for the overall system, an approach followed, e.g., in [989] and implemented in the tool COSMICS [997]. In a recent approach, witness subsystems for MDPs have been shown to provide useful insights using *Farkas certificates* [998].

Feature-oriented systems

In software engineering, *features* are defined as first-class abstractions of an optional or incremental unit of functionality [999]. Feature-oriented systems [1000,1001] follow the concept of features and are well-established, e.g., in the area of Software Product Lines (SPLs) [1002], where each software product corresponds to a combination of features. Feature combinations of a valid product in a feature-oriented system are given by *feature models*, e.g., feature diagrams [1000] and probabilistic feature models [1003] in the static case, and reconfiguration graphs [1004,1005] in adaptive systems with dynamic feature switches. An example feature diagram of a feature-oriented robot system is depicted in Fig. 13.4, on the left. For probabilistic and dynamic feature-oriented systems, Dubslaff et al. [1005,1006] introduced a compositional modeling and analysis framework. Here, operational behaviors of features are described through MDPs with further feature information. The dynamic activation and deactivation of features can be used to describe adaptive behaviors of feature-oriented systems, e.g., to model adaptive heterogeneous hardware systems [1005] and context-dependent systems [1007,1008].

Since the number of feature combinations is exponential in the number of features, they might be not explicitly representable in memory, as it is the case for large product lines as the Linux kernel product line with more than 20 000 features [1009] or redundancy systems [1010]. To this end, the concise representation of sets of feature combinations is important when reasoning about features-oriented systems. Usually, one has given a feature model that provides the domain of valid feature combinations $\mathfrak{V}$, and a set of feature combinations $\mathfrak{C} = \{C_1, \ldots, C_n\}$, for which one seeks a suitable representation. Several methods have been presented to tackle this problem for both, the nonprobabilistic [1011] and probabilistic setting [1006]. Most commonly, a two-level approach first interprets features as propositional logic variables and generates a representation of $\mathfrak{C}$ in Disjunctive Normal Form (DNF) $\bigvee_{i=1}^{n} \bigwedge_{f \in C_i} f$ that is then minimized, e.g., by applying Quine-McCluskey's minimization algorithm [1012,1013]. When n is large, the DNF for $\mathfrak{C}$ might be too big to be explicitly constructed. In this case, partitioning the DNF and applying minimization separately could help. A different approach circumventing the latter drawback is provided through using ordered reduced Binary Decision Diagrams (BDDs). BDDs are tree-like structures, where on a distinct feature is assigned to each depth level, standing for a decision whether the assigned feature is included in the feature combination or not. A BDD that represents a set of feature combinations $\mathfrak{C}$ having a minimal number of nodes can be obtained by successively adding elements of $\mathfrak{C}$ and applying the standard reduction rules. Excluding the description of valid feature combinations $\mathfrak{V}$, i.e., to represent a set of feature combinations $\mathfrak{C}'$ with $\mathfrak{C} \cap \mathfrak{V} = \mathfrak{C}'$, might further reduce the size of the representation of $\mathfrak{C}$. This method has been successfully applied to both, propositional logical and BDD representations [1006,1011].

13.6.2 Key challenges

For the realization of systems with integrated explanation services that meet the requirements of Tactile Internet applications, several challenges have to be addressed. Specifically, explanation facilities have to respect the low-latency, low-energy, and high robustness requirements to be applicable in the Tactile Internet setting. This requires novel programming paradigms that are easy to use, fast, efficient, and, most importantly, interactive to support a Human-in-the-Loop integration. To this end, the software-development infrastructure for Tactile Internet applications (see Section 13.5) has to provide a proper degree of abstraction and hide serve implementation details. Choosing the right level of abstraction, while maintaining functionality and control, and still providing explanations that actually help developers and users, is the first challenge to face when developing explanation facilities. The second challenge is to extract the required information on the chosen level of abstraction, i.e., enabling to reason about causes and effects of events on different time scales. The aim is to extract as much information as possible to support humans and other machine components in, e.g., (*i*) being continuously well-informed about the current situation, (*ii*) understanding reasons for automatically triggered events, and (*iii*) capturing potential risks of own actions, decisions, and possible critical situations. All these scenarios are challenging on their own, but are all required for a tight integration of a Human-in-the-Loop system with explanation facilities.

13.6.3 Beyond state of the art

Well-established in software development, the abstraction level of *features* [999, 1001] seems to be an appropriate starting point for developing explanation facilities in the Tactile Internet application setting. Features encapsulate and structure system functionalities usually provided by the developer and, hence, also have a meaning that could help to explain events during both, design-time and run-time of the developed software. To the best of our knowledge, exposing combinations of active features during run-time as self-explanation method has not yet been considered in the literature. As formal basis for model-based self-explanation, we propose to use stochastic operational models, such as Markov decision processes [557] and their feature-oriented counterparts [1005]. These models are eligible for reasoning methods that lead to cause-effect relations [976], which could be exploited to yield explanation facilities that meet the key challenges posed in the last section.

Feature-based explanations

To support the software-development infrastructure developed for Tactile Internet applications, we now describe self-explanation techniques for dynamic and probabilistic feature-oriented systems [1005]. Based on existing work on their modeling and analysis [1005,1006], such systems are well suited to be integrated into the novel context- and role-oriented programming technology [1014,1015], detailed in Section 13.5 of this chapter, and for human-inspired models for tactile computing (see Chapter 8).

13.6.3.1 Towards feature-annotated applications

At the beginning of Tactile Internet application development, a domain analysis identifies those functionalities of the application that should be encapsulated into features. For instance, in a robotic setting, the developer could image the following classes of features:

Human–machine coadaptation basic functionalities of the application that are of interest for a human, e.g., a knot tying feature or a feature providing assisted haptic guidance for a learning task;

Context-awareness contextual elements that influence the behavior of the machine, e.g., an industrial or surgical environment, testbed or mission-critical context;

State exposure information about the current state important for other components and applications, e.g., whether the communication is security-important (security features) or latency-critical (performance features).

The domain analysis also structures features into a feature model, formalizing constraints on which features can be simultaneously active. To connect features on the abstract level to actual behaviors, code fragments are annotated with those features they belong to. The applications developed then provide hooks and interfaces for explanation services to expose the features that are active during run-time, i.e., make the feature code annotations accessible for other components of the Tactile Internet system.

Example 13.1. Let us consider a feature-oriented robot application that has an *encryption* feature to encrypt the communication to other robots, and a *latency* feature that optimizes the operation of the robot for reducing latency, e.g., to meet the requirements imposed by the Tactile Internet communication. Fig. 13.4 shows an extract of the feature model as feature diagram on the left, including the hierarchical dependencies between communication and performance features. Here, the *latency* feature and *encryption* feature are both optional (indicated by the dots above the features) and could be either active or inactive. Based on the domain analysis that leads to the identification of features and the feature model, actual code of the feature-oriented system is annotated, as illustrated by a snippet of code in the middle of Fig. 13.4. The code implements the processing of an expression, which is only decorated with further formatting in case the *latency* feature is not active, i.e., the application does not need to focus on execution time, and encrypted in case the *encryption* feature is active.

The annotated feature information on the code level and the structural properties provided through feature models pave the way for generating feature-based explanations. A first technique that implements feature-based explanations of feature-annotated Tactile Internet applications during run-time is the exposure of the collection of or a selection of features currently active. This might already provide a

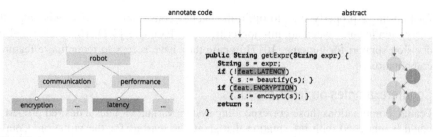

Fig. 13.4

Feature diagram, feature-annotated code, and model-based feature abstraction.

meaningful explanation of general system functionality. Similar as in the case of compact representations for sets of feature combinations (see Section 13.6.1), omitting those features that are necessarily active due to the feature model further reduces the explanation size, and thus could enhance understandability.

In the following sections, we describe further techniques that could be useful for a self-explanation framework on feature-annotated applications:

13.6.3.2 Model-based feature analysis

Since the software-engineering paradigm, as detailed in Section 13.5, follows a model-based approach, formal models for software are at hand. These could be used to support the generation of abstracted models of feature-annotated applications, while maintaining a mapping from the states at run-time to the abstracted model. In Example 13.1, Fig. 13.4 (on the right), illustrates a possible abstraction of the behavior of the *getExpr* method into a state-based operational model with feature annotations. Such models are eligible for a feature-oriented analysis, e.g., using family-based analysis techniques [1005,1006,1016], that could provide insights about the properties of feature behaviors and the whole Tactile Internet application. In particular, techniques described to reason about causality, counterexamples, and witnesses as discussed in Section 13.6.1 could be used and adapted for the feature-oriented setting.

Precomputation of analyses

In the classical setting of feature-oriented analysis, the analysis provides sets of feature combinations that fulfill a desired property, e.g., whether the expected latency is less than 1 ms. Using techniques for a compact feature-combination representation (see Section 13.6.1), these sets of feature combinations could be concisely stored in the explanation facilities and used to reason about feature combinations. For instance, a robot might expose the set of feature combinations that could compromise the latency constraint of Tactile Internet applications. One could also apply counterfactual reasoning to determine the most influential feature with regard to a property, i.e., determining a feature as cause of violation or satisfaction. To this end, the explanation facilities could also suggest a set of feature combinations, to which one could switch for fulfilling a property, e.g., to fulfill the latency constraint by partly switching off encryption features when not mission-critical (see Example 13.1). This includes also decision support for humans [1017] in case they have access to reconfigure feature combinations.

Counterexamples on fixed products

Feature combinations those corresponding system variant violates a desired property could be enriched with information that show the reasons for the violation. Counterexamples in the operational behavior of the system provide a sample run that violates the requirement. Such counterexamples are usually obtained from a formal analysis, e.g., using model-checking techniques. Counterexample annotations could permit other components and humans to retrace undesired behaviors with more detail, accompanied with feature-oriented annotations.

Dynamic feature-based causality

When feature combinations change during run-time, e.g., automatically triggered for system adaptations or actively imposed by humans, causal reasoning on the temporal evolution of feature combinations could reveal features or feature combinations as causes for certain effects.

13.6.4 Synergistic links

The formal basis for model-based self-explanation we proposed relies on stochastic operational models and their feature-oriented counterparts. This aligns with the model-based context-aware software-engineering approach, illustrated in Section 13.5, and the human-inspired models for tactile computing, introduced in Chapter 8. Hence, our methods could be directly applied in the targeted primary research field TP5 and the key technologies and methods K4. Furthermore, all research areas that rely on human–machine coadaptation will benefit from self-explanation facilities developed to ease the timely and efficient communication and coordination in Tactile Internet applications.

13.6.5 Intermediate conclusion

For the development of a self-explanation framework, it seems to be reasonable to address research questions on the abstraction level of feature-based explanations. To tackle such questions, methods on determining causality relations on operational models to determine reasons for certain behaviors, and specialized feature-oriented explanation techniques are key elements. Based on evaluations of the self-explanation framework reliance on the abstraction level of features, feasibility of the overall approach for future programming concepts could be evaluated: Can causal dependency relations identify reasons for feature switches and provide refinement indications in case reasons are on a different level of abstraction than the level of features? Is the abstraction level of features appropriate to provide useful explanations, e.g., to serve as self-explanation facility or to reason about changing system properties that lead to violating Tactile Internet requirements (latency, energy consumption, etc.)? Are the notions of feature dependencies, reasons, and feature-based explanation appropriate to be included in the software-development process of Tactile Internet applications?

13.7 **Conclusion and outlook**

In this chapter we have presented our visions and goals towards a tactile computing infrastructure. We have identified several research questions in different domains, such as safe, secure, and scalable computing infrastructures; world capturing and modeling; and context-adaptive software and self-explanation facilities. These open questions and motivating challenges will drive our research journey for the upcoming years. Thereby we will evaluate our results with the help of the use-cases realized by the research groups for medical applications (U1) (see Chapter 2), industrial applications (U2) (see Chapter 3), and Internet of Skills (U3) (see Chapter 4).

Technological standards and the public

4

Outline

The last part of the book deals with three different areas of research in the field of trace files for tactile applications, standardization in the field of Tactile Internet, and the related societal questions of public opinions and technology transfer.

Traces for the Tactile Internet: Architecture, concepts, and evaluations

14

Patrick Seeling[a], Martin Reisslein[b], and Frank H.P. Fitzek[c]

[a]*Central Michigan University, Mount Pleasant, MI, United States*
[b]*Arizona State University, Tempe, AZ, United States*
[c]*Technische Universität Dresden, Dresden, Germany*

Kohn's second law: An experiment is reproducible until another laboratory tries to repeat it.
— Alexander Kohn, 1989

14.1 Introduction

The Tactile Internet's multimodal (auditory/visual/haptic) application scenarios of human-to-machine and machine-to-human feedback and control (and in combining the two, human-to-human) are bound to permeate all levels of society [1]. Considerations that focus on industrial scenarios of the Tactile Internet are discussed in [1018], especially with a view on integrating existing, commonly utilized and emerging industrial communications standards. However, the inherent Human-in-the-Loop approach that is part of the Tactile Internet with Human-in-the-Loop (TaHiL) applications requires concepts that explicitly accommodate humans in-the-loop to leverage human and machine assistance, e.g., for interventions at the human–machine interface.

Current research efforts already focus on investigating the details of the human–machine interplay from the views of the various disciplines that join at this interface. For example, considerations for the haptic modality of ultralow latency communications approach are discussed in [430], with approaches to encoding and compression in [9]. Underlying investigations into optimizing the encoding and compression approaches date back several years and commonly include human perception optimizations [381]. One common approach that is employed in the perceptual encoding of haptic signals is the reliance on just noticeable differences [378]. This principle has found broad adoption in the Quality-of-Experience (QoE) domain, where relationships between Quality-of-Service (QoS) and QoE have been found to follow logarithmic relationships [1019]. The Weber–Fechner law has subsequently evolved as additional means for comparing the impact across communicated media (audio, video, or even web browsing), e.g., [1020].

Tactile Internet. https://doi.org/10.1016/B978-0-12-821343-8.00027-7

321

As outlined in [1021], bidirectional communications are at the forefront of TaHiL. In their contribution, the authors evaluate several existing studies with such a setup and varying degrees of freedom on the remote side of a human-controlled networked robot. Depending on the applied perceptual coding parameters, the authors find that the data from prior studies [434,1022] exhibit a broader range of packet inter-arrival times than prior assumptions for the Tactile Internet traffic would have provided [1023]. As detailed further in [1021], other inter-arrival and packet delivery characteristics than those previously assumed will likely be found in future Tactile Internet scenarios.

Different settings and use-cases from employed equipment as well as human behavior from TaHiL scenarios will have significant impact on the resulting networked traffic. There is no immediate shortage of envisioned use-cases, where tactile communications can permeate into existing systems: There is a fundamental shortage in overall defined TaHiL scenarios and datasets that enable researchers at the intersection of the human–machine interface to conduct reproducible research. Such future datasets will not only support networking research, but also perceptual coding efforts and other research. The remainder of this paper provides the following contributions: (*i*) In the following Section 14.2, we define a universal terminology and presentation format for Tactile Internet traces using remote teleoperation with TaHiL as example. This newly designed format will be generally applicable beyond the example scenario. (*ii*) In Section 14.3, we describe the content of communications on the network and what type of data can be assumed to be in trace files for the Tactile Internet using a popular robot implementation as example. (*iii*) In Section 14.4, we describe near- and far-term application scenarios, for which Tactile Internet applications will emerge with a high likelihood, and for which we consider providing trace data for other researchers. We conclude in Section 14.5 with an outlook on our ongoing and future trace generation efforts.

14.2 Tactile traces generic system overview

To derive a universally applicable approach to the generation and presentation of Tactile Traces (TT), we initially provide a high-level overview and definition. We consider a generic Human Operator (HO) or Robot Operator (RO) at the core of a Tactile Internet (TI) implementation scenario.

14.2.1 General control operations

Though details of control loops and their utilization in the context of TaHiL for the Tactile Internet can be found described in greater detail in, e.g., [1024], we provide a general high-level overview in the context of tactile traces in what follows. We initially provide a schematic view of the single control loop in Fig. 14.1.

In this scenario, the HO performs actions to remote-control a robot, which results in commands being sent from the sensor(s). The content of the commands is gath-

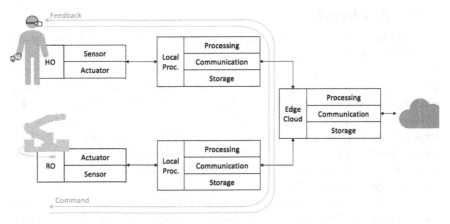

Fig. 14.1

Example illustrating a single control loop common for the Tactile Internet. While various mutations of this scenario are common, the HO is performing the main decisions and sending those as commands towards the RO. The RO performs the commands and generates feedback information, which the HO employs to make the next decision. In-between conversions of sensing/actuation data, processing, and network communications, all increase the computing overhead. Typically control loops (command and feedback, including all system-inherent delays) are required for Tactile Internet experiences that are in the range of 1 ms.

ered via, e.g., wearable sensors, and requires initial local computing. This computing could be initial dimension reduction, compression, or network packetization, to send the data further to a local edge cloud service. In case that no edge cloud service is needed to increase computing resources, the command packets could be forwarded directly to the receiving side. Similarly, if significant amounts of additional resources are needed, the command data could be sent further into the network. On the receiver side, the incoming data will be processed and subsequently utilized to steer the RO via actuators. Monitoring of the RO through sensors will result in data to be sent as feedback to the HO. Again, local computing resources as well as edge cloud or further networked resources will be required to perform operations on the feedback data. Once received on the HO side, the data will be processed to provide feedback information to the HO, e.g., as force feedback or visual feedback in virtual reality environments.

Given the ultralow latency requirements of the Tactile Internet, trade-off decisions and algorithm development need to take both HO as well as RO performance and abilities into account. Furthermore, the parameter space for success and significance of modifications, such as through lossy compression, can vary widely between different applications scenarios, e.g., remote part sorting and remote surgery have significantly different impacts on human well-being. In turn, a given scenario will require significant amounts of experimentation and source data, examples of which we describe throughout this contribution.

14.2.2 Data formats

The Institute of Electrical and Electronics Engineers (IEEE) Robotics and Automation Society has several ongoing standardization initiatives, e.g., [1025], which followed the Core Ontologies for Robotics and Automation (CORA) IEEE standard 1872–2015 [1026]. Other approaches that capture the automation and plant engineering information for exchange are AutomationML [1027,1028], itself based on eXtensible Markup Language (XML) [1029]. Other domains experience similar challenges, oftentimes rooted in the amounts of data to be processed. Smart Cities [1030] and the Internet of Things [1031] are two intertwined examples for these challenges. Solutions across domains commonly make use of the Semantic Web's approaches to real-world representations [1032].

Most recent implementations utilizing data analytics on top of these existing frameworks incorporate time-stamped evaluations of captured data following existing description frameworks. In, e.g., [1033], the authors combine the approaches from the automation and semantic web within time-series data bases [1034] to overcome the current limitations in data representation of the underlying standards. Whereas these standards describe the robotic ontology, features, and programming efforts well, our generic view on the networked communications characteristics for any type of operator requires an additional abstraction while following the general principles described in these standards to facilitate ease of reuse and implementation.

14.2.3 Generic hierarchy for tactile traces

In what follows, we describe our overall design approach based on prior work, such as IEEE 1872–2015 [1026]. We note that our approach differentiates itself from those that are native to the actual implementation domain, as our focus is on capturing the networked data traffic in support of network performance evaluations and related research efforts. We also note that while the majority of data are commonly provided in generic control environment approaches for human readability (including Javascript Object Notation (JSON) and XML), the overhead of these formats will likely be prohibitive for actual network transport and data to be recoded appropriately.

We initially start with the point of view of a generic operator (HO/RO) in our scenario. In our generic scenario, multiple on-operator communication layers and networked communication paths might exist. We consider these paths to generally carry data required for logical command/feedback control loops. Based on the Tactile Internet definition, human-to-machine and machine-to-human interactions incorporate a Human-in-the-Loop configuration, i.e., are geared towards human interaction within the overall control loops. At one or multiple instances, the operator entity under consideration is assumed to exchange information with one or multiple nodes at the network edge (edge node), considered to be the next logical control loop communication partner. We note that there is no requirement for the edge node to logically close the entire or parts of the control loop, as it could just transparently forward data.

Each operator has zero to S Service Access Points (SAPs), which can be considered individual modules comprised of arbitrary combinations of directly attached integrated circuits for actuating/sensing, processing, storing, and communicating. Communications typically consist of command or feedback information that is required for the successful performance of the generic operator. From an SAP point of view, this control loop communication can *logically* be (*i*) internally (*lint-loop*), i.e., to directly attached sensors/actuators; (*ii*) incorporating other local SAPs (*linc-loop*), i.e., to/from other SAPs on the same HO/RO; or (*iii*) to the edge (*ledge-loop*), i.e., externally communicating with one or mode local edge nodes. Furthermore, multiple sensing or actuating components can be combined at each SAP. Each of these components typically will be characterized by their type, frequency, and amount of generated raw or preprocessed data. SAPs could additionally be enabled to store the information exchanged beyond their modular boundaries at different time scales and differently processed degrees. At a minimum, each SAP requires some level of communicating and processing means.

At the base aggregation and layering level $l = 0$, one or more SAPs on an operator will be enabled to communicate externally, e.g., to a local edge node, utilizing edge loops (*ledge-loops* as illustrated at the bottom of Fig. 14.2) for feedback/command information. We denote these modules as External Service Access Points (eSAPs). (We note that this notation is for separating the communications capabilities only, and it does not preclude any other capabilities, i.e., eSAPs could have all other outlined capabilities.) For an initial example, one could think of a single eSAP for a robotic arm that is directly connected to all the different actuators and sensors. This eSAP directly provides all necessary communication and processing. This includes the internal control loops (*lint-loops*) that are employed for the directly attached sensors and actuators. This particular *lint-loop* feedback/control information is internal to the SAP under consideration and not sent over the ledge-loops. We illustrate the configuration with a single eSAP in Fig. 14.2 for an exemplary robot arm with several joints.

Each of the joints here is directly attached to the eSAP, which includes processing for internal control loops. Remote control and feedback data can similarly be stored and processed before sending or after receiving from the ledge-loop. For this scenario, TT could be generated at the eSAP for individual lint-loops (should programmatic access to the eSAP be provided) and ledge-loop (either through programmatic access or by capturing the networked traffic) and include the specific content outlined in Section 14.3. A direct extension of this single eSAP scenario consists of multiple externally connected SAPs, illustrated in Fig. 14.3.

Following the robotic arm scenario, individual joints could have their own external data connection (or, e.g., the actuator could be separately controlled). Lastly, we consider the scenario illustrated in Fig. 14.4.

Within a local environment, such as a robotic arm operator, SAPs can be organized into hierarchical layers. For example, one could consider the individual joint of a robotic arm a single SAP if the joint (with its actuators and sensors) can communicate directly on-operator to another SAP that aggregates the higher-level SAPs

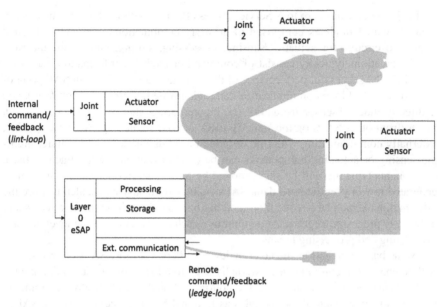

Fig. 14.2

Example illustrating a single eSAP connected with three joints on a robot arm. All command and feedback communication internal to this eSAP occurs on local, internal loops (*lint-loops*). All processing and external communications to the next edge node(s) take place through the single eSAP employing *ledge-loops*.

using incorporating control loops (*linc-loops*). In this scenario, we find all different command and feedback loops, namely (*i*) lint-loops for directly SAP-attached components, (*ii*) linc-loops for locally exchanged command/feedback information with other SAPs, and (*iii*) ledge-loops for external communication. We denote the individual SAP_i^l as the ith SAP on layer l. Whereas layer 0 SAPs always communicate externally (i.e., SAPs at layer 0 are always eSAPs), higher layer SAPs could communicate externally as well. For internal representation purposes, these SAPs are represented in a tree or as graph structure, which allows representing an arbitrary configuration beyond the robotic arm example we considered here. This approach will similarly enable the fine-grained data retrieval at different aggregation and processing levels for tactile trace generation and subsequent evaluation. As an extension to the robotic arm RO example we employed so far, one could consider a human tele-operator remotely controlling the robotic arm HO, illustrated in Fig. 14.5.

In this extended scenario, one could assume that one eSAP (e.g., $eSAP_0^0$) would provide visual interface modalities for the remote robot interaction. This could entail a virtual reality environment as pictured, should the interaction scenario allow for free head movements. Alternatively, nonimmersive visual feedback could be provided, such as commonly employed in medical robotic scenarios. Where needed, computer-generated visual interface information can be computed by the local edge node and

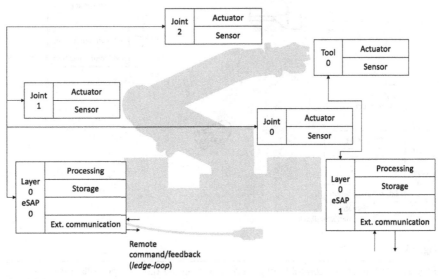

Fig. 14.3

Example that extends Fig. 14.2 and illustrates the possibility for multiple eSAPs on the same layer, here employing two eSAPs.

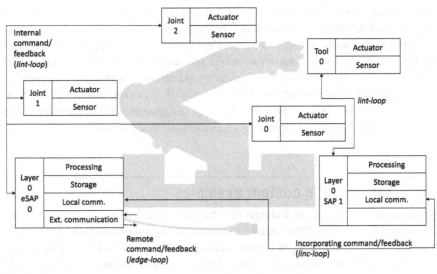

Fig. 14.4

Example showcasing the localized communication between SAPs to incorporate information through *linc-loops*.

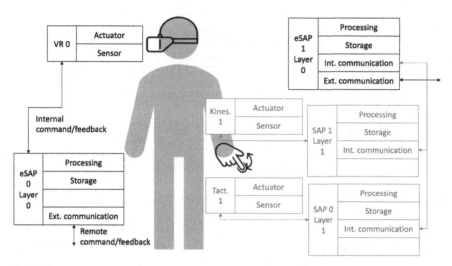

Fig. 14.5

Example for the remote control through a human operator (i.e., the potential remote side to our prior example). Multiple sensing/actuating components and SAPs on different layers exchange information using *lint-* and *linc-loops* before employing the *ledge-loops* for remote communications.

provided through a specific ledge-loop for the visual representation—in contrast to, e.g., sending compressed video. Similarly, a second ledge-loop could be employed (e.g., with $eSAP_1^0$) directly controlling other interaction devices, such as virtual reality controls or a haptic control interface. Depending on the configuration, the latter eSAP (i.e., $eSAP_1^0$) could control the individual components of a haptic interface device, such as tactile and kinesthetic actuators and sensors. In this last configuration example, $eSAP_1^0$ could connect to SAPs on higher layers using linc-loops. (For example, this connection could be to SAP_0^1, which is responsible for tactile information only, and SAP_1^1, which correspondingly handles kinesthetic information.)

14.3 Tactile trace content examples

Traces commonly represent a textual (or otherwise encoded) representation of real-world data. For scenarios that can take a significant range of detailed implementations as source to create traces, significant metadata should be included to provide context as well as enable researchers to perform detailed evaluations and extrapolations. To further this goal, we consider significant details to be included in the traces as descriptive metadata for the individual components that comprise a trace, with more details about these components and their potential future logical organizations provided in Section 14.2.3.

14.3.1 Franka Emika cobot arm

We initially consider the Franka Emika robot arm with a gripper as actuator. For details, we refer the interested reader to, e.g., [1035]. This particular robot arm implementation is ready for industrial applications and not solely geared toward research and education (however, it features a well-described programming interface making it useful in the latter scenarios). It has been used in prior research [1024] in the context of the Tactile Internet and deployed at the Technische Universität Dresden (TUD). To derive usable trace information, logging for trace generation purposes can be performed employing the built-in library functionalities. However, the content thus logged is different from the content sent over the network, which requires a translation. (We note that the content sent over the network could additionally be encrypted in other implementation scenarios. We additionally note that such approach would likely have an impact on processing delays and payload sizes.) The generation of feedback is performed at the rate of 1 kHz, albeit with significant deviations in the time that packets were captured on the wire, as noted previously in [1021].

14.3.1.1 Command and feedback formats

Overall, the format and size of the content sent over the network is well defined in the research libraries provided by the manufacturer and provides the following network traffic.

Commands sent to robot

A *RobotCommand* data structure is sent, which is comprised of a *message_id* (eight bytes), a *MotionGeneratorCommand* and a *ControllerCommand*. The *MotionGeneratorCommand* contains 38 double values and two Boolean values, for a total of 306 bytes (assuming no stuffing of both Boolean values, resulting in one byte per bool). The *ControllerCommand* contains a single array of double for 56 bytes. In total, the required bytes without any other encoding per command would be 370, which is corroborated by the number of bytes captured on the wire.

Feedback received from robot

To complete the control loop, feedback messages are sent from the robot as *RobotState* return message bytes, again headed by the *message_id* (eight bytes). The remaining content consists of several status variables, totaling 250 doubles (2000 bytes) and two boolean arrays with a combined total of 74 bytes (again, assuming no stuffing). In addition, three status enumerations are sent as well, though these are specifically coded as 1-byte versions. Overall, the total number of bytes for the raw feedback message is 2085, which again is corroborated by the number of bytes captured on the wire.

14.3.1.2 Demonstration trace results

Results will be highly dependent on several implementation and external factors influencing each application scenario slightly differently, especially for ultralow latency levels. In what follows, we provide initial insights into some of the potential outcomes

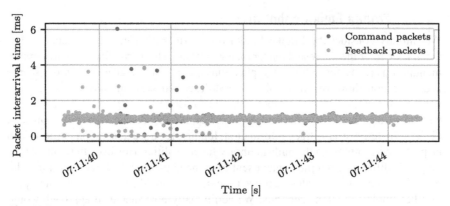

Fig. 14.6

Command and feedback packet interarrival times for a basic arm movement of the Franka Emika robot arm.

at this time scale. We initially consider an example for simple movement of the robot arm in a controlled, direct setting. In this setup, the robot arm was directly connected to the experiment-controlling laptop without external network traffic. We illustrate the resulting packet interarrival times in Fig. 14.6.

We observe that overall, Command (CMD) and Feedback (FB) packets exhibit the tendency to be received around 1 ms apart from each other, as the overall robot configuration is based on 1 kHz control time frames. Whereas this Inter-Arrival Time (IAT) holds for some packets, we observe in Fig. 14.6 that a large number of packets *stray* from this general target, with outliers reaching up to 6 ms in delay on the upper end. Similarly, low-end IATs can reach sub-ms times. As direct connection between controlling laptop and robot arm exists, there is little networking overhead that could be attributed with this deviation. We furthermore note that the IAT variation can be observed for both packet types, i.e., for packets originating from either end of the direct physical network connection (laptop and robot controller, or, in our generic approach, the local edge node and the local eSAP connected via their direct ledge-link).

Due to the likely presence of outliers in this and future traces, visible for the demonstration trace in Fig. 14.6, we note that we applied outlier removal based on [1036,1037], which is an extension of Tukey's box plot approach [1038] to account for skewed distributions. We illustrate the resulting probability densities in Fig. 14.7, noting that we chose the number of bins in the histogram following the guidelines of Freedman–Diaconis rule, described in greater detail in, e.g., [1039].

We initially observe a fairly broad distribution around the 1 ms target for CMD and FB packets in Figs. 14.7 (top) and 14.7 (bottom), respectively. Even after outlier deletion, a significant spread and double-sided heavy tail remains for the IATs observed. As visually noted in Fig. 14.6 initially, this characteristic is independent of the packet type. This is of interest, as the size of FB packets causes Internet Protocol (IP) fragmentation to be utilized. In turn, one could expect this particular packet

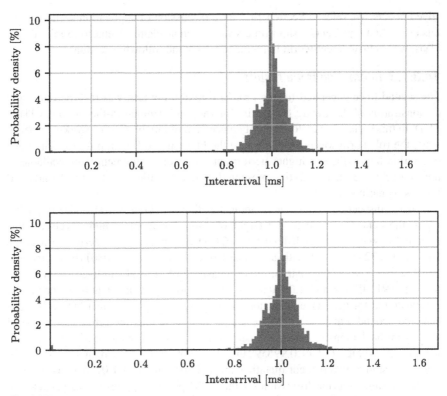

Fig. 14.7

Probability densities for command and feedback packet interarrival times for a basic arm movement of the Franka Emika robot arm.

type to exhibit visibly higher levels of variability of the determined IATs. In contrast, CMD and FB packet types both feature main spreads between around 0.8 ms and 1.2 ms.

Specifically, we note that for the 1994 CMD packet measurements remaining after outlier treatment, the average IAT is 0.99715 ms (median 1.00112 ms), with a standard deviation of $\sigma = 0.10015$ ms. The five-percent quantiles are at 0.888824 ms and 1.102924 ms, respectively. In other words, one can realistically assume that based on our scenario, about 10% of (already outlier-removed) packets will exhibit a IAT deviation from the desired 1 ms of around 10%. Alternative results for the 2472 FB packets indicate an average IAT of 0.996047 ms with $\sigma = 0.114584$ ms (median 1.002789 ms). Correspondingly, the five-percent quantiles lie at IATs of 0.91815 ms and 1.083136 ms, respectively.

Though this level of deviation could be acceptable in some application scenarios, the potential for significant impacts on the common Human-in-the-Loop scenarios associated with the implementation of the Tactile Internet could have significant negative impacts. We refer to Section 14.4 for a discussion of application scenarios. As this evaluation represents a short time scale for an academic scenario to initially fix ideas, we evaluate a real-world implementation in the following section:

14.3.1.3 Interaction trace results

The second trace generated for the Franka Emika robot arm was recorded during a demonstration of the Centre for Tactile Internet with Human-in-the-Loop (CeTI) at TUD. In this scenario, a child randomly placed a building block on a working surface while the robot arm was utilized to place the block in a different position. The trace generation here utilized a higher performance network equipment than available in the shorter, initial demonstration presented earlier. We illustrate the results attained for this scenario in Fig. 14.8.

Visual inspection immediately indicates that for both CMD and FB packet types, the center is close to the target of 1 ms, but exhibits high levels of outliers and deviations. Indeed, for 68 541 CMD-type packets, we derive an average IAT of 0.997235 ms, but with a standard deviation of $\sigma = 0.080116$ ms (median 1.00112 ms). The overall distribution of measured IATs results in five-percent quantiles of 0.915051 ms and 1.080036 ms on the lower and upper tails, respectively. As illustrated for the IAT distribution in Fig. 14.8 (center), the distribution is well-centered around 1 ms, but can have significant deviations therefrom, as observed already in the demonstration case illustrated in Fig. 14.7. For the 68 570 FB packets, we note an average IAT of 0.996592 ms, $\sigma = 0.082631$ ms (median 1.002073 ms). The comparative five-percent quantiles are 0.921011 ms and 1.074076 ms, respectively. Overall, the prior discussion for the CMD packets applies to this packet type as well, though the FB distribution is slightly narrower than the CMD distribution.

Comparing both, in-laboratory and in-world evaluations, we note that the extended time period of the real-world scenario increased the initially observed deviations from the 1 ms target time are resembled in the longer time span with slightly lower overall standard deviations, but increased outliers. In turn, it is a future research direction to identify the in-depth deviations between both scenarios that will enable the generation of data in controlled environments to be adequate representations of real-world scenarios.

14.3.2 Kinesthetic reference dataset

Currently available traces for the TI include tactile and kinesthetic traces, which are available from the Technical University of Munich (TUM). This particular dataset is described in greater detail in [9], with the framework for generation detailed in [1040]. We initially note that as in the previous case, data is commonly fixed in the traces with a timestamp (e.g., captured as long data type) that is followed by 9 float or double values (e.g., typically 8 bytes in case of a double). All data is provided in full millisecond intervals for 25 seconds, resulting in 25 000 samples.

We initially consider the complexity of the contained data as general view on compressibility without deeper context, should this be required by an application

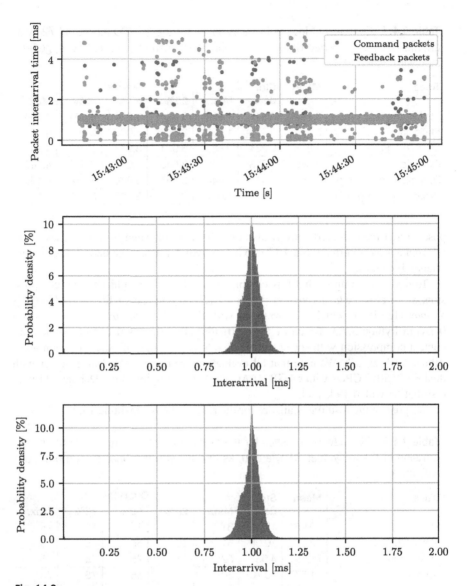

Fig. 14.8

Visualization of packet inter-arrival time results (all, command, feedback) obtained for a real-world interaction with a human device operator of the Franka Emika robot arm.

scenario. We provide an overview of the entropy (based on $\log_e$ for the byte-converted data, obtained with SciPy2001) in Table 14.1.

We initially note that the entropy observed on average is higher than the ideal scenario, indicating that there is a general opportunity for compression. We additionally

Table 14.1 Overview of $\log_e$ entropy statistics for the 25 000 individual 72 byte chunks of byte-converted data contained in the kinesthetic tactile traces, generated with SciPy2001.

Trace	Mean	Std. dev.	Quartiles				
			Min.	25%	50%	75%	Max.
Static	3.732	0.517	1.650	3.874	3.954	4.004	4.154
Static, tapping	3.404	0.476	1.609	3.048	3.095	3.966	4.166
Static, dragging	3.918	0.234	2.886	3.931	3.973	4.012	4.142
Dynamic	3.346	0.513	1.609	3.043	3.087	3.939	4.147
Dynamic, pressing top	3.750	0.416	1.544	3.857	3.946	3.988	4.135
Dynamic, pushing side	3.376	0.513	1.609	3.029	3.083	3.916	4.103

observe that the observed entropy also exhibits a broad range for most cases, overall from a 1.54 minimum to a 4.17 maximum, which reflects the underlying content (value) dynamics.

To derive the impact that this complexity of the data would have on compression, we employ the *libsnappy-dev* (version 1.1.7-1) compression library for Google's *Snappy* algorithm, which was wrapped with the *python-snappy* package (version 0.5.4) in Python 3.7.3. We chose this particular algorithm as it features some non-context compression with speed well-known to exceed that of other common algorithms, such as *zlib*. We note that all evaluations were performed on a server with dual E5-2650v2 CPUs with 64 GB RAM. The server employed the Debian 10 Linux distribution with 4.19 kernel.

We present the resulting statistics for the resulting sizes in Table 14.2.

Table 14.2 Overview of statistics for the sizes in bytes resulting from the compression of the individual 72 byte chunks of data with the *Snappy* compression algorithm.

Trace	Mean	Std. dev.	Quartiles				
			Min.	25%	50%	75%	Max.
Static	69.95	10.43	31	75	75	75	75
Static, tapping	62.50	10.05	31	55	55	75	75
Static, dragging	73.69	4.93	52	75	75	75	75
Dynamic	61.57	10.70	30	55	55	75	75
Dynamic, pressing top	70.39	8.82	31	75	75	75	75
Dynamic, pushing side	62.77	10.95	31	55	55	75	75

We initially note that the overall compression rate is fairly low, as the incoming 72-byte chunks of data are maximally compressed to an average of 61.57 bytes. Interestingly, the variation of the source data is responsible for compression in the median case being either at 55 bytes or already nonfeasible, resulting in a 3-byte overhead, as illustrated in Fig. 14.9 (top). There are, however, several chunks, for which the resulting compression yields compressed sizes of 30 bytes. For these chunks, the resulting smaller sizes could be beneficial for transmission. We additionally restate that the utilized algorithm is a general purpose noncontext compression approach that can likely be traded for a more specialized one in the future. However, the reduction in size typically comes at a trade-off in delay, as computational time is required to achieve the compression outcome.

Subsequently, we focus on the results for the required compression times in Table 14.3 and provide the illustration corresponding to the *Dynamic* scenario in Fig. 14.9 (center).

Table 14.3 Overview of statistics for the computational time in milliseconds required to compress the individual 72-byte chunks of data with the *Snappy* compression algorithm.

Trace	Mean	Std. dev.	Quartiles				
			Min.	25%	50%	75%	Max.
Static	1.340	0.073	1.258	1.331	1.336	1.341	2.097
Static,tapping	1.349	0.079	1.276	1.330	1.336	1.353	2.123
Static, dragging	1.346	0.074	1.297	1.334	1.338	1.344	2.103
Dynamic	1.332	0.074	1.261	1.315	1.322	1.338	2.119
Dynamic, pressing top	1.347	0.073	1.269	1.336	1.342	1.346	2.119
Dynamic, pushing side	1.333	0.073	1.261	1.314	1.324	1.338	2.092

We initially note that the overall average for the time needed to compress the chunks of data is approximately capped around 1.35 ms in general, though extreme cases above 2 ms appear every several hundred milliseconds. The cause of these outliers is subject to further investigation, as the overall Pearson correlation between the two below 0.2 indicates no immediate causality. We provide a view on the overall resulting time to compress distribution in Fig. 14.9 (bottom), which immediately showcases (*i*) multiple distribution peaks and (*ii*) a heavy tail. The latter observation is explained through the periodic observed extreme cases for the compression times throughout the timespan of observed measurements. The remaining distribution peaks of time variations, on the other hand, are likely related to the content and compression algorithm and subject to further investigations.

Overall, we note for this particular scenario that the delay for the compression with a general-purpose compression algorithm clearly increases the time needed to send the data from an eSAP to above 1 ms. A trade-off might exist if multihop transmission paths are considered after the ledge-loop—here, the potential savings in size that can be attained are not high enough to justify the compression times. Nevertheless, an investigation into specialized compression approaches that are more context-dependent or hardware-specialized might yield significant better compression times and, subsequently, might become beneficial due to savings of bytes-in-air or bytes-on-wire.

We also note that lossless compression is typically not the main goal for kinesthetic data. A reduction of the high packet rate is typically much more important. Also, the 72-bytes samples could be compressed in a lossy manner, e.g., by simply

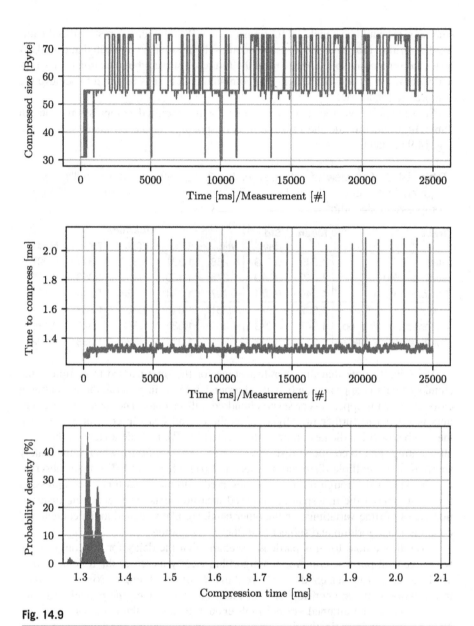

Fig. 14.9

Snappy compression results (compressed sizes, compression times, compression time distribution) from *Dynamic* interaction reference trace data.

mapping the floating point numbers to 16 bit words. Such a compression of samples has not been studied in detail; and, in particular, the combination of sample compression with a packet rate reduction scheme, such as the deadband approach [1040], is an open area for future research. This research topic has not received much attention so far, mainly because of the overall small raw bitrate of the kinesthetic data compared to, e.g., the bitrate of video streams.

14.3.3 Tactile reference dataset

For the tactile data, one can consider two scenarios, namely (*i*) one where the data is transferred beforehand as it is known (e.g., as part of a local installation for a game), and (*ii*) a scenario where data is required to be streamed in real-time with the smallest additional delay introduction possible. For the first scenario, one can consider the kinesthetic information driving the selection of the local installation of available datasets, such as the appropriate selection of the adequate speed (close to the real-world one) and force application (such as the tapping or dragging scenarios) we considered in Section 14.3.2. This data can readily be obtained from existing trace files with the generation process described in [1041].

In the second scenario, however, the immediate streaming of data might require the immediate compression of available information with the smallest overheads possible. As the TUM data available compresses a selection of data after sampling for a significant amount of time, the provided data is not suitable for the TI ultralow latency scenario. Two further approaches appear as interesting avenues, namely (*i*) the direct transmission of gathered x, y, and z accelerometer data in uncompressed or directly compressed form, or (*ii*) initial dimensional reduction applied as in application of, e.g., the DFT321 [1042]. We consider the second scenario in the remainder of this section.

We initially apply the DFT321 to 13 x, y, z value pairs, which were generated through sampling at 2.8 kHz in the original TUM dataset. Smaller numbers of samples are prohibitive with respect to performing the frequency domain translations as part of the dimensionality reduction applied. To remain within the tight boundaries imposed by the TI application scenario, we consider throughout and implement a sliding window over the entire number of samples that moves the window on a per-sample basis. This approach could therefore be utilized to continuously stream data at sample rate as new x, y, z samples are added at the 2.8 kHz rate. Following our prior approaches, we illustrate the results in Fig. 14.10 for the entropy of the sliding window values and the time required to compress them utilizing the *Snappy* compression algorithm. (Future work could consider compressing the tactile information with perceptual lossy compression algorithms [418].) We note that we omit the actual sizes, as they are constant at 107 bytes, i.e., no compression was performed and the 3-byte overhead for the algorithm was added.

We initially observe that the overall entropy is between 3.9 and 4.3, indicating there is opportunity for compression. Similarly, we observe that the overall entropy distribution of the values is fairly normal. However, the efforts of the *Snappy* compression algorithm do not yield any compression, but force a delay. The delays incurred here are higher when compared to prior observations with values above 1.4 ms. They also exhibit the various spikes encountered before, which are currently

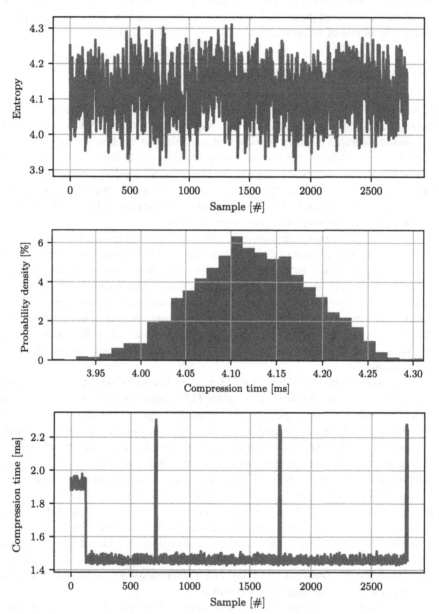

Fig. 14.10

Example results for the *Direct-1spike Probe-aluminumGrid-fast* interaction dataset and compression with *Snappy* ($\log_e$ entropy, entropy distribution, compression times).

under investigation, but likely the result of a combination of hardware and software implementations. We also note that the increased compression time is explained by the compression algorithm working on larger chunks of data to be compressed when comparing to the kinesthetic results. Though it is common that tactile information is exchanged in an open-loop fashion with less stringent latency requirements, scenarios could arise that require similar ultralow latencies, as for kinesthetic information. For these scenarios, specialized complexity reduction and compression approaches are needed to support the ultra-low-latency requirements of the Tactile Internet.

14.3.4 Medical teleoperation

Medical robot assistance for telemedical operations can be considered one of the important future application scenarios of the Tactile Internet. Current common implementations for telesurgery, for example, do not necessarily have the fine-grained data capture and networking facilities required in those future applications. However, an initial extrapolation from courser-grained datasets can enable future research and implementation efforts towards the commonly assumed Tactile Internet granularity of 1 ms and below. In the following, we provide two interpolated data traces gathered from different teleoperation research efforts as first examples for this particular application domain:

14.3.4.1 TUD robot surgery training dataset

We initially employ research data from the Speidel research group at TUD, where users performed different trial scenarios with remote-controlled robot actuators, namely *Circle*, *Gallbladder*, *Peg Transfer*, and *Stitching*. Here, we focus on the first available trial to present ideas and describe our approach, noting that the remaining trials are captured by repeating the overall scenario with different human subjects. In each unmodified source case, a textual content line corresponds to one captured time instance at a rate of about 20 Hz, which corresponds to video frame capturing of the devices and their positions as part of the trials. The tools included in these traces are *Endoscope*, *Grasper1*, *Grasper2*, *Needleholder1*, *Needleholder2*, *Pointer*, *Scissors*, and *Trainer*.

As the frequency of measurements is not sufficient for the ultra-low-latency requirements of the Tactile Internet, we explode the original measurements to milliseconds via interpolation. Specifically, we initially round the source trace timestamps to full millisecond boundaries, followed by linear interpolation at millisecond boundaries. We note that in comparison to the previously evaluated traces, the accuracy here is less, as we overall approximate the generation of the data for future evaluation scenarios (i.e., here the first evaluation consideration lies on the type of data and later processing approaches, rather than an evaluation of real capture fidelity at ultra-low-latency times).

We employ the overall setup we previously introduced for the TUM traces, with the difference that the source data for experiments does not include the tool descriptions and only the values needed (8 values per tool, for all tools). Fig. 14.11

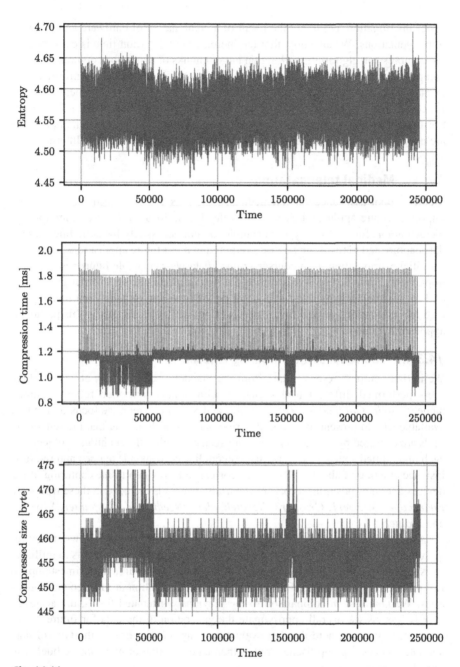

Fig. 14.11

Example results for the *Circle* interaction dataset and compression with *Snappy* (entropy, compression times, compression sizes).

illustrates the overall results, whereas Table 14.4 provides the resulting $\log_e$ entropy value, *Snappy* compression sizes, and *Snappy* compression time statistics.

Table 14.4 Overview statistics of $\log_e$ entropy and *Snappy* compression size in bytes and computational time in milliseconds for TUD interpolated trace data for Subject 1. Note that the number of samples is derived through interpolation onto the millisecond timescale.

Trace	Samples	Entropy		Snappy sizes		Snappy times	
		Mean	Std. dev.	Mean	Std. dev.	Mean	Std. dev.
Circle	245 096	4.569	0.025	452.059	3.455	1.184	0.0704
Gallbladder	334 546	4.576	0.026	455.613	4.169	1.113	0.072
Peg transfer	245 987	4.579	0.025	457.328	3.17	1.124	0.074
Stitching	274 831	4.536	0.027	454.049	5.049	1.138	0.07

Falling in line with prior observations, we observe average entropy values above 4.5, albeit here with a fairly narrow standard deviation. This indicates that the overall values contained in the data are highly variable, and compression, in turn, could be difficult. Indeed, the compression yield is fairly low, with averages above 452, accounting for a compression ratio of around 12 percent. The cost of this compression is just below 1.2 ms on average, which would consume the complete time for the entire control loop considering the Tactile Internet scenario with Human-in-the-Loop.

14.3.4.2 Interpolated JHU-ISI Gesture and Skill Assessment Working Set (JIGSAWS) reference dataset

The publicly available JIGSAWS dataset [1043] contains medical tool information, similar to the TUD traces we discussed prior in this section. The data for the JIGSAWS dataset was gathered directly from a DaVinci telesurgery robot through a dedicated interface. The available dataset was captured at a fixed frequency and does not contain the otherwise available timestamps. Similar to the TUD medical robot interaction traces, we extrapolate from the source data into smaller time intervals through linear interpolation. The main difference between the two approaches lies in the initial timing consideration (which here is assumed fixed at 30 Hz in opposition to the detailed timestamps available in the TUD source data) and the number of values (here 76 float-type values per provided measurement).

We initially illustrate the three different interactions contained in the dataset in Figs. 14.12–14.14, namely *knot tying*, *needle passing*, and *suturing*. These interactions jointly exhibit entropies around five, which results in variable compression values that are attained by compression computational times commonly above 1.6 ms. The detailed results for the subject evaluation *B001* are listed in Table 14.5.

Jointly, our results indicate that the *JIGSAWS* medical robot traces exhibit little general lossless compressibility without additional context. Kinesthetic information reduction is commonly geared towards a reduction of the high packet rate, rather than content (sample) compression. A combined approach of both packet rate reduction

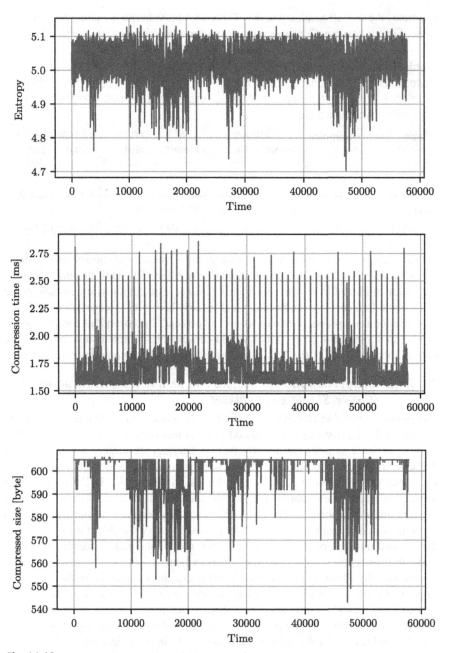

Fig. 14.12

Example results from the *B001* datasets for the *JIGSAW knot tying* interaction and compression with *Snappy* (entropy, compression times, compression sizes).

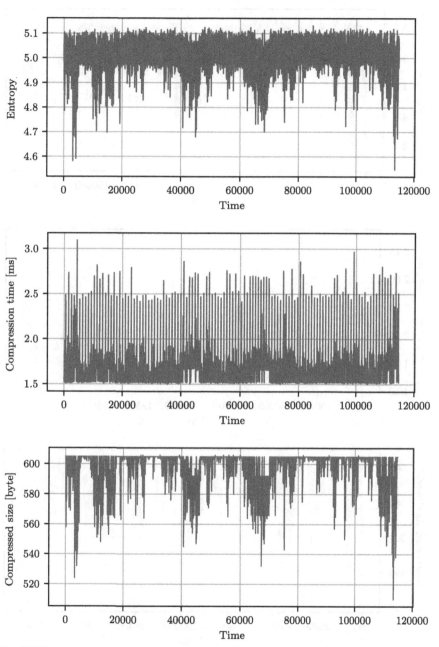

Fig. 14.13

Example results from the *B001* datasets for the *JIGSAW needle passing* interaction and compression with *Snappy* (entropy, compression times, compression sizes).

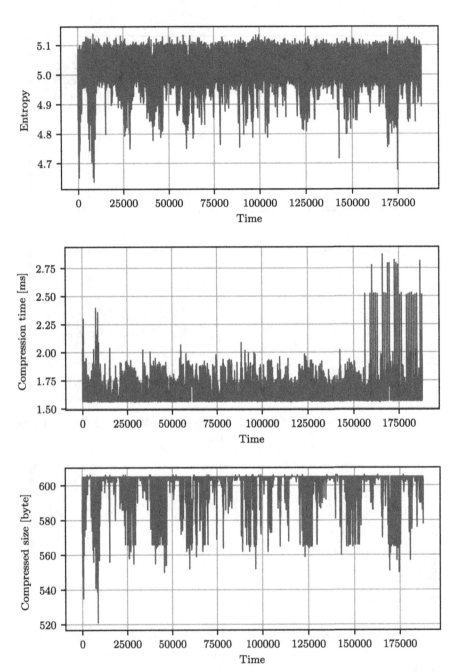

Fig. 14.14

Example results from the *B001* datasets for the *JIGSAW suturing* interaction and compression with *Snappy* (entropy, compression times, compression sizes).

Table 14.5 Overview of statistics derived from interpolated *JIGSAWS* traces for *B001*. Note that the number of samples is derived through interpolation onto the [ms] timescale.

Trace	Samples	Entropy		*Snappy* sizes		*Snappy* times	
		Mean	Std. dev.	Mean	Std. dev.	Mean	Std. dev.
Knot tying	57 801	5.026	0.039	601.633	7.584	1.671	0.114
Needle passing	114 567	5.017	0.049	600.170	9.560	1.624	0.125
Suturing	187 801	5.023	0.041	600.912	8.463	1.633	0.082

and sample compression, however, still requires more than common noncontextual compression approaches in the future.

14.4 Application scenarios

In this section, we outline several application considerations for the Tactile Internet that include Human-in-the-Loop configurations in the context of tactile traces.

14.4.1 Inclusion of metainformation

The prior examples in Section 14.2 showcased the utilization of different data in the communications between a controller and executing components utilizing an eSAP/SAP architecture, as described in Section 14.2.3. In addition, details concerning the individual components comprising the actuating/sensing/communicating/computing facets are required to enable meaningful utilization of the trace data by the broadest possible audience. As a simple continuation of our example, one could consider the differences in fidelity between sensing/actuating, whereby a sensor could feature significantly higher levels of accuracy than a stepping motor that is employed, utilizing the values from the sensor. If there is a (significant) mismatch, an immediate approach to compression without any loss of necessary information would be to reduce the fidelity of the data to be transmitted to a level that matches the one from the stepping motor.

In turn, users of our traces may have a need to access the individual component details (i.e., details about all type of SAP components). Similarly, they would have a need to have access to raw as well as computationally processed data that is generated along the way from source/destination SAPs to eSAPs and over the network. Part of this detail is the information about the frequency of data generation/consumption that is part of the overall setup, for which traces were generated. This, subsequently, also requires a description of the dependency/tree of components comprising the various SAPs as to enable potential research into the combinations of different data flows.

Lastly, to determine the potential impacts of modifications of the content of the traces (e.g., ascertaining the effects of lossy compression approaches or added delays), it would be highly beneficial for researchers utilizing such information to have time or fidelity constraints provided as part of any tactile trace offering.

14.4.2 Consideration for utilization

Without a more detailed knowledge of the sensor and actuator characteristics, one would still be enabled to perform optimization from the beginning and to the end of the local computing endpoints illustrated in Fig. 14.1.

14.4.3 Use-cases

We now highlight these considerations in several application scenarios. We categorize these scenarios into near-term (i.e., these are technically almost ready for production/implementation) and far-term (i.e., these scenarios require additional research and development efforts).

14.4.3.1 Near-term use-cases

Initial implementations that incorporate the TI have begun to percolate into prototypical to full-scale implementations beyond research. We provide some examples here, noting that significant developments for more are currently underway.

For example, initial implementations can be directly geared towards implementations in industrial control, either directly by controlling collaborative robots or other robots in real-time, or through fine-grained programming. The latter approach for industrial control environments can make direct utilization of body-worn sensors that allow an intuitive programming of these collaborative robots and other machines. An example for implementations that are already moving into the marketplace is the Wandelbots platform, which allows even laypersons to program and interact with robots.

An additional approach that can be derived is in human performance monitoring, where in sports and training environments, live athlete performance can be monitored in a fine-grained fashion. In this scenario, comparisons with past or other athlete performances can be derived in real-time, processed, and feedback provided to the athletes. This can include, for example, the capturing of human movements in real environments through wearable sensors, which can help the trainers to provide guidance on how movements can be improved.

14.4.3.2 Far-term use-cases

Additional implementations for the TI are manifold; we consider two examples in the following to highlight the long-term implications and changes that will derive from the TI and the need to provide researchers in these domains with TT as a shareable resource, as a reference, and for repeatability.

Robot-assisted surgery is already commonplace and gaining in momentum, as the level of invasion is reduced and the control is more precise. Telesurgery, on the other hand, is not yet at this level, as the network support to perform live surgery on human patients is not yet deployed. In turn, current robotic surgery is commonly performed by surgeons in the same or adjacent room in a hospital and not by, e.g., a specialist that is several miles away in a different hospital. In the future, the TI will enable telesurgery within the physical distance limits. The Speidel Group as part of CeTI is actively researching this particular domain.

Other use-cases, where the Tactile Internet can readily be employed are military (including combat) scenarios, where direct communication between soldiers and localized command and control infrastructure could be performed, e.g., soldier performance and injury sensing, feedback over potentially very low individual bandwidth links, or even aggregated by platoon.

14.5 Conclusion and outlook

14.5.1 Summary

We analyzed several traces from standard operating procedures and standard equipment to custom-based solutions for the overall robot control and transmission of tactile information for TaHiL scenarios. Our analyses indicate that overall, the tactile data to be transmitted over communication networks is generated at regular intervals, independent of the application scenario. Furthermore, the data is typically of a standardized format and of a fixed, relatively small size. For example, the tactile traces data in Section 14.3.2 are generated for 2.8 kHz source data, which are subsequently dimensionality-compressed over a 30 value window. Reversing the process yields 3 float-type values that would need to be transmitted for each sample or, if reduced to 16-bit floats, about 1 Mbps. This is readily achievable with today's networks. Considering other examples, such as the robot traces, one notices an increase in the data amounts, in addition to variations in the timing of arrival, such as described and modeled in [1021]. However, the common delay variations are comparatively small, especially when considering the use-cases for their employment.

14.5.2 Haptic traces: Quo vadis?

Thus one wonders whether haptic traces that include both kinesthetic and tactile information characterizations matter for the research community. Given that the current tactile data sources are, overall, similar to Constant Bit Rate (CBR) video in their nature, provisioning might be focused on the bandwidth and simple latency optimization. On the other hand, kinesthetic sources might pose greater challenges, due to their high temporal update rate of typically 1000 Hz, which may result in ultra-low-latency requirements. One can additionally argue that the trade-off that could stem from reducing the amount of data, e.g., through additional source coding, will increase the potential variability as well as overall latency, and is currently not beneficial to perform and evaluate. The potential trade-off between required compression time and reduced bandwidth from shaving off one byte from a 12 bytes source will likely not be worthwhile. This becomes more intuitive when considering robot application scenarios, where the actuators and sensors are additionally part of the end-to-end delays.

In direct control scenarios, either fast actuation is required, and compression optimization will increase latency further (emergency stop of an industrial robot), or high precision is required and movements are scaled with respect to precision anyways (surgical robots). We thus arrive at the following future research perspectives for tactile traces:

14.5.3 Future research perspectives
14.5.3.1 Focus on low-latency network transport

One perspective on future research is to focus on the reliable low-latency network transport of CBR haptic data flows with relatively modest bitrates. For this research perspective, it seems more beneficial to target multistream optimizations, protocol design, and latency optimizations across the protocol stack, rather than focusing on the tactile data itself. Extensive network protocol research has already established an extensive foundation for low-latency transport, e.g., [1044–1052]. Moreover, the time-sensitive data processing and decision-making in computing infrastructures close to the location of the tactile Internet application in so-called multiaccess edge computing or similar structures has begun to be extensively investigated, e.g., [8,462,471,1053–1055]. Related adaptive intelligent communication control that can steer tactile traffic flows so as to ensure low-latency routing and processing has been investigated in the context of software-defined networking Software Defined Network (SDN) [1056–1058]. Future research on low-latency network transport and processing of Tactile Internet data should further enhance these approaches and protocols to efficiently and effectively support Tactile Internet data flows.

Similarly, future research should further advance the control paradigms for Tactile Internet applications. For instance, robotic control can operate in a stable and accurate manner with relatively long latencies by decoupling of the control loops, e.g., so-called *digital twin* models of the controlled robot [1024,1059–1061]. These approaches can be further enhanced, e.g., through optimizations based on near-real-time context updates provided by the advancing low-latency communication and multiaccess edge computing approaches, as well as through emerging machine learning approaches.

14.5.3.2 Expansion of tactile-sensing capabilities

A complementary future research perspective is that Tactile Internet data sensing and processing is currently in its infancy. The current status of the Tactile Internet data sensing and processing may be roughly equivalent to the early CBR video streams encoded into H.261 for constant bitrate transmission over Integrated Services Digital Network (ISDN) channels in the early 1990s [1062]. The area of video coding and network transport has greatly evolved over the past three decades, resulting in advanced coding approaches that result in highly variable bitrates as the texture and motion content of the videos varies [1063–1066], as well as video network transport approaches that accommodate the variable video bitrates [1067–1071].

We anticipate that the field of Tactile Internet data sensing will flourish in a similar manner in the coming decades, as the video coding and streaming field has over the past three decades. Indeed, there is a rapidly growing research area on wearables with integrated tactile sensors [1072–1075]. Also, the CeTI team is currently working on an eGlove with 40 gyristors. As more and more tactile data is collected by sensors on a particular device or human, there will be a growing need to judiciously compress the data to enable transmission to the Internet at large. Such compression will likely lead to complex tactile data traffic characteristics that will need to be extensively collected in traces, and then thoroughly examined and modeled. The resulting tactile traffic traces and models will then need to be employed as traffic traces and models in the design and evaluation of low-latency communication protocols and computing approaches.

Tactile Internet standards of the IEEE P1918.1 Working Group

15

Meryem Simsek[a], Sharief Oteafy[b], Zaher Dawy[c], Mohamad Eid[d], Oliver Holland[e], and Eckehard Steinbach[f]

[a]*International Computer Science Institute, Berkeley, CA, United States*
[b]*DePaul University, Chicago, IL, United States*
[c]*American University of Beirut, Beirut, Lebanon*
[d]*New York University Abu Dhabi, Abu Dhabi, United Arab Emirates*
[e]*Advanced Wireless Technology Group, Ltd., London, United Kingdom*
[f]*Technical University of Munich, Munich, Germany*

If you think of standardization as the best that you know today, but which is to be improved tomorrow; you get somewhere.
– Henry Ford

15.1 Introduction

Mobile communication plays a key role in the modern economy and society. At the same time, the current Internet has created a key infrastructure component for our modern world, touching almost every aspect of our daily lives. The Internet enabled access to information and has allowed emerging economies to participate in the global economy. The next big wave of the Internet innovation is now approaching: the Tactile Internet (TI). The TI pushed boundaries of Internet-based applications to remote physical interaction, networked control of highly dynamic processes, and the communication of touch experiences (e.g., [1,1076]). Hereby, the Tactile Internet with Human-in-the-Loop (TaHiL) will yield to innovation in quasi-real-time human–machine interactions in real, virtual, mixed, and remote environments enabling a broad range of new applications in various fields. Whereas senses like hearing (audio) and sight (visual), or a combination of them (audiovisual) are relatively less challenging to transmit, touch (haptics) has much stricter communication requirements. One reason for this is that stable and ultra-low-latency interaction needs to be guaranteed if the intention is to achieve sensorimotor control over the communication medium in perceived real-time. This interaction can be with a Human-in-the-Loop or with any virtual or real object, e.g., a machine.

Central to the TI is the more general realization of new realms of communication applications not only requiring the ultra-low-latency touch interaction, but also

Tactile Internet. https://doi.org/10.1016/B978-0-12-821343-8.00028-9

ultra-high reliability, security, and availability, such as industrial control (pertaining to *Industry 4.0* scenarios [1077]). Ultrahigh reliability might also be required in many other TI scenarios, both in TaHiL as well as machine-in-the-loop cases.

Within the context of the TI, components that comprise a TI system as well as the dedicated haptic human-interaction hardware, typically use different and often proprietary communication/interaction formats and means. Moreover, elements and structures of End-to-End (E2E) TI deployment might significantly vary and even be conflicting in different solutions. It is therefore of utmost importance to standardize aspects of the TI to harmonize such essentials. This will allow TI components to freely interact with each other directly out-of-the-box, without requiring custom/proprietary communication design that is dependent on the scenario and specific set of equipment used. Such standardization will also facilitate other aspects of the network supporting the TI to be deployed in a consistent way, such as network-side processing.

To this end, the Institute of Electrical and Electronics Engineers (IEEE) P1918.1 TI Standards Working Group (WG) [1078] was formulated initially out of the IEEE ComSoc Standards Development Board (COM/SDB) Fifth Generation (5G) Rapid Rapid Reaction Standardization Initiative (RRSI) as a collaborative effort of King's College London and Technical University of Dresden (the latter having originated the concept of the TI) to bring a proposal for TI standardization to a RRSI meeting in Santa Clara, CA, USA, in November 2015.

15.2 Definition of the Tactile Internet

The TI is one key example of the benefits of some of the pioneering capabilities argued for 5G-and-beyond communication systems, specifically the Ultra-Reliable Low-Latency Communication (URLLC). However, the TI standardization cannot redefine the standards that are being developed to realize 5G or other networks, or combinations of networks, on which it might run. In most realistic scenarios the TI must simply operate on top of them. With this in mind, the IEEE P1918.1 TI standards WG is intended to complement and identify what is missing in 5G and other appropriate networks that might serve the TI, such as haptic communication protocols/codecs, and, e.g., network-side support for the TI, emulating remote physical environments. The IEEE P1918.1 WG and its standards, particularly the IEEE P1918.1 baseline standard, also intend to define which functionalities and functional entities have to be present in which locations, the relationships between them, how they are interfaced, and how the overall network is invoked among other considerations.

The TI has undergone different interpretations from different adopters, each having different objectives for the use of the technology. It is therefore key to define the TI to precisely understand and define the terminology. To this end, the definition of the Tactile Internet has been introduced within the IEEE P1918.1 WG.

It is also crucial to define the context of the TI's operation and interactions. Building on the above definition, seven core aspects of the TI are detailed as basic assumptions of the WG: (*i*) The TI provides a medium for remote physical interaction, which often requires the exchange of haptic information. (*ii*) This interaction may be among humans or machines, or humans and machines. (*iii*) In the context of TI operation, the term *object* refers to any form of physical entity, including humans. Machines may include robots, networked functions, software, or any other connected entity. (*iv*) Scenarios encompassing Human-in-the-Loop physical interaction with haptic feedback are often referred to as bilateral haptic teleoperation. The goal of TI in such scenarios is that humans should not be able to distinguish between locally executing a manipulation task compared to remotely performing the same task across the TI. (*v*) The results of machine-in-the-loop physical interactions will ideally be the same as if the machines were interacting with objects directly at—or close to—the locations of those objects. (*vi*) There are two broad categories of haptic information, namely tactile or kinesthetic. There may also be a combination of both. Tactile information refers to the perception of information by the various receptors (mechanoreceptor, thermoreceptor, and nociceptor) of the human skin, such as surface texture, friction, temperature, and pain. Kinesthetic information refers to the information perceived by the mechanoreceptors in the skeleton, muscles, and tendons of the human body, such as force, torque, position, and velocity. (*vii*) The definition of perceived real-time may differ for humans and machines, and is therefore use-case specific.

Though the purpose of this section is to clearly define the TI and identify the scope of its interactions, it is ongoing work to ensure a common understanding of the metrics and Key Performance Indicator (KPI) used, contrasting, for example, with the definitions of latency that are being adopted under the 3rd-Generation Partnership Project (3GPP) [1079,1080]. Further work is also continuing around the definitions of functions for the TI, together with the selection or creation of definitions of basic as well as composite concepts that are repeatedly used in the standard. The exhaustive list of such definitions will be included in the baseline standard.

15.3 IEEE P1918.1 Tactile Internet Standards Working Group

The IEEE P1918.1 TI Standards WG was approved in November 2015. The approval from the RRSI meeting led to the development of a Project Authorization Request (PAR), itself also thereafter being approved by the COM/SDB to be submitted to the IEEE Standards Association (IEEE-SA) New Standards Committee (NesCom) for their consideration. This process, including all the stages in-between, led to the approval of the PAR by NesCom and the wider IEEE-SA in March 2016, with the project to develop the baseline standard being authorized to operate until the end of 2020.

According to the PAR of IEEE P1918.1 [1081], the scope of the baseline standard is to define a framework for the TI, including descriptions of its application scenarios, definitions and terminology, necessary functions involved, and technical assumptions. This also fundamentally includes the definition of a reference model and architecture for the TI, comprising the detailing of common architectural entities, interfaces between those entities, and the definition and mapping of functions to those entities. Moreover, in contrast to the general perception that the TI focuses only on ultra-low-latency use-cases, it is noted that the TI encompasses mission-critical applications, as well as noncritical applications. The developed standard therefore takes into account for the high reliability, security, and availability as well as low latency that apply in some TI use-cases. At the same time, however, it must also be compatible with a considerable relaxation of such aspects even towards relatively high latency, low reliability TI scenarios.

The standard also includes core baseline work on terminology for the TI, such that the standard's instructions can be consistently followed for the TI to be compatibly realized and understood among the manufacturers, operators, end-users, and others that might take advantage of or implement the service, or that might in some other way be stakeholders.

Expanding on the PAR, the IEEE P1918.1 WG and its baseline standard aim to serve as a foundation for the TI in general: a toolbox of items needed to invoke TI services from a network architecture and functionalities point of view. This includes the definition of entities that have to be involved in the end-to-end communication and interaction, the mapping of functions to those entities, the interfacing of those entities/functions, and the core additional likely higher-layer new functionalities that support the TI, such as Network Processing Support (likely termed a *Support Engine* in the WG's context) providing various functionalities, e.g., emulating the remote environment. Furthermore, the standard defines various use-cases for the TI, the ultimate objective being that the selection among those use-cases will be made at the invocation of the TI service, and the network will be configured accordingly. Use-cases in IEEE P1918.1 are therefore defined in a codified way.

In addition to the above, the IEEE P1918.1 WG and its baseline standard aim to serve as a foundation for further standards adding extra capabilities, functionalities, or other complementary aspects related to the TI. An example of this is illustrated in Fig. 15.1. It can be seen that there can be *additional standards within the WG*, which serve as standards in their own rights, i.e., they might operate alone, likely in conjunction with the baseline standard and its associated assumptions, but not as a requirement. One reason for encompassing this form of standard is to maximize flexibility and impact. These additional standards are numbered IEEE P1918.1.X, *X* being a numerical designation in the same temporal order in the PAR for the standards that are approved. Examples of such standards related to the TI are illustrated in the bottom row in Fig. 15.1. The IEEE P1918.1.1 *Haptic Codecs for the TI* standard is an already initiated example of this, where the codecs being developed therein will be operable on the IEEE P1918.1 baseline standard scenarios and architecture, but will also be usable in far removed contexts outside of P1918.1, e.g., over a range of other networks.

The other form of standards is an *amendment standard*, which might add specific functionalities to the TI baseline standard building its capabilities. These are numbered IEEE P1918.1a, IEEE P1918.1b, etc., *a* and *b* being the first and second standards in terms of order of PAR approval. They are illustrated on the right

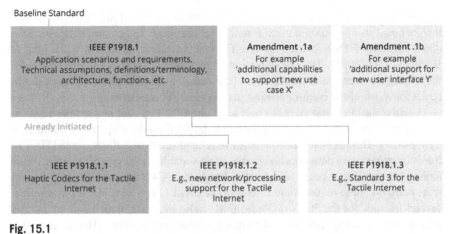

Fig. 15.1

The TI Standards WG and its baseline standard as a foundation for further TI standards. Note, all standards projects indicated are possible examples, except for IEEE P1918.1 and IEEE P1918.1.1 which are already initiated and for which work is ongoing.

side of the top row in Fig. 15.1. Within the TI WG context some examples of such amendment standards might be the addition of new use-cases and potentially entire protocols (as optional modes of operation) to support the use-cases, and perhaps new architectural entities supporting the TI, among others.

Differentiating IEEE P1918.1 from other standardization groups The purpose of this section is to provide some background on completed or ongoing standardization activities of other standardization groups, either directly covering or related to the TI and its associated capabilities. The ultimate goal is to position and differentiate the IEEE TI WG and its standards.

At the top of the hierarchy in an international regulatory-standards sense, International Telecommunication Union Standardization Sector (ITU-T) has defined the TI as a (*Technology Watch*) area, and prepared an associated report. The report covers aspects such as the TI's applications both in mission-critical and noncritical scopes, its benefits for society, implications for equipment, and other areas [1082]. Many of the covered aspects serve for the purpose of affirming the importance of this new technology as well as its introductorily definition and scope. These are all initial steps in assessing the need and potential for standardization. At a similar international level, the International Standards Organization (ISO) has prepared a standard covering aspects of human-system interaction, and specifically in this case haptic/tactile interaction [1083].

The Society of Motion Picture and Television Engineers (SMPTE) has defined a standard aiming to capture the essence of haptic/tactile information, as well as what needs to be communicated and how it is represented, for the purpose of broadcasting haptic/tactile information together with audiovisual information [1084]. This provides an interesting new viewpoint, given the unidirectional nature of broadcasting and associated implications for reliability and latency (and the flexibility in both thereof), and its *open-loop* nature. Finally, the European Telecommunications Standards Institute (ETSI) focused on the TI standardization through a work item on IPv6-based TI [1085]. The consideration of the higher layers is essential to realize the TI performance requirements in an E2E sense, given the involvement of the Internet for much of the communication path in many TI scenarios.

Again at the international regulatory level, the ITU-T has defined requirements for 5G communication systems [1086]. Here, the URLLC mode of operation could serve TI use-cases. This is essential in the scope of related standards efforts, since ITU-T refines communication networks and makes them suitable for carrying TI/haptic traffic, among other traffic types. On the other hand, the 3GPP is standardizing the systems realizing these requirements. To address URLLC services, 3GPP has specified several features for the 5G New Radio (NR) radio interface, which can be grouped into latency-reducing features and reliability-enhancing features [1087], as well as has defined a novel architecture, entities, and interfaces to support such use-cases (simultaneously). In addition, the 5G communication systems embrace several new communication paradigms that are beneficial for TI [1076,1087]. One major change is the transformation of the network from a hardware-based to a software-based network as well as the virtualization of network functions. Within such a network, not only network functionality can be realized, but also application functions can be placed and executed on this distributed cloud platform. This allows putting applications at locations that provide best performance, such as edge computing at the base station to minimize latency.

3GPP has also specified reliability-enhancing features for 5G NR, so that the data transmissions over the radio interface can be guaranteed with a defined latency bound. These features include the definition of highly robust transmission modes, including robust coding and modulation schemes, robust multiantenna transmission modes, and dual-connectivity, where the data is duplicated at the transmitter and simultaneously transmitted to the receiver via different radio links.

A major challenge remains when TI services are applied over longer distances. This challenge alludes to a first differentiating aspect of the TI and the IEEE P1918.1 TI WG standards effort compared with 5G URLLC, for example. Such capability can be achieved through network-side support functions built into the TI architecture, as envisioned through the standards work in IEEE P1918.1 [1078].

Furthermore, the TI can be framed as an application, with unique characteristics implied by that application and with the expectation that the application can be deployed as an overlay network on top of almost any network or combination of networks—not intended to apply only in the context of 5G URLLC as the underlying communication means. Noting the greatly increased network flexibility in 5G and beyond contexts through softwarization of network functions, the IEEE P1918.1 TI standards effort aims to be able to invoke the E2E TI service on top of such capabilities, conveying the constraints to configure network entities, interfaces, and other factors based on the specific use-case in the context of such fully flexible networks. This is acknowledging, however, that the TI and IEEE P1918.1

standard will also deal with the mapping of entities, interfaces, etc., to the hardware deployed in cases or portions of the utilized networks, where there is less flexibility or no flexibility. Indeed, the developed architecture aims to act as a bootstrapping of such an overlay network, providing the means to rendezvous and negotiate/configure requirements/capabilities over each link towards the realization of the required architectural components/entities and overall communication path(s) to invoke the E2E use-case and associated E2E requirements, using whichever appropriate communication means, or combination of means, available. Depending on the deployment scenario, such a bootstrapping might be combined with the utilized haptic codec negotiation or mode of operation information exchange being covered under the scope of IEEE P1918.1.1 standards effort.

15.4 IEEE P1918.1 architecture

The IEEE P1918.1 working group defines an architecture for the TI. Some key aspects are detailed in the following subsections:

15.4.1 System and functional architecture

The IEEE P1918.1 standard is based on a reference system architecture that has been designed to meet the requirements of TI use-cases. The architecture is based on a set of key principles that include the following: (i) high performance capability to meet the stringent delay and reliability targets of TI services, (ii) flexibility to support a wide range of TI use-cases with varying quality requirements, (iii) scalability to handle a large density of simultaneous TI connections, (iv) adaptability to optimize performance dynamically and in real-time (based on traffic and network conditions), (v) interoperability to facilitate connectivity over multiple existing network technologies and with external application service providers, and (vi) intelligence to benefit from the promising potential of artificial intelligence advances, combined with edge caching and computing techniques. The key components of the reference architecture are summarized in Figs. 15.2 and 15.3. Tactile Device (TD) communicate tactile/haptic information over end-to-end connections that span tactile edge and core network domain elements. Each TD shall have the following characteristics: (i) support functions such as sensing, actuation, haptic feedback, or control; (ii) be equipped with at least one network connectivity interface, normally wireless; (iii) perform monitoring, management, and optimization functionalities to achieve the target quality of service requirements.

The tactile edge consists of one or multiple TDs, which can either be physical devices (e.g., sensors or haptic devices) or a virtual implementation thereof that emulates their functions and interfaces. TDs within a given tactile edge may communicate among each other over peer-to-peer links. In addition, they may cooperate among each other to provide data relaying in addition to distributed storage and computing functionalities.

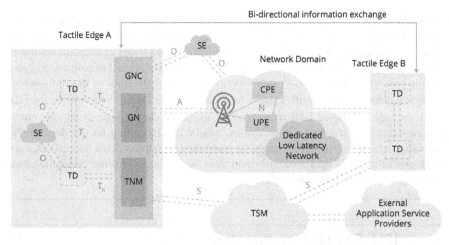

Fig. 15.2

TI reference architecture with the GNC residing in the tactile edge.

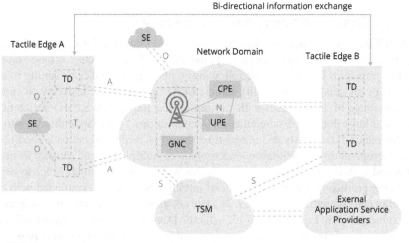

Fig. 15.3

TI reference architecture with the GNC residing in the network domain.

The TDs connect to the core network domain through a Gateway Node Controller (GNC), which consists of a data-plane Gateway Node (GN) and a control-plane Tactile Network Manager (TNM). The GNC is an entity with enhanced networking/communications and management capabilities, and resides at the interface between the TDs in the tactile edge and the core network domain. The TNM manages functions, such as admission control, resource allocation, service provisioning, mobility management, and connection management. The GNC can reside either in the tactile edge (see Fig. 15.2) or the network domain (see Fig. 15.3), depending on the deployment scenario design and implementation. Each of these two options has its pros and cons; for example, having the GNC part of the tactile edge allows for more customized and advanced functionality closer to the TDs, which can ultimately provide enhanced performance. On the other hand, having the GNC part of the core network domain, such as 5G core, can benefit from existing infrastructure and the management and orchestration functionalities that are already therein.

The core network domain acts as the backbone to connect two or more tactile edges to each other. It shall be able to provide an adequate level of performance under certain conditions to meet the end-to-end performance requirements. The reference architecture is generic to support multiple possible core network domain technologies [1076,1088,1089], including: (*i*) shared wireless network (e.g., cellular 5G and beyond radio access and core networks); (*ii*) shared wired network (e.g., Internet core network routers); (*iii*) dedicated wireless network (e.g., proprietary networks or point-to-point microwave or millimeter wave links); (*iv*) dedicated wired network (e.g., point-to-point fiber optic leased line links).

The core network domain normally includes User Plane Entity (UPE) and Control Plane Entity (CPE) to handle data forwarding and control management functionalities. In addition, TDs in two tactile edges can communicate with each other over multiple network domains simultaneously, which can enhance reliability due to redundancy and reduce latency due to traffic splitting [1090–1092].

As noted above, one of the key principles behind the design of the reference architecture is intelligence to provide computing, analytics and storage resources for improving the performance of the tactile edges to meet the strict delay and reliability requirements. This is realized in the architecture via the Subjective Equality (SE) node. The SE will be equipped with advanced computing and storage resources to run Artificial Intelligence (AI), predictive learning, and distributed computing algorithms. This has a plethora of potential impactful applications that include haptic rendering, motion-trajectory prediction, sensory compensation [419], edge caching [1093], workload assignment [1094], and network load estimation, among others. The SE can reside either within the tactile edge or remotely in the cloud, and it can be based on either centralized or distributed architecture. It can be even deployed on demand using drones to complement existing infrastructure-based resources [1095].

Finally, the Tactile Service Manager (TSM) supports service-level functions, such as device registration and authentication, and serves as a gateway to Application Service Provider (ASP). In general, TI applications can be developed and offered either by TI network operators or external ASP.

15.4.2 Interfaces

A number of basic interfaces have been defined to serve interactions among the key entities in the TI architecture, as shown in Figs. 15.2 and 15.3. The key identified physical interfaces include the following:

Access (A) interface This interface provides connectivity between the tactile edge and the network domain. It is the main reference point for the user-plane and the control-plane information exchange between the network domain and the tactile edge. The end points of the A are dependent on the variant of the reference architecture. In the first scenario depicted in Fig. 15.2, the A connects the GN and the Network Controller (NC) to a network domain entity, e.g., as a cellular base station. In the second scenario depicted in Fig. 15.3, the A connects a tactile device to the GN and the NC; both GN and NC belong to the network domain.

Tactile (T) interface This interface provides connectivity between entities within the tactile edge. It is the main reference point for the user-plane and the control-plane information exchange between the entities of the tactile edge. The T Interface is divided into two subinterfaces Ta and Tb to support different modes of TD connectivity, whereby the Ta interface is used for TD-to-TD communications and the Tb interface is used for TD-to-GNC communications when the GNC resides in the tactile edge.

Open (O) interface This interface provides connectivity between any architectural entity and the support engine. The O provides connectivity between the support engine and any entity related to the tactile edge, e.g., the TD or the GN and the NC. In another case, the O provides connectivity between the support engine and defined network domain entities.

Service (S) interface This interface provides connectivity between the TSM and the GNC. The S interface carries control-plane information only.

Network-side (N) interface This interface refers to any interface providing internal connectivity between network domain entities. The N is the main reference point for the user-plane and control-plane information exchange between the network domain entities.

In addition to the physical interfaces mentioned above, the reference architecture has identified various logical interfaces denoted by L followed by a number, as follows: (*i*) The L0 interface interconnects the GN and the NC. (*ii*) The L1 interface interconnects a TD and the TSM. (*iii*) The L2 interface interconnects a TD and the CPE. (*iv*) The L3 interface interconnects a TD and the UPE.

In terms of performance requirements, meeting the end-to-end quality of service targets for active TI sessions imposes specific requirements on each of the interfaces along the path from source to destination TDs. The relationship between the end-to-end requirements and the per-interface requirements is complex, due to statistical variability per interface and interdependence among the different interfaces.

The KPIs for each interface include the following: (*i*) The *availability* of an interface is defined as the probability that a given interface is available to be utilized by any user-plane or control-plane connectivity service. (*ii*) The *reliability* of an interface measures its packet delivery performance. It is defined as the capability of transmitting a fixed-size Protocol Data Unit (PDU) within a predefined time duration with high success probability. (*iii*) The *latency* of an interface is a measure of its responsiveness. It is defined as the capability to successfully deliver a protocol layer packet from a transmitter to the same protocol layer receiver point to satisfy the end-to-end latency requirements. The end-to-end latency is defined as the one-way delay to successfully deliver an application layer packet from a tactile device in tactile edge A to a tactile device in tactile edge B. (*iv*) The *scalability* of an interface describes its capability to cope and perform under an increased number of devices. It is defined as the maximum number of devices that can be supported without deteriorating the availability, reliability, and latency requirements. (*v*) The *security* of an interface describes its capability to perform secure TD connectivity, including data integrity, user authentication, and data encryption capabilities. (*vi*) The *privacy* of an interface is a measure to indicate the user data transmission and storage is according to the service access level policy defined by the TSM. Typical requirements per interface are summarized in [1096], where two grades of service are defined: normal-grade and ultragrade. This is to better capture the variability in requirements among different use-cases.

15.4.3 Bootstrapping of the Tactile Internet service and architecture instantiation

The TI is envisioned to connect a multitude of devices, potentially across continents. One of the challenges of establishing the TI architecture is facilitating rapid E2E connections, and coordinating the operational mandates of the TI formed by its communicating Tactile Edges (TEs). Over the span of a TI session, it is pivotal to determine how TI communication will be invoked, the TI devices involved, and the potential paradigms under which TI communication would operate, and terminate. This is a challenging undertaking, given the volatility of network resources over time, and more critically the URLLC mandates that dictate careful selection and maintenance of sessions over TI devices and networks.

The E2E connection establishment is referred to as TI bootstrapping, after which a TI session would be instantiated, maintained, and eventually terminated. Prudently, choosing a single bootstrapping paradigm would be stifling. The involvement of different architectural components, both for rendezvous and session management, has been carefully considered in presenting TI bootstrapping techniques.

Three paradigms for TI bootstrapping are presented, building on different operational mandates. Our goal is to have two TI edges bootstrap their communication and remote interaction. The design and execution of these three paradigms depend on a number of factors, encompassing latency, reliability, availability of TI resources, spatial configurations and distance between TDs. The variation between these paradigms stems from the ubiquity and cost of utilizing different network backbones, as well as the slicing of resources dedicated to TI operation, both at the edge and core of the network [1076]. These three paradigms are described in the following together with their requirements and limitations.

15.4.3.1 Omnipresent Tactile Internet paradigm

This paradigm builds on a ubiquitous deployment of TI components, which form a readily-accessible TI core topology. This core is built on the TSM and CPE, in addition to a redundant distribution of strategically placed NC modules over all regions of TI operation. Ultimately, this design enables a quick *latching* of TDs onto the omnipresent TI core infrastructure, facilitating rapid bootstrapping. These access points, mostly represented as NCs, could be realized as dedicated TI hardware that is strategically placed in high-traffic TI zones or manifested on Software Defined Network (SDN) components.

This paradigm would enable quick TI bootstrapping, where the first task of a TD aiming to initiate a TI session would be to identify the optimal latching NC. This decision will likely factor distance, reliability of connection to the NC, the capability of the NC, its resources and load, among other factors. However, building such an omnipresent TI infrastructure would present a significant cost for the deployment and sustenance of such NCs and TI core components. This cost is compounded by the fact that many of these TI components would overlay on other existing network topologies, which have varying operational mandates and resilience measures.

15.4.3.2 Ad hoc Tactile Internet paradigm

Many envisioned TI deployments will have geographic or URLLC restrictions, that mandate a unique approach to TI bootstrapping. These TI networks will mostly benefit from shorter-range networks, which build on ad hoc topologies and restricted hop-counts. Thus, for such cases, there is neither the support nor the need for long-range communication across TI components. For example, TI networks that are used for automation and control in emergency settings, need not resort to established omnipresent topologies.

To address this subset of TI networks, the ad hoc TI communication paradigm is built, which forgoes the omnipresent infrastructure, and delegates TI connection setup from the edge to the GNC. Under this paradigm, if the TI session is confined within a single TI edge, then the GNC will solely assume the task of establishing, maintaining and terminating the connection with other TDs on that edge. However, if the TI session requests bootstrapping TDs beyond a single TE, then the initiating GNC will solicit the cooperation of a TSM module, from a predesignated set of TSM modules that are frequently updated in the GNCs registry. If the registry is empty, then a TSM-discovery protocol will be invoked to populate it with potential TSMs. Thereafter, the chosen TSM will take over the task of orchestrating end-to-end TI bootstrapping, which includes recruiting/soliciting the services of all required TI network components.

While this paradigm is inherently minimalistic, it is less resource-hungry in setup, and would present a viable *starting* point for TI architectures. Evidently, TI bootstrapping here would take significantly longer time than in the omnipresent paradigm, especially in the recruitment phase of the TSM looking for E2E resources.

15.4.3.3 Hybrid Tactile Internet paradigm

As more TI deployments are realized, it is fair to assume that a number of TI providers would invest in geographically distributed *focal* TI anchors, that could enable faster TI bootstrapping. These focal points would be expected to remain online, with the inherent task of facilitating E2E bootstrapping.

Thus, a hybrid TI paradigm is proposed that would capitalize on abundantly available rendezvous points, that would be pre-deployed (potentially with distributed surrogates) to facilitate TI anchoring across regions of TI operation. These anchor points would manage incoming TI bootstrapping requests, and maintain a list of accessible TI resources and their duty cycles. As with the case of NCs, such rendezvous points could be virtualized in Network Function Virtualization (NFV) modules, on already deployed network topologies.

Under this paradigm, TI bootstrapping is formed in a two-step process: finding the anchor point and then triggering it to start a cascading protocol to tether with other rendezvous points from E2E. The rendezvous component could be incorporated into the operation of the TSM.

15.4.3.4 Contrasting Tactile Internet bootstrapping paradigms

As evident in this book, TI use cases span a multitude of scenarios, with varying levels of involvement from humans and machines. TI bootstrapping is inherently complicated, and does not conform to a single operation model. Boostrapping is subject to varying network conditions, such as fluctuations in KPIs, which often present hurdles in assuring URLLC. Although recent efforts have attempted to leverage multiple network interfaces to boost URLLC performance [1097], they are not directly applicable to the inherently variable challenge of TI.

TI communication is ultimately edge-driven, yet its maintenance and operation is core-managed. That is, TI devices at the TE initiate the sessions with a deterministic set of operational mandates (such as reliability, expected latency, etc.) and attempt to leverage core-resources to establish E2E communication. The choice of the bootstrapping paradigm is influenced by the assumptions made on apriori setup of URLLC paths, which includes leveraging network resources [1098].

Recent developments in a number of networking domains are further realizing core functionalities that could leverage TI operation. This includes recent developments in SDNs and network coding, which would improve access and communication latency [2], thus facilitating resilient E2E access under more stringent time mandates. This would enable elastic operation under the hybrid and ad hoc bootstrapping paradigms. Moreover, capitalizing on fine-tuned offloading, and leveraging cloud variants [1099] will aid both core and edge TI network components in maintaining and improving TI operation. Moreover, building on Edge technologies could improve URLLC delivery, and sustain rapid access to TDs under resource provisioning schemes that capitalize on mobile edge computing [1100]. Moreover, TI bootstrapping could exploit nearby Internet of Things (IoT) resources to leverage their sensing, connectivity and actuation capabilities, especially in systems that are designed with reliability at their core [1101]. In fact, the potential for IoT and Edge technologies in the realm of TI operation has been expanded upon in [1102].

The past few years have witnessed significant leaps in resource discovery at the network edge. Fog computing architectures will enable rapid resource discovery at the edge, potentiating a previously untapped pool of resources. These resources could be utilized even in mobile settings, with agile decision making on the cost and latency of offloading [1103]. More recently, developments in Fog Computing based Radio Access Network (F-RAN) [1104] present new frontiers of performance gain both in edge and network domain bootstrapping procedures.

15.4.4 Tactile Internet operational states

The operation of a TD needs careful management to ensure predictable behavior and measurable KPIs. A deterministic Finite State Machine is presented that defines the operation of a TD across its operational life cycle. That is encompassed from the instant the TD attempts to start a TI session, then bootstrapping, and throughout its authentication, operation, and termination states.

The remainder of this subsection details the operational states of a generic TD as it invokes a TI session, and carries out the operations required to start, maintain, recover, and terminate a TI session. These states are summarized in Table 15.1. Furthermore, the deterministic transitions between these states are depicted in Fig. 15.4.

The starting point for each TD would be the registration phase, defined as the task of establishing communication with (latching on) the TI architecture. In the omnipresent TI bootstrapping paradigm, the TD would register with a GNC, which may also involve communication with the TSM. Thereafter, the *latching* point of the TD to initiate registration will be referred to as the TI Anchor. The is the first stage, where the TD is attempting to invoke E2E communication, and thereby is not allowed to carry out any functions beyond latching onto the TI backbone, if present. In both the ad hoc and the hybrid models, this step will involve the TSM to establish registration, often via the GNC in the former model.

The next state depends on the type of the TD. If it is a lower-end SN/AN, then the TD will have a designated *parent* in its close proximity, with which the TD will need to associate with first. This parent TI node will thereafter ensure reliable operation and assist in connection establishment and error recovery. If a TD device operates independently, then this would be an optional step.

Some mission-critical TDs, as well as new ones, may need to be authenticated prior to being allowed to join/start a TI session. The third phase, is an optional state in which a TD would communicate with the authenticating agent in the TI infrastructure to carry out authentication. The TSM is the main module that could carry out this task, perhaps with assistance from the SE when needed, or with significant amounts of traffic.

The TD will then commence its E2E control synchronization, where it will probe and establish a link to the end tactile edge. At this state, the TD is not allowed to communicate operational data, yet would focus on relaying connection setup and

Table 15.1 Describing the operational states of a general TD, when initiating communication with another TD.

State: mandatory or optional	Functional mandate	Functional capacity	TI node connected to
Registration	Register with TI architecture	TI *probing*	TI anchor
Association	Pair with *parent* TI node	TI association	TI parent node
Authentication	Authenticate with designated TI network-domain component	Exchange authentication messages	TI authentication system
Control synchronization	Establish E2E connection (set communication parameters)	Communicate with other TI nodes for *control only*	TI nodes (edge and core)
Haptic synchronization	Establish E2E haptic-specific parameters (session, codecs, etc.)	Communicate with other TI nodes for *control only*	TI nodes (edge and core)
Operation	Carry out normal TI operation	Communicate with other TI nodes for *data only*	TI nodes (edge and core)
Recovery (network)	Recover from network failure	Communicate with other TI nodes for *control only*	TI nodes (edge and core)
Recovery (operational)	Recover from operational error or data/encoding errors	Communicate with other TI nodes for *control only*	TI nodes (edge and core)
Termination	Terminate connection	Tear down connection with TI edge and core	TI nodes (edge, core, local parent)

maintenance parameters. This may include setting the parameters for the interfaces along the E2E path, which will aid the network domain in selecting the optimal path throughout the network to deliver the requested connection parameters. This is a critical state, as it encompasses the path establishment and route selection phases of TI operation. More importantly, it will typically involve multiple tiers of the TI architecture, which will communicate to ensure that a path that meets the minimum requirements set in the *setup* message is indeed available and reserved.

If the TD engaging in a TI session is targeting haptic communication, then the next state would encompass the specific communication and establishment of haptic-specific information, still before actual data communication. This state is pivotal in deciding on the codecs, session parameters, and messaging formats specific to this current TI session. While different use cases may mandate different haptic exchange frequencies, it is expected that every haptic communication will start with the haptic

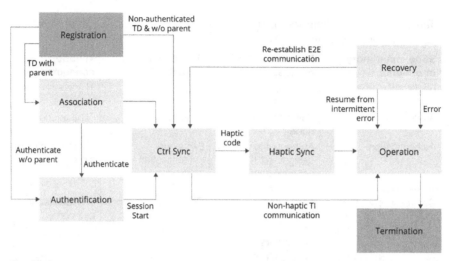

Fig. 15.4

IEEE P1918.1 architecture with the GN and the NC residing as part of the network domain.

synchronization state to establish initial parameters. Future changes to codecs and other haptic parameters will then be handled as data communication in the *operation* state. This is critical to ensuring that all haptic communication will enforce an initial setup, regardless of future updates to the parameters which may be included in operational data payloads.

All TD components will then transition to the operational state. At this state, the E2E path has been established, it has met all connection setup requirements, and the TD are ready to exchange TI information. This is expected to be the most time-dominant state, as it will encompass all TI data communication.

During operation in this state, one TD may detect an intermittent network error, in which case the TD will transition into *recovery* mode, in which designated protocols will take over error checking and potential correction mechanisms to attempt to re-establish reliable communication. If the error proves to be intermittent and is resolved, then the TD will transition back to the operational state. If for any reason the error perseveres, then the TD will transition back to control synchronization, and re-discover whether or not an E2E path is indeed available under the operational requirements set out by the edge user.

Finally, once the TI operation is successfully completed, the TD will transition to *termination* phase, in which all the resources that were previously dedicated to this TD are released back to the TI management plane. If that was initially handled by the NC, then the resources return to it. Most typically, the TSM would be involved in the provisioning of TI resources.

The transitions across all these states are depicted in the Finite State Machine (FSM) in Fig. 15.4. It is important to note that these transitions represent the states and transitions of a single TD, which has transitioned from dormant/off to the initial registration phase. The paradigm of communication detailed above will dictate the TI entity this TD will communicate with, and the overall expectation for latency and reliability in establishing this E2E communication. This FSM is not meant to capture the protocol each entity would invoke under each state, however it is kept generic to capture different types of TI systems.

15.4.5 Interface messages

TI devices need a common interface model to communicate, with a predetermined header format and clearly defined fields. To these ends, a detailed messaging standard is defined that will encompass the key TI data communication, particularly concentrating on the parameters requested by the initiating TD to establish the expected E2E path. In Fig. 15.5, an ASN.1 is presented based on the definition of a message being sent from one TD node to another TI component in the TI network [1105]. The messages are designed to be generic enough to capture the various types of TI communication. However, as a TD transitions into operational state, it might negotiate with the receiver at the end of the E2E path to commit to a lighter-weight version of these messages, in order to reduce the size of headers. These messages are depicted in Fig. 15.5, and elaborated upon below.

For the most part, the message definitions are self-explanatory, spanning typical headers that identify the sender and receiver, for example. The specific *Mode* component is designed to detail the expected mode of operation for the current TD initiating/maintaining communication. This includes the operational parameters (*opParams*) which identify the expected thresholds for each of the four performance metrics explained in the interfaces section. The *Compensation* component identifies whether or not the current TD is engaging any AI-based compensation techniques, such as those adopted to compensate for inevitable delay/lag in communication. Lower-end nodes, which would require a parent node to operate through, would have the *controllerADDR* component set to that parent node and the *controllerType* set to *Pnode*. If there were no such parent (i.e., a higher end TD node), these components would be set to *null*. The list of TI components that could carry out the task of a parent node are detailed under the *Pnode* field.

In the transmission of each message, the TD will identify the current state it is operating under to ease the bootstrapping phase and expedite path establishment. This is captured under the *State* message component. Finally, the *Stack* component is designed to capture different protocol stacks that might serve as the foundational communication infrastructure upon which the TI would operate. That is, an agile architecture design for TI operation is pivotal to allow for message definitions that would highlight whether the operation is on the typical TCP/IP protocol stack, or on a potentially more scalable architecture such as Information Centric Networkss (ICNs). Any such new type would be listed under *StackType* which would also detail the *LayerName* if it is operating on a different architecture from TCP/IP.

The lighter weight version of these TI messages should be employed in the *Operation* state, and could remove headers that detail the type of Node, the *Pnode*, as

```
TIMessage ::= SEQUENCE {
currentMode    Mode,
pduLayer       Stack,
sourceAddr     ADDR,
destAddr       ADDR,
sourceType     Node,
destType       Node,
header         OctetString,
payload        DATA }
```

```
Mode ::= SEQUENCE {
opParams       Params,
state          State,
Compensation   BOOLEAN,
controllerAddr ADDR OPTIONAL,
controllerType Pnode OPTIONAL }
```

```
Params ::= SEQUENCE {
maxLatency     REAL(1.00..1000.00),
minReli        REAL(80.0..99.99999),
minAvail       REAL(80.0..99.99999),
minCapacity    REAL(1.00..9999999.99) }
```

```
ADDR ::= SEQUENCE {
port       Integer(0..65535),
netAddr    NumericString( IP_ADDR ) }
```

Comments on fields:

currentMode identifies the current operational mode in which this message is sent, which is used to decide on message priorities, triggering bootstrapping, recovery actions; in context of current state

pduLayer identifies the current layer on which this PDU is sent, to extend beyond the Internet stack in future network infrastructures, which is defined under Stack

header is a generic field kept to capture auxiliary information that are not captured under other headers

opParams dictate TD/Edge settings that are expected for this communication & are typically set in the bootstrapping phase

Compensation identifies whether the contents of this message have been manipulated by an AI-compensation system.

ADDR captures a tuple that uniquely identifies a communicating process, interpreted in context of Stack, yet currently using IP terms for clarity

```
State ::= SEQUENCE {
stateCode  INTEGER(1..8),
stateName  VisibleString( "Registration"  | "Auth" | "Association"
              | "ctrl_sync" | "haptic_sync" | "operation"
              | "recovery"  | "termination") }
```

```
Node ::= SEQUENCE {
typeId     INTEGER(1..14),
typeName   VisibleString("CPE"|"UPE"|"TSM"|"SE_EDGE"|"SE_CORE"|
                 "GN"|"NC"|"TD"|"CN"|"HN"|"AN"|"SN"|"A"|"S") }
```

```
PNode ::= SEQUENCE {
typeId     INTEGER(1..6),
typeName   VisibleString("CN"|"GN"|"SG"|"AG"|"AN"|"SN")  }
```

```
Stack ::= SEQUENCE {
layerId    INTEGER(1..5),
stackType  VisibleString("TCP/IP" | "other"),
layerName  VisibleString(SIZE (2..30)) OPTIONAL }
```

Fig. 15.5

ASN.1 based definitions of TI messages, exchanged between any two entities in the TI architecture. Evidently, some fields may be superfluous, depending on the interaction. Some fields are intentionally left as OPTIONAL, rendering them TI-component specific.

well as the *Stack* component, since these would typically not change once the initial control and haptic synchronization have taken place.

15.5 **IEEE P1918.1 use cases**

The IEEE P1918.1 WG identifies and defines the use cases that must be supported by the Tactile Internet, and the basic requirements for each of those use cases in terms of key characteristics and performance measures. It is still under discussion if the use cases and their requirements are going to be normative or informative. *Normative*, in this context, is defined by IEEE as information that is required to implement the standard. It is therefore officially part of the standard. *Informative*, on the other hand, is considered as information only and is therefore not officially part of the standard. The current viewpoint on the list of use cases for the IEEE P1918.1 baseline standard is as follows:

Teleoperation Teleoperation allows human users to immerse into a distant or inaccessible environment to perform complex tasks. A typical teleoperation system comprises a master and a slave device, which exchange haptic signals (forces, torques, position, velocity, vibration, etc.), video signals, and audio signals over a (communication) network. In particular, the communication of haptic information imposes strong demands on the network as it closes a global control loop between the master and the slave device. The Quality-of-Service (QoS) requirements and the capabilities of teleoperation systems vary considerably with the dynamics of the remote environment where the teleoperator is placed.

Automotive Vehicular networks are currently standardized within IEEE 802.1 and consider trends in automotive high speed networks and ultra-low latency requirements that are driven by adding new applications such as high-resolution cameras. While IEEE 802.1 focuses on latency reduction between Electronic Control Units (ECU), Audio-Video Bridging (AVB) and Time Sensitive Networks (TSNs) are soon to be standardized and will allow new enhanced applications for remote control of driving functions. The upper boundary of intra-vehicular network delay is targeted below 1 ms [1106]. To support this, the edge unit may be relevant to support local decision making among cars. New haptic applications might target the remote driving support of shuttles, trucks, and road machines in areas which are hard to serve or difficult to maintain. Remote driving requires spontaneous feedback, including haptic events, to make reliable decisions in life-critical situations. In addition, teledriving can be considered as a fallback solution, if autonomy fails.

Immersive virtual reality Immersive Virtual Reality (IVR) describes the case of a human which interacts with virtual entities in a remote environment such that the perception of an interaction with a real physical world/objects is achieved. IVR systems have already been applied or have enormous potential to be utilized in the numerous areas including education, health care, and skills-transfer such as training drivers, pilots, and surgeons, among many others. Here, humans are supposed to perceive all five senses (vision, sound, touch, smell, taste) for full immersion in the virtual environment. A key point of interest to the Tactile Internet as a platform for IVR is latency. Hence, the TI with ultra-low latency is an appropriate platform for IVR systems.

Internet of drones With the increase of the usage of drone delivery systems, traffic management for delivery drones will be necessary. Collisions and other conflicts between drones will be inevitable with increasing number of drones, in particular, if operated by different companies. As a result, it will be necessary to transmit data from the drones to a control center for dynamic route allocation. Built on the TI, it will be possible to guarantee the required ultra-low latency, efficiency, reliability, and overall safety of the drone delivery system. In future, drones might be used for different tasks and humans rather than machines might act as masters with drones acting as slaves devices. Consequently, not only GPS, audio, and video data will be involved, but most-likely haptic (kinesthetic and tactile) information will be transmitted through the communication network.

Interpersonal communication Haptic Interpersonal Communication (HIC) aims to facilitate mediated touch (kinesthetic or tactile cues) over a computer network to feel the presence of a remote user and to perform social interactions. A typical HIC system comprises a local user, a remote participant, a remote participant model at the local environment, and a local user model at the remote environment. Maintaining a human model for remote usage involves the exchange of haptic data (position, velocity, interaction forces, etc.) and non-haptic data (gestures, head movements and posture, eye contact, facial expressions, etc.). The HIC system supports two types of interactions: *dialog interaction*, affecting the remote participant presence, and *observation interaction* perceiving the remote participant presence. Note that the human models (remote participant or local user) can be either a physical entity (such as a social robot) or a virtual representation of a human.

Live haptic-enabled broadcast The live haptic-enabled broadcasting aims to enable the ability to let the viewer actually *feel*, *sense* or *perceive* on-screen action creating a truly immersive and personalized experience. Haptic-tactile broadcasting is the E2E use of technology to capture, encode, broadcast – transmit, transport, by any means – decode, convert and deliver the *feeling* or *impact* or *motion* of a live event such that a remote viewer can experience the same haptic-tactile experience of the live event at a remote location. The addition of haptics, in addition to the capture and transmission of the audio and video essences makes haptic-tactile broadcasting different from traditional broadcasting or streaming.

Cooperative automated driving Without cooperation, in the context of self-driving vehicles, the field of perception of the vehicle is limited to the local coverage of the on-board sensors. Cooperation greatly enhances the field of perception, and will be enabled by the TI for Vehicle-to-Vehicle/Vehicle-to-Infrastructure (V2V/V2I) or Vehicle-to-Any (V2X) communications. TI V2X enables fast and reliable exchange of highly-detailed sensor data between vehicles, along with haptic information on driving trajectories, opening the door to cooperative perception and maneuvering functionalities [1107,1108]. Although the existing V2X standards (i.e., DSRC/ITS=G5 over IEEE 802.11p) support driver assistance and partial automation services, they are not able to cover the requirements for higher levels of automation – which call for new network architectures interconnecting vehicles and infrastructure utilizing ultra-low-latency networks based on the TI for cooperative driving services.

Based on these identified use cases and their analysis, which is ongoing work, the IEEE P1918.1 TI standards WG captures almost the complete range of expectations and performance requirements of URLLC applications in 5G, without specifically focusing on the 5G capacity and data rates expectations for them. At one end of the extreme requirements of TI, for example, consider the teleoperation use case. This use case demands extremely high reliability requirements to avoid any risk of significant (expensive) damage – or potentially even risk of injury or to life. At the intermediate range of requirements is the Interpersonal Communication case. In this use case, the requirements are generally more relaxed in terms of reliability and end-to-end latency (5 ms or more will be acceptable given the human user element and the associated slower reactivity). Finally, highly relaxed requirements are apparent in the Live Haptic-Enabled Broadcast use case. Here, the communication is unidirectional, and as long as the various forms of media (e.g., video, audio, haptic) are well synchronized – noting that the content might also be heavily buffered in such cases to assist – latencies of hundreds of milliseconds, or even seconds, might be acceptable.

15.6 IEEE P1918.1 haptic codecs

The Task Group IEEE P1918.1.1 [1109] (also known as Haptic Codecs Task Group (HCTG)) is currently working towards the standardization of the first generation of haptic codecs for the Tactile Internet. This project within the IEEE P1918.1 Working Group has been active since December 2016 and is planning to finalize the first standard by the end of 2020. The scope of this standardization activity is summarized in its PAR [1110]: *This standard defines haptic codecs for the TI. These codecs address TI application scenarios where the human is in the loop (i.e., teleoperation or remote touch applications) as well as scenarios that rely on machine remote control. The standard defines (perceptual) data reduction algorithms and schemes for both closed-loop (kinesthetic information exchange) and open-loop (tactile information exchange) communication. These codecs are designed such that they can be combined with stabilizing control and local communication architectures for time-delayed teleoperation. Further, the standard also specifies mechanisms and protocols for the exchange of the capabilities (e.g., workspace, the number of degrees of freedom, amplitude range, temporal and spatial resolution, etc.) of the haptic de-vices.*

The HCTG is proceeding in several phases. Initially, the requirements for haptic codec technology in the context of the TI were identified. For this, a call for contributions was issued. The result of this requirement analysis phase is summarized in detail in [9]. Then, based on these results, a first call for haptic codec technology was published. In this first codec development phase, the HCTG concentrated on the standardization of a perceptual codec for the exchange of kinesthetic information. This covers the encoding of forces, torques, and velocity streams for a single point of contact in multiple dimensions (typically three-dimensional). It is assumed that the corresponding information is captured using appropriate sensors and will be exchanged between different TI nodes. The main goal of the corresponding codec is to lower the average packet rate between the two entities without affecting the quality of experience of the user. The selected codec is based on the perceptual deadband-based kinesthetic data reduction scheme described in [380,381,1111]. The codec exploits Weber's law of just noticeable differences, which applies for example to the perception of force, torque and velocity changes. Changes in the kinesthetic data streams, which are below the change detection threshold, can be ignored from a perceptual point of view and are hence not transmitted. This leads to an irregular subsampling of the original sensor data stream, which normally operates with an update rate of at least 1 kHz. By decreasing or increasing the sensitivity threshold in the codec, i.e., by changing the corresponding deadband, the average packet rate can be controlled. A larger deadband means that fewer updates need to transmitted. This comes, however, at the price of lower perceptual transparency, where full transparency means that the user cannot distinguish between local and remote interactions. The resulting kinesthetic codec is able to reduce the average packet rate by about 80%–95% depending on the specific setup. The defined codec is suitable for interaction scenarios where the two TI nodes are close in terms of the experienced communication latency. Stable interaction is typically possible for communication latencies below 5–10 ms depending on the used kinesthetic interfaces and the impedance of the remote environment where the physical interaction takes place. This part of the standard has been finalized.

In the second codec development phase, which is currently still ongoing, the compression of tactile information moved into the focus of the HCTG. Tactile data streams carry information about surface material properties such as roughness, stiffness, friction and thermal conductivity. These properties are sensed by the various receptors in the human skin. So far, only vibrotactile information has been considered within IEEE P1918.1.1 as this is the most widely used tactile information as it directly relates to the micro-roughness and friction of surfaces. These submodalities of tactile perception currently receive substantial interest in the context of so-called surface haptics, a research field which tries to capture and display the roughness and friction of surfaces to users using haptic interfaces. Vibrotactile information is typically sensed using accelerometers attached to a tool or directly to the finger while interacting (tapping, sliding) with an object surface. The requirements for a tactile codec are fundamentally different from the aforementioned kinesthetic codec. For a tactile codec, not the reduction of the average packet rate, but the compression of the raw data is the goal. This is because tactile information is exchanged in an open-loop manner between two TI nodes. Hence, delay is less of an issue and segments of tactile samples can be processed and compressed jointly. Hence, the compression of tactile information is conceptually similar to audio or speech coding. The HCTG has received two competing proposals for tactile codec technology. The first one is based on a Discrete Wavelet Transform, a perceptual quantizer and the Set Partitioning In Hierarchical Trees (SPIHT) coding algorithm. The perceptual quantizer uses a model of human vibrotactile perception which in addition to frequency-dependent sensitivity thresholds also considers masking effects. The second proposal uses Sparse Linear Prediction in combination with a perceptual quantizer that is also based on sensitivity thresholds followed by standard entropy coding. Both approaches are described in detail in Chapter 9.

The third activity extends the kinesthetic codec from the first development phase towards the exchange of kinesthetic information in the presence of communication delay. To this end, the kinesthetic packet rate reduction approach has to be combined with control approaches, which stabilize the networked physical interaction when the delay becomes larger than the aforementioned 5–10 ms. Several algorithms have been proposed in the literature, which stabilize kinesthetic interaction in the presence of delay. Some of these control approaches have already been combined with kinesthetic data reduction. An overview about the state of the art in this field and the delay-tolerant kinesthetic codec currently under investigation within the HCTG is given in Chapter 9.

A parallel activity to the haptic codec development (kinesthetic and tactile, without and with delay) proposes a TIM to make haptic interfaces and applications self-described. The TIM metadata describes the haptic interface and data (media), in addition to the intended quality of experience (quality of service and user experience). An overview of the TIM metadata is shown in Fig. 15.6. Furthermore, this activity develops a handshake protocol to provide seamless exchange of the TIM metadata in order to make haptic communication a *plug-and-play* regardless of the haptic interface used by these applications. The haptic handshake/communication protocol is shown in Fig. 15.7. The implementation of the TIM metadata and the handshake/communication protocol is detailed in [1112].

15.7 Conclusion and outlook

The IEEE P1918.1 TI standards WG has a baseline standard development schedule which foresees to have a complete standard draft by the end of 2020 to start a letter ballot such that IEEE Standards Association can facilitate the development of a global standard for the TI. This standard document will contain the definition of interfaces, definition of functional capabilities, the functional architecture, and the TI use cases including TaHiL applications. Additionally, the haptic codecs standardization will

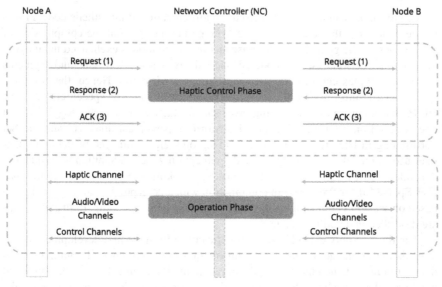

Fig. 15.6

Overview of the Tactile Internet Metadata (TIM) metadata.

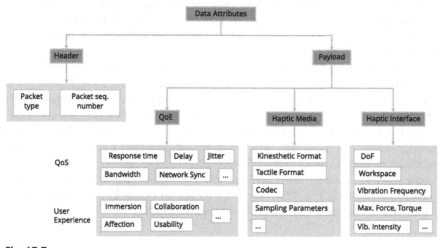

Fig. 15.7

Haptic handshake and communication protocol.

be completed. For the future, the IEEE P1918.1 TI standards WG aims to focus on intelligent solutions for the TI.

Public opinion and the Tactile Internet

16

Sven Engesser, Lisa Weidmüller, and Lutz M. Hagen

Technische Universität Dresden, Dresden, Germany

Technology is neither good nor bad; nor is it neutral.
– Melvin Kranzberg

16.1 Introduction

To understand the nexus of science, technology, and society, and to provide solutions for challenges arising from it, the mutual exchange between engineering sciences, natural sciences, and social sciences has proven crucial. However, it is not sufficient for the scientific disciplines to merely coexist and to simultaneously pursue different questions within a common research project. Instead, it is necessary to integrate natural sciences, engineering sciences, and social sciences from the beginning and to continuously foster actual cooperation. This is what Guston and Sarewitz called *real-time* Technology Assessment (TA) [1113]. It has become an essential part of the Responsible Research and Innovation (RRI) paradigm at European and national levels [1114–1116]. Therefore we deem it highly advisable to integrate real-time TA into the emerging research field of Tactile Internet with Human-in-the-Loop (TaHiL).

This approach also incorporates important tenets of transdisciplinarity—an overarching concept combining efforts by scholars from different disciplines to create theoretical, methodological, and translational innovations that integrate and move beyond discipline-specific perspectives to address a common problem with great societal impact and to generate practical knowledge [1117].

One of the most important aspects of real-time TA is Communication and Early Warning (CEW) [1113], which aims at *enhancing the quality of communication* by anticipating potential *conflict, opposition, and backlash*, both within the research project and in relation to the wider society. A methodological pillar of CEW is *survey research*, which allows mapping public opinion and track its changes over time. It is important to note that CEW does not only focus on knowledge and rational thoughts, but also takes into account *emotions and affect* of the public, which are *an underappreciated dimension of risk judgment that is particularly important with respect to technologies* [1113]. This applies particularly to research on the Tactile Internet (TI) with Human-in-the-Loop.

For effective real-time TA, it is required to know the public opinion on technology and to understand how it is formed. Accordingly, we follow a combination of

Tactile Internet. https://doi.org/10.1016/B978-0-12-821343-8.00029-0

exploratory and confirmatory approaches and formulate the following three Research Questions (RQs):

RQ1 What is the public opinion on the TI in terms of awareness, knowledge, benefits, risks, concerns, support, and intended use?

RQ2 Which factors influence (a) the support and (b) the intended use of the TI?

RQ3 Which role do (a) reasoning and (b) intuition play for the formation of attitudes toward the TI?

16.2 Theoretical background

When analyzing the public opinion toward new technologies, core concepts are attitudes and use intentions. Thus our study builds upon general models of attitude formation as well as specific technology acceptance models that we will explicate in this section. Furthermore, this section contains a brief overview of findings from prior research on public opinion toward technology and the factors that influence it.

16.2.1 Models of attitude formation

There is a wide consensus in the research literature that, at the individual level, attitude formation can be described as dual process (Fig. 16.1): one is automatic, unconscious, and fast, whereas the other one is controlled, conscious, and slow [1118,1119]. The former has been labeled *peripheral* [1120], *heuristic* [1121], *experiential* [1122], *intuition* [1123] or, generically, *System 1* [1124], and the latter *central* [1120], *systematic* [1121], *rational* [1122], *reasoning* [1123], or *System 2* [1124].

Intuition primarily draws on personal experience [1122] and heuristics [1121]. It is also associated with emotions and affect [1118,1122]. In contrast, reasoning involves rational assessment, the weighing up of arguments, and deliberation, which requires access to the central working memory [1118,1125]. The motivation and cognitive capacity of an individual determine which process is more dominant. Individuals with higher involvement regarding an information about a new technology tend to rely on reasoning, whereas individuals with lower involvement rely on intuition [1121]. Attitudes formed through intuition are considered less stable than those formed through reasoning [1120]. Although it has become evident that intuition and reasoning are far from completely separate in reality [1126], dual-process models have proven highly valuable from an analytical point of view.

As mentioned above, we know that emotions and affect play an important role for intuition. Moreover, affect cannot only influence the way heuristics are consulted during intuitive processes but can also act as heuristic itself [1127–1129].

In the field of technology, individuals are particularly inclined to intuitive attitude formation, because they frequently experience little personal relevance and little competence with regard to the technology in question [1130,1131]. For emerging technologies, such as the TI, this effect may be even stronger. At early stages of

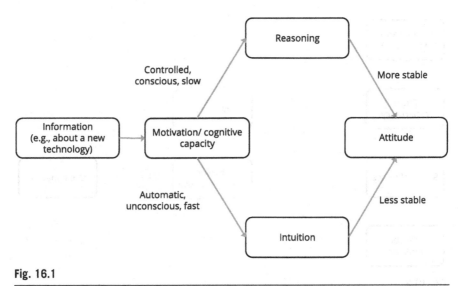

Fig. 16.1

Dual-process model of attitude formation (adapted from [1118,1120–1123]).

development, when heuristics from other sources (e.g., personal experience, mass media) are still scarce, individuals tend to rely on their emotions as heuristics [1132].

16.2.2 Models of technology use

In the field of technology communication, it is not only relevant to what extent the public supports the technology under study. It is also important to know if the public uses the technology or at least intends to do so (which is a good proxy of actual use [1133]). The most widely established instrument to explain the behavioral intention to use a technology is the Technology Acceptance Model (TAM) [1134] and its follow-up, the Unified Theory of Acceptance and Use of Technology (UTAUT) [1133] (Fig. 16.2).

The strongest predictors of behavioral intention to use were initially labeled *perceived usefulness* and *perceived ease of use* [1134] and have been renamed *performance expectancy* and *effort expectancy* [1133]. Both are assumed to be channeled through the *attitude* toward the technology. Subsequent extensions of the TAM have added *social influence* and *facilitating conditions* as predictors. *Age, gender,* and *experience* were included as moderators. All predictors were confirmed in a recent metaanalysis [1135]. In previous studies, the TAM explained 45%–56% of the variance in behavioral intention to use [1133,1135].

16.2.3 Empirical findings related to public opinion on technology

In Germany, 60% of the population feel constrained by technology in general, and an additional 33% are undecided concerning this question [1136]. In terms of specific

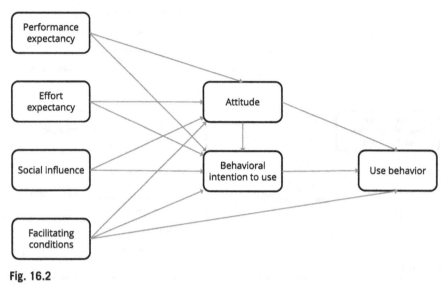

Fig. 16.2

Technology use model (based on [1133–1135]).

technologies, Gnambs and Appel [1137] identified increasing concerns toward robots throughout the European Union between 2012 and 2017, in particular with regard to assistance at the workplace.

The authors in [1131] demonstrated that individuals rely on heuristics from the mass media to form their attitudes toward nanotechnology. Lee and colleagues [1132] showed that both cognitive *and* affective factors influence the support of nanotechnology among the population. They also identified an interaction effect of knowledge and negative emotions. Druckman and Bolsen [1138] found that *general* knowledge is a stronger predictor of support for less familiar technologies, whereas *specific* knowledge is more relevant concerning technologies that are more familiar.

Priest [1139] found that trust in institutions predicted the support of biotechnology. The support of nanotechnology was found to be positively influenced by trust in scientists [1132]. The importance of public trust as predictor of the intention to use automated driving was underlined by Liu and colleagues [1140].

16.3 Survey on the public opinion regarding the Tactile Internet

To answer our research questions, we conducted an empirical study to collect data about the perceptions of the TI in the German population. This section contains detailed information about how the study was carried out.

16.3.1 **Sample and procedure**

Data was collected via Computer-Assisted Telephone Interviews (CATI) among the German population (18 years and older) using a random-digit dialing technique, including both landline and cell phone numbers (dual frame).

Overall, 784 interviews were completed (5.7% minimum response rate) over a time period of 30 days at the end of 2019 and the beginning of 2020. Respondents were slightly more often male (52%) and, on average, 53 years old (SD = 17.12), which is ten years older than the average of the German population [1141]. The majority of the sample had received higher education (51.4% had a college or university degree). This implies a slight overrepresentation of male, older, and higher educated citizens compared to the population. In general, the higher educated and those with a higher interest in the topic of a survey are slightly overrepresented in telephone surveys [1142]. This raises the suspicion that those who were more interested in technology had a higher probability to participate in our study, which presumably contributed to the overrepresentation of males.

To ensure that all respondents shared a similar understanding of the TI and to enable those who had not yet heard of it to answer the questions, the following short description was included in the interview:

> *Regardless of how the population conceives of the Tactile Internet, scholars frequently define it as a network that allows the real-time wireless transmission of information. This allows the collaboration between humans and machines in numerous areas, such as autonomous driving, robotics, and telemedicine. For example, a robotic arm can be controlled from a distance and without any perceptible delay through a data glove. The technological basis for this is 5G, the next generation of mobile communication.*

16.3.2 **Measures**

Awareness of the TI was measured by asking how much the respondents had heard of the TI before the interview (four-point scale) [1143] and through which channels they had heard of it (e.g., conversations, mass media, social media). Additionally, we asked whether respondents were aware of the Center for TaHiL Research at TU Dresden (0 = no, 1 = yes) and measured respondents' *perceived knowledge* about the TI by asking them how well informed they felt on a five-point scale ranging from 1 *(very badly)* to 5 *(very well informed)*.

To identify *perceived risks and benefits*, two methodological paths were taken: First, respondents were openly asked to name a maximum of three potential risks and benefits after having received a short description of the TI (see above). Second, we asked them about the *social implications* of the TI. The latter bears on the respondents indicating how much they thought the TI would change certain aspects of society, such as social inequality, data security, health, or quality of life, either positively (very, rather), negatively (very, rather), or not at all. These aspects were selected based on a literature review of perceived risks and benefits regarding the TI [1144], other emerging technologies, such as the Internet of Things or nanotechnology [1132], and regarding technology in general [1136,1145,1146].

Support of the TI was measured on a five-point scale ranging from 1 *(greatly oppose)* to 5 *(greatly support)* [1132]. Additionally, we asked for respondents' acceptance of five *application areas* of the TI. The areas relate to the three TaHiL use-cases: medicine, industry, and Internet of Skills.

Respondents were also asked to indicate their *intention to use* the TI on a five-point scale from 1 *(not at all)* to 5 *(as soon as possible)*.

Additionally, we included several factors that had been found to influence the support and intention to use a technology in previous research.

Cognitive factors Aside from perceived knowledge, we included a general five-point item to measure respondents' *risk-benefit assessment*, where they had to indicate whether they believed the risks/benefits of the TI greatly/slightly outweighed its benefits/risks or if they were in balance (adapted from [1143]). This item was inspired by dual-process models of attitude formation and, together with perceived knowledge, was assumed to measure how inclined respondents were toward System 2 or reasoning. However, at this early stage of TI development respondents' weighing of risks against benefits might not be completely free of emotion.

Emotional factors The most relevant emotional factors were *concerns toward technology* in general and *concerns toward the TI* in particular (adapted from [1132]). Another important emotional factor found in prior research was *trust in scientists* (three items, five-point scale, adapted from [1132]). These three measures were also inspired by dual-process models and were assumed to measure System 1 or intuition. Unexpectedly, the internal consistency of the *trust in scientists* scale was low (Cronbach's Alpha = 0.10), which means that the findings have to be interpreted with caution.

TAM From the TAM, we included *perceived usefulness* (performance expectancy, Cronbach's Alpha = 0.83), *ease of use* (effort expectancy, Cronbach's Alpha = 0.78), and *social influence* measured by agreement with three items, each on a five-point scale (1 – *disagree completely* to 5 – *agree completely*, adapted from [1147] and [1148]). The internal consistency for *social influence* was low (Cronbach's Alpha = 0.38), which means that this predictor has to be interpreted with caution. We refrained from including measures for facilitating conditions from the TAM, because we assumed that those would be difficult to assess at this early stage of the TI's development. Sample items for perceived usefulness included *The Tactile Internet is useful for me* and *The Tactile Internet will simplify the completion of my daily tasks.* For perceived ease of use, sample items were *I would be able to use the Tactile Internet in everyday life* and *It would be easy for me to become skilled in using the Tactile Internet.* Social influence was represented by items, such as *My decision to use the Tactile Internet depends on whether people in my personal environment use it* or . . . *whether the media support its use.*

Personal attributes As *personal attributes*, we measured whether respondents were biased toward supporting and intending to use the TI due to a higher general *innova-*

tiveness [1149] by including three items for which respondents indicated agreement on a five-point scale (1 – *disagree completely* to 5 – *agree completely*). Sample items included *Other people come to me for advice on new technologies* or *I think it is easier for me than other people to familiarize myself with a new technology* and showed good internal consistency (Cronbach's Alpha = 0.81). Additionally, their *interest in technology* was measured on a five-point scale ranging from 1 (*not interested at all*) to 5 (*very interested*) [1136]. *Demographic variables* included *age*, *gender*, and *education*.

16.4 Survey results

After data was collected through the survey, we conducted statistical analyses to find out what the German population knows and thinks about the TI. This section contains the key findings and answers to our research questions. The first four subsections are dedicated to answering RQ1, whereas subsection five focuses on RQ2 and RQ3.

16.4.1 Awareness and knowledge

As could be expected at this early point of technological development, awareness of the TI was generally low: Three out of four respondents had not yet heard anything about it. Among those who had, the majority had heard *just a little* (Fig. 16.3). Even less surprising at this point of time: Only one out of seventeen respondents claimed to have heard about the Center for TaHiL Research at TU Dresden.

Question: How much have you heard of the Tactile Internet before today?

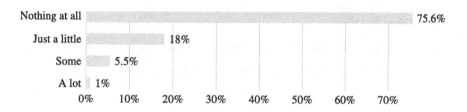

Fig. 16.3

Public awareness of the Tactile Internet (n = 835).

In accordance with the low degree of awareness, only fewer than one sixth of the sample felt that they were well or even very well informed about the TI. However, more than four in ten claimed to be moderately informed (Fig. 16.4).

Traditional mass media were the main source of information (Fig. 16.5). This referred to generally informing oneself about technology, as well as to information about the TI and TaHiL in particular. The more specific the issue, the more important

Question: How informed do you feel about the Tactile Internet?

Fig. 16.4

Perceived knowledge of the Tactile Internet (n = 779).

Questions: Which sources do you use to inform yourself about technology? Do you remember from which sources you have heard about the Tactile Internet/TaHiL?

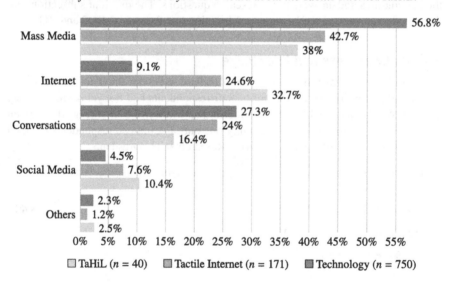

Fig. 16.5

Sources of information concerning technology in general, the Tactile Internet and TaHiL.

conversations became as a source. The opposite applied to the Internet. Social media still played a minor—albeit not insignificant—role.

16.4.2 Benefits and risks

When asked openly about possible benefits, health issues came to the minds of the respondents—way ahead of anything else: More than 40% named either *medicine/surgery* or *health care* (Fig. 16.6). *Automated driving* was named second most often, but by less than one tenth of the population. Closely related, benefits for mobility ranked third. No other benefit was named by more than four out of one hundred respondents. Among those who named benefits, most felt confident to name not more than one (out of three possible). This may mirror the vague conception people hold in mind at this early stage of the TI's development and may be indicative of a contemporary low involvement. Attitudes toward the TI seem to be weakly defined at this point, not yet allowing for profound judgment on benefits by most of the people.

Question: Which benefits come to your mind spontaneously that this technology could offer?

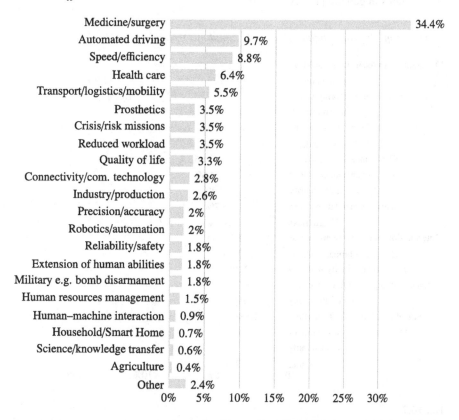

Fig. 16.6

Perceived benefits of the Tactile Internet (open answers, only benefits that were named first, categorized, n = 544).

Two other observations support this conclusion: First, 544 (69.4%) of all respondents named at least one benefit, but only 244 (31.1%) managed to mention a second, and a mere 69 (8.8%) managed to mention a third benefit. Second, the most frequently named benefits mirrored two aspects (i.e., autonomous driving and telemedicine), which were already included in the definition of the TI provided to the respondents prior to the question about benefits (see Method section).

Even fewer respondents were able to name risks (named one: 552, 70.4%; named two: 166, 21.2%; named three: 31, 4.0%). Additionally, the vague nature of risks that were voiced by the respondents gives further weight to the interpretation that people only have a weakly defined conception of the TI and its implications at this point: Vulnerability, loss of control, and other safety issues lead the list (Fig. 16.7). Fear of unemployment ranks next, which is frequent (and often justifiable) for new technologies in general [1150].

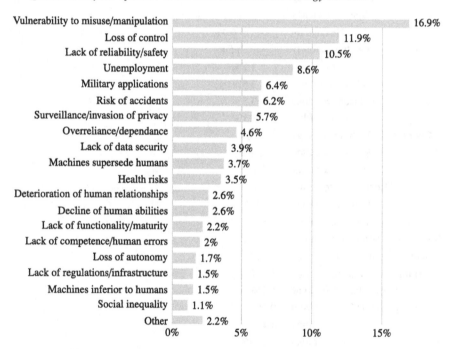

Question: In your opinion, which risks could this technology involve?

Fig. 16.7

Perceived risks of the Tactile Internet (open answers, only risks that were named first, categorized, n = 545).

All in all, benefits seemed to dominate the minds of most respondents (37.2%), whereas almost as many were undecided (35.1%) and a considerable proportion of the sample regarded risks as dominant (27.7%, Fig. 16.8).

Question: All in all, with regard to the Tactile Internet what would you say?

Fig. 16.8

Risk-benefit assessment of the Tactile Internet (n = 776).

16.4.3 Concerns, application areas, and social implications

In accordance with the general risk-benefit assessment, the share of respondents who indicated they did not worry about the development of the TI clearly outweighed the share of those who did, whereas a large part stayed undecided (Fig. 16.9).

Regarding specific support for the TI in several application areas relevant to TaHiL research, our analysis shows that emotions toward the TI depend on the specific application (Fig. 16.10). Only the implementation of exoskeletons was evaluated clearly positively. Deploying robots for medical and economic purposes predominantly lead to positive feelings, but these varied between respondents. Concerning automated traffic and Virtual Reality (VR) learning, the public seemed to react with rather mixed feelings. Reasons for these divergent affective predispositions remain unclear at this point and should be dealt with in consecutive research.

Another indicator for the fact that most people had not made up their minds in a definite manner can be seen in the stark variation of the answers: The standard deviation for most items comes up for a bit more than one fourth of the entire scale.

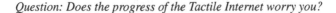

Question: Does the progress of the Tactile Internet worry you?

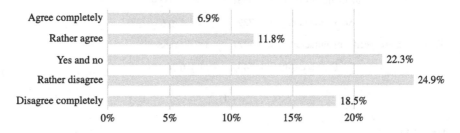

Fig. 16.9

Concerns toward the Tactile Internet (n = 757).

Question: Please tell me for each of the following applications whether you find the notion of it very pleasant (+2), rather pleasant (+1), partly pleasant, partly unpleasant (0), rather unpleasant (-1), or very unpleasant (-2).

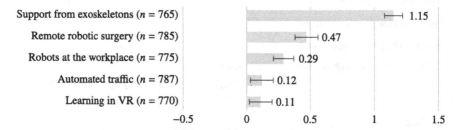

Fig. 16.10

Assessment of application areas of the Tactile Internet (means and 95%-confidence intervals of means).

Question: I will now read to you a couple of aspects that could possibly change induced by the Tactile Internet. For each aspect, please tell me if you believe that the Tactile Internet will change it in a very positive (+2), rather positive (+1), rather negative (-1), very negative (-2) way, or if it will not change it at all (0).

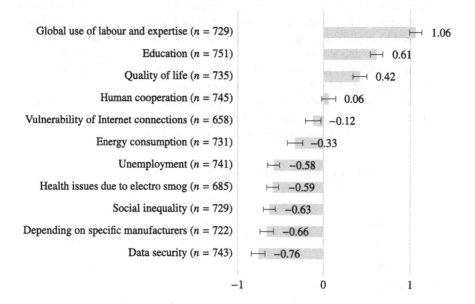

Fig. 16.11

Evaluation of social implications of the Tactile Internet (means and 95%-confidence intervals of means).

Asking for assessments of social implications resulted in a predominantly negative picture (Fig. 16.11). A majority seemed to expect the TI to generally improve the quality of life, to enhance education and, above all, to facilitate the availability and connection of expertise and labor on a global level. However, human cooperation in general was not clearly perceived as positively affected. All other implications tended to generate unpleasant feelings. Among them were mostly general ones, such as rising unemployment, deterioration of health, increasing social inequality, economic dependence, and a decrease of data security.

Again, the rather large standard deviations may indicate that attitudes were probably in flow and still need to consolidate. The wording of the questionnaire may also have influenced the answers: With the exception of *data security*, all of the more negatively assessed social implications were also referred to by negative terms (e.g., *unemployment*, *inequality*).

16.4.4 Support and intention to use

The TI can rely on the general support of almost one half of the sample (Fig. 16.12). A large proportion of respondents, roughly two fifth, was undecided, whereas only one out of ten was in opposition to this technology.

Question: All in all, would you say that you greatly support, rather support, rather oppose, or greatly oppose the development of the Tactile Internet, or are you undecided?

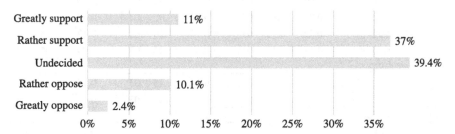

Fig. 16.12

Public support for the Tactile Internet (n = 789).

When it came to the question whether they would intend to use the TI themselves, the respondents were cautious. Roughly one out of four rather intended to do so, whereas almost two fifth would not, and slightly more than one third was undecided (Fig. 16.13).

16.4.5 Predictors of support and intention to use

To identify factors that influence whether people support and intend to use the TI (*RQ2*) and to analyze the role reasoning (cognitive factors) and intuition (emotional

Question: If you had the opportunity, to what extent would you use the Tactile Internet? Answers ranged from -2 "I do not want to use the TI at all" to +2 "I would like to use the TI as soon as possible".

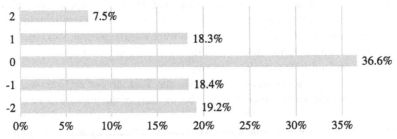

Fig. 16.13

Intention to use the Tactile Internet (n = 787).

Table 16.1 Regression on support for the Tactile Internet.

Variable	Model 1			Model 2			Model 3		
	B	**β**	**SE**	**B**	**β**	**SE**	**B**	**β**	**SE**
Constant	2.41***		.20	1.21***		.19	1.83***		.24
Socio-demographics/personal attributes									
Gender (female)	-0.18*	-.10	.07	-0.09	-.05	.06	-0.04	-.02	.06
Age	0.00	-.05	.00	0.00	-.02	.00	0.00	-.01	.00
Education	0.05*	.08	.02	0.03	.06	.02	0.02	.03	.02
Innovativeness	0.22***	.27	.03	0.16***	.20	.03	0.12***	.15	.03
Interest in technology	0.11**	.13	.04	0.05	.06	.03	0.04	.05	.03
Cognitive factors									
Knowledge about TI				0.14***	.15	.03	0.14***	.15	.03
Risk vs. benefit				0.36***	.41	.03	0.27***	.30	.03
Emotional factors									
Trust in scientists							0.12**	.10	.04
Tech. concerns							-0.04	-.06	.02
TI concerns							-0.18***	-.23	.03
R²		.18***			.38***			.45***	
ΔR²					.20***			.07***	

Note: N = 692, *p < .05, **p < .01, ***p < .001; residuals very much resemble a normal distribution (visual check), model is not impaired by multicollinearity (all tolerances higher than 0.8)

factors) play for the formation of these attitudes (*RQ3*), we conducted two series of ordinary least squares (OLS) regressions (Tables 16.1 and 16.2).

The analyses reveal that cognitive factors, such as perceived knowledge about the TI (2% of total variance explained) and risk-benefit assessment (9%), have a posi-

Table 16.2 Regression on intention to use the Tactile Internet.

Variable	Model 1			Model 2			Model 3		
	B	β	SE	B	β	SE	B	β	SE
Constant	1.65***		.20	0.22		.25	-0.28		.25
Socio-demographics/personal attributes									
Gender (female)	-0.06	-.03	.08	0.03	.01	.07	0.06	.03	.07
Age	-0.01***	-.21	.00	-0.01**	-.10	.00	-0.01***	-.12	.00
Education	0.05*	.06	.03	0.04	.05	.02	0.02	.03	.02
Innovativeness	0.36***	.34	.04	0.16***	.15	.04	0.13***	.13	.04
Interest in technology	0.18**	.16	.04	0.13**	.12	.04	0.11**	.10	.04
Predictors from TAM									
Usefulness				0.51***	.45	.04	0.37***	.32	.04
Ease of Use				0.12**	.10	.04	0.10*	.09	.04
Social Influence				0.01	.01	.04	0.02	.01	.04
Attitude									
Support for TI							0.33***	.26	.04
R^2		.30***			.49***			.53***	
ΔR^2					.19***			.04***	

Note: N = 715, *p < .05, **p < .01, ***p < .001; residuals very much resemble a normal distribution (visual check), model not impaired by multicollinearity (all tolerances higher than 0.6)

tive effect on the support of the TI (Table 16.1). Among the emotional factors, trust in scientists (1%) positively predicts, and concerns toward the TI (5%) negatively predict the support of the TI. Concerns toward technology in general have no significant effect. In total, cognitive factors explain more variance (11%) than emotional factors (6%).

Perceived usefulness (10%), perceived ease of use (1%), and the attitude toward the TI measured as general support (7%) positively predict behavioral intention (Table 16.2). Social influence does not have a significant effect on intention to use.

Innovativeness has a positive effect on the support of the TI (2%) and the intention to use it (2%), whereas age (1%) negatively, and interest in technology (1%) positively predict behavioral intention (Tables 16.1 and 16.2). There is no significant effect of gender nor of education on support or intention to use.

16.5 **Discussion**

Public awareness and perceived knowledge of the TI are still low. Accordingly, public opinion can be assumed to be weakly defined and to leave many degrees of freedom, which can be shaped by communication measures and public relations. This applies even more to the specific case of the Center for TaHiL Research at TU Dresden, which is still unknown to the majority of the public. Those who have heard of TaHiL have done so through the mass media or personal conversations. This would suggest that communication measures should target both leading media outlets and scientific opinion leaders. However, this strategy should be complemented by including non-established actors and social media to increase the diversity of perspectives.

Regardless of knowledge and awareness, a third of the population already supports the technological development of the TI, but another 40% are still undecided, which leaves much room for potential supporters or opponents. Three quarters of the population are still rather hesitant to use TI applications.

Both cognitive factors and emotional factors influence the support of the TI. Reasoning and intuition seem to complement each other. This is in line with previous studies on nanotechnology [1132] and emphasizes the validity of dual-process models [1118] for the formation of attitudes. Future surveys may show how the relation between cognitive and emotional factors changes when awareness and knowledge of the TI increase.

Among the cognitive factors, rational risk-benefit assessment exerts the largest influence in our study. This suggests that communication measures should clearly illustrate the risks and benefits of the TI, and carefully weight them up. Although cognitive factors are more influential than emotional factors for the support of the TI, the latter should not be underestimated. Concerns toward the TI play an important role in attitude formation and should be sufficiently addressed by communication measures.

It should be considered, however, that emotions toward the TI depend on the context. Risks are mostly associated with rather general issues like manipulation, loss of control, rising unemployment, social inequality, economic dependence, and lack of data security. Concrete application areas, such as exoskeletons and robotic assistants, generate more positive feelings. Communication measures could focus on these cases, present them in an appealing manner, fill the cognitive voids, and draw the emotions toward the positive.

Resonating previous studies [1139,1140], our analysis showed that trust in scientists positively predicts the support of technology. Accordingly, it may be worthwhile for TaHiL researchers to invest in trust-building measures. Although it was not part of our analysis, communication measures should include personalized messages, e.g., by focusing on outstanding principal investigators, their personality, and achievements.

Perceived usefulness is the strongest predictor of the intention to use the TI, whereas ease of use has only a minimal effect and social influence has no significant effect at all, which is presumably due to the early stage of development, where social influences are still difficult to grasp. This implies that communication measures should elaborate upon the usefulness of the TI and its potential applications. Applications in the areas of medicine/health, mobility/automated driving, and exoskeletons/prosthetics seem to be particularly attractive, which may also be due to the current media coverage. However, TaHiL research should not shy away from promoting other promising but less well-known applications.

The Technology Acceptance Model [1133,1134] is only partially supported in our study. This observation may be related to the early developmental stage of the TI, in which ease of use and social influence are still difficult to imagine for the public. Future surveys may uncover whether the effects of ease of use and social influence increase over time.

16.6 Conclusion and outlook

The findings of this study do not need to be regarded as *early warning*. There is no widespread opposition toward the TI among the population, and there is no indication of conflicts. However, it cannot be denied that the combination of low awareness, undecided opinions, and general concerns bears the risk of rising opposition toward the TI. Therefore all these issues should be addressed by TaHiL research as early as possible to reduce the probability of any backlash. In particular, concerns regarding data security, economic dependence, and loss of control should be taken seriously. Appropriate communication measures should be designed and implemented. Apart from this, survey research should continue to monitor public opinion and identify potential problems.

Accordingly, two follow-up surveys are planned within the next years, which will include vignettes to specifically investigate TaHiL research use-cases. So far, 577 participants from the survey reported on in this chapter have been recruited for these follow-up surveys. This allows examining how awareness, attitudes, and intention to use TI develop over time on an aggregate (societal) level as well as on an individual (personal) level. Additionally, we will be able to investigate how the relative importance of cognitive and emotional factors for opinion formation about the TI changes with increasing integration of TI applications into everyday life and available information about the technology, for example, from TaHiL communication measures.

Bibliography

[1] G.P. Fettweis, The Tactile Internet: Applications and challenges, IEEE Vehicular Technology Magazine 9 (1) (2014) 64–70.

[2] D. Szabo, A. Gulyás, F.H.P. Fitzek, D.E. Lucani Rötter, Towards the Tactile Internet: Decreasing communication latency with network coding and software defined networking, in: Proceedings of the European Wireless Conference (EW), 2015.

[3] G. Schirner, D. Erdogmus, K. Chowdhury, T. Padir, The future of Human-in-the-Loop cyber-physical systems, Computer 46 (1) (2013) 36–45.

[4] Y. Yang, A.M. Zador, Differences in sensitivity to neural timing among cortical areas, The Journal of Neuroscience 32 (43) (2012) 15142–15147.

[5] M. Dohler, T. Mahmoodi, M.A. Lema, M. Condoluci, F. Sardis, K. Antonakoglou, H. Aghvami, Internet of skills, where robotics meets AI, 5G and the Tactile Internet, in: Proceedings of the European Conference on Networks and Communications (EuCNC), 2017.

[6] A.J. Bremner, D.J. Lewkowicz, C. Spence, Multisensory Development, Oxford University Press, 2012.

[7] S.-C. Li, U. Lindenberger, B. Hommel, G. Aschersleben, W. Prinz, P.B. Baltes, Transformations in the couplings among intellectual abilities and constituent cognitive processes across the life span, Psychological Science 15 (3) (2004) 155–163.

[8] Z. Xiang, F. Gabriel, E. Urbano, G.T. Nguyen, M. Reisslein, F.H.P. Fitzek, Reducing latency in virtual machines: Enabling Tactile Internet for human–machine co-working, IEEE Journal on Selected Areas in Communications 37 (5) (2019) 1098–1116.

[9] E. Steinbach, M. Strese, M. Eid, X. Liu, A. Bhardwaj, Q. Liu, M. Al-Ja'afreh, T. Mahmoodi, R. Hassen, A. El Saddik, O. Holland, Haptic codecs for the Tactile Internet, Proceedings of the IEEE 107 (2) (2019) 447–470.

[10] I. Rahwan, M. Cebrian, N. Obradovich, J. Bongard, J.-F. Bonnefon, C. Breazeal, J.W. Crandall, N.A. Christakis, I.D. Couzin, M.O. Jackson, N.R. Jennings, E. Kamar, I.M. Kloumann, H. Larochelle, D. Lazer, R. McElreath, A. Mislove, D.C. Parkes, A. Pentland, M.E. Roberts, A. Shariff, J.B. Tenenbaum, M. Wellman, Machine behaviour, Nature 568 (7753) (2019) 477–486.

[11] M.R. Frank, D. Autor, J.E. Bessen, E. Brynjolfsson, M. Cebrian, D.J. Deming, M. Feldman, M. Groh, J. Lobo, E. Moro, D. Wang, H. Youn, I. Rahwan, Toward understanding the impact of artificial intelligence on labor, Proceedings of the National Academy of Sciences of the United States of America 116 (14) (2019) 6531–6539.

[12] E. Awad, S. Dsouza, R. Kim, J. Schulz, J. Henrich, A. Shariff, J.-F. Bonnefon, I. Rahwan, The moral machine experiment, Nature 563 (7729) (2018) 59–64.

[13] H. Nakawala, G. Ferrigno, E. De Momi, Toward a knowledge-driven context-aware system for surgical assistance, Journal of Medical Robotics Research 2 (3) (2017) 1740007, pp. 1–14.

[14] V. Gabler, T. Stahl, G. Huber, O. Oguz, D. Wollherr, A game-theoretic approach for adaptive action selection in close proximity human-robot-collaboration, in: Proceedings of the IEEE International Conference on Robotics and Automation (ICRA), 2017.

[15] S.S. Vedula, M. Ishii, G.D. Hager, Objective assessment of surgical technical skill and competency in the operating room, Annual Review of Biomedical Engineering 19 (2017) 301–325.

[16] D. Azari, C. Greenberg, C. Pugh, D. Wiegmann, R. Radwin, In search of characterizing surgical skill, Journal of Surgical Education 76 (5) (2019) 1348–1363.

[17] A. Madani, M.C. Vassiliou, Y. Watanabe, B. Al-Halabi, M.S. Al-Rowais, D.L. Deckelbaum, G.M. Fried, L.S. Feldman, What are the principles that guide behaviors in the operating room?, Annals of Surgery 265 (2) (2017) 255–267.

[18] R.J. Seidel, K.C. Perencevich, A.L. Kett, From Principles of Learning to Strategies for Instruction, Springer, 2007, pp. 75–113, chapter Psychomotor domain.

[19] B. Burbach, S. Barnason, S.A. Thompson, Using "Think aloud" to capture clinical reasoning during patient simulation, International Journal of Nursing Education Scholarship 12 (1) (2015) 1–7.

[20] L. Corrin, M. Olmos, Capturing clinical experiences: Supporting medical education through the implementation of an online clinical log, in: Proceedings of the Ascilite Conference (ASCILITE), 2010.

[21] A. Rafiq, J.A. Moore, X. Zhao, C.R. Doarn, R.C. Merrell, Digital video capture and synchronous consultation in open surgery, Annals of Surgery 239 (4) (2004) 567–573.

[22] S. Shackelford, E. Garofalo, V. Shalin, K. Pugh, H. Chen, J. Pasley, B. Sarani, S. Henry, M. Bowyer, C.F. Mackenzie, Development and validation of trauma surgical skills metrics: Preliminary assessment of performance after training, Journal of Trauma and Acute Care Surgery 79 (1) (2015) 105–110.

[23] V. Belagiannis, X. Wang, H.B.B. Shitrit, K. Hashimoto, R. Stauder, Y. Aoki, M. Kranzfelder, A. Schneider, P. Fua, S. Ilić, H. Feußner, N. Navab, Parsing human skeletons in an operating room, Machine Vision and Applications 27 (7) (2016) 1035–1046.

[24] T. Beyl, P. Nicolai, M.D. Comparetti, J. Raczkowsky, E. De Momi, H. Wörn, Time-of-flight-assisted Kinect camera-based people detection for intuitive human robot cooperation in the surgical operating room, International Journal of Computer Assisted Radiology and Surgery 11 (7) (2016) 1329–1345.

[25] V. Srivastav, T. Issenhuth, A. Kadkhodamohammadi, M. de Mathelin, A. Gangi, N. Padoy, MVOR: A multi-view RGB-D operating room dataset for 2D and 3D human pose estimation, CoRR, arXiv:1808.08180, 2019.

[26] A.P. Twinanda, E.O. Alkan, A. Gangi, M. de Mathelin, N. Padoy, Data-driven spatio-temporal RGBD feature encoding for action recognition in operating rooms, International Journal of Computer Assisted Radiology and Surgery 10 (6) (2015) 737–747.

[27] G. Birgand, C. Azevedo, S. Rukly, R. Pissard-Gibollet, G. Toupet, J.-F. Timsit, J.-C. Lucet, et al., Motion-capture system to assess intraoperative staff movements and door openings: Impact on surrogates of the infectious risk in surgery, Infection Control & Hospital Epidemiology 40 (5) (2019) 566–573.

[28] D. Bouget, M. Allan, D. Stoyanov, P. Jannin, Vision-based and marker-less surgical tool detection and tracking: A review of the literature, Medical Image Analysis 35 (2017) 633–654.

[29] A. Sánchez, O. Rodríguez, R. Sánchez, G. Benítez, R. Pena, O. Salamo, V. Baez, Laparoscopic surgery skills evaluation: Analysis based on accelerometers, Journal of the Society of Laparoscopic & Robotic Surgeons 18 (4) (2014) e2014.00234, pp. 1–5.

[30] S. Speidel, G. Sudra, J. Senemaud, M. Drentschew, B.P. Müller-Stich, C. Gutt, R. Dillmann, Recognition of risk situations based on endoscopic instrument tracking and

knowledge based situation modeling, in: M.I. Miga, K.R. Cleary (Eds.), Medical Imaging 2008: Visualization, Image-Guided Procedures, and Modeling, in: Proceedings of SPIE, vol. 6918, SPIE, 2008, pp. 326–333.

[31] V. Lahanas, C. Loukas, E. Georgiou, A simple sensor calibration technique for estimating the 3D pose of endoscopic instruments, Surgical Endoscopy 30 (3) (2016) 1198–1204.

[32] S. Bodenstedt, N. Padoy, G. Hager, Learned partial automation for shared control in tele-robotic manipulation, in: Proceedings of the AAAI Fall Symposium Series, 2012.

[33] N. Padoy, G.D. Hager, Human-machine collaborative surgery using learned models, in: Proceedings of the IEEE International Conference on Robotics and Automation (ICRA), 2011.

[34] L. Zappella, B. Béjar, G. Hager, R. Vidal, Surgical gesture classification from video and kinematic data, Medical Image Analysis 17 (7) (2013) 732–745.

[35] R. Mudunuri, O. Burgert, T. Neumuth, Ontological modelling of surgical knowledge, in: S. Fischer, E. Mähle, R. Reischuk (Eds.), INFORMATIK 2009 – Im Focus das Leben, in: Lecture Notes in Informatics, vol. P154, Gesellschaft für Informatik, 2009, pp. 1044–1054.

[36] P. Jannin, X. Morandi, Surgical models for computer-assisted neurosurgery, NeuroImage 37 (3) (2007) 783–791.

[37] A. Uciteli, J. Neumann, K. Tahar, K. Saleh, S. Stucke, S. Faulbrück-Röhr, A. Kaeding, M. Specht, T. Schmidt, T. Neumuth, A. Besting, D. Stegemann, F. Portheine, H. Herre, Ontology-based specification, identification and analysis of perioperative risks, Journal of Biomedical Semantics 8 (1) (2017) 36, pp. 1–14.

[38] D. Katić, C. Julliard, A.-L. Wekerle, H.G. Kenngott, B.P. Müller-Stich, R. Dillmann, S. Speidel, P. Jannin, B. Gibaud, LapOntoSPM: An ontology for laparoscopic surgeries and its application to surgical phase recognition, International Journal of Computer Assisted Radiology and Surgery 10 (9) (2015) 1427–1434.

[39] D.A. Nagy, I.J. Rudas, T. Haidegger, OntoFlow, a software tool for surgical workflow recording, in: Proceedings of the IEEE World Symposium on Applied Machine Intelligence and Informatics (SAMI), 2018.

[40] B. Gibaud, G. Forestier, C. Feldmann, G. Ferrigno, P. Gonçalves, T. Haidegger, C. Julliard, D. Katić, H.G. Kenngott, L. Maier-Hein, K. März, E. De Momi, D.Á. Nagy, H. Nakawala, J. Neumann, T. Neumuth, J. Rojas Balderrama, S. Speidel, M. Wagner, P. Jannin, Toward a standard ontology of surgical process models, International Journal of Computer Assisted Radiology and Surgery 13 (9) (2018) 1397–1408.

[41] S. Bodenstedt, D. Rivoir, A. Jenke, M. Wagner, M. Breucha, B. Müller-Stich, S.T. Mees, J. Weitz, S. Speidel, Active learning using deep Bayesian networks for surgical workflow analysis, International Journal of Computer Assisted Radiology and Surgery 14 (6) (2019) 1079–1087.

[42] Y. Gur, M. Moradi, H. Bulu, Y. Guo, C. Compas, T. Syeda-Mahmood, Towards an efficient way of building annotated medical image collections for big data studies, in: M.J. Cardoso, T. Arbel, S.-L. Lee, V. Cheplygina, S. Balocco, D. Mateus, G. Zahnd, L. Maier-Hein, S. Demirci, E. Granger, L. Duong, M.-A. Carbonneau, S. Albarqouni, G. Carneiro (Eds.), Intravascular Imaging and Computer Assisted Stenting, and Large-Scale Annotation of Biomedical Data and Expert Label Synthesis, in: Lecture Notes in Computer Science, vol. 10552, Springer, 2017, pp. 87–95.

[43] G. Lecuyer, M. Ragot, N. Martin, L. Launay, P. Jannin, Assisted phase and step annotation for surgical videos, International Journal of Computer Assisted Radiology and Surgery 15 (4) (2020) 673–680.

[44] T.S. Kim, A. Malpani, A. Reiter, G.D. Hager, S. Sikder, S.S. Vedula, Crowdsourcing annotation of surgical instruments in videos of cataract surgery, in: D. Stoyanov, Z. Taylor, S. Balocco, R. Sznitman, A. Martel, L. Maier-Hein, L. Duong, G. Zahnd, S. Demirci, S. Albarqouni, S.-L. Lee, S. Moriconi, V. Cheplygina, D. Mateus, E. Trucco, E. Granger, P. Jannin (Eds.), Intravascular Imaging and Computer Assisted Stenting and Large-Scale Annotation of Biomedical Data and Expert Label Synthesis, in: Lecture Notes in Computer Science, vol. 11043, Springer, 2018, pp. 121–130.

[45] T.S. Lendvay, L. White, T. Kowalewski, Crowdsourcing to assess surgical skill, JAMA Surgery 150 (11) (2015) 1086–1087.

[46] L. Maier-Hein, D. Kondermann, T. Roß, S. Mersmann, E. Heim, S. Bodenstedt, H.G. Kenngott, A. Sanchez, M. Wagner, A. Preukschas, A.-L. Wekerle, S. Helfert, K. März, A. Mehrabi, S. Speidel, C. Stock, Crowdtruth validation: A new paradigm for validating algorithms that rely on image correspondences, International Journal of Computer Assisted Radiology and Surgery 10 (8) (2015) 1201–1212.

[47] L. Maier-Hein, T. Roß, J. Gröhl, B. Glocker, S. Bodenstedt, C. Stock, E. Heim, M. Götz, S. Wirkert, H.G. Kenngott, S. Speidel, K. Maier-Hein, Crowd-algorithm collaboration for large-scale endoscopic image annotation with confidence, in: S. Ourselin, L. Joskowicz, M.R. Sabuncu, G. Unal, W. Wells (Eds.), Medical Image Computing and Computer-Assisted Intervention – MICCAI 2016, in: Lecture Notes in Computer Science, vol. 9901, Springer, 2016, pp. 616–623.

[48] T. Vercauteren, M. Unberath, N. Padoy, N. Navab, CAI4CAI: The rise of contextual artificial intelligence in computer-assisted interventions, Proceedings of the IEEE 108 (1) (2019) 198–214.

[49] T. Blum, N. Padoy, H. Feußner, N. Navab, Modeling and online recognition of surgical phases using hidden Markov models, in: D. Metaxas, L. Axel, G. Fichtinger, G. Székely (Eds.), Medical Image Computing and Computer-Assisted Intervention – MICCAI 2008, in: Lecture Notes in Computer Science, vol. 5242, Springer, 2008, pp. 627–635.

[50] N. Padoy, T. Blum, S.-A. Ahmadi, H. Feußner, M.-O. Berger, N. Navab, Statistical modeling and recognition of surgical workflow, Medical Image Analysis 16 (3) (2012) 632–641.

[51] I. Funke, A. Jenke, S.T. Mees, J. Weitz, S. Speidel, S. Bodenstedt, Temporal coherence-based self-supervised learning for laparoscopic workflow analysis, in: D. Stoyanov, Z. Taylor, D. Sarikaya, J. McLeod, M.A. González Ballester, N.C.F. Codella, A. Martel, L. Maier-Hein, A. Malpani, M.A. Zenati, S. De Ribaupierre, L. Xiongbiao, T. Collins, T. Reichl, K. Drechsler, M. Erdt, M.G. Linguraru, C.O. Laura, R. Shekhar, S. Wesarg, M.E. Celebi, K. Dana, A. Halpern (Eds.), OR 2.0 Context-Aware Operating Theaters, Computer Assisted Robotic Endoscopy, Clinical Image-Based Procedures, and Skin Image Analysis, in: Lecture Notes in Computer Science, vol. 11041, Springer, 2018, pp. 85–93.

[52] A.P. Twinanda, S. Shehata, D. Mutter, J. Marescaux, M. de Mathelin, N. Padoy, EndoNet: A deep architecture for recognition tasks on laparoscopic videos, IEEE Transactions on Medical Imaging 36 (1) (2016) 86–97.

[53] A.P. Twinanda, Vision-based approaches for surgical activity recognition using laparoscopic and RBGD videos, PhD thesis, University of Strasbourg, France, 2017.

[54] N. Padoy, T. Blum, H. Feußner, M.-O. Berger, N. Navab, On-line recognition of surgical activity for monitoring in the operating room, in: Proceedings of the AAAI Conference on Artificial Intelligence (AAAI), 2008.

[55] R. DiPietro, C. Lea, A. Malpani, N. Ahmidi, S.S. Vedula, G.I. Lee, M.R. Lee, G.D. Hager, Recognizing surgical activities with recurrent neural networks, in: S. Ourselin, L. Joskowicz, M.R. Sabuncu, G. Unal, W. Wells (Eds.), Medical Image Computing and Computer-Assisted Intervention – MICCAI 2016, in: Lecture Notes in Computer Science, vol. 9900, Springer, 2016, pp. 551–558.

[56] C.E. Reiley, H.C. Lin, B. Varadarajan, B. Vagvolgyi, S. Khudanpur, D.D. Yuh, G.D. Hager, Automatic recognition of surgical motions using statistical modeling for capturing variability, in: J.D. Westwood, R.S. Haluck, H.M. Hoffman, G.T. Mogel, R. Phillips, R.A. Robb, K.G. Vosburgh (Eds.), Medicine Meets Virtual Reality 16 – Parallel, Combinatorial, Convergent: NextMed by Design, in: Studies in Health Technology and Informatics, vol. 132, IOS Press, 2008, pp. 396–401.

[57] C. Lea, G.D. Hager, R. Vidal, An improved model for segmentation and recognition of fine-grained activities with application to surgical training tasks, in: Proceedings of the IEEE Winter Conference on Applications of Computer Vision (WACV), 2015.

[58] L. Tao, L. Zappella, G.D. Hager, R. Vidal, Surgical gesture segmentation and recognition, in: K. Mori, I. Sakuma, Y. Sato, C. Barillot, N. Navab (Eds.), Medical Image Computing and Computer-Assisted Intervention – MICCAI 2013, in: Lecture Notes in Computer Science, vol. 8151, Springer, 2013, pp. 339–346.

[59] M. Pfeiffer, I. Funke, M.R. Robu, S. Bodenstedt, L. Strenger, S. Engelhardt, T. Roß, M.J. Clarkson, K. Gurusamy, B.R. Davidson, L. Maier-Hein, C. Riediger, T. Welsch, J. Weitz, S. Speidel, Generating large labeled data sets for laparoscopic image processing tasks using unpaired image-to-image translation, in: D. Shen, T. Liu, T.M. Peters, L.H. Staib, C. Essert, S. Zhou, P.-T. Yap, A. Khan (Eds.), Medical Image Computing and Computer Assisted Intervention – MICCAI 2019, in: Lecture Notes in Computer Science, vol. 11768, Springer, 2019, pp. 119–127.

[60] K.A. Ericsson, Deliberate practice and the acquisition and maintenance of expert performance in medicine and related domains, Academic Medicine 79 (10) (2004) S70–S81.

[61] J.D. Birkmeyer, J.F. Finks, A. O'Reilly, M. Oerline, A.M. Carlin, A.R. Nunn, J. Dimick, M. Banerjee, N.J.O. Birkmeyer, Surgical skill and complication rates after bariatric surgery, New England Journal of Medicine 369 (15) (2013) 1434–1442.

[62] J.D. Birkmeyer, A.E. Siewers, E.V.A. Finlayson, T.A. Stukel, F.L. Lucas, I. Batista, H.G. Welch, D.E. Wennberg, Hospital volume and surgical mortality in the United States, New England Journal of Medicine 346 (15) (2002) 1128–1137.

[63] S. Yule, R. Flin, S. Paterson-Brown, N. Maran, Non-technical skills for surgeons in the operating room: A review of the literature, Surgery 139 (2) (2006) 140–149.

[64] J.H. Peters, G.M. Fried, L.L. Swanstrom, N.J. Soper, L.F. Sillin, B. Schirmer, K. Hoffman, et al., Development and validation of a comprehensive program of education and assessment of the basic fundamentals of laparoscopic surgery, Surgery 135 (1) (2004) 21–27.

[65] R. Aggarwal, K. Moorthy, A. Darzi, Laparoscopic skills training and assessment, British Journal of Surgery 91 (12) (2004) 1549–1558.

[66] H.W.R. Schreuder, R. Wolswijk, R.P. Zweemer, M.P. Schijven, R.H.M. Verheijen, Training and learning robotic surgery, time for a more structured approach: A systematic review, International Journal of Obstetrics & Gynaecology 119 (2) (2012) 137–149.

[67] N.E. Seymour, A.G. Gallagher, S.A. Roman, M.K. O'Brien, V.K. Bansal, D.K. Andersen, R.M. Satava, Virtual reality training improves operating room performance:

Results of a randomized, double-blinded study, Annals of Surgery 236 (4) (2002) 458–464.

[68] C. Våpenstad, E.F. Hofstad, L.E. Bø, E. Kuhry, G. Johnsen, R. Mårvik, T. Langø, T.N. Hernes, Lack of transfer of skills after virtual reality simulator training with haptic feedback, Minimally Invasive Therapy & Allied Technologies 26 (6) (2017) 346–354.

[69] K. Gurusamy, R. Aggarwal, L. Palanivelu, B.R. Davidson, Systematic review of randomized controlled trials on the effectiveness of virtual reality training for laparoscopic surgery, British Journal of Surgery 95 (9) (2008) 1088–1097.

[70] C.R. Larsen, J.L. Sørensen, T.P. Grantcharov, T. Dalsgaard, L. Schouenborg, C. Ottosen, T.V. Schroeder, B.S. Ottesen, Effect of virtual reality training on laparoscopic surgery: Randomised controlled trial, British Medical Journal 338 (7705) (2009) 1253–1256.

[71] L.J. Moore, M.R. Wilson, E. Waine, R.S.W. Masters, J.S. McGrath, S.J. Vine, Robotic technology results in faster and more robust surgical skill acquisition than traditional laparoscopy, Journal of Robotic Surgery 9 (1) (2015) 67–73.

[72] P. Yohannes, P. Rotariu, P. Pinto, A.D. Smith, B.R. Lee, Comparison of robotic versus laparoscopic skills: Is there a difference in the learning curve?, Urology 60 (1) (2002) 39–45.

[73] E. Boyle, M. Al-Akash, A.G. Gallagher, O. Traynor, A.D.K. Hill, P.C. Neary, Optimising surgical training: Use of feedback to reduce errors during a simulated surgical procedure, Postgraduate Medical Journal 87 (1030) (2011) 524–528.

[74] T.P. Grantcharov, S. Schulze, V.B. Kristiansen, The impact of objective assessment and constructive feedback on improvement of laparoscopic performance in the operating room, Surgical Endoscopy 21 (12) (2007) 2240–2243.

[75] A. Trehan, A. Barnett-Vanes, M.J. Carty, P. McCulloch, M. Maruthappu, The impact of feedback of intraoperative technical performance in surgery: A systematic review, British Medical Journal Open 5 (6) (2015) e006759, pp. 1–5.

[76] K. Moorthy, Y. Munz, S.K. Sarker, A. Darzi, Objective assessment of technical skills in surgery, British Medical Journal 327 (7422) (2003) 1032–1037.

[77] J.A. Martin, G. Regehr, R. Reznick, H. Macrae, J. Murnaghan, C. Hutchison, M. Brown, Objective structured assessment of technical skill (OSATS) for surgical residents, British Journal of Surgery 84 (2) (1997) 273–278.

[78] C.E. Reiley, H.C. Lin, D.D. Yuh, G.D. Hager, Review of methods for objective surgical skill evaluation, Surgical Endoscopy 25 (2) (2011) 356–366.

[79] Z. Lin, M. Uemura, M. Zecca, S. Sessa, H. Ishii, M. Tomikawa, M. Hashizume, A. Takanishi, Objective skill evaluation for laparoscopic training based on motion analysis, IEEE Transactions on Biomedical Engineering 60 (4) (2013) 977–985.

[80] L. Richstone, M.J. Schwartz, C. Seideman, J. Cadeddu, S. Marshall, L.R. Kavoussi, Eye metrics as an objective assessment of surgical skill, Annals of Surgery 252 (1) (2010) 177–182.

[81] N. Ahmidi, G.D. Hager, L. Ishii, G. Fichtinger, G.L. Gallia, M. Ishii, Surgical task and skill classification from eye tracking and tool motion in minimally invasive surgery, in: T. Jiang, N. Navab, J.P.W. Pluim, M.A. Viergeve (Eds.), Medical Image Computing and Computer-Assisted Intervention – MICCAI 2010, in: Lecture Notes in Computer Science, vol. 6363, Springer, 2010, pp. 295–302.

[82] F. Pérez-Escamirosa, A. Alarcón-Paredes, G.A. Alonso-Silverio, I. Oropesa, O. Camacho-Nieto, D. Lorias-Espinoza, A. Minor-Martínez, Objective classification of psychomotor laparoscopic skills of surgeons based on three different approaches, International Journal of Computer Assisted Radiology and Surgery 15 (1) (2020) 27–40.

[83] K.-F. Kowalewski, J.D. Hendrie, M.W. Schmidt, C.R. Garrow, T. Bruckner, T. Proctor, S. Paul, D. Adigüzel, S. Bodenstedt, A. Erben, H.G. Kenngott, Y. Erben, S. Speidel, B.P. Müller-Stich, F. Nickel, Development and validation of a sensor- and expert model-based training system for laparoscopic surgery: The iSurgeon, Surgical Endoscopy 31 (5) (2017) 2155–2165.

[84] H. Ismail Fawaz, G. Forestier, J. Weber, L. Idoumghar, P.-A. Müller, Evaluating surgical skills from kinematic data using convolutional neural networks, in: A.F. Frangi, J.A. Schnabel, C. Davatzikos, C. Alberola-López, G. Fichtinger (Eds.), Medical Image Computing and Computer Assisted Intervention – MICCAI 2018, in: Lecture Notes in Computer Science, vol. 11073, Springer, 2018, pp. 214–221.

[85] Z. Wang, A. Majewicz Fey, Deep learning with convolutional neural network for objective skill evaluation in robot-assisted surgery, International Journal of Computer Assisted Radiology and Surgery 13 (12) (2018) 1959–1970.

[86] H. Min, D.R. Morales, D. Orgill, D.S. Smink, S. Yule, Systematic review of coaching to enhance surgeons' operative performance, Surgery 158 (5) (2015) 1168–1191.

[87] S.J. Cole, H. Mackenzie, J. Ha, G.B. Hanna, D. Mišković, Randomized controlled trial on the effect of coaching in simulated laparoscopic training, Surgical Endoscopy 28 (3) (2014) 979–986.

[88] P. Singh, R. Aggarwal, M. Tahir, P.H. Pucher, A. Darzi, A randomized controlled study to evaluate the role of video-based coaching in training laparoscopic skills, Annals of Surgery 261 (5) (2015) 862–869.

[89] E.M. Bonrath, N.J. Dedy, L.E. Gordon, T.P. Grantcharov, Comprehensive surgical coaching enhances surgical skill in the operating room, Annals of Surgery 262 (2) (2015) 205–212.

[90] S. Wijewickrema, P. Piromchai, Y. Zhou, I. Ioannou, J. Bailey, G. Kennedy, S. O'Leary, Developing effective automated feedback in temporal bone surgery simulation, Otolaryngology – Head and Neck Surgery 152 (6) (2015) 1082–1088.

[91] T. Horeman, S.P. Rodrigues, J.J. van den Dobbelsteen, F.-W. Jansen, J. Dankelman, Visual force feedback in laparoscopic training, Surgical Endoscopy 26 (1) (2012) 242–248.

[92] D. Black, C. Hansen, A. Nabavi, R. Kikinis, H. Hahn, A survey of auditory display in image-guided interventions, International Journal of Computer Assisted Radiology and Surgery 12 (10) (2017) 1665–1676.

[93] R. Sigrist, G. Rauter, R. Riener, P. Wolf, Augmented visual, auditory, haptic, and multimodal feedback in motor learning: A review, Psychonomic Bulletin & Review 20 (1) (2013) 21–53.

[94] G. Islam, K. Kahol, B. Li, M. Smith, V.L. Patel, Affordable, web-based surgical skill training and evaluation tool, Journal of Biomedical Informatics 59 (2016) 102–114.

[95] T. Yamaguchi, R. Nakamura, Laparoscopic training using a quantitative assessment and instructional system, International Journal of Computer Assisted Radiology and Surgery 13 (9) (2018) 1453–1461.

[96] A. Malpani, S.S. Vedula, H.C. Lin, G.D. Hager, R.H. Taylor, Real-time teaching cues for automated surgical coaching, CoRR, arXiv:1704.07436, 2017.

[97] S. Wijewickrema, X. Ma, P. Piromchai, R. Briggs, J. Bailey, G. Kennedy, S. O'Leary, Providing automated real-time technical feedback for virtual reality based surgical training: Is the simpler the better?, in: Proceedings of the International Conference on Artificial Intelligence in Education (AIED), 2018.

[98] N. Ahmidi, G.D. Hager, L. Ishii, G.L. Gallia, M. Ishii, Robotic path planning for surgeon skill evaluation in minimally-invasive sinus surgery, in: N. Ayache, H. Delingette, P. Golland, K. Mori (Eds.), Medical Image Computing and Computer-Assisted Intervention – MICCAI 2012, in: Lecture Notes in Computer Science, vol. 7510, Springer, 2012, pp. 471–478.

[99] S. Wijewickrema, Y. Zhou, J. Bailey, G. Kennedy, S. O'Leary, Provision of automated step-by-step procedural guidance in virtual reality surgery simulation, in: Proceedings of the ACM Conference on Virtual Reality Software and Technology (VRST), 2016.

[100] N. Oppermann, C. Yang, J. Weitz, C. Reissfelder, S.T. Mees, Establishment of multidimensional structured training curriculum in surgery, Zentralblatt für Chirurgie 6 (2019) 536–542.

[101] I. Funke, S.T. Mees, J. Weitz, S. Speidel, Video-based surgical skill assessment using 3D convolutional neural networks, International Journal of Computer Assisted Radiology and Surgery 14 (7) (2019) 1217–1225.

[102] D.D. Woods, E.S. Patterson, E.M. Roth, Can we ever escape from data overload? A cognitive systems diagnosis, Cognition, Technology & Work 4 (1) (2002) 22–36.

[103] D. Katić, A.-L. Wekerle, J. Görtler, P. Spengler, S. Bodenstedt, S. Röhl, S. Suwelack, H.G. Kenngott, M. Wagner, B.P. Müller-Stich, R. Dillmann, S. Speidel, Context-aware augmented reality in laparoscopic surgery, Computerized Medical Imaging and Graphics 37 (2) (2013) 174–182.

[104] L. Maier-Hein, S.S. Vedula, S. Speidel, N. Navab, R. Kikinis, A. Park, M. Eisenmann, H. Feußner, G. Forestier, S. Giannarou, M. Hashizume, D. Katić, H.G. Kenngott, M. Kranzfelder, A. Malpani, K. März, T. Neumuth, N. Padoy, C. Pugh, N. Schoch, D. Stoyanov, R. Taylor, M. Wagner, G.D. Hager, P. Jannin, Surgical data science for next-generation interventions, Nature Biomedical Engineering 1 (9) (2017) 691–696.

[105] S. Bernhardt, S.A. Nicolau, L. Soler, C. Doignon, The status of augmented reality in laparoscopic surgery as of 2016, Medical Image Analysis 37 (2017) 66–90.

[106] E. De Momi, L. Kranendonk, M. Valenti, N. Enayati, G. Ferrigno, A neural network-based approach for trajectory planning in robot–human handover tasks, Frontiers in Robotics and AI 3 (2016) 34, pp. 1–10.

[107] N. Navab, J. Traub, T. Sielhorst, M. Feuerstein, C. Bichlmeier, Action- and workflow-driven augmented reality for computer-aided medical procedures, IEEE Computer Graphics and Applications 27 (5) (2007) 10–14.

[108] D.B. Douglas, C.A. Wilke, J.D. Gibson, J.M. Boone, M. Wintermark, Augmented reality: Advances in diagnostic imaging, Multimodal Technologies and Interaction 1 (4) (2017) 29, pp. 1–12.

[109] D. Guha, N.M. Alotaibi, N. Nguyen, S. Gupta, C. McFaul, V.X.D. Yang, Augmented reality in neurosurgery: A review of current concepts and emerging applications, Canadian Journal of Neurological Sciences 44 (3) (2017) 235–245.

[110] M. Peterhans, A. vom Berg, B. Dagon, D. Inderbitzin, C. Baur, D. Candinas, S. Weber, A navigation system for open liver surgery: Design, workflow and first clinical applications, International Journal of Medical Robotics and Computer Assisted Surgery 7 (1) (2011) 7–16.

[111] A.L. Simpson, T.P. Kingham, Current evidence in image-guided liver surgery, Journal of Gastrointestinal Surgery 20 (6) (2016) 1265–1269.

[112] G. Quero, A. Lapergola, L. Soler, M. Shabaz, A. Hostettler, T. Collins, J. Marescaux, D. Mutter, M. Diana, P. Pessaux, Virtual and augmented reality in oncologic liver surgery, Surgical Oncology Clinics 28 (1) (2019) 31–44.

[113] A. Hughes-Hallett, E.K. Mayer, H.J. Marcus, T.P. Cundy, P.J. Pratt, A.W. Darzi, J.A. Vale, Augmented reality partial nephrectomy: Examining the current status and future perspectives, Urology 83 (2) (2014) 266–273.

[114] M.N. van Oosterom, H.G. van der Poel, N. Navab, C.J.H. van de Velde, F.W.B. van Leeuwen, Computer-assisted surgery: Virtual- and augmented-reality displays for navigation during urological interventions, Current Opinion in Urology 28 (2) (2018) 205–213.

[115] L. Guerriero, G. Quero, M. Diana, L. Soler, V. Agnus, J. Marescaux, F. Corcione, Virtual reality exploration and planning for precision colorectal surgery, Diseases of the Colon & Rectum 61 (6) (2018) 719–723.

[116] L. Qian, J.Y. Wu, S.P. DiMaio, N. Navab, P. Kazanzides, A review of augmented reality in robotic-assisted surgery, IEEE Transactions on Medical Robotics and Bionics 2 (1) (2020) 1–16.

[117] B. Fida, F. Cutolo, G. di Franco, M. Ferrari, V. Ferrari, Augmented reality in open surgery, Updates in Surgery 70 (3) (2018) 389–400.

[118] J.T. Doswell, A. Skinner, Augmenting human cognition with adaptive augmented reality, in: D.D. Schmorrow, C.M. Fidopiastis (Eds.), Foundations of Augmented Cognition – Advancing Human Performance and Decision-Making through Adaptive Systems, in: Lecture Notes in Computer Science, vol. 8534, Springer, 2014, pp. 104–113.

[119] M.A. Zenati, L. Kennedy-Metz, R.D. Dias, Cognitive engineering to improve patient safety and outcomes in cardiothoracic surgery, Seminars in Thoracic and Cardiovascular Surgery 32 (1) (2020) 1–7.

[120] R.D. Dias, M.C. Ngo-Howard, M.T. Boskovski, M.A. Zenati, S.J. Yule, Systematic review of measurement tools to assess surgeons' intraoperative cognitive workload, British Journal of Surgery 105 (5) (2018) 491–501.

[121] L. Maier-Hein, S. Speidel, E. Stenau, E.C. Chen, B. Ma, Registration, in: Mixed and Augmented Reality in Medicine, CRC Press, 2020, pp. 29–45.

[122] S. Speidel, S. Bodenstedt, F. Vasconcelos, D. Stoyanov, Interventional imaging: Vision, in: S.K. Zhou, D. Rückert, G. Fichtinger (Eds.), Handbook of Medical Image Computing and Computer Assisted Intervention, Academic Press, 2020, pp. 721–745.

[123] M. Pfeiffer, C. Riediger, J. Weitz, S. Speidel, Learning soft tissue behavior of organs for surgical navigation with convolutional neural networks, International Journal of Computer Assisted Radiology and Surgery 14 (7) (2019) 1147–1155.

[124] J.-N. Brunet, A. Mendizabal, A. Petit, N. Golse, E. Vibert, S. Cotin, Physics-based deep neural network for augmented reality during liver surgery, in: D. Shen, T. Liu, T.M. Peters, L.H. Staib, C. Essert, S. Zhou, P.-T. Yap, A. Khan (Eds.), Medical Image Computing and Computer Assisted Intervention – MICCAI 2019, in: Lecture Notes in Computer Science, vol. 11768, Springer, 2019, pp. 137–145.

[125] P. Reipschläger, S. Engert, R. Dachselt, Augmented displays: Seamlessly extending interactive surfaces with head-mounted augmented reality, in: Proceedings of the ACM Conference on Human Factors in Computing Systems (CHI), Extended Abstracts, 2020.

[126] M. Spindler, W. Büschel, R. Dachselt, Use your head: Tangible windows for 3D information spaces in a tabletop environment, in: Proceedings of the ACM International Conference on Interactive Tabletops and Surfaces (ITS), 2012.

[127] W. Büschel, P. Reipschläger, R. Dachselt, Improving 3D visualizations: Exploring spatial interaction with mobile devices, in: Proceedings of the ACM Companion on Interactive Surfaces and Spaces (ISS Companion), 2016.

[128] K. Klamka, A. Siegel, S. Vogt, F. Göbel, S. Stellmach, R. Dachselt, Look & pedal: Hands-free navigation in zoomable information spaces through gaze-supported foot input, in: Proceedings of the ACM International Conference on Multimodal Interaction (ICMI), 2015.

[129] B. Hatscher, M. Luz, L.E. Nacke, N. Elkmann, V. Müller, C. Hansen, GazeTap: Towards hands-free interaction in the operating room, in: Proceedings of the ACM International Conference on Multimodal Interaction (ICMI), 2017.

[130] N. Nestorov, P. Hughes, N. Healy, N. Sheehy, N. O'Hare, Application of natural user interface devices for touch-free control of radiological images during surgery, in: Proceedings of the IEEE International Symposium on Computer-Based Medical Systems (CBMS), 2016.

[131] A.V. Reinschlüssel, T. Münder, V. Uslar, D. Weyhe, A. Schenk, R. Malaka, Tangible organs: Introducing 3D printed organ models with VR to interact with medical 3D models, in: Proceedings of the ACM Conference on Human Factors in Computing Systems (CHI), Extended Abstracts, 2019.

[132] I. Avellino, G. Bailly, G. Canlorbe, J. Belghiti, G. Morel, M.-A. Vitrani, Impacts of telemanipulation in robotic assisted surgery, in: Proceedings of the ACM Conference on Human Factors in Computing Systems (CHI), 2019.

[133] A. Üneri, M.A. Balicki, J. Handa, P. Gehlbach, R.H. Taylor, I. Iordachita, New steady-hand eye robot with micro-force sensing for vitreoretinal surgery, in: Proceedings of the IEEE RAS/EMBS International Conference on Biomedical Robotics and Biomechatronics (BioRob), 2010.

[134] X. He, D. Roppenecker, D. Gierlach, M. Balicki, K. Olds, P. Gehlbach, J. Handa, R. Taylor, I. Iordachita, Toward clinically applicable steady-hand eye robot for vitreoretinal surgery, in: Proceedings of the ASME International Mechanical Engineering Congress and Exposition (IMECE), 2012.

[135] S. Yang, R.A. MacLachlan, C.N. Riviere, Manipulator design and operation of a six-degree-of-freedom handheld tremor-canceling microsurgical instrument, IEEE/ASME Transactions on Mechatronics 20 (2) (2014) 761–772.

[136] B.C. Becker, R.A. MacLachlan, L.A. Lobes Jr., C.N. Riviere, Position-based virtual fixtures for membrane peeling with a handheld micromanipulator, in: Proceedings of the IEEE International Conference on Robotics and Automation (ICRA), 2012.

[137] B. Fallahi, M. Waine, C. Rossa, R. Sloboda, N. Usmani, M. Tavakoli, An integrator-backstepping control approach for three-dimensional needle steering, IEEE/ASME Transactions on Mechatronics 24 (5) (2019) 2204–2214.

[138] P. Moreira, N. Zemiti, C. Liu, P. Poignet, Viscoelastic model based force control for soft tissue interaction and its application in physiological motion compensation, Computer Methods and Programs in Biomedicine 116 (2) (2014) 52–67.

[139] M. Bowthorpe, M. Tavakoli, Generalized predictive control of a surgical robot for beating-heart surgery under delayed and slowly-sampled ultrasound image data, IEEE Robotics and Automation Letters 1 (2) (2016) 892–899.

[140] M. Bowthorpe, M. Tavakoli, Ultrasound-based image guidance and motion compensating control for robot-assisted beating-heart surgery, Journal of Medical Robotics Research 1 (1) (2016) 1640002, pp. 1–11.

[141] N.A. Patronik, C.N. Riviere, S. El Qarra, M.A. Zenati, The HeartLander: A novel epicardial crawling robot for myocardial injections, International Congress Series 1281 (2005) 735–739.

[142] J. Carriere, J. Fong, T. Meyer, R. Sloboda, S. Husain, N. Usmani, M. Tavakoli, An admittance-controlled robotic assistant for semi-autonomous breast ultrasound scanning, in: Proceedings of the International Symposium on Medical Robotics (ISMR), 2019.

[143] S. Billings, N. Deshmukh, H.J. Kang, R. Taylor, E.M. Boctor, System for robot-assisted real-time laparoscopic ultrasound elastography, in: D.R. Holmes III, K.H. Wong (Eds.), Medical Imaging 2012: Image-Guided Procedures, Robotic Interventions, and Modeling, in: Proceedings of SPIE, vol. 8316, SPIE, 2012, pp. 589–596.

[144] L. Yu, H. Li, L. Zhao, S. Ren, Q. Gu, Automatic guidance of laparoscope based on the region of interest for robot assisted laparoscopic surgery, Computer Assisted Surgery 21 (Sup. 1) (2016) 17–21.

[145] O. Weede, H. Mönnich, B. Müller, H. Wörn, An intelligent and autonomous endoscopic guidance system for minimally invasive surgery, in: Proceedings of the IEEE International Conference on Robotics and Automation (ICRA), 2011.

[146] A. Bihlmaier, H. Wörn, Automated endoscopic camera guidance: A knowledge-based system towards robot assisted surgery, in: Proceedings of the International Symposium on Robotics (ISR), 2014.

[147] H. Mayer, I. Nagy, A. Knoll, E.U. Braun, R. Lange, R. Bauernschmitt, Adaptive control for human–robot skilltransfer: Trajectory planning based on fluid dynamics, in: Proceedings of the IEEE International Conference on Robotics and Automation (ICRA), 2007.

[148] H. Mayer, I. Nagy, D. Burschka, A. Knoll, E.U. Braun, R. Lange, R. Bauernschmitt, Automation of manual tasks for minimally invasive surgery, in: Proceedings of the International Conference on Autonomic and Autonomous Systems (ICAS), 2008.

[149] H. Mayer, F. Gomez, D. Wierstra, I. Nagy, A. Knoll, J. Schmidhuber, A system for robotic heart surgery that learns to tie knots using recurrent neural networks, Advanced Robotics 22 (13–14) (2008) 1521–1537.

[150] J. van den Berg, S. Miller, D. Duckworth, H. Hu, A. Wan, X.-Y. Fu, K. Goldberg, P. Abbeel, Superhuman performance of surgical tasks by robots using iterative learning from human-guided demonstrations, in: Proceedings of the IEEE International Conference on Robotics and Automation (ICRA), 2010.

[151] T. Osa, N. Sugita, M. Mitsuishi, Online trajectory planning in dynamic environments for surgical task automation, in: Proceedings of the Conference on Robotics: Science and Systems (RSS), 2014.

[152] G.P. Mylonas, P. Giataganas, M. Chaudery, V. Vitiello, A. Darzi, G.-Z. Yang, in: Proceedings of the IEEE/RSJ International Conference on Intelligent Robots and Systems (IROS), 2013.

[153] P. Giataganas, V. Vitiello, V. Simaiaki, E. Lopez, G.-Z. Yang, Cooperative in situ microscopic scanning and simultaneous tissue surface reconstruction using a compliant robotic manipulator, in: Proceedings of the IEEE International Conference on Robotics and Automation (ICRA), 2013.

[154] B. Kehoe, G. Kahn, J. Mahler, J. Kim, A. Lee, A. Lee, K. Nakagawa, S. Patil, W.D. Boyd, P. Abbeel, K. Goldberg, Autonomous multilateral debridement with the raven surgical robot, in: Proceedings of the IEEE International Conference on Robotics and Automation (ICRA), 2014.

[155] A. Murali, S. Sen, B. Kehoe, A. Garg, S. McFarland, S. Patil, W.D. Boyd, S. Lim, P. Abbeel, K. Goldberg, Learning by observation for surgical subtasks: Multilateral cutting of 3D viscoelastic and 2D orthotropic tissue phantoms, in: Proceedings of the IEEE International Conference on Robotics and Automation (ICRA), 2015.

[156] F. Richter, R.K. Orosco, M.C. Yip, Open-sourced reinforcement learning environments for surgical robotics, CoRR, arXiv:1903.02090, 2019.

[157] M. Tang, F. Perazzi, A. Djelouah, I. Ben Ayed, C. Schroers, Y. Boykov, On regularized losses for weakly-supervised CNN segmentation, in: Proceedings of the European Conference on Computer Vision (ECCV), 2018.

[158] R.R. Selvaraju, A. Das, R. Vedantam, M. Cogswell, D. Parikh, D. Batra, Grad-CAM: Why did you say that? Visual explanations from deep networks via gradient-based localization, CoRR, arXiv:1610.02391, 2016.

[159] H. Laga, A survey on deep learning architectures for image-based depth reconstruction, CoRR, arXiv:1906.06113, 2019.

[160] J.-R. Chang, Y.-S. Chen, Pyramid stereo matching network, in: Proceedings of the IEEE Conference on Computer Vision and Pattern Recognition (CVPR), 2018.

[161] M. Ye, E. Johns, A. Handa, L. Zhang, P. Pratt, G.-Z. Yang, Self-supervised Siamese learning on stereo image pairs for depth estimation in robotic surgery, CoRR, arXiv: 1705.08260, 2017.

[162] G. Yang, J. Manela, M. Happold, D. Ramanan, Hierarchical deep stereo matching on high-resolution images, in: Proceedings of the IEEE Conference on Computer Vision and Pattern Recognition (CVPR), 2019.

[163] O. Ronneberger, P. Fischer, T. Brox, U-net: Convolutional networks for biomedical image segmentation, in: N. Navab, J. Hornegger, W.M. Wells, A.F. Frangi (Eds.), Medical Image Computing and Computer-Assisted Intervention – MICCAI 2015, in: Lecture Notes in Computer Science, vol. 9351, Springer, 2015, pp. 234–241.

[164] S. Ren, K. He, R. Girshick, J.S. Faster, R-CNN: Towards real-time object detection with region proposal networks, IEEE Transactions on Pattern Analysis and Machine Intelligence 39 (6) (2016) 1137–1149.

[165] K. Klamka, R. Dachselt, J. Steimle, Rapid iron-on user interfaces: Hands-on fabrication of interactive textile prototypes, in: Proceedings of the ACM Conference on Human Factors in Computing Systems (CHI), 2020.

[166] S. Haddadin, L. Johannsmeier, F.D. Ledezma, Tactile robots as a central embodiment of the Tactile Internet, Proceedings of the IEEE 107 (2) (2018) 471–487.

[167] U. Aßmann, D. Grzelak, J. Mey, D. Pukhkaiev, R. Schöne, C. Werner, G. Püschel, Cross-layer adaptation in multi-layer autonomic systems (invited talk), in: B. Catania, R. Královič, J. Nawrocki, G. Pighizzini (Eds.), SOFSEM 2019: Theory and Practice of Computer Science, in: Lecture Notes in Computer Science, vol. 11376, Springer, 2019, pp. 1–20.

[168] S. Bouchard, Lean Robotics: A Guide to Making Robots Work in Your Factory, Samuel Bouchard, 2017.

[169] R.P. Van Til, S. Sengupta, R.J. Srodawa, M.A. Latcha, Robotic assembly cell, in: Proceedings of the Frontiers in Education Conference (FIE), 2003.

[170] F. Chen, K. Sekiyama, J. Huang, B. Sun, H. Sasaki, T. Fukuda, An assembly strategy scheduling method for human and robot coordinated cell manufacturing, International Journal of Intelligent Computing and Cybernetics 4 (4) (2011) 487–510.

[171] F. Chen, K. Sekiyama, F. Cannella, T. Fukuda, Optimal subtask allocation for human and robot collaboration within hybrid assembly system, IEEE Transactions on Automation Science and Engineering 11 (4) (2014) 1065–1075.

[172] B. Sadrfaridpour, Y. Wang, Collaborative assembly in hybrid manufacturing cells: An integrated framework for human–robot interaction, IEEE Transactions on Automation Science and Engineering 15 (3) (2018) 1178–1192.

[173] V.V. Unhelkar, P.A. Lasota, Q. Tyroller, R.-D. Buhai, L. Marceau, B. Deml, J.A. Shah, Human-aware robotic assistant for collaborative assembly: Integrating human motion prediction with planning in time, IEEE Robotics and Automation Letters 3 (3) (2018) 2394–2401.

[174] H.M. Do, T.Y. Choi, C. Park, D.I. Park, J.H. Kyung, Design of robotic cell with multi dual-arm robots, in: Proceedings of the International Conference on Ubiquitous Robots and Ambient Intelligence (URAI), 2014.

[175] U. Aßmann, C. Piechnick, G. Püschel, M. Piechnick, J. Falkenberg, S. Werner, Modelling the world of a smart room for robotic co-working, in: L.F. Pires, S. Hammoudi, B. Selić (Eds.), Model-Driven Engineering and Software Development, in: Communications in Computer and Information Science, vol. 880, Springer, 2017, pp. 484–506.

[176] F.J. Furrer, Future-Proof Software-Systems – A Sustainable Evolution Strategy, Springer, 2019.

[177] E.M. Clarke, O. Grumberg, S. Jha, Y. Lu, H. Veith, Counterexample-guided abstraction refinement for symbolic model checking, Journal of the ACM 50 (5) (2003) 752–794.

[178] S.A. Seshia, N. Sharygina, S. Tripakis, Modeling for verification, in: E.M. Clarke, T.A. Henzinger, H. Veith, R. Bloem (Eds.), Handbook of Model Checking, Springer, 2018, pp. 75–105.

[179] G. Engels, L. Groenewegen, Object-oriented modeling: A roadmap, in: Proceedings of the Conference on the Future of Software Engineering (FOSE), 2000.

[180] J. Rumbaugh, I. Jacobson, G. Booch, Unified Modeling Language Reference Manual, Pearson Higher Education, 2004.

[181] M. Hause, The SysML modelling language, in: Proceedings of the European Systems Engineering Conference, 2006.

[182] K. Henricksen, J. Indulska, A. Rakotonirainy, Modeling context information in pervasive computing systems, in: F. Mattern, M. Naghshineh (Eds.), Pervasive Computing, in: Lecture Notes in Computer Science, vol. 2414, Springer, 2002, pp. 167–180.

[183] M. Basel, A framework for evaluating model-driven self-adaptive software systems, CoRR, arXiv:1901.04020, 2019.

[184] C.A. Petri, Communication with automata, PhD thesis, Universität, Hamburg, Germany, 1966.

[185] J.C.M. Baeten, D.A. van Beek, J.E. Rooda, Process algebra, in: P.A. Fishwick (Ed.), Handbook of Dynamic System Modeling, in: Chapman & Hall/CRC Computer and Information Science Series, vol. 19, Chapman & Hall/CRC Press, 2007, pp. 1–21.

[186] L. Aceto, A. Ingólfsdóttir, K.G. Larsen, J. Srba, Reactive Systems: Modelling, Specification and Verification, Cambridge University Press, 2007.

[187] D. René, H. Alla, Discrete, Continuous, and Hybrid Petri Nets, Springer, 2010.

[188] T. Stahl, M. Völter, S. Efftinge, Modellgetriebene Softwareentwicklung. Techniken, Engineering, Management, Dpunkt-Verlag, 2007.

[189] A. Schürr, F. Klar, 15 years of triple graph grammars, in: H. Ehrig, R. Heckel, G. Rozenberg, G. Taentzer (Eds.), Graph Transformations, in: Lecture Notes in Computer Science, vol. 5214, Springer, 2008, pp. 411–425.

[190] F. Jouault, F. Allilaire, J. Bézivin, I. Kurtev, P. Valduriez, ATL: A QVT-like transformation language, in: Proceedings of the Companion to the ACM Symposium on Object-Oriented Programming Systems, Languages, and Applications (OOPSLA), 2006.

[191] D.S. Kolovos, R.F. Paige, F.A.C. Polack, The Epsilon transformation language, in: A. Vallecillo, J. Gray, A. Pierantonio (Eds.), Theory and Practice of Model Transformations, in: Lecture Notes in Computer Science, vol. 5063, Springer, 2008, pp. 46–60.

[192] A. Uwe, Invasive Software Composition, Springer, 2003.

[193] M. Eysholdt, H. Behrens, Xtext: Implement your language faster than the quick and dirty way, in: Proceedings of the ACM International Conference Companion on Object Oriented Programming Systems Languages and Applications Companion (OOPSLA), 2010.

[194] L.C.L. Kats, E. Visser, The Spoofax language workbench: Rules for declarative specification of languages and IDEs, in: Proceedings of the ACM International Conference on Object Oriented Programming Systems Languages and Applications (OOPSLA), 2010.

[195] H. Krahn, B. Rumpe, S. Völkel, MontiCore: Modular development of textual domain specific languages, in: R.F. Paige, B. Meyer (Eds.), Objects, Components, Models and Patterns, in: Lecture Notes in Business Information Processing, vol. 11, Springer, 2008, pp. 297–315.

[196] R. Salay, J. Mylopoulos, S.M. Easterbrook, Using macromodels to manage collections of related models, in: P. van Eck, J. Gordijn, R. Wieringa (Eds.), Advanced Information Systems Engineering, in: Lecture Notes in Computer Science, vol. 5565, Springer, 2009, pp. 141–155.

[197] C. Atkinson, R. Gerbig, K. Markert, M. Zrianina, A. Egurnov, F. Kajzar, Towards a deep, domain specific modeling framework for robot applications, in: A. Uwe, G. Wagner (Eds.), Proceedings of the International Workshop on Model-Driven Robot Software Engineering (MORSE) co-located with International Conference on Software Technologies: Applications and Foundations (STAF), in: CEUR Workshop Proceedings, vol. 1319, CEUR-WS.org, 2014, pp. 4–15.

[198] C. Atkinson, D. Stoll, P. Bostan, Orthographic software modeling: A practical approach to view-based development, in: L.A. Maciaszek, C. González-Pérez, S. Jablonski (Eds.), Evaluation of Novel Approaches to Software Engineering, in: Communications in Computer and Information Science, vol. 69, Springer, 2010, pp. 206–219.

[199] M. Wimmer, A. Schauerhuber, G. Kappel, W. Retschitzegger, W. Schwinger, E. Kapsammer, A survey on UML-based aspect-oriented design modeling, ACM Computing Surveys 43 (4) (2011) 28, pp. 1–59.

[200] U. Aßmann, S. Götz, J.-M. Jézéquel, B. Morin, M. Trapp, A reference architecture and roadmap for models@run.time systems, in: N. Bencomo, R. France, B.H.C. Cheng, U. Aßmann (Eds.), Models@run.time – Foundations, Applications, and Roadmaps, in: Lecture Notes in Computer Science, vol. 8378, Springer, 2014, pp. 1–18.

[201] I. Crnković, Component-based software engineering – New challenges in software development, Software Focus 2 (4) (2001) 127–133.

[202] L. Chen, Microservices: Architecting for continuous delivery and DevOps, in: Proceedings of the IEEE International Conference on Software Architecture (ICSA), 2018.

[203] C. Mai, R. Schöne, J. Mey, T. Kühn, U. Aßmann, Adaptive Petri nets – A Petri net extension for reconfigurable structures, in: Proceedings of the International Conference on Adaptive and Self-Adaptive Systems and Applications (ADAPTIVE), 2018.

[204] R. Hirschfeld, P. Costanza, M. Haupt, An introduction to context-oriented programming with ContextS, in: R. Lämmel, J. Visser, J. Saraiva (Eds.), Generative and Transformational Techniques in Software Engineering II, in: Lecture Notes in Computer Science, vol. 5235, Springer, 2007, pp. 396–407.

[205] F. Steimann, On the representation of roles in object-oriented and conceptual modelling, Data & Knowledge Engineering 35 (1) (2000) 83–106.

[206] T. Kühn, S. Böhme, S. Götz, U. Aßmann, A combined formal model for relational context-dependent roles, in: Proceedings of the ACM International Conference on Software Language Engineering (SLE), 2015.

[207] N. Bencomo, S. Götz, H. Song, Models@run.time: A guided tour of the state of the art and research challenges, Software and Systems Modeling 18 (5) (2019) 3049–3082.

[208] An architectural blueprint for autonomic computing, Autonomic Computing, White Paper, IBM, 2006.

[209] S. Kounev, P. Lewis, K.L. Bellman, N. Bencomo, J. Camara, A. Diaconescu, L. Esterle, K. Geihs, H. Giese, S. Götz, P. Inverardi, J.O. Kephart, A. Zisman, The notion of self-aware computing, in: S. Kounev, J.O. Kephart, A. Milenkoski, X. Zhu (Eds.), Self-Aware Computing Systems, Springer, 2017, pp. 3–16.

[210] M. Quigley, K. Conley, B. Gerkey, J. Faust, T. Foote, J. Leibs, R. Wheeler, A.Y. Ng, ROS: An open-source robot operating system, 2009.

[211] A.K. Dey, Understanding and using context, Personal and Ubiquitous Computing 5 (1) (2001) 4–7.

[212] S. Proß, B. Bachmann, An advanced environment for hybrid modeling of biological systems based on Modelica, Journal of Integrative Bioinformatics 8 (1) (2011) 152, pp. 1–34.

[213] S. Haddadin, M. Suppa, S. Fuchs, T. Bodenmüller, A. Albu-Schäffer, G. Hirzinger, Towards the robotic co-worker, in: C. Pradalier, R. Siegwart, G. Hirzinger (Eds.), Robotics Research, in: Springer Tracts in Advanced Robotics, vol. 70, Springer, 2011, pp. 261–282.

[214] C.-M. Huang, B. Mutlu, Anticipatory robot control for efficient human-robot collaboration, in: Proceedings of the ACM/IEEE International Conference on Human-Robot Interaction (HRI), 2016.

[215] Y. Maruyama, S. Kato, T. Azumi, Exploring the performance of ROS2, in: Proceedings of the International Conference on Embedded Software (EMSOFT), 2016.

[216] M. Derakhshanmanesh, J. Ebert, M. Grieger, G. Engels, Model-integrating development of software systems: A flexible component-based approach, Software and Systems Modeling 18 (4) (2019) 2557–2586.

[217] P. Klint, R. Lämmel, C. Verhoef, Toward an engineering discipline for grammarware, ACM Transactions on Software Engineering and Methodology 14 (3) (2005) 331–380.

[218] D.E. Knuth, Semantics of context-free languages, Mathematical Systems Theory 2 (2) (1968) 127–145.

[219] C. Bürger, Reference attribute grammar controlled graph rewriting: Motivation and overview, in: Proceedings of the ACM International Conference on Software Language Engineering (SLE), 2015.

[220] G. Hedin, Reference attributed grammars, Informatica 24 (3) (2000) 301–317.

[221] J. Mey, R. Schöne, G. Hedin, E. Söderberg, T. Kühn, N. Fors, J. Öqvist, U. Aßmann, Relational reference attribute grammars: Improving continuous model validation, Journal of Computer Languages 57 (2020) 100940, pp. 1–21.

[222] R. Schöne, J. Mey, B. Ren, U. Aßmann, Bridging the gap between smart home platforms and machine learning using relational reference attribute grammars, in: Proceedings of the ACM/IEEE International Conference on Model Driven Engineering Languages and Systems Companion (MODELS-C), 2019.

[223] R. Milner, The Space and Motion of Communicating Agents, Cambridge University Press, 2009.

[224] R. Bruni, U. Montanari, G. Plotkin, D. Terreni, On hierarchical graphs: Reconciling bigraphs, Gs-monoidal theories and Gs-graphs, Fundamenta Informaticae 134 (2014) 287–317.

[225] L. Birkedal, S. Debois, E. Elsborg, T. Hildebrandt, H. Niss, Bigraphical models of context-aware systems, in: L. Aceto, A. Ingólfsdóttir (Eds.), Foundations of Software Science and Computation Structures, in: Lecture Notes in Computer Science, vol. 3921, Springer, 2006, pp. 187–201.

[226] A. Mansutti, M. Miculan, M. Peressotti, Multi-agent systems design and prototyping with bigraphical reactive systems, in: K. Magoutis, P. Pietzuch (Eds.), Proceedings of the 14th IFIP WG 6.1 International Conference on Distributed Applications and Interoperable Systems, in: Lecture Notes in Computer Science, vol. 8460, Springer, 2014, pp. 201–208.

[227] E. Pereira, C.M. Kirsch, R. Sengupta, J.B. de Sousa, BigActors – A model for structure-aware computation, in: Proceedings of the ACM/IEEE International Conference on Cyber-Physical Systems (ICCPS), 2013.

[228] S. Debois, Computation in the informatic jungle, Technical report, IT University of Copenhagen, 2011.

[229] G. Perrone, Domain-specific modelling languages in bigraphs, PhD thesis, IT University of Copenhagen, Denmark, 2013.

[230] G. Hirzinger, N. Sporer, A. Albu-Schäffer, M. Hähnle, R. Krenn, A. Pascucci, M. Schedl, DLR's torque-controlled light weight robot III – Are we reaching the technological limits now?, in: Proceedings of the IEEE International Conference on Robotics and Automation (ICRA), 2002.

[231] A. Albu-Schäffer, S. Haddadin, C. Ott, A. Stemmer, T. Wimböck, G. Hirzinger, The DLR lightweight robot: Design and control concepts for robots in human environments, Industrial Robot 34 (5) (2007) 376–385.

[232] G. Hirzinger, A. Albu-Schäffer, M. Hähnle, I. Schäfer, N. Sporer, On a new generation of torque controlled light-weight robots, in: Proceedings of the IEEE International Conference on Robotics and Automation (ICRA), 2001.

[233] A. Albu-Schäffer, C. Ott, U. Frese, G. Hirzinger, Cartesian impedance control of redundant robots: Recent results with the DLR-light-weight-arms, in: Proceedings of the IEEE International Conference on Robotics and Automation (ICRA), 2003.

[234] C. Ott, Cartesian Impedance Control of Redundant and Flexible-Joint Robots, Springer, 2008.

[235] B. Siciliano, L. Villani, Robot Force Control, Springer, 2012.

[236] C. Schindlbeck, S. Haddadin, Unified passivity-based Cartesian force/impedance control for rigid and flexible joint robots via task-energy tanks, in: Proceedings of the IEEE International Conference on Robotics and Automation (ICRA), 2015.

[237] S. Haddadin, A. Albu-Schäffer, G. Hirzinger, Requirements for safe robots: Measurements, analysis and new insights, The International Journal of Robotics Research 28 (11–12) (2009) 1507–1527.

[238] H. Sami, Towards Safe Robots – Approaching Asimov's 1st Law, Springer, 2013.

[239] S. Haddadin, A. De Luca, A. Albu-Schäffer, Robot collisions: A survey on detection, isolation, and identification, IEEE Transactions on Robotics 33 (6) (2017) 1292–1312.

[240] S. Haddadin, S. Haddadin, A. Khoury, T. Rokahr, S. Parusel, R. Burgkart, A. Bicchi, A. Albu-Schäffer, On making robots understand safety: Embedding injury knowledge into control, The International Journal of Robotics Research 31 (13) (2012) 1578–1602.

[241] N. Mansfeld, M. Hamad, M. Becker, A.G. Marin, S. Haddadin, Safety map: A unified representation for biomechanics impact data and robot instantaneous dynamic properties, IEEE Robotics and Automation Letters 3 (3) (2018) 1880–1887.

[242] M. Safeea, P. Neto, KUKA sunrise toolbox: Interfacing collaborative robots with MATLAB, IEEE Robotics & Automation Magazine 26 (1) (2018) 91–96.

[243] B. Akan, Human robot interaction solutions for intuitive industrial robot programming, PhD thesis, Mälardalen University, Sweden, 2012.

[244] L. Johannsmeier, S. Haddadin, A hierarchical human-robot interaction-planning framework for task allocation in collaborative industrial assembly processes, IEEE Robotics and Automation Letters 2 (1) (2016) 41–48.

[245] R. Deimel, O. Brock, A novel type of compliant and underactuated robotic hand for dexterous grasping, The International Journal of Robotics Research 35 (1–3) (2016) 161–185.

[246] M.V. Liarokapis, A.M. Dollar, Yale OpenHand Project: Optimizing open-source hand designs for ease of fabrication and adoption, IEEE Transactions on Automation Science and Engineering 15 (2) (2016) 456–467.

[247] C. Piazza, G. Grioli, M.G. Catalano, A. Bicchi, A century of robotic hands, Annual Review of Control, Robotics, and Autonomous Systems 2 (2019) 1–32.

[248] A. Bicchi, M. Gabiccini, M. Santello, Modelling natural and artificial hands with synergies, Philosophical Transactions of the Royal Society B: Biological Sciences 366 (1581) (2011) 3153–3161.

[249] M.G. Catalano, G. Grioli, E. Farnioli, A. Serio, C. Piazza, A. Bicchi, Adaptive synergies for the design and control of the Pisa/IIT SoftHand, The International Journal of Robotics Research 33 (5) (2014) 768–782.

[250] M. Hemmi, R. Morita, Y. Hirota, K. Inoue, H. Nabae, G. Endo, K. Suzumori, Development of hydraulic tough motors with high power density and their application to a 7-axis robotic arm, in: Proceedings of the IEEE/SICE International Symposium on System Integration (SII), 2019.

[251] B.U. Rehman, M. Focchi, M. Frigerio, J. Goldsmith, D.G. Caldwell, C. Semini, Design of a hydraulically actuated arm for a quadruped robot, in: Proceedings of the International Conference on Climbing and Walking Robots and the Support Technologies for Mobile Machine (CLAWAR), 2015.

[252] K. Izawa, S.-H. Hyon, Design of arm for a hydraulic humanoid robot, in: Proceedings of the JSME Conference on Robotics and Mechatronics (ROBOMECH), 2014.

[253] D. Peleg, E. Braiman, E. Yom-Tov, G.F. Inbar, Classification of finger activation for use in a robotic prosthesis arm, IEEE Transactions on Neural Systems and Rehabilitation Engineering 10 (4) (2002) 290–293.

[254] P.D. Marasco, K. Kim, J.E. Colgate, M.A. Peshkin, T.A. Kuiken, Robotic touch shifts perception of embodiment to a prosthesis in targeted reinnervation amputees, Brain 134 (3) (2011) 747–758.

[255] M. Minsky, Telepresence, OMNI Magazine 6 (1980) 44–52.

[256] G. Niemeyer, J.-J.E. Slotine, Stable adaptive teleoperation, IEEE Journal of Oceanic Engineering 16 (1) (1991) 152–162.

[257] G. Niemeyer, J.-J.E. Slotine, Telemanipulation with time delays, The International Journal of Robotics Research 23 (9) (2004) 873–890.

[258] E. Nuño, R. Ortega, N. Barabanov, L. Basañez, A globally stable PD controller for bilateral teleoperators, IEEE Transactions on Robotics 24 (3) (2008) 753–758.

[259] C. Seo, J.-P. Kim, J. Kim, H.-S. Ahn, J. Ryu, Robustly stable bilateral teleoperation under time-varying delays and data losses: An energy-bounding approach, Journal of Mechanical Science and Technology 25 (8) (2011) 2089–2100.

[260] J. Artigas, J. Vilanova, C. Preusche, G. Hirzinger, Time domain passivity control-based telepresence with time delay, in: Proceedings of the IEEE/RSJ International Conference on Intelligent Robots and Systems (IROS), 2006.

[261] J. Artigas, C. Preusche, G. Hirzinger, in: Proceedings of the IEEE/RSJ International Conference on Intelligent Robots and Systems (IROS), 2007.

[262] A.S. Voloshina, S.H. Collins, Lower limb active prosthetic systems – Overview, in: Wearable Robotics: Systems and Applications, Academic Press, 2020, pp. 469–486.

[263] J. Mattila, J. Koivum, D.G. Caldwell, C. Semini, A survey on control of hydraulic robotic manipulators with projection to future trends, IEEE/ASME Transactions on Mechatronics 22 (2) (2017) 669–680.

[264] D.A. Lawrence, Stability and transparency in bilateral teleoperation, IEEE Transactions on Robotics and Automation 9 (5) (1993) 624–637.

[265] L. Baccelliere, N. Kashiri, L. Muratore, A. Laurenzi, M. Kamedula, A. Margan, S. Cordasco, J. Malzahn, N.G. Tsagarakis, Development of a human size and strength compliant bi-manual platform for realistic heavy manipulation tasks, in: Proceedings of the IEEE/RSJ International Conference on Intelligent Robots and Systems (IROS), 2017.

[266] M. Grebenstein, M. Chalon, G. Hirzinger, R. Siegwart, A method for hand kinematics designers 7 billion perfect hands, in: Proceedings of the International Conference on Applied Bionics and Biomechanics (ICABB), 2010.

[267] S. Patel, T. Sobh, Manipulator performance measures – A comprehensive literature survey, Journal of Intelligent & Robotic Systems 77 (3–4) (2015) 547–570.

[268] T. Chen, M. Haas-Heger, M. Ciocarlie, Underactuated hand design using mechanically realizable manifolds, in: Proceedings of the IEEE International Conference on Robotics and Automation (ICRA), 2018.

[269] Y. Nitta, S. Tamura, H. Takase, A study on introducing FPGA to ROS based autonomous driving system, in: Proceedings of the International Conference on Field-Programmable Technology (FPT), 2018.

[270] I. Kuon, J. Rose, Measuring the gap between FPGAs and ASICs, IEEE Transactions on Computer-Aided Design of Integrated Circuits and Systems 26 (2) (2007) 203–215.

[271] E. Nurvitadhi, J. Sim, D. Sheffield, A. Mishra, S. Krishnan, D. Marr, Accelerating recurrent neural networks in analytics servers: Comparison of FPGA, CPU, GPU, and ASIC, in: Proceedings of the International Conference on Field-Programmable Logic and Applications (FPL), 2016.

[272] L. Kalms, H. Ibrahim, D. Göhringer, Full-HD accelerated and embedded feature detection video system with 63 fps using ORB for FREAK, in: Proceedings of the International Conference on ReConFigurable Computing and FPGAs (ReConFig), 2018.

[273] L. Kalms, A. Podlubne, D. Göhringer, HiFlipVX: An open source high-level synthesis FPGA library for image processing, in: C. Hochberger, B. Nelson, A. Koch, R. Woods, P. Diniz (Eds.), Applied Reconfigurable Computing, in: Lecture Notes in Computer Science, vol. 11444, Springer, 2019, pp. 149–164.

[274] A. Suardi, E.C. Kerrigan, G.A. Constantinides, Fast FPGA prototyping toolbox for embedded optimization, in: Proceedings of the European Control Conference (ECC), 2015.

[275] A. Elkady, T. Sobh, Robotics middleware: A comprehensive literature survey and attribute-based bibliography, Journal of Robotics 2012 (2012) 959013, pp. 1–15.

[276] H. Bruyninckx, Open robot control software: The OROCOS project, in: Proceedings of the IEEE International Conference on Robotics and Automation (ICRA), 2001.

[277] G. Metta, P. Fitzpatrick, L. Natale, YARP: Yet another robot platform, International Journal of Advanced Robotic Systems 3 (1) (2006) 43–48.

[278] W.H. Wolf, Hardware-software co-design of embedded systems, Proceedings of the IEEE 82 (7) (1994) 967–989.

[279] J. Teich, Hardware/software codesign: The past, the present, and predicting the future, Proceedings of the IEEE 100 (Special Centennial Issue) (2012) 1411–1430.

[280] K. Yamashina, T. Ohkawa, K. Ootsu, T. Yokota, Proposal of ROS-compliant FPGA component for low-power robotic systems-case study on image processing application, in: Proceedings of the International Workshop on FPGAs for Software Programmers (FSP), 2015.

[281] K. Yamashina, H. Kimura, T. Ohkawa, K. Ootsu, T. Yokota, cReComp: Automated design tool for ROS-compliant FPGA component, in: Proceedings of the IEEE International Symposium on Embedded Multicore/Many-core Systems-on-Chip (MCSOC), 2016.

[282] Y. Sugata, T. Ohkawa, K. Ootsu, T. Yokota, Acceleration of publish/subscribe messaging in ROS-compliant FPGA component, in: Proceedings of the International Symposium on Highly Efficient Accelerators and Reconfigurable Technologies (HEART), 2017.

[283] A. Podlubne, D. Göhringer, FPGA-ROS: Methodology to augment the robot operating system with FPGA designs, in: Proceedings of the International Conference on ReConFigurable Computing and FPGAs (ReConFig), 2019.

[284] V. Pradeep, K. Konolige, E. Berger, Calibrating a multi-arm multi-sensor robot: A bundle adjustment approach, in: O. Khatib, V. Kumar, G. Sukhatme (Eds.), Experimental Robotics, in: Springer Tracts in Advanced Robotics, vol. 79, Springer, 2014, pp. 211–225.

[285] L. Kalms, D. Göhringer, Exploration of OpenCL for FPGAs using SDAccel and comparison to GPUs and multicore CPUs, in: Proceedings of the International Conference on Field Programmable Logic and Applications (FPL), 2017.

[286] T. Ohkawa, Y. Sugata, H. Watanabe, N. Ogura, K. Ootsu, T. Yokota, High level synthesis of ROS protocol interpretation and communication circuit for FPGA, in: Proceedings of the International Workshop on Robotics Software Engineering (RoSE), 2019.

[287] M. Spittle, Motor Learning and Skill Acquisition: Applications for Physical Education and Sport, Palgrave Macmillan, 2013.

[288] J.J. Gibson, The Ecological Approach to Visual Perception, Houghton Mifflin, 1979.

[289] A. Chemero, An outline of a theory of affordances, Ecological Psychology 15 (2) (2003) 181–195.

[290] D. Araújo, R. Hristovski, L. Seifert, J. Carvalho, K. Davids, Ecological cognition: Expert decision-making behaviour in sport, International Review of Sport and Exercise Psychology 12 (1) (2019) 1–25.

[291] A.M. Williams, P.R. Ford, N.J. Hodges, P. Ward, Expertise in sport: Specificity, plasticity, and adaptability in high-performance athletes, in: A.M. Williams, A. Kozbelt, K.A. Ericsson, R.R. Hoffman (Eds.), The Cambridge Handbook of Expertise and Expert Performance, Cambridge University Press, 2018, pp. 653–674.

[292] J.A. Adams, Historical review and appraisal of research on the learning, retention, and transfer of human motor skills, Psychological Bulletin 101 (1) (1987) 41–74.

[293] R.A. Schmidt, T.D. Lee, Motor Control and Learning: A Behavioral Emphasis, Human Kinetics, 2011.

[294] A. Newell, P.S. Rosenbloom, Mechanisms of skill acquisition and the law of practice, in: J.R. Anderson (Ed.), Cognitive Skills and Their Acquisition, in: Carnegie Mellon Symposia on Cognition Series, Erlbaum, 1981, pp. 1–55.

[295] N.J. Evans, S.D. Brown, D.J.K. Mewhort, A. Heathcote, Refining the law of practice, Psychological Review 125 (4) (2018) 592–605.

[296] R.A. Magill, D.I. Anderson, Motor Learning and Control: Concepts and Applications, McGraw-Hill, 2017.

[297] D. Araújo, K. Davids, What exactly is acquired during skill acquisition?, Journal of Consciousness Studies 18 (2011) 7–23.

[298] N.C. Soderstrom, R.A. Bjork, Learning versus performance: An integrative review, Perspectives on Psychological Science 10 (2) (2015) 176–199.

[299] J.A. Adams, A closed-loop theory of motor learning, Journal of Motor Behavior 3 (2) (1971) 111–150.

[300] A.W. Salmoni, R.A. Schmidt, C.B. Walter, Knowledge of results and motor learning: A review and critical reappraisal, Psychological Bulletin 95 (3) (1984) 355–386.

[301] G. Wulf, C. Shea, R. Lewthwaite, Motor skill learning and performance: A review of influential factors, Medical Education 44 (1) (2010) 75–84.

[302] G. Wulf, Attentional focus and motor learning: A review of 15 years, International Review of Sport and Exercise Psychology 6 (1) (2013) 77–104.

[303] G. Wulf, W. Prinz, Directing attention to movement effects enhances learning: A review, Psychonomic Bulletin & Review 8 (4) (2001) 648–660.

[304] G. Wulf, An external focus of attention is a *conditio sine qua non* for athletes: A response to Carson, Collins, and Toner (2015), Journal of Sports Sciences 34 (13) (2016) 1293–1295.

[305] H. Rohbanfard, L. Proteau, Learning through observation: A combination of expert and novice models favors learning, Experimental Brain Research 215 (3) (2011) 183–197.

[306] C.M. Janelle, J.D. Champenoy, S.A. Coombes, M.B. Mousseau, Mechanisms of attentional cueing during observational learning to facilitate motor skill acquisition, Journal of Sports Sciences 21 (10) (2003) 825–838.

[307] S.P. Swinnen, Information feedback for motor learning: A review, in: H.N. Zelaznik (Ed.), Advances in Motor Learning and Control, Human Kinetics, 1996, pp. 37–66.

[308] H.P. Crowell, I.S. Davis, Gait retraining to reduce lower extremity loading in runners, Clinical Biomechanics 26 (1) (2011) 78–83.

[309] G. Wulf, C.H. Shea, S. Matschiner, Frequent feedback enhances complex motor skill learning, Journal of Motor Behavior 30 (2) (1998) 180–192.

[310] P. van Vliet, G. Wulf, Extrinsic feedback for motor learning after stroke: What is the evidence?, Disability and Rehabilitation 28 (13–14) (2006) 831–840.

[311] M. Agethen, D. Krause, Effects of bandwidth feedback on the automatization of an arm movement sequence, Human Movement Science 45 (2016) 71–83.

[312] Y. Aoyagi, E. Ohnishi, Y. Yamamoto, N. Kado, T. Suzuki, H. Ohnishi, N. Hokimoto, N. Fukaya, Feedback protocol of 'fading knowledge of results' is effective for prolonging motor learning retention, Journal of Physical Therapy Science 31 (8) (2019) 687–691.

[313] G. Wulf, Self-controlled practice enhances motor learning: Implications for physiotherapy, Physiotherapy 93 (2) (2007) 96–101.

[314] A.M. Gentile, A working model of skill acquisition with application to teaching, Quest 17 (1) (1972) 3–23.

[315] K.M. Newell, C.B. Walter, Kinematic and kinetic parameters as information feedback in motor skill acquisition, Journal of Human Movement Studies 7 (4) (1981) 235–254.

[316] M.M. Pacheco, C.W. Lafe, K.M. Newell, Search strategies in the perceptual-motor workspace and the acquisition of coordination, control, and skill, Frontiers in Psychology 10 (2019) 1874, pp. 1–24.

[317] K.M. Newell, L.R. Morris, D.M. Scully, Augmented information and the acquisition of skill in physical activity, Exercise and Sport Sciences Reviews 13 (1985) 235–261.

[318] C.A. Fowler, M.T. Turvey, Skill acquisition: An event approach with special reference to searching for the optimum of a function of several variables, in: G.E. Stelmach (Ed.), Information Processing in Motor Control and Learning, Academic Press, 1978, pp. 1–40.

[319] R. Robertson, L.St. Germain, D.M. Ste-Marie, The effects of self-observation when combined with a skilled model on the learning of gymnastics skills, Journal of Motor Learning and Development 6 (1) (2018) 18–34.

[320] J.F. Dyer, P. Stapleton, M. Rodger, Mapping sonification for perception and action in motor skill learning, Frontiers in Neuroscience 11 (2017) 463, pp. 1–4.

[321] K.M. Newell, Constraints on the development of coordination, in: M.G. Wade, H.T.A. Whiting (Eds.), Motor Development in Children: Aspects of Coordination and Control, Martinus Nijhoff, 1986, pp. 341–360.

[322] J.A.S. Kelso, Dynamic Patterns: The Self-Organization of Brain and Behavior, MIT Press, 1995.

[323] I. Renshaw, J.Y. Chow, A constraint-led approach to sport and physical education pedagogy, Physical Education and Sport Pedagogy 24 (2) (2019) 103–116.

[324] T. Buszard, M. Reid, R.S.W. Masters, D. Farrow, Scaling the equipment and play area in children's sport to improve motor skill acquisition: A systematic review, Sports Medicine 46 (6) (2016) 829–843.

[325] T. Buszard, M. Reid, R.S.W. Masters, D. Farrow, Scaling tennis racquets during PE in primary school to enhance motor skill acquisition, Research Quarterly for Exercise and Sport 87 (4) (2016) 414–420.

[326] D. Farrow, M. Reid, The effect of equipment scaling on the skill acquisition of beginning tennis players, Journal of Sports Sciences 28 (7) (2010) 723–732.

[327] L. Oppici, D. Panchuk, F.R. Serpiello, D. Farrow, The influence of a modified ball on transfer of passing skill in soccer, Psychology of Sport and Exercise 39 (2018) 63–71.

[328] L. Oppici, D. Panchuk, F.R. Serpiello, D. Farrow, Long-term practice with domain-specific task constraints influences perceptual skills, Frontiers in Psychology 8 (2017) 1387, pp. 1–9.

[329] D. Conte, T.G. Favero, M. Niederhausen, L. Capranica, A. Tessitore, Effect of different number of players and training regimes on physiological and technical demands of ball-drills in basketball, Journal of Sports Sciences 34 (8) (2016) 780–786.

[330] A.L. Owen, D.P. Wong, M. McKenna, A. Dellal, Heart rate responses and technical comparison between small- vs. large-sided games in elite professional soccer, Journal of Strength & Conditioning Research 25 (8) (2011) 2104–2110.

[331] A.O. Effenberg, U. Fehse, G. Schmitz, B. Krüger, H. Mechling, Movement sonification: Effects on motor learning beyond rhythmic adjustments, Frontiers in Neuroscience 10 (2016) 219, pp. 1–18.

[332] R. Sigrist, G. Rauter, L. Marchal-Crespo, R. Riener, P. Wolf, Sonification and haptic feedback in addition to visual feedback enhances complex motor task learning, Experimental Brain Research 233 (3) (2015) 909–925.

[333] N. Konttinen, K. Mononen, J. Viitasalo, T. Mets, The effects of augmented auditory feedback on psychomotor skill learning in precision shooting, Journal of Sport and Exercise Psychology 26 (2) (2004) 306–316.

[334] T. Horeman, F. van Delft, M.D. Blikkendaal, J. Dankelman, J.J. van den Dobbelsteen, F.W. Jansen, Learning from visual force feedback in box trainers: Tissue manipulation in laparoscopic surgery, Surgical Endoscopy 28 (6) (2014) 1961–1970.

[335] C. Deußer, S. Passmann, T. Strufe, Browsing unicity: On the limits of anonymizing web tracking data, in: Proceedings of the IEEE Symposium on Security and Privacy (S&P), 2020.

[336] C. Kuhn, M. Beck, T. Strufe, Covid notions: Towards formal definitions – and documented understanding – of privacy goals and claimed protection in proximity-tracing services, CoRR, arXiv:2004.07723, 2020.

[337] S. Seiler-Hwang, P. Arias-Cabarcos, A. Marín, F. Almenares, D. Díaz-Sánchez, C. Becker, "I don't see why I would ever want to use it": Analyzing the usability of popular smartphone password managers, in: Proceedings of the ACM Conference on Computer and Communications Security (CCS), 2019.

[338] D. Rebollo-Monedero, J. Parra-Arnau, C. Diaz, J. Forne, On the measurement of privacy as an attacker's estimation error, International Journal of Information Security 12 (2) (2013) 129–149.

[339] P. Arias-Cabarcos, C. Krupitzer, C. Becker, A survey on adaptive authentication, ACM Computing Surveys 52 (4) (2019) 80, pp. 1–30.

[340] W. Meng, D.S. Wong, S. Furnell, J. Zhou, Surveying the development of biometric user authentication on mobile phones, IEEE Communications Surveys & Tutorials 17 (3) (2014) 1268–1293.

[341] A. Alzubaidi, J. Kalita, Authentication of smartphone users using behavioral biometrics, IEEE Communications Surveys & Tutorials 18 (3) (2016) 1998–2026.

[342] A. Dantcheva, P. Elia, A. Ross, What else does your biometric data reveal? A survey on soft biometrics, IEEE Transactions on Information Forensics and Security 11 (3) (2015) 441–467.

[343] F. Armknecht, T. Strufe, An efficient distributed privacy-preserving recommendation system, in: Proceedings of the IFIP Mediterranean Ad Hoc Networking Workshop (Med-Hoc-Net), 2011.

[344] S. Roos, M. Beck, T. Strufe, Anonymous addresses for efficient and resilient routing in F2F overlays, in: Proceedings of the IEEE International Conference on Computer Communications (INFOCOM), 2016.

[345] D.L. Quoc, M. Beck, P. Bhatotia, R. Chen, C.W. Fetzer, T. Strufe, Privacy-preserving stream analytics, in: Proceedings of the USENIX Annual Technical Conference (USENIX ATC), 2017.

[346] B. Abernethy, K. Zawi, Pickup of essential kinematics underpins expert perception of movement patterns, Journal of Motor Behavior 39 (5) (2007) 353–367.

[347] L. Vilar, D. Araujo, K. Davids, B. Travassos, R. Duarte, J. Parreira, Interpersonal coordination tendencies supporting the creation/prevention of goal scoring opportunities in futsal, European Journal of Sport Science 14 (1) (2014) 28–35.

[348] D. Farrow, S. Robertson, Development of a skill acquisition periodisation framework for high-performance sport, Sports Medicine 47 (6) (2017) 1043–1054.

[349] M.A. Guadagnoli, T.D. Lee, Challenge point: A framework for conceptualizing the effects of various practice conditions in motor learning, Journal of Motor Behavior 36 (2) (2004) 212–224.

[350] J. Fagerberg, Innovation: A guide to the literature, in: J. Fagerberg, D.C. Mowery, R.R. Nelson (Eds.), The Oxford Handbook of Innovation, Oxford University Press, 2009, pp. 1–26.

[351] R. Garcia, R. Calantone, A critical look at technological innovation typology and innovativeness terminology: A literature review, Journal of Product Innovation Management 19 (2) (2002) 110–132.

[352] V. Chiesa, F. Frattini, Commercializing technological innovation: Learning from failures in high-tech markets, Journal of Product Innovation Management 28 (4) (2011) 437–454.

[353] J.M. Utterback, Design-Inspired Innovation, World Scientific, 2007.

[354] R. Verganti, Design Driven Innovation: Changing the Rules of Competition by Radically Innovating What Things Mean, Harvard Business Review Press, 2009.

[355] M. Hassenzahl, N. Tractinsky, User experience – A research agenda, Behaviour & Information Technology 25 (2) (2006) 91–97.

[356] P.A. Hancock, A.A. Pepe, L.L. Murphy, Hedonomics: The power of positive and pleasurable ergonomics, Ergonomics in Design 13 (1) (2005) 8–14.

[357] C. Wölfel, J. Krzywinski, F. Drechsel, Knowing, reasoning and visualizing in industrial design, The Knowledge Engineering Review 28 (3) (2013) 287–302.

[358] T. Bobbe, J. Krzywinski, C. Wölfel, A comparison of design process models from academic theory and professional practice, in: Proceedings of the International Design Conference (IDC), 2016.

[359] J. Krzywinski, C. Wölfel, Industriedesign und nutzerzentrierte Produktentwicklung, in: B. Bender, K. Gericke (Eds.), Pahl/Beitz Konstruktionslehre – Methoden und Anwendung erfolgreicher Produktentwicklung, Springer, 2020.

[360] A. Cooper, R. Reimann, D. Cronin, About Face 3: The Essentials of Interaction Design, Wiley, 2007.

[361] M. Mazzucato, Mission-Oriented Research & Innovation in the European Union: A problem-solving approach to fuel innovation-led growth, Publications Office of the European Union, 2018.

[362] C.P. Hoffmann, S. Lennerts, C. Schmitz, W. Stölzle, F. Uebernickel, Business Innovation: Das St. Galler Modell, Springer, 2016.

[363] J.T. Gourville, Eager sellers and stony buyers: Understanding the psychology of new-product adoption, Harvard Business Review 84 (6) (2006) 98–106.

[364] L. Lüneburg, E. Papp, J. Krzywinski, The potential of wearable demonstrators introducing innovative technologies, in: Proceedings of the International Design Conference (IDC), 2020.

[365] J.N. Katz, Lumbar disc disorders and low-back pain: Socioeconomic factors and consequences, Journal of Bone & Joint Surgery 88-A (Supplement 2) (2006) 21–24.

[366] B.L. Pinto, S.M. Beaudette, S.H.M. Brown, Tactile cues can change movement: An example using tape to redistribute flexion from the lumbar spine to the hips and knees during lifting, Human Movement Science 60 (2018) 32–39.

[367] V. Lorenzoni, J. Staley, T. Marchant, K.E. Onderdijk, P.J. Maes, M. Leman, The sonic instructor: A music-based biofeedback system for improving weightlifting technique, PLOS One 14 (8) (2019) e0220915, pp. 1–19.

[368] M. Boocock, Y. Naudé, S. Taylor, J. Kilby, G. Mawston, Influencing lumbar posture through real-time biofeedback and its effects on the kinematics and kinetics of a repetitive lifting task, Gait & Posture 73 (2019) 93–100.

[369] J.Y. Yoon, M.H. Kang, J.S. Oh, Effects of visual biofeedback using a laser beam on the EMG ratio of the medial and lateral vasti muscles and kinematics of hip and knee joints during a squat exercise, Journal of Physical Therapy Science 23 (4) (2011) 559–563.

[370] E. Papp, C. Wölfel, J. Krzywinski, Acceptance and user experience of wearable assistive devices for industrial purposes, in: Proceedings of the International Design Conference (IDC), 2020.

[371] C. Barriault, D. Pearson, Assessing exhibits for learning in science centers: A practical tool, Visitor Studies 13 (1) (2010) 90–106.

[372] S.J. Lederman, R.L. Klatzky, Haptic perception: A tutorial, Attention, Perception, & Psychophysics 71 (7) (2009) 1439–1459.

[373] L.A. Jones, Kinesthetic sensing, in: Human and Machine Haptics, MIT Press, 2000 (should have had appeared in 2000), http://citeseerx.ist.psu.edu/viewdoc/summary?doi=10.1.1.133.5356.

[374] S. Okamoto, H. Nagano, Y. Yamada, Psychophysical dimensions of tactile perception of textures, IEEE Transactions on Haptics 6 (1) (2013) 81–93.

[375] A. Noll, B. Gülecyüz, A. Hofmann, E. Steinbach, A rate-scalable perceptual wavelet-based vibrotactile codec, in: Proceedings of the IEEE Haptics Symposium (HAPTICS), 2020.

[376] Z. Shao, J. Wu, Q. Ouyang, C. He, Z. Cao, Multi-layered perceptual model for haptic perception of compliance, Electronics 8 (12) (2019) 1497.

[377] G. Gescheider, Psychophysics: The Fundamentals, Psychology Press, 2013.

[378] E.H. Weber, Die Lehre vom Tastsinne und Gemeingefühle auf Versuche gegründet, Friedrich Vieweg und Sohn, 1851.

[379] S. Hirche, M. Buss, Human-oriented control for haptic teleoperation, Proceedings of the IEEE 100 (3) (2012) 623–647.

[380] P. Hinterseer, E. Steinbach, S. Hirche, M. Buss, A novel, psychophysically motivated transmission approach for haptic data streams in telepresence and teleaction systems, in: Proceedings of the IEEE International Conference on Acoustics, Speech, and Signal Processing (ICASSP), 2005.

[381] P. Hinterseer, S. Hirche, S. Chaudhuri, E. Steinbach, M. Buss, Perception-based data reduction and transmission of haptic data in telepresence and teleaction systems, IEEE Transactions on Signal Processing 56 (2) (2008) 588–597.

[382] P. Hinterseer, E. Steinbach, A psychophysically motivated compression approach for 3D haptic data, in: Proceedings of the IEEE Symposium on Haptic Interfaces for Virtual Environment and Teleoperator Systems (HAPTICS), 2006.

[383] D.M. Wolpert, Z. Ghahramani, M.I. Jordan, An internal model for sensorimotor integration, Science 269 (5232) (1995) 1880–1882.

[384] H.Z. Tan, F. Barbagli, K. Salisbury, C. Ho, C. Spence, Force-direction discrimination is not influenced by reference force direction (short paper), Haptics-e 4 (1) (2006) 1–6.

[385] F. Barbagli, K. Salisbury, C. Ho, C. Spence, H.Z. Tan, Haptic discrimination of force direction and the influence of visual information, ACM Transactions on Applied Perception 3 (2) (2006) 125–135.

[386] H. Pongrac, Vibrotactile perception: Examining the coding of vibrations and the just noticeable difference under various conditions, Multimedia Systems 13 (2008) 297–307.

[387] A. Torabi, M. Khadem, K. Zareinia, G.R. Sutherland, M. Tavakoli, Application of a redundant haptic interface in enhancing soft-tissue stiffness discrimination, IEEE Robotics and Automation Letters 4 (2) (2019) 1037–1044.

[388] K. Higashi, S. Okamoto, Y. Yamada, H. Nagano, M. Konyo, Hardness perception based on dynamic stiffness in tapping, Frontiers in Psychology 9 (2019) 2654, pp. 1–9.

[389] O. Caldiran, H.Z. Tan, C. Basdogan, Visuo-haptic discrimination of viscoelastic materials, IEEE Transactions on Haptics 12 (4) (2019) 438–450.

[390] S.J. Bensmaïa, M. Hollins, The vibrations of texture, Somatosensory & Motor Research 20 (1) (2003) 33–43.

[391] J.C. Makous, R.M. Friedman, C.J. Vierck, A critical band filter in touch, The Journal of Neuroscience 15 (4) (1995) 2808–2818.

[392] R.T. Verrillo, Vibrotactile sensitivity and the frequency response of the Pacinian corpuscle, Psychonomic Science 4 (1) (1966) 135–136.

[393] R.T. Verrillo, Vibrotactile thresholds measured at the finger, Perception & Psychophysics 9 (1971) 329–330.

[394] M. Rothenberg, R.T. Verrillo, S.A. Zahorian, S. Bolanowski, Vibrotactile frequency for encoding a speech parameter, The Journal of the Acoustical Society of America 62 (4) (1977) 1003–1012.

[395] G. Gescheider, S. Bolanowski, J. Pope, R.T. Verrillo, A four-channel analysis of the tactile sensitivity of the fingertip: Frequency selectivity, spatial summation, and temporal summation, Somatosensory & Motor Research 19 (2) (2002) 114–124.

[396] S. Bolanowski, G. Gescheider, R.T. Verrillo, C. Checkosky, Four channels mediate the mechanical aspects of touch, The Journal of the Acoustical Society of America 84 (1988) 1680–1694.

[397] R. Schubert, S. Haufe, F. Blankenburg, A. Villringer, G. Curio, Now you'll feel it, now you won't: EEG rhythms predict the effectiveness of perceptual masking, Journal of Cognitive Neuroscience 21 (12) (2009) 2407–2419.

[398] G. Gescheider, A. Valetutti, M. Padula, R.T. Verrillo, Vibrotactile forward masking as a function of age, The Journal of the Acoustical Society of America 91 (3) (1992) 1690–1696.

[399] R.T. Verrillo, G. Gescheider, B. Calman, C. Doren, Vibrotactile masking: Effects of one- and two-site stimulation, Perception & Psychophysics 33 (1983) 379–387.

[400] R. Chaudhari, C. Schuwerk, M. Danaei, E. Steinbach, Perceptual and bitrate-scalable coding of haptic surface texture signals, IEEE Journal of Selected Topics in Signal Processing 9 (3) (2014) 462–473.

[401] A. Noll, B. Gülecyüz, E. Steinbach, Vibrotactile perceptual codec based on DWT and SPIHT, in: IEEE P1918.1 Haptics Codecs Task Group – DC: HC NGS-19-1-r0, 2019.

[402] R. Hassen, B. Gülecyüz, E. Steinbach, Perceptual vibrotactile codec based on sparse linear prediction, in: IEEE P1918.1 Haptics Codecs Task Group – DC: HC HGS-19-1-r0, 2019.

[403] L.A. Jones, N. Sarter, Tactile displays: Guidance for their design and application, Human Factors 50 (2008) 90–111.

[404] S.J. Bensmaïa, J. Yau, M. Hollins, Vibrotactile intensity and frequency information in the Pacinian system: A psychophysical model, Perception & Psychophysics 5 (67) (2005) 828–841.

[405] N. Cao, H. Nagano, M. Konyo, S. Okamoto, S. Tadokoro, A pilot study: Introduction of time-domain segment to intensity-based perception model of high-frequency vibration, in: D. Prattichizzo, H. Shinoda, H.Z. Tan, E. Ruffaldi, A. Frisoli (Eds.), Haptics: Science, Technology, and Applications, in: Lecture Notes in Computer Science, vol. 10893, Springer, 2018, pp. 321–332.

[406] S.-C. Li, M. Jordanova, U. Lindenberger, From good senses to good sense: A link between tactile information processing and intelligence, Intelligence 26 (2) (1998) 99–122.

[407] S. Master, M. Larue, F. Tremblay, Characterization of human tactile pattern recognition performance at different ages, Somatosensory & Motor Research 27 (2) (2010) 60–67.

[408] L. Tamè, E. Azañón, M.R. Longo, A conceptual model of tactile processing across body features of size, shape, side, and spatial location, Frontiers in Psychology 10 (2019) 291, pp. 1–19.

[409] N. Weisz, A. Wühle, G. Monittola, G. Demarchi, J. Frey, T. Popov, C. Braun, Prestimulus oscillatory power and connectivity patterns predispose conscious somatosensory perception, Proceedings of the National Academy of Sciences of the United States of America 111 (2014) E417–E425.

[410] J. Misselhorn, U. Friese, A.K. Engel, Frontal and parietal alpha oscillations reflect attentional modulation of cross-modal matching, Scientific Reports 9 (2019) 5030, pp. 1–11.

[411] C. Shahabi, A. Ortega, M.R. Kolahdouzan, A comparison of different haptic compression techniques, in: Proceedings of the IEEE International Conference on Multimedia and Expo (ICME), 2002.

[412] Y. You, M.Y. Sung, Haptic data transmission based on the prediction and compression, in: Proceedings of the IEEE International Conference on Communications (ICC), 2008.

[413] H. Tanaka, K. Ohnishi, Haptic data compression/decompression using DCT for motion copy system, in: Proceedings of the IEEE International Conference on Mechatronics (ICM), 2009.

[414] A. Kuzu, E.A. Baran, S. Bogosyan, M. Gökaşan, A. Şabanoviç, Wavelet packet transform-based compression for teleoperation, Journal of Systems and Control Engineering 229 (7) (2015) 639–651.

[415] P.G. Otanez, J.R. Moyne, D.M. Tilbury, Using deadbands to reduce communication in networked control systems, in: Proceedings of the American Control Conference (ACC), 2002.

[416] S. Hirche, P. Hinterseer, E. Steinbach, M. Buss, Transparent data reduction in networked telepresence and teleaction systems – Part I: Communication without time delay, Presence: Teleoperators and Virtual Environments 16 (5) (2007) 523–531.

[417] S. Hirche, P. Hinterseer, E. Steinbach, M. Buss, Transparent data reduction in networked telepresence and teleaction systems – Part II: Time-delayed communication, Presence: Teleoperators and Virtual Environments 16 (5) (2007) 532–542.

[418] E. Steinbach, S. Hirche, J. Kammerl, I. Vittorias, R. Chaudhari, Haptic data compression and communication, IEEE Signal Processing Magazine 28 (1) (2011) 87–96.

[419] E. Steinbach, S. Hirche, M. Ernst, F. Brandi, R. Chaudhari, J. Kammerl, I. Vittorias, Haptic communications, Proceedings of the IEEE 100 (4) (2012) 937–956.

[420] N. Sakr, N. Georganas, J. Zhao, X. Shen, Motion and force prediction in haptic media, in: Proceedings of the IEEE International Conference on Multimedia and Expo (ICME), 2007.

[421] F. Guo, C. Zhang, Y. He, Haptic data compression based on a linear prediction model and quadratic curve reconstruction, Journal of Software 9 (11) (2014) 2796–2803.

[422] H. Pongrac, B. Färber, P. Hinterseer, J. Kammerl, E. Steinbach, Limitations of human 3D force discrimination, in: Proceedings of the International Workshop on Human-Centered Robotics Systems (HCRS), 2006.

[423] J. Kammerl, R. Chaudhari, E. Steinbach, Exploiting directional dependencies of force perception for lossy haptic data reduction, in: Proceedings of the IEEE International Symposium on Haptic Audio-Visual Environments and Games (HAVE), 2010.

[424] D.A. Lawrence, L.Y. Pao, A.M. Dougherty, M.A. Salada, Y. Pavlou, Rate-hardness: A new performance metric for haptic interfaces, IEEE Transactions on Robotics and Automation 16 (4) (2000) 357–371.

[425] R.J. Anderson, M.W. Spong, Bilateral control of teleoperators with time delay, IEEE Transactions on Automatic Control 34 (5) (1989) 494–501.

[426] B. Hannaford, J. Ryu, Time-domain passivity control of haptic interfaces, IEEE Transactions on Robotics and Automation 18 (1) (2002) 1–10.

[427] J. Ryu, J. Artigas, C. Preusche, A passive bilateral control scheme for a teleoperator with time-varying communication delay, Mechatronics 20 (7) (2010) 812–823.

[428] G. Hirzinger, K. Landzettel, D. Reintsema, C. Preusche, A. Albu-Schäffer, B. Rebele, M. Turk, ROKVISS – Robotics component verification on ISS, in: Proceedings of the International Symposium on Artificial Intelligence, Robotics and Automation in Space (i-SAIRAS), 2005.

[429] M. Panzirsch, J. Ryu, M. Ferre, Reducing the conservatism of the time domain passivity approach through consideration of energy reflection in delayed coupled network systems, Mechatronics 58 (2019) 58–69.

[430] K. Antonakoglou, X. Xu, E. Steinbach, T. Mahmoodi, M. Dohler, Toward haptic communications over the 5G Tactile Internet, IEEE Communications Surveys & Tutorials 20 (4) (2018) 3034–3059.

[431] I. Vittorias, J. Kammerl, S. Hirche, E. Steinbach, Perceptual coding of haptic data in time-delayed teleoperation, in: Proceedings of the IEEE World Haptics Conference (WHC), 2009.

[432] N. Chopra, M.W. Spong, S. Hirche, M. Buss, Bilateral teleoperation over the Internet: The time varying delay problem, in: Proceedings of the American Control Conference (ACC), 2003.

[433] R. Balachandran, J. Artigas, U. Mehmood, J.-H. Ryu, Performance comparison of wave variable transformation and time domain passivity approaches for time-delayed teleoperation: Preliminary results, in: Proceedings of the IEEE/RSJ International Conference on Intelligent Robots and Systems (IROS), 2016.

[434] X. Xu, C. Schuwerk, B. Çizmeci, E. Steinbach, Energy prediction for teleoperation systems that combine the time domain passivity approach with perceptual deadband-based haptic data reduction, IEEE Transactions on Haptics 9 (4) (2016) 560–573.

[435] X. Xu, B. Çizmeci, C. Schuwerk, E. Steinbach, Haptic data reduction for time-delayed teleoperation using the time domain passivity approach, in: Proceedings of the IEEE World Haptics Conference (WHC), 2015.

[436] X. Xu, B. Çizmeci, C. Schuwerk, E. Steinbach, Model-mediated teleoperation: Toward stable and transparent teleoperation systems, IEEE Access 4 (2016) 425–449.

[437] X. Xu, B. Çizmeci, A. Al-Nuaimi, E. Steinbach, Point cloud-based model-mediated teleoperation with dynamic and perception-based model updating, IEEE Transactions on Instrumentation and Measurement 63 (11) (2014) 2558–2569.

[438] X. Xu, B. Çizmeci, E. Steinbach, Point-cloud-based model-mediated teleoperation, in: Proceedings of the IEEE International Symposium on Haptic Audio-Visual Environments and Games (HAVE), 2013.

[439] X. Xu, J. Kammerl, R. Chaudhari, E. Steinbach, Hybrid signal-based and geometry-based prediction for haptic data reduction, in: Proceedings of the IEEE International Symposium on Haptic Audio-Visual Environments and Games (HAVE), 2011.

[440] X. Xu, E. Steinbach, Towards real-time modeling and haptic rendering of deformable objects for point cloud-based model-mediated teleoperation, in: Proceedings of the IEEE International Workshop on Hot Topics in 3D (Hot3D), 2014.

[441] X. Xu, S. Chen, E. Steinbach, Model-mediated teleoperation for movable objects: Dynamics modeling and packet rate reduction, in: Proceedings of the IEEE International Symposium on Haptic Audio-Visual Environments and Games (HAVE), 2015.

[442] X. Xu, Q. Liu, E. Steinbach, Haptic data reduction for time-delayed teleoperation using the input-to-state stability approach, in: Proceedings of the IEEE World Haptics Conference (WHC), 2019.

[443] S. Okamoto, Y. Yamada, Perceptual properties of vibrotactile material texture: Effects of amplitude changes and stimuli beneath detection thresholds, in: Proceedings of the IEEE/SICE International Symposium on System Integration (SII), 2010.

[444] S. Okamoto, Y. Yamada, Lossy data compression of vibrotactile material-like textures, IEEE Transactions on Haptics 6 (1) (2012) 69–80.

[445] R. Chaudhari, B. Çizmeci, K.J. Kuchenbecker, S. Choi, E. Steinbach, Low bitrate source-filter model based compression of vibrotactile texture signals in haptic teleoperation, in: Proceedings of the ACM International Conference on Multimedia (MM), 2012.

[446] D. Pan, A tutorial on mpeg/audio compression, IEEE MultiMedia 2 (2) (1995) 60–74.

[447] G.A. Gescheider, R.T. Verrillo, C.L. Van Doren, Prediction of vibrotactile masking functions, The Journal of the Acoustical Society of America 72 (5) (1982) 1421–1426.

[448] A. Said, W.A. Pearlman, A new, fast, and efficient image codec based on set partitioning in hierarchical trees, IEEE Transactions on Circuits and Systems for Video Technology 6 (3) (1996) 243–250.

[449] R. Hassen, E. Steinbach, Vibrotactile signal compression based on sparse linear prediction and human tactile sensitivity function, in: Proceedings of the IEEE World Haptics Conference (WHC), 2019.

[450] G. Gescheider, S. Bolanowski, K. Hardick, The frequency selectivity of information-processing channels in the tactile sensory system, Somatosensory & Motor Research 18 (3) (2001) 191–201.

[451] R.T. Verillo, Effect of contactor area on the vibrotactile threshold, The Journal of the Acoustical Society of America 35 (12) (1963) 1962–1966.

[452] X. Xu, M. Panzirsch, L. Qian, E. Steinbach, Integrating haptic data reduction with energy reflection-based passivity control for time-delayed teleoperation, in: Proceedings of the IEEE Haptics Symposium (HAPTICS), 2020.

[453] M. Panzirsch, H. Singh, T. Kruger, C. Ott, A. Albu-Schäffer, Safe interactions and kinesthetic feedback in high performance earth-to-moon teleoperation, in: Proceedings of the IEEE Aerospace Conference (AeroConf), 2020.

[454] H. Singh, A. Jafari, J. Ryu, Enhancing the force transparency of time domain passivity approach: Observer-based gradient controller, in: Proceedings of the IEEE International Conference on Robotics and Automation (ICRA), 2019.

[455] H. Singh, D. Janetzko, A. Jafari, B. Weber, C. Lee, J. Ryu, Enhancing the rate-hardness of haptic interaction: Successive force augmentation approach, IEEE Transactions on Industrial Electronics 67 (1) (2019) 809–819.

[456] A. Coelho, H. Singh, T. Muskardin, R. Balachandran, K. Kondak, Smoother position-drift compensation for time domain passivity approach based teleoperation, in: Proceedings of the IEEE/RSJ International Conference on Intelligent Robots and Systems (IROS), 2018.

[457] A. Coelho, C. Ott, H. Singh, F. Lizarralde, K. Kondak, Multi-DoF time domain passivity approach based drift compensation for telemanipulation, in: Proceedings of the International Conference on Advanced Robotics (ICAR), 2019.

[458] V. Chawda, H. Van Quang, M.K. O'Malley, J. Ryu, Compensating position drift in time domain passivity approach based teleoperation, in: Proceedings of the IEEE Haptics Symposium (HAPTICS), 2014.

[459] F.E. van Beek, The effect of damping on the perception of hardness, in: Making Sense of Haptics, Springer, 2017, pp. 81–101.

[460] H. Singh, A. Jafari, J. Ryu, Increasing the rate-hardness of haptic interaction: Successive force augmentation approach, in: Proceedings of the IEEE World Haptics Conference (WHC), 2017.

[461] F.H.P. Fitzek, F. Granelli, P. Seeling, Computing in Communication Networks: From Theory to Practice, Academic Press, 2020.

[462] J.A. Cabrera Guerrero, R. Schmoll, G.T. Nguyen, S. Pandi, F.H.P. Fitzek, Softwarization and network coding in the mobile edge cloud for the Tactile Internet, Proceedings of the IEEE 107 (2) (2019) 350–363.

[463] A. Osman, A. Wasicek, S. Köpsell, T. Strufe, Transparent microsegmentation in smart home IoT networks, in: Proceedings of the USENIX Workshop on Hot Topics in Edge Computing (HotEdge), 2020.

[464] T. Li, H. Salah, X. Ding, T. Strufe, F.H.P. Fitzek, S. Santini, INFAS: In-network flow management scheme for SDN control plane protection, in: Proceedings of the IEEE Symposium on Integrated Network and Service Management (IM), 2019.

[465] T. Li, H. Salah, M. He, T. Strufe, S. Santini, REMO: Resource efficient distributed network monitoring, in: Proceedings of the IEEE/IFIP Network Operations and Management Symposium (NOMS), 2018.

[466] S. Hanisch, A. Osman, T. Li, T. Strufe, Security for mobile edge cloud, in: Computing in Communication Networks: From Theory to Practice, Academic Press, 2020, pp. 391–407, chapter 23.

[467] M. Taghouti, Compressed sensing, in: Computing in Communication Networks: From Theory to Practice, Academic Press, 2020, pp. 207–227, chapter 10.

[468] M. Taghouti, M. Höweler, In-network compressed sensing, in: Computing in Communication Networks: From Theory to Practice, Academic Press, 2020, pp. 379–389, chapter 22.

[469] J. Rischke, Z. Xiang, Network coding for transport, in: Computing in Communication Networks: From Theory to Practice, Academic Press, 2020, pp. 357–367, chapter 20.

[470] S. Wunderlich, F. Gabriel, S. Pandi, F.H.P. Fitzek, M. Reisslein, Caterpillar RLNC (CRLNC): A practical finite sliding window RLNC approach, IEEE Access 5 (2017) 20183–20197.

[471] F. Gabriel, S. Wunderlich, S. Pandi, F.H.P. Fitzek, M. Reisslein, Caterpillar RLNC with Feedback (CRLNC-FB): Reducing delay in selective repeat ARQ through coding, IEEE Access 6 (2018) 44787–44802.

[472] S. Wunderlich, F. Gabriel, S. Pandi, F.H.P. Fitzek, We don't need no generation – A practical approach to sliding window RLNC, in: Proceedings of the Wireless Days Conference (WD), 2017.

[473] S. Pandi, F. Gabriel, J.A. Cabrera Guerrero, S. Wunderlich, M. Reisslein, F.H.P. Fitzek, PACE: Redundancy engineering in RLNC for low-latency communication, IEEE Access 5 (2017) 20477–20493.

[474] S. Pandi, S. Wunderlich, F.H.P. Fitzek, Reliable low latency wireless mesh networks – From myth to reality, in: Proceedings of the IEEE Consumer Communications & Networking Conference (CCNC), 2018.

[475] R. Torre, T.V. Doan, H. Salah, Mobile edge cloud, in: Computing in Communication Networks: From Theory to Practice, Academic Press, 2020, pp. 81–97, chapter 4.

[476] Z. Xiang, C. Collmann, P. Seeling, Realizing mobile edge clouds, in: Computing in Communication Networks: From Theory to Practice, Academic Press, 2020, pp. 291–301, chapter 15.

[477] S. Wunderlich, J.A. Cabrera Guerrero, F.H.P. Fitzek, M. Reisslein, Network coding in heterogeneous multicore IoT nodes with DAG scheduling of parallel matrix block operations, IEEE Internet of Things Journal 4 (4) (2017) 917–933.

[478] S. Wunderlich, F.H.P. Fitzek, M. Reisslein, Progressive multicore RLNC decoding with online DAG scheduling, IEEE Access 7 (2019) 161184–161200.

[479] J. Acevedo, R. Scheffel, S. Wunderlich, M. Hasler, S. Pandi, J. Cabrera, F.H.P. Fitzek, G.P. Fettweis, M. Reisslein, Hardware acceleration for RLNC: A case study based on the Xtensa processor with the tensilica instruction-set extension, Electronics 7 (9) (2018) 180.

[480] R.L. Gregory, The Oxford Companion to the Mind, Oxford University Press, 1987, pp. 598–601, chapter Perception.

[481] E. Ahissar, E. Assa, Perception as a closed-loop convergence process, eLife 5 (2016) e12830, pp. 1–26.

[482] T. Naftali, D. Polani, Information theory of decisions and actions, in: V. Cutsuridis, A. Hussain, J.G. Taylor (Eds.), Perception-Action Cycle: Models, Architectures, and Hardware, in: Springer Series in Cognitive and Neural Systems, vol. 1, Springer, 2011, pp. 601–636.

[483] P. Majaranta, K.-J. Räihä, A. Hyrskykari, O. Špakov, Eye movements and human-computer interaction, in: C. Klein, U. Ettinger (Eds.), Eye Movement Research: An Introduction to its Scientific Foundations and Applications, in: Studies in Neuroscience, Psychology and Behavioral Economics, vol. 7, Springer, 2019, pp. 971–1015.

[484] W. Barfield, C. Hendrix, O. Bjorneseth, K.A. Kaczmarek, W. Lotens, Comparison of human sensory capabilities with technical specifications of virtual environment equipment, Presence: Teleoperators and Virtual Environments 4 (4) (1995) 329–356.

[485] E.P. Gardner, K.O. Johnson, The somatosensory system: Receptors and central pathways, in: Principles of Neural Science, McGraw-Hill, 2012, pp. 475–497.

[486] L. Freina, M. Ott, A literature review on immersive virtual reality in education: State of the art and perspectives, in: Proceedings of the International Scientific Conference on eLearning and Software for Education (eLSE), 2015.

[487] O.A. van der Meijden, M.P. Schijven, The value of haptic feedback in conventional and robot-assisted minimal invasive surgery and virtual reality training: A current review, Surgical Endoscopy 23 (6) (2009) 1180–1190.

[488] A.C. Muller Queiroz, A. Moreira Nascimento, R. Tori, T. Brashear Alejandro, V.V. de Melo, F. de Souza Meirelles, M.I. da Silva Leme, Immersive virtual environments in corporate education and training, in: Proceedings of the Americas Conference on Information Systems (AMCIS), 2018.

[489] C.J. Bohil, B. Alicea, F.A. Biocca, Virtual reality in neuroscience research and therapy, Nature Reviews Neuroscience 12 (12) (2011) 752–762.

[490] M.T.M. Lambooij, W.A. IJsselsteijn, I. Heynderickx, Visual discomfort in stereoscopic displays: A review, in: A.J. Woods, N.A. Dodgson, J.O. Merritt, M.T. Bolas, I.E.

McDowall (Eds.), Stereoscopic Displays and Virtual Reality Systems XIV, in: Proceedings of SPIE, vol. 6490, SPIE, 2007, 64900I, pp. 1–13.

[491] J. Diemer, G.W. Alpers, H.M. Peperkorn, Y. Shiban, A. Mühlberger, The impact of perception and presence on emotional reactions: A review of research in virtual reality, Frontiers in Psychology 6 (2015) 26, pp. 1–9.

[492] R.S. Renner, B.M. Velichkovsky, J.R. Helmert, The perception of egocentric distances in virtual environments – A review, ACM Computing Surveys 46 (2) (2013) 23, pp. 1–40.

[493] C. Boletsis, The new era of virtual reality locomotion: A systematic literature review of techniques and a proposed typology, Multimodal Technologies and Interaction 1 (4) (2017) 24, pp. 1–17.

[494] J. Dargahi, S. Najarian, Human tactile perception as a standard for artificial tactile sensing – A review, International Journal of Medical Robotics and Computer Assisted Surgery 1 (1) (2004) 23–35.

[495] J.P. Bresciani, K. Drewing, M.O. Ernst, Human haptic perception and the design of haptic-enhanced virtual environments, in: A. Bicchi, M. Buss, M.O. Ernst, A. Peer (Eds.), The Sense of Touch and Its Rendering, in: Springer Tracts in Advanced Robotics, vol. 45, Springer, 2008, pp. 61–106.

[496] L.A. Hall, D.I. McCloskey, Detections of movements imposed on finger, elbow and shoulder joints, The Journal of Physiology 335 (1) (1983) 519–533.

[497] U. Proske, Kinesthesia: The role of muscle receptors, Muscle & Nerve 34 (5) (2006) 545–558.

[498] A.K. Wise, J.E. Gregory, U. Proske, The effects of muscle conditioning on movement detection thresholds at the human forearm, Brain Research 735 (1) (1996) 125–130.

[499] J.A. Taylor, D.I. McCloskey, Detection of slow movements imposed at the elbow during active flexion in man, Journal of Physiology 457 (1) (1992) 503–513.

[500] U. Proske, S.C. Gandevia, The proprioceptive senses: Their roles in signaling body shape, body position and movement, and muscle force, Physiological Reviews 92 (4) (2012) 1651–1697.

[501] G. Baud-Bovy, E. Gatti, Hand-held object force direction identification thresholds at rest and during movement, in: A.M.L. Kappers, J.B.F. van Erp, W.M. Bergmann Tiest, F.C.T. van der Helm (Eds.), Haptics: Generating and Perceiving Tangible Sensations, in: Lecture Notes in Computer Science, vol. 6192, Springer, 2010, pp. 231–236.

[502] F.E. van Beek, W.M. Bergmann Tiest, A.M. Kappers, G. Baud-Bovy, Integrating force and position: Testing model predictions, Experimental Brain Research 234 (11) (2016) 3367–3379.

[503] L.A. Jones, Perception of force and weight: Theory and research, Psychological Bulletin 100 (1) (1986) 29–42.

[504] L.D. Walsh, J.L. Taylor, S.C. Gandevia, Overestimation of force during matching of externally generated forces, Journal of Physiology 589 (3) (2011) 547–557.

[505] K. Myles, M.S. Binseel, The tactile modality: A review of tactile sensitivity and human tactile interfaces, Technical report, Army Research Laboratory – Aberdeen Proving Ground, 2007.

[506] K.A. Kaczmarek, J.G. Webster, P. Bach-y-Rita, W.J. Tompkins, Electrotactile and vibrotactile displays for sensory substitution systems, IEEE Transactions on Biomedical Engineering 38 (1) (1991) 1–15.

[507] K. Driggs-Campbell, V. Shia, R. Bajcsy, Improved driver modeling for Human-in-the-Loop vehicular control, in: Proceedings of the IEEE International Conference on Robotics and Automation (ICRA), 2015.

[508] T. Sugiyama, S.L. Liew, The effects of sensory manipulations on motor behavior: From basic science to clinical rehabilitation, Journal of Motor Behavior 49 (1) (2017) 67–77.

[509] M. Turk, Multimodal interaction: A review, Pattern Recognition Letters 36 (2014) 189–195.

[510] C.K. Williams, H. Carnahan, Motor learning perspectives on haptic training for the upper extremities, IEEE Transactions on Haptics 7 (2) (2014) 240–250.

[511] C.E. Sherrick, R.W. Cholewiak, Cutaneous sensitivity, in: K.R. Boff, L. Kaufman, J.P. Thomas (Eds.), Handbook of Perception and Human Performance, Wiley, 1986, pp. 12:1–58.

[512] E.R. Kandel, T.M. Jessell, Touch, in: Principles of Neural Science, Elsevier, 1991, pp. 349–414.

[513] J.B.F. van Erp, J.J. van den Dobbelsteen, On the Design of Tactile Displays, TNO Human Factors, 1998.

[514] J.B.F. van Erp, I.M.L.C. Vogels, Vibrotactile Perception: A Literature Review, TNO Human Factors, 1998.

[515] L. Venkatesan, S.M. Barlow, D. Kieweg, Age- and sex-related changes in vibrotactile sensitivity of hand and face in neurotypical adults, Somatosensory & Motor Research 32 (1) (2015) 44–50.

[516] T.J. Allen, U. Proske, Effect of muscle fatigue on the sense of limb position and movement, Experimental Brain Research 170 (1) (2006) 30–38.

[517] J.M. Saxton, P.M. Clarkson, R. James, M. Miles, M. Westerfer, S. Clark, A.E. Donnelly, Neuromuscular dysfunction following eccentric exercise, Medicine and Science in Sports and Exercise 27 (8) (1995) 1185–1193.

[518] L.D. Walsh, C.W. Hesse, D.L. Morgan, U. Proske, Human forearm position sense after fatigue of elbow flexor muscles, Journal of Physiology 558 (2) (2004) 705–715.

[519] N. Weerakkody, P. Percival, D.L. Morgan, J.E. Gregory, U. Proske, Matching different levels of isometric torque in elbow flexor muscles after eccentric exercise, Experimental Brain Research 149 (2) (2003) 141–150.

[520] M.F. Bruce, The relation of tactile thresholds to histology in the fingers of elderly people, Journal of Neurology, Neurosurgery and Psychiatry 43 (1980) 730–734.

[521] J. Decorps, J.L. Saumet, P. Sommer, D. Sigaudo-Roussel, B. Fromy, Effect of ageing on tactile transduction processes, Ageing Research Reviews 13 (2014) 90–99.

[522] T. Iwasaki, The aging of human Meissner's corpuscles as evidenced by parallel sectioning, Okajimas Folia Anatomica Japonica 79 (6) (2003) 185–189.

[523] Y.K. Dillon, J. Haynes, M. Henneberg, The relationship of the number of Meissner's corpuscles to dermatoglyphic characters and finger size, Journal of Anatomy 199 (2001) 577–584.

[524] R.M. Peters, E. Hackeman, D. Goldreich, Diminutive digits discern delicate details: Fingertip size and the sex difference in tactile spatial acuity, The Journal of Neuroscience 29 (50) (2009) 15756–15761.

[525] A. Abdouni, M. Djaghloul, C. Thieulin, R. Vargiolu, C. Pailler-Mattei, H. Zahouani, Biophysical properties of the human finger for touch comprehension: Influences of ageing and gender, Royal Society Open Science 4 (8) (2017) 170321, pp. 1–14.

[526] D.R. Proffitt, M. Bhalla, R. Gossweiler, J. Midgett, Perceiving geographical slant, Psychonomic Bulletin & Review 2 (4) (1995) 409–428.

[527] F. Gemperle, T. Hirsch, A. Goode, J. Pearce, D.P. Siewiorek, A. Smailigić, Wearable vibro-tactile display, Technical report, Carnegie Mellon University, CMU Wearable Group, 2003.

[528] F.A. Geldard, C.E. Sherrick, The cutaneous "rabbit": A perceptual illusion, Science 178 (4057) (1972) 178–179.

[529] H.L. Hollingworth, The inaccuracy of movement – with special reference to constant errors, Archives of Psychology 13 (1909) 1–87.

[530] C. Pacchierotti, S. Sinclair, M. Solazzi, A. Frisoli, V. Hayward, D. Prattichizzo, Wearable haptic systems for the fingertip and the hand: Taxonomy, review, and perspectives, IEEE Transactions on Haptics 10 (4) (2017) 580–600.

[531] K. Salisbury, D. Brock, T. Massie, N. Swarup, C. Zilles, Haptic rendering, in: Proceedings of the Symposium on Interactive 3D graphics (SI3D), 1995.

[532] D. Wang, M. Song, A. Naqash, Y. Zheng, W. Xu, Y. Zhang, Toward whole-hand kinesthetic feedback: A survey of force feedback gloves, IEEE Transactions on Haptics 12 (2) (2019) 189–204.

[533] M.A. McEvoy, N. Correll, Materials that couple sensing, actuation, computation, and communication, Science 347 (6228) (2015) 1261689.

[534] J.C. Craig, K.O. Johnson, The two-point threshold: Not a measure of tactile spatial resolution, Current Directions in Psychological Science 9 (1) (2000) 29–32.

[535] M.S. Gandhi, R. Sesek, R. Tuckett, S.J. Bamberg, Progress in vibrotactile threshold evaluation techniques: A review, Journal of Hand Therapy 24 (3) (2011) 240–256.

[536] A. Schwarz, C.M. Kanzler, O. Lambercy, A.R. Luft, J.M. Veerbeek, Systematic review on kinematic assessments of upper limb movements after stroke, Stroke 50 (3) (2019) 718–727.

[537] R.A. Stevenson, D. Ghose, J.K. Fister, D.K. Sarko, N.A. Altieri, A.R. Nidiffer, L.R. Kurela, J.K. Siemann, T.W. James, M.T. Wallace, Identifying and quantifying multisensory integration: A tutorial review, Brain Topography 27 (6) (2014) 707–730.

[538] M.E. Altinsoy, Auditory-Tactile Interaction in Virtual Environments, Shaker, 2006.

[539] I. Koch, E. Poljac, H. Müller, A. Kiesel, Cognitive structure, flexibility, and plasticity in human multitasking – An integrative review of dual-task and task-switching research, Psychological Bulletin 144 (6) (2018) 557–583.

[540] R.S. Sutton, A.G. Barto, Reinforcement Learning: An Introduction, MIT Press, 2018.

[541] Y. Bengio, A. Courville, P. Vincent, Representation learning: A review and new perspectives, IEEE Transactions on Pattern Analysis and Machine Intelligence 35 (8) (2013) 1798–1828.

[542] L.J. Byom, B. Mutlu, Theory of mind: Mechanisms, methods, and new directions, Frontiers in Human Neuroscience 7 (2013) 413, pp. 1–12.

[543] N. Kolling, M. Wittmann, M.F.S. Rushworth, Multiple neural mechanisms of decision making and their competition under changing risk pressure, Neuron 81 (5) (2014) 1190–1202.

[544] N.D. Daw, J.P. O'Doherty, P. Dayan, B. Seymour, R.J. Dolan, Cortical substrates for exploratory decisions in humans, Nature 441 (7095) (2006) 876–879.

[545] N.D. Daw, Y. Niv, P. Dayan, Uncertainty-based competition between prefrontal and dorsolateral striatal systems for behavioral control, Nature Neuroscience 8 (12) (2005) 1704–1711.

[546] J. Gläscher, N. Daw, P. Dayan, J.P. O'Doherty, States versus rewards: Dissociable neural prediction error signals underlying model-based and model-free reinforcement learning, Neuron 66 (4) (2010) 585–595.

[547] P. Schwartenbeck, T.H.B. FitzGerald, C. Mathys, R. Dolan, K. Friston, The dopaminergic midbrain encodes the expected certainty about desired outcomes, Cerebral Cortex 25 (10) (2015) 3434–3445.

[548] D. Lee, H. Seo, M.W. Jung, Neural basis of reinforcement learning and decision making, Annual Review of Neuroscience 35 (2012) 287–308.

[549] K. Friston, F. Rigoli, D. Ognibene, C. Mathys, T.H.B. FitzGerald, G. Pezzulo, Active inference and epistemic value, Cognitive Neuroscience 6 (4) (2015) 187–214.

[550] D. Cuevas Rivera, F. Ott, D. Marković, A. Strobel, S.J. Kiebel, Context-dependent risk aversion: A model-based approach, Frontiers in Psychology 9 (2018) 2053, pp. 1–17.

[551] D.A. Simon, N.D. Daw, Neural correlates of forward planning in a spatial decision task in humans, The Journal of Neuroscience 31 (14) (2011) 5526–5539.

[552] B.B. Doll, K.D. Duncan, D.A. Simon, D. Shohamy, N.D. Daw, Model-based choices involve prospective neural activity, Nature Neuroscience 18 (5) (2015) 767–772.

[553] W. Wood, D. Rünger, Psychology of habit, Annual Review of Psychology 67 (1) (2016) 289–314.

[554] P. Lally, C.H.M. van Jaarsveld, H.W.W. Potts, J. Wardle, How are habits formed: Modelling habit formation in the real world, European Journal of Social Psychology 40 (6) (2010) 998–1009.

[555] J.R. Anderson, Learning and Memory: An Integrated Approach, Wiley, 2000.

[556] E.L. Thorndike, Animal Intelligence: An Experimental Study of the Associative Processes in Animals, Macmillan, 1898.

[557] M. Puterman, Markov Decision Processes: Discrete Stochastic Dynamic Programming, Wiley, 1994.

[558] P. Dayan, Improving generalization for temporal difference learning: The successor representation, Neural Computation 5 (4) (1993) 613–624.

[559] C.J.C.H. Watkins, P. Dayan, Q-learning, Machine Learning 8 (3) (1992) 279–292.

[560] S.W. Lee, S. Shimojo, J.P. O'Doherty, Neural computations underlying arbitration between model-based and model-free learning, Neuron 81 (3) (2014) 687–699.

[561] K. Friston, T. FitzGerald, F. Rigoli, P. Schwartenbeck, J. O'Doherty, G. Pezzulo, Active inference and learning, Neuroscience & Biobehavioral Reviews 68 (2016) 862–879.

[562] R.S. Sutton, A.G. Barto, Toward a modern theory of adaptive networks: Expectation and prediction, Psychological Review 88 (2) (1981) 135–170.

[563] S. Schwöbel, D. Marković, M.N. Smolka, S.J. Kiebel, Balancing control: A Bayesian interpretation of habitual and goal-directed behavior, Journal of Mathematical Psychology 100 (2021) 102472, pp. 1–22.

[564] S. Monsell, Task switching, Trends in Cognitive Sciences 7 (3) (2003) 134–140.

[565] S. Maren, K.L. Phan, I. Liberzon, The contextual brain: Implications for fear conditioning, extinction and psychopathology, Nature Reviews Neuroscience 14 (6) (2013) 417–428.

[566] S.J. Gershman, Y. Niv, Learning latent structure: Carving nature at its joints, Current Opinion in Neurobiology 20 (2) (2010) 251–256.

[567] M. Toussaint, A. Storkey, Probabilistic inference for solving discrete and continuous state Markov decision processes, in: Proceedings of the International Conference on Machine Learning (ICML), 2006.

[568] K. Friston, P. Schwartenbeck, T. FitzGerald, M. Moutoussis, T. Behrens, R.J. Dolan, The anatomy of choice: Dopamine and decision-making, Philosophical Transactions of the Royal Society of London B: Biological Sciences 369 (1655) (2014) 20130481, pp. 1–12.

[569] R. Kaplan, K.J. Friston, Planning and navigation as active inference, Biological Cybernetics 112 (4) (2018) 323–343.

[570] M. Botvinick, M. Toussaint, Planning as inference, Trends in Cognitive Sciences 16 (10) (2012) 485–488.

[571] H.C. Barron, R.J. Dolan, T.E.J. Behrens, Online evaluation of novel choices by simultaneous representation of multiple memories, Nature Neuroscience 16 (10) (2013) 1492–1498.

[572] G.E. Wimmer, D. Shohamy, Preference by association: How memory mechanisms in the hippocampus bias decisions, Science 338 (6104) (2012) 270–273.

[573] Y. Liu, R.J. Dolan, Z. Kurth-Nelson, T.E.J. Behrens, Human replay spontaneously reorganizes experience, Cell 178 (3) (2019) 640–652.

[574] J. Filar, K. Vrieze, Competitive Markov Decision Processes, Springer, 1996.

[575] D.A. Peled, Formal methods, in: S. Cha, R.N. Taylor, K. Kang (Eds.), Handbook of Software Engineering, Springer, 2019, pp. 193–222.

[576] S. Ali, H. Sun, Y. Zhao, Model learning: A survey on foundation, tools and applications, CoRR, arXiv:1901.01910, 2019.

[577] M. Isberner, B. Steffen, An abstract framework for counterexample analysis in active automata learning, in: Proceedings of the International Conference on Grammatical Inference (ICGI), 2014.

[578] T. Brázdil, K. Chatterjee, M. Chmelik, A. Fellner, J. Kretínský, Counterexample explanation by learning small strategies in Markov decision processes, in: Proceedings of the International Conference on Computer Aided Verification (CAV), 2015.

[579] X. Huang, M. Kwiatkowska, S. Wang, M. Wu, Safety verification of deep neural networks, in: R. Majumdar, V. Kunčak (Eds.), Computer Aided Verification, in: Lecture Notes in Computer Science, vol. 10426, Springer, 2017, pp. 3–29.

[580] T. Brázdil, K. Chatterjee, M. Chmelík, V. Forejt, J. Křetínský, M. Kwiatkowska, D. Parker, M. Ujma, Verification of Markov decision processes using learning algorithms, in: F. Cassez, J.-F. Raskin (Eds.), Automated Technology for Verification and Analysis, in: Lecture Notes in Computer Science, vol. 8837, Springer, 2014, pp. 98–114.

[581] D. Angluin, Learning regular sets from queries and counterexamples, Information and Computation 75 (2) (1987) 87–106.

[582] D. Peled, M.Y. Vardi, M. Yannakakis, Black box checking, in: J. Wu, S.T. Chanson, Q. Gao (Eds.), Formal Methods for Protocol Engineering and Distributed Systems, in: IFIP Advances in Information and Communication Technology, vol. 28, Springer, 1999, pp. 225–240.

[583] E.M. Clarke Jr., O. Grumberg, D.A. Peled, Model Checking, MIT Press, 1999.

[584] C. Baier, J.-P. Katoen, Principles of Model Checking, MIT Press, 2008.

[585] C. de la Higuera, Grammatical Inference: Learning Automata and Grammars, Cambridge University Press, 2010.

[586] F.W. Vaandrager, Model learning, Communications of the ACM 60 (2) (2017) 86–95.

[587] R.E. Bellman, Dynamic Programming, Princeton University Press, 1957.

[588] R. Howard, Dynamic Programming and Markov Processes, MIT Press, 1960.

[589] F. Pedregosa, G. Varoquaux, A. Gramfort, V. Michel, B. Thirion, O. Grisel, M. Blondel, P. Prettenhofer, R. Weiss, V. Dubourg, J. Vanderplas, A. Passos, D. Cournapeau, M. Brucher, M. Perrot, É. Duchesnay, Scikit-learn: Machine learning in Python, Journal of Machine Learning Research 12 (85) (2011) 2825–2830.

[590] T.M. Mitchell, Machine Learning, McGraw-Hill, 1997.

[591] S. Liu, A. Panangadan, A. Talukder, C.S. Raghavendra, Compact representation of coordinated sampling policies for body sensor networks, in: Proceedings of the IEEE Global Communications Conference (GLOBECOM), Workshop on Advances in Communication and Networks (M9): Smart Homes for Tele-Health, 2010.

[592] T. Brázdil, K. Chatterjee, J. Křetínský, V. Toman, Strategy representation by decision trees in reactive synthesis, in: D. Beyer, M. Huisman (Eds.), Tools and Algorithms for the Construction and Analysis of Systems, in: Lecture Notes in Computer Science, vol. 10805, Springer, 2018, pp. 385–407.

[593] J. Kretínský, T. Meggendorfer, Of cores: A partial-exploration framework for Markov decision processes, in: W. Fokkink, R. van Glabbeek (Eds.), 30th International Conference on Concurrency Theory (CONCUR), in: Leibniz International Proceedings in Informatics (LIPIcs), vol. 140, Schloss Dagstuhl – Leibniz-Zentrum für Informatik, 2019, pp. 5:1–17.

[594] P. Ashok, K. Chatterjee, J. Křetínský, M. Weininger, T. Winkler, Approximating values of generalized-reachability stochastic games, in: Proceedings of the ACM/IEEE Symposium on Logic in Computer Science (LICS), 2020.

[595] P. Ashok, M. Jackermeier, P. Jagtap, J. Křetínský, M. Weininger, M. Zamani, dtControl: Decision tree learning algorithms for controller representation, in: Proceedings of the International Conference on Hybrid Systems: Computation and Control (HSCC), 2020.

[596] W. Damm, M. Fränzle, A. Lüdtke, J.W. Rieger, A. Trende, A. Unni, Integrating neurophysiological sensors and driver models for safe and performant automated vehicle control in mixed traffic, in: Proceedings of the IEEE Intelligent Vehicles Symposium (IV), 2019.

[597] L.P. Kaelbling, M.L. Littman, A.W. Moore, Reinforcement learning: A survey, Journal of Artificial Intelligence Research 4 (1996) 237–285.

[598] H.E. Robbins, Some aspects of the sequential design of experiments, Bulletin of the American Mathematical Society 58 (5) (1952) 527–535.

[599] T.L. Lai, H. Robbins, Asymptotically efficient adaptive allocation rules, Advances in Applied Mathematics 6 (1) (1985) 4–22.

[600] R.S. Sutton, Learning to predict by the methods of temporal differences, Machine Learning 3 (1) (1988) 9–44.

[601] U. Castiello, The neuroscience of grasping, Nature Reviews Neuroscience 6 (9) (2005) 726–736.

[602] H.J. Foley, M.W. Matlin, Sensation and Perception, Routledge, 2016.

[603] K. Jasmin, C.F. Lima, S.K. Scott, Understanding rostral-caudal auditory cortex contributions to auditory perception, Nature Reviews Neuroscience 20 (7) (2019) 425–434.

[604] R.S. Johansson, J.R. Flanagan, Coding and use of tactile signals from the fingertips in object manipulation tasks, Nature Reviews Neuroscience 10 (5) (2009) 345–359.

[605] S. Preusser, S.D. Thiel, C. Rook, E. Roggenhofer, A. Kosatschek, B. Draganski, F. Blankenburg, J. Driver, A. Villringer, B. Pleger, The perception of touch and the ventral somatosensory pathway, Brain 138 (3) (2015) 540–548.

[606] Y. Sasaki, J.E. Nanez, T. Watanabe, Advances in visual perceptual learning and plasticity, Nature Reviews Neuroscience 11 (1) (2010) 53–60.

[607] B.E. Stein, The New Handbook of Multisensory Processing, MIT Press, 2019.

[608] R. Romo, R. Rossi-Pool, Turning touch into perception, Neuron 105 (1) (2020) 16–33.

[609] M. Haller, J. Case, N.E. Crone, E.F. Chang, D. King-Stephens, K.D. Laxer, P.B. Weber, J. Parvizi, R.T. Knight, A.Y. Shestyuk, Persistent neuronal activity in human prefrontal cortex links perception and action, Nature Human Behaviour 2 (1) (2018) 80–91.

[610] A. Herwig, Linking perception and action by structure or process? Toward an integrative perspective, Neuroscience & Biobehavioral Reviews 52 (2015) 105–116.

[611] S. Schütz-Bosbach, W. Prinz, Perceptual resonance: Action-induced modulation of perception, Trends in Cognitive Sciences 11 (8) (2007) 349–355.

[612] H. von Helmholtz, Über die Natur der menschlichen Sinnesempfindungen, Königs-berger Naturwissenschaftliche Unterhaltungen 3 (1) (1854) 1–20.

[613] H. von Helmholtz, R. Kahl, Selected writings of Hermann von Helmholtz, Wesleyan University Press, 1971.

[614] M.O. Ernst, H.H. Bulthoff, Merging the senses into a robust percept, Trends in Cognitive Sciences 8 (4) (2004) 162–169.

[615] S.J. Sober, P.N. Sabes, Flexible strategies for sensory integration during motor planning, Nature Neuroscience 8 (4) (2005) 490–497.

[616] N. van Atteveldt, M.M. Murray, G. Thut, C.E. Schroeder, Multisensory integration: Flexible use of general operations, Neuron 81 (6) (2014) 1240–1253.

[617] M. Di Luca, B. Knorlein, M.O. Ernst, M. Harders, Effects of visual-haptic asynchronies and loading-unloading movements on compliance perception, Brain Research Bulletin 85 (5) (2011) 245–259.

[618] N. Gurari, A.M. Okamura, K.J. Kuchenbecker, Perception of force and stiffness in the presence of low-frequency haptic noise, PLOS One 12 (6) (2017) e0178605, pp. 1–26.

[619] R. Gau, U. Noppeney, How prior expectations shape multisensory perception, NeuroImage 124 (Part A) (2016) 876–886.

[620] T. Rohe, A.C. Ehlis, U. Noppeney, The neural dynamics of hierarchical Bayesian causal inference in multisensory perception, Nature Communications 10 (2019) 1907, pp. 1–17.

[621] J. Limanowski, K. Friston, Attentional modulation of vision versus proprioception during action, Cerebral Cortex 30 (3) (2020) 1637–1648.

[622] K. Friston, S.J. Kiebel, Predictive coding under the free-energy principle, Philosophical Transactions of the Royal Society B: Biological Sciences 364 (1521) (2009) 1211–1221.

[623] K. Friston, T. FitzGerald, F. Rigoli, P. Schwartenbeck, G. Pezzulo, Active inference: A process theory, Neural Computation 29 (1) (2017) 1–49.

[624] Z. Shi, D. Burr, Predictive coding of multisensory timing, Current Opinion in Behavioral Sciences 8 (2016) 200–206.

[625] S.-C. Li, A. Rieckmann, Neuromodulation and aging: Implications of aging neuronal gain control on cognition, Current Opinion in Neurobiology 29 (2014) 148–158.

[626] G. Calvert, C. Spence, B.E. Stein (Eds.), The Handbook of Multisensory Processes, MIT Press, 2004.

[627] D. Purves, G.J. Augustine, D. Fitzpatrick, W.C. Hall, A.S. Lamantia, J.O. McNamara, S.M. Williams, Neuroscience, Sinauer Associates, 2004.

[628] D.O. Hebb, The Organization of Behavior: A Neuropsychological Theory, Wiley, 1949.

[629] N. Caporale, Y. Dan, Spike timing-dependent plasticity: A Hebbian learning rule, Annual Review of Neuroscience 31 (2008) 25–46.

[630] B.E. Stein, T.R. Stanford, B.A. Rowland, Development of multisensory integration from the perspective of the individual neuron, Nature Reviews Neuroscience 15 (8) (2014) 520–535.

[631] C.S. Choe, R.B. Welch, R.M. Gilford, J.F. Juola, The "ventriloquist effect": Visual dominance or response bias?, Perception & Psychophysics 18 (1) (1975) 55–60.

[632] H. McGurk, J. MacDonald, Hearing lips and seeing voices, Nature 264 (5588) (1976) 746–748.

[633] M. Botvinick, J. Cohen, Rubber hands 'feel' touch that eyes see, Nature 391 (6669) (1998) 756.

[634] M. Tsakiris, P. Haggard, The rubber hand illusion revisited: Visuotactile integration and self-attribution, Journal of Experimental Psychology: Human Perception and Performance 31 (1) (2005) 80–91.

[635] M.T. Wallace, B.E. Stein, Sensory and multisensory responses in the newborn monkey superior colliculus, The Journal of Neuroscience 21 (22) (2001) 8886–8894.

[636] P. Redgrave, V. Coizet, E. Comoli, J.G. McHaffie, M. Leriche, N. Vautrelle, L.M. Hayes, P. Overton, Interactions between the midbrain superior colliculus and the basal ganglia, Frontiers in Neuroanatomy 4 (2010) 132, pp. 1–8.

[637] M.I. Sereno, R.S. Huang, Multisensory maps in parietal cortex, Current Opinion in Neurobiology 24 (2014) 39–46.

[638] L. Yu, J. Xu, B.A. Rowland, B.E. Stein, Multisensory plasticity in superior colliculus neurons is mediated by association cortex, Cerebral Cortex 26 (3) (2016) 1130–1137.

[639] S. Everling, K. Johnston, Control of the superior colliculus by the lateral prefrontal cortex, Philosophical Transactions of the Royal Society B: Biological Sciences 368 (1628) (2013) 20130068, pp. 1–11.

[640] E. Macaluso, J. Driver, Multisensory spatial interactions: A window onto functional integration in the human brain, Trends in Neurosciences 28 (5) (2005) 264–271.

[641] M. Avillac, S.B. Hamed, J.-R. Duhamel, Multisensory integration in the ventral intraparietal area of the macaque monkey, The Journal of Neuroscience 27 (8) (2007) 1922–1932.

[642] G. Gentile, V.I. Petkova, H.H. Ehrsson, Integration of visual and tactile signals from the hand in the human brain: An fMRI study, Journal of Neurophysiology 105 (2) (2011) 910–922.

[643] B.E. Stein, M.A. Meredith, The Merging of the Senses, MIT Press, 1993.

[644] E. Macaluso, J. Driver, Spatial attention and crossmodal interactions between vision and touch, Neuropsychologia 39 (12) (2001) 1304–1316.

[645] A. Amedi, R. Malach, S. Peled, E. Zohary, Visuo-haptic object-related activation in the ventral visual pathway, Nature Neuroscience 4 (3) (2001) 324–330.

[646] A. Amedi, G. Jacobson, T. Hendler, R. Malach, E. Zohary, Convergence of visual and tactile shape processing in the human lateral occipital complex, Cerebral Cortex 12 (11) (2002) 1202–1212.

[647] M.S. Beauchamp, N.E. Yasar, R.E. Frye, T. Ro, Touch, sound and vision in human superior temporal sulcus, NeuroImage 41 (3) (2008) 1011–1020.

[648] F. Bremmer, A. Schlack, N.J. Shah, O. Zafiris, M. Kubischik, K.-P. Hoffmann, K. Zilles, G.R. Fink, Polymodal motion processing in posterior parietal and premotor cortex: A human fMRI study strongly implies equivalencies between humans and monkeys, Neuron 29 (1) (2001) 287–296.

[649] M.S.A. Graziano, Where is my arm? The relative role of vision and proprioception in the neuronal representation of limb position, Proceedings of the National Academy of Sciences of the United States of America 96 (18) (1999) 10418–10421.

[650] H.H. Ehrsson, C. Spence, R.E. Passingham, That's my hand! Activity in premotor cortex reflects feeling of ownership of a limb, Science 305 (5685) (2004) 875–877.

[651] J. Limanowski, F. Blankenburg, Network activity underlying the illusory self-attribution of a dummy arm, Human Brain Mapping 36 (6) (2015) 2284–2304.

[652] J. Limanowski, F. Blankenburg, Integration of visual and proprioceptive limb position information in human posterior parietal, premotor, and extrastriate cortex, The Journal of Neuroscience 36 (9) (2016) 2582–2589.

[653] V. de Lafuente, R. Romo, Neural correlate of subjective sensory experience gradually builds up across cortical areas, Proceedings of the National Academy of Sciences of the United States of America 103 (39) (2006) 14266–14271.

[654] S.A. Hillyard, E.K. Vogel, S.J. Luck, Sensory gain control (amplification) as a mechanism of selective attention: Electrophysiological and neuroimaging evidence, Philosophical Transactions of the Royal Society of London, Series B: Biological Sciences 353 (1373) (1998) 1257–1270.

[655] J.H. Reynolds, R. Desimone, The role of neural mechanisms of attention in solving the binding problem, Neuron 24 (1) (1999) 19–29.

[656] Y.B. Saalmann, S. Kastner, Gain control in the visual thalamus during perception and cognition, Current Opinion in Neurobiology 19 (4) (2009) 408–414.

[657] A. Lajtha, S. Vizi, Handbook of Neurochemistry and Molecular Neurobiology: Neurotransmitter Systems, Springer, 2008.

[658] S. Bao, V.T. Chan, M.M. Merzenich, Cortical remodelling induced by activity of ventral tegmental dopamine neurons, Nature 412 (6842) (2001) 79–83.

[659] V. de Lafuente, R. Romo, Dopamine neurons code subjective sensory experience and uncertainty of perceptual decisions, Proceedings of the National Academy of Sciences of the United States of America 108 (49) (2011) 19767–19771.

[660] W. Schultz, Updating dopamine reward signals, Current Opinion in Neurobiology 23 (2) (2013) 229–238.

[661] S. Sarno, V. de Lafuente, R. Romo, N. Parga, Dopamine reward prediction error signal codes the temporal evaluation of a perceptual decision report, Proceedings of the National Academy of Sciences of the United States of America 114 (48) (2017) E10494–E10503.

[662] T.G. Fechner, Elemente der Psychophysik, Breitkopf und Hartel, 1860.

[663] M.E. Diamond, Perceptual uncertainty, PLOS Biology 17 (8) (2019) e3000430, pp. 1–7.

[664] M.O. Ernst, M.S. Banks, Humans integrate visual and haptic information in a statistically optimal fashion, Nature 415 (24) (2002) 429–433.

[665] J.F. Ferreira, M. Castelo-Branco, J. Dias, A hierarchical Bayesian framework for multimodal active perception, Adaptive Behavior 20 (3) (2012) 172–190.

[666] D.C. Knill, A. Pouget, The Bayesian brain: The role of uncertainty in neural coding and computation, Trends in Neurosciences 27 (12) (2004) 712–719.

[667] W.J. Ma, J.M. Beck, P.E. Latham, A. Pouget, Bayesian inference with probabilistic population codes, Nature Neuroscience 9 (11) (2006) 1432–1438.

[668] A. Huk, E. Hart, Parsing signal and noise in the brain, Science 364 (6437) (2019) 236–237.

[669] O. Deroy, C. Spence, U. Noppeney, Metacognition in multisensory perception, Trends in Cognitive Sciences 20 (10) (2016) 736–747.

[670] S.C. Yang, D.M. Wolpert, M. Lengyel, Theoretical perspectives on active sensing, Current Opinion in Behavioral Sciences 11 (2018) 100–108.

[671] M. Obrist, E. Gatti, E. Maggioni, C.T. Vi, C. Velasco, Multisensory experiences in HCI, IEEE MultiMedia 24 (2) (2017) 9–13.

[672] C. Bachhuber, E.G. Steinbach, A system for precise end-to-end delay measurements in video communication, CoRR, arXiv:1510.01134, 2015.

[673] R.S. Johansson, I. Birznieks, First spikes in ensembles of human tactile afferents code complex spatial fingertip events, Nature Neuroscience 7 (2) (2004) 170–177.

[674] Y. Sato, K. Aihara, A Bayesian model of sensory adaptation, PLOS One 6 (4) (2011) e19377, pp. 1–7.

[675] K.W. Latimer, D. Barbera, M. Sokoletsky, B. Awwad, Y. Katz, I. Nelken, I. Lampl, A.L. Fairhall, N.J. Priebe, Multiple timescales account for adaptive responses across sensory cortices, The Journal of Neuroscience 39 (50) (2019) 10019–10033.

[676] Z. Ghahramani, Probabilistic machine learning and artificial intelligence, Nature 521 (7553) (2015) 452–459.

[677] M.M. Murray, D.J. Lewkowicz, A. Amedi, M.T. Wallace, Multisensory processes: A balancing act across the lifespan, Trends in Neurosciences 39 (8) (2016) 567–579.

[678] A. Jucaite, H. Forssberg, P. Karlsson, C. Halldin, L. Farde, Age-related reduction in dopamine D1 receptors in the human brain: From late childhood to adulthood – A positron emission tomography study, Neuroscience 167 (1) (2010) 104–110.

[679] E.R. Sowell, B.S. Peterson, P.M. Thompson, S.E. Welcome, A.L. Henkenius, A.W. Toga, Mapping cortical change across the human life span, Nature Neuroscience 6 (3) (2003) 309–315.

[680] S. Passow, M. Müller, R. Westerhausen, K. Hugdahl, I. Wartenburger, H.R. Heekeren, U. Lindenberger, S.-C. Li, Development of attentional control of verbal auditory perception from middle to late childhood: Comparisons to healthy aging, Developmental Psychology 49 (10) (2013) 1982–1993.

[681] T.M. Dekker, H. Ban, B. van der Velde, M.I. Sereno, A.E. Welchman, M. Nardini, Late development of cue integration is linked to sensory fusion in cortex, Current Biology 25 (21) (2015) 2856–2861.

[682] B. Hommel, K.Z. Li, S.-C. Li, Visual search across the life span, Developmental Psychology 40 (4) (2004) 545–558.

[683] M. Gori, M. Del Viva, G. Sandini, D.C. Burr, Young children do not integrate visual and haptic form information, Current Biology 18 (9) (2008) 694–698.

[684] U. Lindenberger, P.B. Baltes, Sensory functioning and intelligence in old age: A strong connection, Psychology and Aging 9 (3) (1994) 339–355.

[685] P.B. Baltes, U. Lindenberger, Emergence of a powerful connection between sensory and cognitive functions across the adult life span: A new window to the study of cognitive aging?, Psychology and Aging 12 (1) (1997) 12–21.

[686] N. Deshpande, E.J. Metter, S. Ling, R. Conwit, L. Ferrucci, Physiological correlates of age-related decline in vibrotactile sensitivity, Neurobiology of Aging 29 (5) (2008) 765–773.

[687] N. Raz, U. Lindenberger, K.M. Rodrigue, K.M. Kennedy, D. Head, A. Williamson, C. Dahle, D. Gerstorf, J.D. Acker, Regional brain changes in aging healthy adults: General trends, individual differences and modifiers, Cerebral Cortex 15 (11) (2005) 1676–1689.

[688] C. Grady, Trends in neurocognitive aging, Nature Reviews Neuroscience 13 (7) (2012) 491–505.

[689] D. Servan-Schreiber, H. Printz, J.D. Cohen, A network model of catecholamine effects: Gain, signal-to-noise ratio, and behavior, Science 249 (4971) (1990) 892–895.

[690] S.-C. Li, U. Lindenberger, S. Sikström, Aging cognition: From neuromodulation to representation, Trends in Cognitive Sciences 5 (11) (2001) 479–486.

[691] S.-C. Li, T. von Oertzen, U. Lindenberger, A neurocomputational model of stochastic resonance and aging, Neurocomputing 69 (13–15) (2006) 1553–1560.

[692] S. Passow, F. Thurm, S.-C. Li, Activating developmental reserve capacity via cognitive training or non-invasive brain stimulation: Potentials for promoting fronto-parietal and hippocampal-striatal network functions in old age, Frontiers in Aging Neuroscience 9 (2017) 33, pp. 1–20.

[693] G. Papenberg, D. Hämmerer, V. Müller, U. Lindenberger, S.-C. Li, Lower theta inter-trial phase coherence during performance monitoring is related to higher reaction time variability: A lifespan study, NeuroImage 83 (2013) 912–920.

[694] S. Passow, R. Westerhausen, K. Hugdahl, I. Wartenburger, H.R. Heekeren, U. Lindenberger, S.-C. Li, Electrophysiological correlates of adult age differences in attentional control of auditory processing, Cerebral Cortex 24 (1) (2014) 249–260.

[695] V.S. Störmer, S.-C. Li, H.R. Heekeren, U. Lindenberger, Normal aging delays and compromises early multifocal visual attention during object tracking, Journal of Cognitive Neuroscience 25 (2) (2013) 188–202.

[696] D. Hämmerer, S.-C. Li, V. Müller, U. Lindenberger, An electrophysiological study of response conflict processing across the lifespan: Assessing the roles of conflict monitoring, cue utilization, response anticipation, and response suppression, Neuropsychologia 48 (11) (2010) 3305–3316.

[697] S.-C. Li, F. Schmiedek, O. Huxhold, C. Rocke, J. Smith, U. Lindenberger, Working memory plasticity in old age: Practice gain, transfer, and maintenance, Psychology and Aging 23 (4) (2008) 731–742.

[698] I.E. Nagel, C. Preuschhof, S.-C. Li, L. Nyberg, L. Backman, U. Lindenberger, H.R. Heekeren, Performance level modulates adult age differences in brain activation during spatial working memory, Proceedings of the National Academy of Sciences of the United States of America 106 (52) (2009) 22552–22557.

[699] D. Hämmerer, S.-C. Li, V. Müller, U. Lindenberger, Life span differences in electrophysiological correlates of monitoring gains and losses during probabilistic reinforcement learning, Journal of Cognitive Neuroscience 23 (3) (2011) 579–592.

[700] B. Eppinger, H.R. Heekeren, S.-C. Li, Age-related prefrontal impairments implicate deficient prediction of future reward in older adults, Neurobiology of Aging 36 (8) (2015) 2380–2390.

[701] D. Ruggles, H. Bharadwaj, B.G. Shinn-Cunningham, Normal hearing is not enough to guarantee robust encoding of suprathreshold features important in everyday communication, Proceedings of the National Academy of Sciences of the United States of America 108 (37) (2011) 15516–15521.

[702] S.-C. Li, S. Passow, W. Nietfeld, J. Schröder, L. Bertram, H.R. Heekeren, U. Lindenberger, Dopamine modulates attentional control of auditory perception: DARPP-32 (PPP1R1B) genotype effects on behavior and cortical evoked potentials, Neuropsychologia 51 (8) (2013) 1649–1661.

[703] A.J. Parker, Binocular depth perception and the cerebral cortex, Nature Reviews Neuroscience 8 (5) (2007) 379–391.

[704] S.J. Westerman, T. Cribbin, Individual differences in the use of depth cues: Implications for computer-and video-based tasks, Acta Psychologica 99 (3) (1998) 293–310.

[705] G.R. Fink, J.C. Marshall, P.W. Halligan, C.D. Frith, J. Driver, R.S.J. Frackowiak, R.J. Dolan, The neural consequences of conflict between intention and the senses, Brain 122 (3) (1999) 497–512.

[706] A.J.M. Foulkes, R.C. Miall, Adaptation to visual feedback delays in a human manual tracking task, Experimental Brain Research 131 (1) (2000) 101–110.

[707] C. Grefkes, A. Ritzl, K. Zilles, G.R. Fink, Human medial intraparietal cortex subserves visuomotor coordinate transformation, NeuroImage 23 (4) (2004) 1494–1506.

[708] J. Limanowski, E. Kirilina, F. Blankenburg, Neuronal correlates of continuous manual tracking under varying visual movement feedback in a virtual reality environment, NeuroImage 146 (2017) 81–89.

[709] S.T. Grafton, P. Schmitt, J. Van Horn, J. Diedrichsen, Neural substrates of visuomotor learning based on improved feedback control and prediction, NeuroImage 39 (3) (2008) 1383–1395.

[710] M.K. Rand, H. Heuer, Visual and proprioceptive recalibrations after exposure to a visuomotor rotation, European Journal of Neuroscience 50 (8) (2019) 3296–3310.

[711] K.J. Friston, J. Daunizeau, J. Kilner, S.J. Kiebel, Action and behavior: A free-energy formulation, Biological Cybernetics 102 (3) (2010) 227–260.

[712] K.J. Friston, R. Rosch, T. Parr, C. Price, H. Bowman, Deep temporal models and active inference, Neuroscience & Biobehavioral Reviews 90 (2018) 486–501.

[713] R. Shadmehr, J.W. Krakauer, A computational neuroanatomy for motor control, Experimental Brain Research 185 (3) (2008) 359–381.

[714] E. Todorov, M.I. Jordan, Optimal feedback control as a theory of motor coordination, Nature Neuroscience 5 (11) (2002) 1226–1235.

[715] S. Vijayakumar, T. Hospedales, A. Haith, Generative probabilistic modeling: Understanding causal sensorimotor integration, in: J. Trommershäuser, K. Kording, M.S. Landy (Eds.), Sensory Cue Integration, in: Oxford Series in Computational Neuroscience, vol. 3, Oxford University Press, 2011, pp. 63–81.

[716] M.E. Altinsoy, S. Merchel, Electrotactile feedback for handheld devices with touch screen and simulation of roughness, IEEE Transactions on Haptics 5 (1) (2011) 6–13.

[717] P. Bakardjiev, A. Richter, M.E. Altinsoy, U. Marschner, J. Troge, Dielectric elastomer loudspeaker driver, in: Proceedings of the International Conference on Electromechanically Active Polymer (EAP) Transducers & Artificial Muscles (EuroEAP), 2017.

[718] K. Klamka, R. Dachselt, IllumiPaper: Illuminated interactive paper, in: Proceedings of the ACM Conference on Human Factors in Computing Systems (CHI), 2017.

[719] A. Rand, The Romantic Manifesto, Penguin Books, 1975, pp. 34–44, chapter 3: Art and sense of life.

[720] S. Komurasaki, H. Kajimoto, H. Ishizuka, Fundamental perceptual characterization of an integrated tactile display with electrovibration and electrical stimuli, Micromachines 10 (5) (2019) 301, pp. 1–12.

[721] T.-H. Yang, S.-Y. Kim, C.H. Kim, D.-S. Kwon, W.J. Book, Development of a miniature pin-array tactile module using elastic and electromagnetic force for mobile devices, in: Proceedings of the IEEE World Haptics Conference (WHC), 2009.

[722] G.-H. Yang, K.-U. Kyung, M.A. Srinivasan, D.-S. Kwon, Quantitative tactile display device with pin-array type tactile feedback and thermal feedback, in: Proceedings of the IEEE International Conference on Robotics and Automation (ICRA), 2006.

[723] R. Velázquez, E.E. Pissaloux, M. Wiertlewski, A compact tactile display for the blind with shape memory alloys, in: Proceedings of the IEEE International Conference on Robotics and Automation (ICRA), 2006.

[724] G. Moy, C. Wagner, R.S. Fearing, A compliant tactile display for teletaction, in: Proceedings of the IEEE International Conference on Robotics and Automation (ICRA), 2000.

[725] K. Minamizawa, S. Fukamachi, H. Kajimoto, N. Kawakami, S. Tachi, Gravity grabber: Wearable haptic display to present virtual mass sensation, in: Proceedings of the ACM Special Interest Group on Computer Graphics and Interactive Techniques Conference (SIGGRAPH), 2007.

[726] S.B. Schorr, A.M. Okamura, Fingertip tactile devices for virtual object manipulation and exploration, in: Proceedings of the ACM Conference on Human Factors in Computing Systems (CHI), 2017.

[727] D. Prattichizzo, F. Chinello, C. Pacchierotti, M. Malvezzi, Towards wearability in fingertip haptics: A 3-DoF wearable device for cutaneous force feedback, IEEE Transactions on Haptics 6 (4) (2013) 506–516.

[728] M. Solazzi, A. Frisoli, M. Bergamasco, Design of a cutaneous fingertip display for improving haptic exploration of virtual objects, in: Proceedings of the IEEE International Symposium on Robot and Human Interactive Communication (RO-MAN), 2010.

[729] M. Gabardi, M. Solazzi, D. Leonardis, A. Frisoli, A new wearable fingertip haptic interface for the rendering of virtual shapes and surface features, in: Proceedings of the IEEE Haptics Symposium (HAPTICS), 2016.

[730] R. Scheibe, M. Moehring, B. Fröhlich, Tactile feedback at the finger tips for improved direct interaction in immersive environments, in: Proceedings of the IEEE Symposium on 3D User Interfaces (3DUI), 2007.

[731] D. De Rossi, F. Carpi, N. Carbonaro, A. Tognetti, E.P. Scilingo, Electroactive polymer patches for wearable haptic interfaces, in: Proceedings of the International Conference of the IEEE Engineering in Medicine and Biology Society (EMBC), 2011.

[732] S. Mun, S. Yun, S. Nam, S.K. Park, S. Park, B.J. Park, J.M. Lim, K.-U. Kyung, Electroactive polymer based soft tactile interface for wearable devices, IEEE Transactions on Haptics 11 (1) (2018) 15–21.

[733] T. Nara, M. Takasaki, S. Tachi, T. Higuchi, An application of SAW to a tactile display in virtual reality, in: Proceedings of the IEEE Ultrasonics Symposium (IUS), 2000.

[734] K. Tsukada, M. Yasumura, ActiveBelt: Belt-type wearable tactile display for directional navigation, in: N. Davies, E.D. Mynatt, I. Siio (Eds.), UbiComp 2004: Ubiquitous Computing, in: Lecture Notes in Computer Science, vol. 3205, Springer, 2004, pp. 384–399.

[735] J.B.F. van Erp, H.A.H.C. van Veen, A multi-purpose tactile vest for astronauts in the International Space Station, in: Proceedings of the Eurohaptics Conference (EuroHaptics), 2003.

[736] S. Schätzle, T. Ende, T. Wüsthoff, C. Preusche, VibroTac: An ergonomic and versatile usable vibrotactile feedback device, in: Proceedings of the IEEE International Symposium on Robot and Human Interactive Communication (RO-MAN), 2010.

[737] H. Culbertson, C.M. Nunez, A. Israr, F. Lau, F. Abnousi, A.M. Okamura, A social haptic device to create continuous lateral motion using sequential normal indentation, in: Proceedings of the IEEE Haptics Symposium (HAPTICS), 2018.

[738] J. Kühn, S. Haddadin, An artificial robot nervous system to teach robots how to feel pain and reflexively react to potentially damaging contacts, IEEE Robotics and Automation Letters 2 (1) (2016) 72–79.

[739] B. Henze, M.A. Roa, C. Ott, Passivity-based whole-body balancing for torque-controlled humanoid robots in multi-contact scenarios, The International Journal of Robotics Research 35 (12) (2016) 1522–1543.

[740] E. Häntzsche, R. Müller, M. Hübner, T. Ruder, R. Unger, A. Nocke, C. Cherif, Manufacturing technology of integrated textile-based sensor networks for in situ monitoring applications of composite wind turbine blades, Smart Materials and Structures 25 (10) (2016) 105012–105022.

[741] U. Vogel, B. Richter, O.R. Hild, P. Wartenberg, K. Fehse, M. Schober, S. Brenner, J. Baumgarten, P. König, B. Beyer, G. Bunk, S. Ulbricht, C. Schmidt, M. Jahnel, E. Bodenstein, S. Saager, C. Metzner, V. Kirchhoff, OLED microdisplays – Enabling advanced near-to-eye displays, sensors, and beyond, in: Proceedings of the SID's International Symposium Display Week, 2016 (invited paper).

[742] A. Silzle, H. Strauss, P. Novo, IKA-SIM: A system to generate auditory virtual environments, in: Proceedings of the Audio Engineering Society Convention, 2004.

[743] P.R. Cook, Real Sound Synthesis for Interactive Applications, CRC Press, 2002.

[744] R.S. Pellegrini, A Virtual Reference Listening Room as an Application of Auditory Virtual Environments, dissertation.de, 2002.

[745] J. Ahrens, R. Rabenstein, S. Spors, The theory of wave field synthesis revisited, in: Proceedings of the Audio Engineering Society Convention, 2008.

[746] G. Theile, H. Wittek, Wave field synthesis: A promising spatial audio rendering concept, Acoustical Science and Technology 25 (6) (2004) 393–399.

[747] D. de Vries, Microphone arrays for measurement and recording, in: Proceedings of the International Congress on Acoustics (ICA), 2007.

[748] H. Wittek, F. Rumsey, G. Theile, Perceptual enhancement of wavefield synthesis by stereophonic means, Journal of the Audio Engineering Society 55 (9) (2007) 723–751.

[749] I. Poupyrev, N.-W. Gong, S. Fukuhara, M.E. Karagozler, C. Schwesig, K.E. Robinson, Project Jacquard: Interactive digital textiles at scale, in: Proceedings of the ACM Conference on Human Factors in Computing Systems (CHI), 2016.

[750] L. Buechley, M. Eisenberg, J. Catchen, A. Crockett, The LilyPad Arduino: Using computational textiles to investigate engagement, aesthetics, and diversity in computer science education, in: Proceedings of the ACM Conference on Human Factors in Computing Systems (CHI), 2008.

[751] L. Buechley, M. Eisenberg, Fabric PCBs, electronic sequins, and socket buttons: Techniques for e-textile craft, Personal and Ubiquitous Computing 13 (2) (2009) 133–150.

[752] K.D.D. Willis, C. Xu, K.-J. Wu, G. Levin, M.D. Gross, Interactive fabrication: New interfaces for digital fabrication, in: Proceedings of the International Conference on Tangible, Embedded, and Embodied Interaction (TEI), 2010.

[753] P. Baudisch, S. Müller, Personal fabrication, Foundations and Trends in Human–Computer Interaction 10 (3–4) (2017) 165–293.

[754] N. Al-huda Hamdan, S. Völker, J. Borchers, Sketch&Stitch: Interactive embroidery for e-textiles, in: Proceedings of the ACM Conference on Human Factors in Computing Systems (CHI), 2018.

[755] A. Khan, J.S. Roo, T. Kraus, J. Steimle, Soft inkjet circuits: Rapid multi-material fabrication of soft circuits using a commodity inkjet printer, in: Proceedings of the ACM Symposium on User Interface Software and Technology (UIST), 2019.

[756] P. Strohmeier, J. Knibbe, S. Boring, K. Hornbæk, zPatch: Hybrid resistive/capacitive eTextile input, in: Proceedings of the International Conference on Tangible, Embedded, and Embodied Interaction (TEI), 2018.

[757] K. Salisbury, F. Conti, F. Barbagli, Haptic rendering: Introductory concepts, IEEE Computer Graphics and Applications 24 (2) (2004) 24–32.

[758] Y. Visell, Tactile sensory substitution: Models for enaction in HCI, Interacting with Computers 21 (1–2) (2008) 38–53.

[759] S. Schätzle, Entwicklung eines vibrotaktilen Armbands zur vielseitigen Informationsübermittlung in Mensch-Maschine-Systemen, PhD thesis, Leibniz Universität Hannover, Germany, 2018.

[760] A. Silzle, Generation of quality taxonomies for auditory virtual environments by means of systematic expert survey, Fortschritte der Akustik 33 (2) (2007) 869.

[761] J. Berg, F. Rumsey, Identification of quality attributes of spatial audio by repertory grid technique, Journal of the Audio Engineering Society 54 (5) (2006) 365–379.

[762] N. Zacharov, K. Koivuniemi, Unravelling the perception of spatial sound reproduction: Language development, verbal protocol analysis and listener training, in: Proceedings of the Audio Engineering Society Convention, 2001.

[763] S. Choisel, F. Wickelmaier, Extraction of auditory features and elicitation of attributes for the assessment of multichannel reproduced sound, Journal of the Audio Engineering Society 54 (9) (2006) 815–826.

[764] S. Bech, N. Zacharov, Perceptual Audio Evaluation – Theory, Method and Application, Wiley, 2007.

[765] S. George, Objective models for predicting selected multichannel audio quality attributes, PhD thesis, University of Surrey, United Kingdom, 2009.

[766] M.D. Husain, R. Kennon, T. Dias, Design and fabrication of temperature sensing fabric, Journal of Industrial Textiles 44 (3) (2014) 398–417.

[767] O. Atalay, W.R. Kennon, E. Demirok, Weft-knitted strain sensor for monitoring respiratory rate and its electro-mechanical modeling, IEEE Sensors Journal 15 (1) (2015) 110–122.

[768] N. Baribina, A. Oks, I. Baltina, P. Eizentals, Comparative analysis of knitted pressure sensors, in: Proceedings of the International Scientific Conference "Engineering for Rural Development", 2018.

[769] C. Gonçalves, A.F. da Silva, J. Gomes, R. Simoes, Wearable e-textile technologies: A review on sensors, actuators and control elements, Inventions 3 (1) (2018) 14, pp. 1–13.

[770] A.-S. Augurelle, A.M. Smith, T. Lejeune, J.-L. Thonnard, Importance of cutaneous feedback in maintaining a secure grip during manipulation of hand-held objects, Journal of Neurophysiology 89 (2) (2003) 665–671.

[771] T. Hulin, M. Rothammer, I. Tannert, S.S. Giri, B. Pleintinger, H. Singh, B. Weber, C. Ott, FingerTac – A wearable tactile thimble for mobile haptic augmented reality applications, in: Proceedings of the International Conference on Human–Computer Interaction (HCI), 2020.

[772] H. Singh, B. Suthar, S.Z. Mehdi, J.-H. Ryu, Ferro-fluid based portable fingertip haptic display and its preliminary experimental evaluation, in: Proceedings of the IEEE Haptics Symposium (HAPTICS), 2018.

[773] H. Uoyama, K. Goushi, K. Shizu, H. Nomura, C. Adachi, Highly efficient organic light-emitting diodes from delayed fluorescence, Nature 492 (7428) (2012) 234–238.

[774] H. Nakanotani, T. Higuchi, T. Furukawa, K. Masui, K. Morimoto, M. Numata, H. Tanaka, Y. Sagara, T. Yasuda, C. Adachi, High-efficiency organic light-emitting diodes with fluorescent emitters, Nature Communications 5 (2014) 4016, pp. 1–7.

[775] T. Wühle, S. Merchel, M.E. Altinsoy, Verfahren zur Beeinflussung einer auditiven Richtungswahrnehmung eines Hörers, German patent, DE 10 2018 108 852.3, 2019.

[776] K. Klamka, T. Horak, R. Dachselt, Watch+Strap: Extending smartwatches with interactive StrapDisplays, in: Proceedings of the ACM Conference on Human Factors in Computing Systems (CHI), 2020.

[777] K. Klamka, R. Dachselt, The future role of visual feedback for unobtrusive E-Textile interfaces, in: Proceedings of the ACM Conference on Human Factors in Computing Systems (CHI), Workshop "(Un)Acceptable!?! – Re-thinking the Social Acceptability of Emerging Technologies", 2018.

[778] K. Klamka, R. Dachselt, ARCord: Visually augmented interactive cords for mobile interaction, in: Proceedings of the ACM Conference on Human Factors in Computing Systems (CHI), Extended Abstracts, 2018.

[779] K. Klamka, P. Reipschläger, R. Dachselt, CHARM: Cord-based haptic augmented reality manipulation, in: J.Y.C. Chen, G. Fragomeni (Eds.), Virtual, Augmented and Mixed Reality – Multimodal Interaction, in: Lecture Notes in Computer Science, vol. 11574, Springer, 2019, pp. 96–114.

[780] A. Peetz, K. Klamka, R. Dachselt, BodyHub: A reconfigurable wearable system for clothing, in: Adjunct Publication of the ACM Symposium on User Interface Software and Technology (UIST), 2019.

[781] S.L. Star, J.R. Griesemer, Institutional ecology, 'translations' and boundary objects: Amateurs and professionals in Berkeley's Museum of Vertebrate Zoology, 1907–39, Social Studies of Science 19 (3) (1989) 387–420.

[782] P. Naghshtabrizi, J.P. Hespanha, Implementation considerations for wireless networked control systems, in: S.K. Mazumder (Ed.), Wireless Networking Based Control, Springer, 2011, pp. 1–27.

[783] G. Walsh, H. Ye, Scheduling of networked control systems, IEEE Control Systems Magazine 21 (1) (2001) 57–65.

[784] D. Carnevale, A. Teel, D. Nešić, A Lyapunov proof of an improved maximum allowable transfer interval for networked systems, IEEE Transactions on Automatic Control 52 (5) (2007) 892–897.

[785] W. Heemels, A. Teel, N. van de Wouw, D. Nešić, Networked control systems with communication constraints: Tradeoffs between transmission intervals, delays and performance, IEEE Transactions on Automatic Control 55 (8) (2010) 1781–1796.

[786] M. Mahmoudu, Control and Estimation Methods over Communication Networks, Springer, 2014.

[787] E. Peters, D. Marelli, D. Quevedo, M. Fu, Predictive control for networked systems affected by correlated packet loss, International Journal of Robust and Nonlinear Control 29 (15) (2019) 5078–5094.

[788] D. Quevedo, D. Nešić, Input-to-state stability of packetized predictive control over unreliable networks affected by packet-dropouts, IEEE Transactions on Automatic Control 56 (2) (2011) 370–375.

[789] X. Tong, G. Zhao, M.A. Imran, Z. Pang, Z. Chen, Minimizing wireless resource consumption for packetized predictive control in real-time cyber physical systems, in: Proceedings of the IEEE International Conference on Communications (ICC), Workshops, 2018.

[790] M. Lemmon, Event-triggered feedback in control, estimation, and optimization, in: A. Bemporad, M. Heemels, M. Johansson (Eds.), Networked Control Systems, in: Lecture Notes in Control and Information Sciences, vol. 406, Springer, 2010, pp. 293–358.

[791] A. González, A. Villamil, N. Franchi, G.P. Fettweis, String stable CACC under LTE-V2V mode 3: Scheduling periods and transmission delays, in: Proceedings of the IEEE 5G World Forum (5G-WF), 2019.

[792] A. Villamil, A. González, N. Franchi, G.P. Fettweis, Observer-based packet drop mitigation for string stable CACC, in: Proceedings of the International Conference on Control and Robot Technology (ICCRT), 2019.

[793] A. Zappone, E. Björnson, L. Sanguinetti, E. Jorswieck, Globally optimal energy-efficient power control and receiver design in wireless networks, IEEE Transactions on Signal Processing 65 (11) (2017) 2844–2859.

[794] M. Butt, E. Jorswieck, A. Mohamed, Energy and bursty packet loss tradeoff over fading channels: A system-level model, IEEE Systems Journal 12 (1) (2018) 527–538.

[795] C. Sun, E. Jorswieck, Y. Yuan, Sum rate maximization for non-regenerative MIMO relay networks, IEEE Transactions on Signal Processing 64 (24) (2016) 6392–6405.

[796] E.J. Candès, J.K. Romberg, T. Tao, Robust uncertainty principles: Exact signal reconstruction from highly incomplete frequency information, IEEE Transactions on Information Theory 52 (2) (2006) 489–509.

[797] D.L. Donoho, Compressed sensing, IEEE Transactions on Information Theory 52 (4) (2006) 1289–1306.

[798] H. Boche, R. Calderbank, G. Kutyniok, J. Vybíral, A survey of compressed sensing, in: H. Boche, R. Calderbank, G. Kutyniok, J. Vybíral (Eds.), Applied and Numerical Harmonic Analysis, in: Compressed Sensing and Its Applications, vol. 67, Springer, 2013, pp. 1–39.

[799] E.J. Candès, T. Tao, The Dantzig selector: Statistical estimation when p is much larger than n, The Annals of Statistics 35 (6) (2007).

[800] R. Prony, Essai expérimental et analytique: Sur les lois de la dilatabilité des fluides élastiques et sur celles de la force expansive de la vapeur de l'eau et de la vapeur de l'alkool, à différentes températures, Journal de l'École Polytechnique Floréal et Plairial 1 (22) (1795) 24–76.

[801] S.L. Marple Jr., Digital Spectral Analysis with Applications, Prentice Hall, 1987.

[802] B.F. Logan, Properties of high-pass signals, PhD thesis, Columbia University, USA, 1965.

[803] L.I. Rudin, S. Osher, E. Fatemi, Nonlinear total variation based noise removal algorithms, Physica D: Nonlinear Phenomena 60 (1–4) (1992) 259–268.

[804] D.L. Donoho, B.F. Logan, Signal recovery and the large sieve, SIAM Journal on Applied Mathematics 52 (2) (1992) 577–591.

[805] R. Tibshirani, Regression shrinkage and selection via the lasso, Journal of the Royal Statistical Society: Series B 58 (1) (1996) 267–288.

[806] S.S. Chen, D.L. Donoho, M.A. Saunders, Atomic decomposition by basis pursuit, SIAM Review 43 (1) (2001) 129–159.

[807] S.G. Mallat, Z. Zhong, Matching pursuits with time-frequency dictionaries, IEEE Transactions on Signal Processing 41 (12) (1993) 3397–3415.

[808] B.K. Natarajan, Sparse approximate solutions to linear systems, SIAM Journal on Computing 24 (2) (1995) 227–234.

[809] E.J. Candès, J.K. Romberg, T. Tao, Stable signal recovery from incomplete and inaccurate measurements, Communications on Pure and Applied Mathematics 59 (8) (2006) 1207–1223.

[810] G. Gasso, A. Rakotomamonjy, S. Canu, Recovering sparse signals with a certain family of non-convex penalties and DC programming, IEEE Transactions on Signal Processing 57 (12) (2009) 4686–4698.

[811] O. Christensen, An Introduction to Frames and Riesz Bases, Birkhäuser, 2003.

[812] M. Elad, Sparse and Redundant Representations – From Theory to Applications in Signal and Image Processing, Springer, 2010.

[813] S.D. Babacan, R. Molina, A.K. Katsaggelos, Variational Bayesian blind deconvolution using a total variation prior, IEEE Transactions on Image Processing 18 (1) (2009) 12–26.

[814] N.P. Galatsanos, A.K. Katsaggelos, Methods for choosing the regularization parameter and estimating the noise variance in image restoration and their relation, IEEE Transactions on Image Processing 1 (3) (1992) 322–336.

[815] C. Hansen, Analysis of discrete ill-posed problems by means of the L-curve, SIAM Review 34 (4) (1992) 561–580.

[816] V.A. Morozov, Methods for Solving Incorrectly Posed Problems, Springer, 1984.

[817] M.G. Shirangi, History matching production data and uncertainty assessment with an efficient TSVD parameterization algorithm, Journal of Petroleum Science and Engineering 113 (2014) 54–71.

[818] D. Needell, R. Vershynin, Uniform uncertainty principle and signal recovery via regularized orthogonal matching pursuit, Foundations of Computational Mathematics 9 (3) (2009) 317–334.

[819] Y.C. Pati, R. Rezaiifar, P.S. Krishnaprasad, Orthogonal matching pursuit: Recursive function approximation with applications to wavelet decomposition, in: Proceedings of the Asilomar Conference on Signals, Systems, and Computers (ACSSC), 1993.

[820] S. Foucart, H. Rauhut, A Mathematical Introduction to Compressive Sensing, Birkhäuser, 2013.

[821] P. Wojtaszczyk, Stability and instance optimality for Gaussian measurements in compressed sensing, Foundations of Computational Mathematics 10 (1) (2010) 1–13.

[822] B. Recht, M. Fazel, P.A. Parrilo, Guaranteed minimum-rank solutions of linear matrix equations via nuclear norm minimization, SIAM Review 52 (3) (2010) 471–501.

[823] E.J. Candès, Y. Plan, Tight oracle bounds for low-rank matrix recovery from a minimal number of random measurements, CoRR, arXiv:1001.0339, 2010.

[824] P. Jain, R. Meka, I.S. Dhillon, Guaranteed rank minimization via singular value projection, in: J.D. Lafferty, C.K.I. Williams, J. Shawe-Taylor, R.S. Zemel, A. Culotta (Eds.), Advances in Neural Information Processing Systems, vol. 23, Curran Associates, 2010, pp. 937–945.

[825] M.-J. Lai, W. Yin, Augmented ℓ_1 and nuclear-norm models with a globally linearly convergent algorithm, SIAM Journal on Imaging Sciences 6 (2) (2013) 1059–1091.

[826] K. Mohan, M. Fazel, New restricted isometry results for noisy low-rank recovery, in: Proceedings of the IEEE International Symposium on Information Theory (ISIT), 2010.

[827] S. Dasgupta, A. Gupta, An elementary proof of a theorem of Johnson and Lindenstrauss, Random Structures & Algorithms 22 (1) (2003) 60–65.

[828] B. Recht, W. Xu, B. Hassibi, Null space conditions and thresholds for rank minimization, Mathematical Programming 127 (1) (2011) 175–202.

[829] B. Bah, J. Tanner, Improved bounds on restricted isometry constants for Gaussian matrices, SIAM Journal on Matrix Analysis and Applications 31 (5) (2010) 2882–2898.

[830] E.J. Candès, Compressive sampling, in: Proceedings of the International Congress of Mathematicians (ICM), 2006.

[831] D. Achlioptas, Database-friendly random projections: Johnson–Lindenstrauss with binary coins, Journal of Computer and System Sciences 66 (4) (2003) 671–687.

[832] W.B. Johnson, J. Lindenstrauss, Extensions of Lipschitz mappings into a Hilbert space, Contemporary Mathematics 26 (1984) 189–206.

[833] R. Baraniuk, M. Davenport, R. DeVore, M. Wakin, A simple proof of the restricted isometry property for random matrices, Constructive Approximation 28 (3) (2008) 253–263.

[834] F. Krahmer, R.A. Ward, New and improved Johnson–Lindenstrauss embeddings via the restricted isometry property, SIAM Journal on Mathematical Analysis 43 (3) (2011) 1269–1281.

[835] H. Rauhut, Compressive sensing and structured random matrices, in: M. Fornasier (Ed.), Theoretical Foundations and Numerical Methods for Sparse Recovery, in: Radon Series on Computational and Applied Mathematics, vol. 9, De Gruyter, 2010, pp. 1–92.

[836] M. Rudelson, R. Vershynin, On sparse reconstruction from Fourier and Gaussian measurements, Communications on Pure and Applied Mathematics 61 (8) (2008) 1025–1045.

[837] F. Krahmer, S. Mendelson, H. Rauhut, Suprema of chaos processes and the restricted isometry property, Communications on Pure and Applied Mathematics 67 (11) (2014) 1877–1904.

[838] H. Rauhut, J. Romberg, J.A. Tropp, Restricted isometries for partial random circulant matrices, Applied and Computational Harmonic Analysis 32 (2) (2012) 242–254.

[839] A. Eftekhari, H.L. Yap, C.J. Rozell, M.B. Wakin, The restricted isometry property for random block diagonal matrices, CoRR, arXiv:1210.3395, 2012.

[840] Y. Zhang, Theory of compressive sensing via ℓ_1-minimization: A non-RIP analysis and extensions, Journal of the Operations Research Society of China 1 (1) (2013) 79–105.

[841] S. Negahban, M.J. Wainwright, Estimation of (near) low-rank matrices with noise and high-dimensional scaling, The Annals of Statistics 39 (2) (2011) 1069–1097.

[842] S. Negahban, B. Yu, M.J. Wainwright, P.K. Ravikumar, A unified framework for high-dimensional analysis of M-estimators with decomposable regularizers, Statistical Science 27 (4) (2012) 538–557.

[843] K.R. Davidson, S.J. Szarek, Local operator theory, random matrices and Banach spaces, in: W.B. Johnson, J. Lindenstrauss (Eds.), Handbook of the Geometry of Banach Spaces, vol. 1, Elsevier, 2001, pp. 317–366.

[844] M. Ledoux, Deviation inequalities on largest eigenvalues, in: V.D. Milman, G. Schechtman (Eds.), Geometric Aspects of Functional Analysis, in: Lecture Notes in Mathematics, vol. 1910, Springer, 2007, pp. 167–219.

[845] R. Qiu, M. Wicks, Cognitive Networked Sensing and Big Data, Springer, 2014.

[846] G. Guillot, R.L. Schilling, E. Porcu, M. Bevilacqua, Validity of covariance models for the analysis of geographical variation, Methods in Ecology and Evolution 5 (4) (2014) 329–335.

[847] S.L.H. Nguyen, A. Ghrayeb, Compressive sensing-based channel estimation for massive multiuser MIMO systems, in: Proceedings of the IEEE Wireless Communications and Networking Conference (WCNC), 2013.

[848] T. Ho, R. Kötter, M. Médard, D.R. Karger, M. Effros, The benefits of coding over routing in a randomized setting, in: Proceedings of the IEEE International Symposium on Information Theory (ISIT), 2003.

[849] F. Gabriel, J. Rischke, F.H.P. Fitzek, M. Mühleisen, T. Lohmar, No plan survives contact with the enemy: On gains of coded multipath over MPTCP in dynamic settings, in: Proceedings of the IEEE Wireless Communications and Networking Conference (WCNC), 2019.

[850] A. Grohmann, F. Gabriel, S. Zimmermann, F.H.P. Fitzek, SourceShift: Resilient routing in highly dynamic wireless mesh networks, in: Proceedings of the IEEE Wireless Communications and Networking Conference (WCNC), 2020.

[851] S. Pandi, F. Gabriel, O. Zhdanenko, S. Wunderlich, F.H.P. Fitzek, MESHMERIZE: An interactive demo of resilient mesh networks in drones, in: Proceedings of the IEEE Consumer Communications & Networking Conference (CCNC), 2019.

[852] M. Taghouti, M. Tömösközi, T. Waurick, A.K. Chorppath, F.H.P. Fitzek, On the joint design of compressed sensing and network coding for wireless communications, Transactions on Emerging Telecommunications Technologies (2019) e3645, pp. 1–17 (early view article).

[853] C.E. Shannon, A mathematical theory of communication, Bell System Technical Journal 27 (3) (1948) 379–423.

[854] D. Slepian, J. Wolf, Noiseless coding of correlated information sources, IEEE Transactions on Information Theory 19 (4) (1973) 471–480.

[855] A. Orlitsky, J. Roche, Coding for computing, in: Proceedings of the IEEE FOCS, 1995.

[856] R. Ahlswede, J. Körner, Source coding with side information and a converse for degraded broadcast channels, IEEE Transactions on Information Theory 21 (6) (1975) 629–637.

[857] J. Körner, K. Marton, How to encode the modulo-two sum of binary sources, IEEE Transactions on Information Theory 25 (2) (1979) 219–221.

[858] V. Doshi, D. Shah, M. Médard, M. Effros, Functional compression through graph coloring, IEEE Transactions on Information Theory 56 (8) (2010) 3901–3917.

[859] S. Feizi, M. Médard, On network functional compression, IEEE Transactions on Information Theory 60 (9) (2014) 5387–5401.

[860] S. Feizi, M. Médard, When do only sources need to compute? On functional compression in tree networks, in: Proceedings of the Allerton Conference on Communication, Control, and Computing (Allerton), 2009.

[861] J. Jornet, I. Akyildiz, Channel modeling and capacity analysis for electromagnetic wireless nanonetworks in the terahertz band, IEEE Transactions on Wireless Communications 10 (10) (2011) 3211–3221.

[862] E. Larsson, O. Edfors, F. Tufvesson, T. Marzetta, Massive MIMO for next generation wireless systems, IEEE Communications Magazine 52 (2) (2014) 186–195.

[863] L. Landau, M. Dörpinghaus, G.P. Fettweis, 1-bit quantization and oversampling at the receiver: Communication over bandlimited channels with noise, IEEE Communications Letters 21 (5) (2017) 1007–1010.

[864] M.-T. Suer, C. Thein, H. Tchouankem, L. Wolf, Multi-connectivity as an enabler for reliable low latency communications – An overview, IEEE Communications Surveys & Tutorials 22 (1) (2020) 156–169.

[865] A. Traßl, L. Scheuvens, T. Hößler, E. Schmitt, N. Franchi, G.P. Fettweis, Outage prediction for URLLC in Rayleigh fading, in: Proceedings of the European Conference on Networks and Communications (EuCNC), 2020.

[866] R. Ahlswede, G. Dueck, Identification via channels, IEEE Transactions on Information Theory 35 (1) (1989) 15–29.

[867] R. Ahlswede, G. Dueck, Identification in the presence of feedback – A discovery of new capacity formulas, IEEE Transactions on Information Theory 35 (1) (1989) 30–36.

[868] H. Boche, C. Deppe, Secure identification for wiretap channels; robustness, superadditivity and continuity, IEEE Transactions on Information Forensics and Security 13 (7) (2018) 1641–1655.

[869] H. Boche, R.F. Schaefer, H.V. Poor, Identification capacity of correlation-assisted discrete memoryless channels: Analytical properties and representations, in: Proceedings of the IEEE International Symposium on Information Theory (ISIT), 2019.

[870] F.H.P. Fitzek, The medium is the message, in: Proceedings of the IEEE International Conference on Communications (ICC), 2006.

[871] X. Zhou, P. Kyritsi, P.C.F. Eggers, F.H.P. Fitzek, The medium is the message: Secure communication via waveform coding in MIMO systems, in: Proceedings of the IEEE Vehicular Technology Conference (VTC Spring), 2007.

[872] H. Kimble, The Quantum Internet, Nature 453 (7198) (2008) 1023–1030.

[873] S. Wehner, D. Elkouss, R. Hanson, Quantum Internet: A vision for the road ahead, Science 362 (6412) (2018) eaam9288, pp. 1–9.

[874] A.S. Cacciapuoti, M. Caleffi, F. Tafuri, F.S. Cataliotti, S. Gherardini, G. Bianchi, Quantum Internet: Networking challenges in distributed quantum computing, IEEE Network 34 (1) (2020) 137–143.

[875] K. Duffy, J. Li, M. Médard, Capacity-achieving guessing random additive noise decoding, IEEE Transactions on Information Theory 65 (7) (2019) 4023–4040.

[876] K. Duffy, M. Médard, Guessing random additive noise decoding with soft detection symbol reliability information SGRAND, in: Proceedings of the IEEE International Symposium on Information Theory (ISIT), 2019.

[877] P. Rodríguez-Vázquez, J. Grzyb, N. Sarmah, B. Heinemann, U.R. Pfeiffer, A 65 Gbps QPSK one meter wireless link operating at a 225–255 GHz tunable carrier in a SiGe HBT technology, in: Proceedings of the IEEE Radio and Wireless Symposium (RWS), 2018.

[878] D. Fritsche, P. Stärke, C. Carta, F. Ellinger, A low-power SiGe BiCMOS 190 GHz transceiver chipset with demonstrated data rates up to 50 Gbit/s using on-chip antennas, IEEE Transactions on Microwave Theory and Techniques 65 (9) (2017) 3312–3323.

[879] M. Taghivand, Y. Rajavi, K. Aggarwal, A.S.Y. Poon, An energy harvesting 2×2 60 GHz transceiver with scalable data rate of 38-to-2450 Mb/s for near range communication, in: Proceedings of the IEEE Custom Integrated Circuits Conference (CICC), 2014.

[880] A. Siligaris, F. Chaix, M. Pelissier, V. Puyal, J. Zevallos, L. Dussopt, P. Vincent, A low power 60 GHz 2.2 Gbps UWB transceiver with integrated antennas for short range communications, in: Proceedings of the IEEE Radio Frequency Integrated Circuits Symposium (RFIC), 2013.

[881] Y. Wang, B. Liu, R. Wu, H. Liu, A.T. Narayanan, J. Pang, N. Li, T. Yoshioka, Y. Terashima, H. Zhang, D. Tang, M. Katsuragi, D. Lee, S. Choi, K. Okada, A. Matsuzawa, A 60 GHz 3.0 Gb/s spectrum efficient BPOOK transceiver for low-power short-range wireless in 65 nm CMOS, IEEE Journal of Solid-State Circuits 54 (5) (2019) 1363–1374.

[882] C.W. Byeon, C.H. Yoon, C.S. Park, A 67 mW 10.7 Gb/s 60 GHz OOK CMOS transceiver for short-range wireless communications, IEEE Transactions on Microwave Theory and Techniques 61 (9) (2013) 3391–3401.

[883] A. Mustapha, D. Cracan, R. Gadhafi, M. Sanduleanu, A V-band transceiver with integrated resonator and receiver/transmitter antenna for near-field IoT, IEEE Transactions on Circuits and Systems II: Express Briefs 65 (10) (2018) 1300–1304.

[884] N. Joram, J. Wagner, A. Strobel, F. Ellinger, Distance measurement and time synchronization using frequency modulated continuous wave (FMCW) radar, Embedded Projects Journal 15 (2012) 26–31.

[885] H. Wang, Y. Huang, S. Chung, Spatial diversity 24 GHz FMCW radar with ground effect compensation for automotive applications, IEEE Transactions on Vehicular Technology 66 (2) (2017) 965–973.

[886] R. Lee, Z. Tsai, C. Lang, C. Chang, S. Chang, A switched-beam FMCW radar for wireless indoor positioning system, in: Proceedings of the European Radar Conference (EuRAD), 2011.

[887] S. Wehrli, R. Gierlich, J. Huttner, D. Barras, F. Ellinger, H. Jäckel, Integrated active pulsed reflector for an indoor local positioning system, IEEE Transactions on Microwave Theory and Techniques 58 (2) (2010) 267–276.

[888] M. Park, J. Park, U. Jeong, Design of conductive composite elastomers for stretchable electronics, Nano Today 9 (2) (2014) 244–260.

[889] D. Brosteaux, F. Axisa, M. Gonzalez, J. Vanfleteren, Design and fabrication of elastic interconnections for stretchable electronic circuits, IEEE Electron Device Letters 28 (7) (2007) 552–554.

[890] M. Gonzalez, F. Axisa, M.V. Bulcke, D. Brosteaux, B. Vandevelde, J. Vanfleteren, Design of metal interconnects for stretchable electronic circuits, Microelectronics Reliability 48 (6) (2008) 825–832.

[891] M. Schubert, L. Rebohle, Y. Wang, M. Fritsch, K. Bock, M. Vinnichenko, T. Schumann, Evaluation of nanoparticle inks on flexible and stretchable substrates for biocompatible application, in: Proceedings of the Electronic System-Integration Technology Conference (ESTC), 2018.

[892] M.S. Sarwar, Y. Dobashi, C. Preston, J.K.M. Wyss, S. Mirabbasi, J.D.W. Madden, Bend, stretch, and touch: Locating a finger on an actively deformed transparent sensor array, Science Advances 3 (3) (2017) e1602200, pp. 1–8.

[893] R.D.P. Wong, J.D. Posner, V.J. Santos, Flexible microfluidic normal force sensor skin for tactile feedback, Sensors and Actuators A: Physical 179 (2012) 62–69.

[894] T. Linz, C. Kallmayer, R. Aschenbrenner, H. Reichl, Embroidering electrical interconnects with conductive yarn for the integration of flexible electronic modules into fabric, in: Proceedings of the IEEE International Symposium on Wearable Computers (ISWC), 2005.

[895] D. Ernst, M. Faghih, R. Liebfried, M. Melzer, D. Karnaushenko, W. Hofmann, O.G. Schmidt, T. Zerna, Packaging of ultrathin flexible magnetic field sensors with polyimide interposer and integration in an active magnetic bearing, IEEE Transactions on Components, Packaging and Manufacturing Technology 10 (1) (2020) 39–43.

[896] N. Palavesam, D. Bonfert, W.W. Hell, C. Landesberger, H. Gieser, C. Kutter, K. Bock, Mechanical reliability analysis of ultra-thin chip-on-foil assemblies under different types of recurrent bending, in: Proceedings of the IEEE Electronic Components and Technology Conference (ECTC), 2016.

[897] N. Palavesam, E. Yacoub-George, W. Hell, C. Landesberger, C. Kutter, K. Bock, Dynamic bending reliability analysis of flexible hybrid integrated chip-foil packages, in: Proceedings of the IEEE Electronics Packaging Technology Conference (EPTC), 2018.

[898] M. Schubert, J. Rasche, M.-M. Laurila, T. Vuorinen, M. Mäntysalo, K. Bock, Printed flexible microelectrode for application of nanosecond pulsed electric fields on cells, Materials 12 (17) (2019) 2713, pp. 1–13.

[899] S. Höppner, Y. Yan, B. Vogginger, A. Dixius, J. Partzsch, F. Neumärker, S. Hartmann, S. Schiefer, S. Scholze, G. Ellguth, L. Cederstroem, M. Eberlein, C. Mayr, S. Temple, L. Plana, L. Garside, S. Davison, D.R. Lester, S. Furber, Dynamic voltage and frequency scaling for neuromorphic many-core systems, in: Proceedings of the IEEE International Symposium on Circuits and Systems (ISCAS), 2017.

[900] B. Moons, R. Uytterhoeven, W. Dehaene, M. Verhelst, 14.5 Envision: A 0.26-to-10 TOPS/W subword-parallel dynamic-voltage-accuracy-frequency-scalable convolutional neural network processor in 28 nm FDSOI, in: Proceedings of the IEEE International Solid-State Circuits Conference (ISSCC), 2017.

[901] Z. Yuan, J. Yue, H. Yang, Z. Wang, J. Li, Y. Yang, Q. Guo, X. Li, M. Chang, H. Yang, Y. Liu, Sticker: A 0.41–62.1 TOPS/W 8 bit neural network processor with multi-sparsity compatible convolution arrays and online tuning acceleration for fully connected layers, in: Proceedings of the IEEE Symposium on VLSI Circuits (VLSIC), 2018.

[902] G. Bellec, D. Kappel, W. Maass, R. Legenstein, Deep rewiring: Training very sparse deep networks, in: Proceedings of the International Conference on Learning Representations (ICLR), 2018.

[903] D. Göhringer, M. Hübner, E.N. Zeutebouo, J. Becker, Operating system for runtime reconfigurable multiprocessor systems, International Journal of Reconfigurable Computing 2011 (2011) 121353, pp. 1–16.

[904] E. Juntunen, D. Dawn, S. Pinel, J. Laskar, A high-efficiency, high-power millimeter-wave oscillator using a feedback class E power amplifier in 45 nm CMOS, IEEE Microwave and Wireless Components Letters 21 (8) (2011) 430–432.

[905] X. An, J. Wagner, F. Ellinger, An efficient ultra-wideband pulse transmitter with automatic on-off functionality for primary radar systems, IEEE Microwave and Wireless Components Letters 30 (4) (2020) 449–452.

[906] R. Tetzlaff, V. Senger, The seizure prediction problem in epilepsy: Cellular nonlinear networks, IEEE Circuits and Systems Magazine 12 (4) (2012) 8–20.

[907] M. Eberlein, R. Hildebrand, R. Tetzlaff, N. Hoffmann, L. Kuhlmann, B. Brinkmann, J. Müller, Convolutional neural networks for epileptic seizure prediction, in: Proceedings of the IEEE International Conference Bioinformatics and Biomedicine (BIBM), 2018.

[908] M. Eberlein, J. Müller, H. Yang, S. Walz, J. Schreiber, R. Tetzlaff, S. Creutz, O. Uckermann, G. Leonhardt, Evaluation of machine learning methods for seizure prediction in epilepsy, Current Directions in Biomedical Engineering 5 (1) (2019) 109–112.

[909] C.W. Fetzer, Building critical applications using microservices, IEEE Security & Privacy 14 (6) (2016) 86–89.

[910] M. Wolf, D. Serpanos, Safety and security in cyber-physical systems and Internet-of-Things systems, Proceedings of the IEEE 106 (1) (2018) 9–20.

[911] K. Murdock, D. Oswald, F.D. Garcia, J. Van Bulck, D. Gruss, F. Piessens, Plundervolt: Software-based fault injection attacks against Intel SGX, in: Proceedings of the IEEE Symposium on Security and Privacy (S&P), 2020.

[912] V. Costan, S. Devadas, Intel SGX explained, Cryptology ePrint Archive, Report 2016/086, 2016.

[913] C.C. Tsai, D.E. Porter, M. Vij, Graphene-SGX: A practical library OS for unmodified applications on SGX, in: Proceedings of the USENIX Annual Technical Conference (USENIX ATC), 2017.

[914] S. Arnautov, B. Trach, F. Gregor, T. Knauth, A. Martin, C. Priebe, J. Lind, D. Muthukumaran, D. O'Keeffe, M.L. Stillwell, D. Goltzsche, D. Eyers, R. Kapitza, P. Pietzuch, C.W. Fetzer, SCONE: Secure Linux containers with Intel SGX, in: Proceedings of the USENIX Symposium on Operating Systems Design and Implementation (OSDI), 2016.

[915] D. Kuvaiskii, C.W. Fetzer, δ-encoding: Practical encoded processing, in: Proceedings of the IEEE/IFIP International Conference on Dependable Systems and Networks (DSN), 2015.

[916] D. Kuvaiskii, R. Faqeh, P. Bhatotia, P. Felber, C.W. Fetzer, HAFT: Hardware-assisted fault tolerance, in: Proceedings of the European Conference on Computer Systems (EuroSys), 2016.

[917] M. Bohman, B. James, M.J. Wirthlin, H. Quinn, J. Goeders, Microcontroller compiler-assisted software fault tolerance, IEEE Transactions on Nuclear Science 66 (1) (2019) 223–232.

[918] C. Chen, G. Eisenhauer, S. Pande, Q. Guan, CARE: Compiler-assisted recovery from soft failures, in: Proceedings of the International Conference for High Performance Computing, Networking, Storage and Analysis (SC), 2019.

[919] C. Chen, G. Eisenhauer, M. Wolf, S. Pande, LADR: Low-cost application-level detector for reducing silent output corruptions, in: Proceedings of the International Symposium on High-Performance Parallel and Distributed Computing (HPDC), 2018.

[920] P. Jagtap, F. Abdi, M. Rungger, M. Zamani, M. Caccamo, Software fault tolerance for cyber-physical systems via full system restart, CoRR, arXiv:1812.03546, 2018.

[921] Y. Shen, G. Heiser, K. Elphinstone, Fault tolerance through redundant execution on COTS multicores: Exploring trade-offs, in: Proceedings of the IEEE/IFIP International Conference on Dependable Systems and Networks (DSN), 2019.

[922] K. Elphinstone, Y. Shen, Increasing the trustworthiness of commodity hardware through software, in: Proceedings of the IEEE/IFIP International Conference on Dependable Systems and Networks (DSN), 2013.

[923] A. Osman, S. Hanisch, T. Strufe, SeCoNetBench: A modular framework for secure container networking benchmarks, in: Proceedings of the European Workshop on Security and Privacy in Edge Computing (EuroSPEC), 2019.

[924] M. Alizadeh, A. Kabbani, T. Edsall, B. Prabhakar, A. Vahdat, M. Yasuda, Less is more: Trading a little bandwidth for ultra-low latency in the data center, in: Proceedings of the USENIX Conference on Networked Systems Design and Implementation (NSDI), 2012.

[925] K. Bhardwaj, J.C. Miranda, A. Gavrilovska, Towards IoT-DDoS prevention using edge computing, in: Proceedings of the USENIX Workshop on Hot Topics in Edge Computing (HotEdge), 2018.

[926] C. Wan, L. Wang, V.V. Phoha, A survey on gait recognition, ACM Computing Surveys 51 (5) (2018) 89, pp. 1–35.

[927] M. Atas, Hand tremor based biometric recognition using leap motion device, IEEE Access 5 (2017) 23320–23326.

[928] M. Abadi, A. Chu, I. Goodfellow, H.B. McMahan, I. Mironov, K. Talwar, L. Zhang, Deep learning with differential privacy, in: Proceedings of the ACM Conference on Computer and Communications Security (CCS), 2016.

[929] R.A. Newcombe, S. Izadi, O. Hilliges, D. Molyneaux, D. Kim, A.J. Davison, P. Kohli, J. Shotton, S. Hodges, A.W. Fitzgibbon, Kinectfusion: Real-time dense surface mapping and tracking, in: Proceedings of the IEEE International Symposium on Mixed and Augmented Reality (ISMAR), 2011.

[930] R.A. Newcombe, D. Fox, S.M. Seitz, DynamicFusion: Reconstruction and tracking of non-rigid scenes in real-time, in: Proceedings of the IEEE Conference on Computer Vision and Pattern Recognition (CVPR), 2015.

[931] T. Whelan, S. Leutenegger, R. Salas-Moreno, B. Glocker, A. Davison, ElasticFusion: Dense SLAM without a pose graph, in: Proceedings of the Conference on Robotics: Science and Systems (RSS), 2015.

[932] T. Whelan, R.F. Salas-Moreno, B. Glocker, A.J. Davison, S. Leutenegger, ElasticFusion: Real-time dense SLAM and light source estimation, The International Journal of Robotics Research 35 (14) (2016) 1697–1716.

[933] A. Dai, M. Nießner, M. Zollhöfer, S. Izadi, C. Theobalt, Bundlefusion: Real-time globally consistent 3D reconstruction using on-the-fly surface reintegration, ACM Transactions on Graphics 36 (3) (2017) 24, pp. 1–18.

[934] H. Zhang, Z.-H. Bo, J.-H. Yong, F. Xu, Interactionfusion: Real-time reconstruction of hand poses and deformable objects in hand-object interactions, ACM Transactions on Graphics 38 (4) (2019) 48, pp. 1–11.

[935] P. Stotko, S. Krumpen, M. Schwarz, C. Lenz, S. Behnke, R. Klein, M. Weinmann, A VR system for immersive teleoperation and live exploration with a mobile robot, CoRR, arXiv:1908.02949, 2019.

[936] C.R. Qi, L. Yi, H. Su, L.J. Guibas, PointNet++: Deep hierarchical feature learning on point sets in a metric space, in: Proceedings of the Conference on Neural Information Processing Systems (NIPS), 2017.

[937] Y. Li, R. Bu, M. Sun, W. Wu, X. Di, B. Chen, PointCNN: Convolution on $\mathcal{X}$-transformed points, in: Proceedings of the Neural Information Processing Systems (NIPS), 2018.

[938] E. Brachmann, F. Michel, A. Krull, M.Y. Yang, S. Gumhold, C. Rother, Uncertainty-driven 6D pose estimation of objects and scenes from a single RGB image, in: Proceedings of the IEEE Conference on Computer Vision and Pattern Recognition (CVPR), 2016.

[939] B. Dong, S. Byna, K. Wu, Prabhat, H. Johansen, J.N. Johnson, N. Keen, Data elevator: Low-contention data movement in hierarchical storage system, in: Proceedings of the IEEE International Conference on High Performance Computing (HiPC), 2016.

[940] M. Weiland, H. Brunst, T. Quintino, N. Johnson, O. Iffrig, S. Smart, C. Herold, A. Bonanni, A. Jackson, M. Parsons, An early evaluation of Intel's Optane DC persistent memory module and its impact on high-performance scientific applications, in: Proceedings of the International Conference for High Performance Computing, Networking, Storage and Analysis (SC), 2019.

[941] M. Asch, T. Moore, R. Badia, M. Beck, P. Beckman, T. Bidot, F. Bodin, F. Cappello, A. Choudhary, B. de Supinski, E. Deelman, J. Dongarra, A. Dubey, G. Fox, H. Fu, S. Girona, W. Gropp, M. Heroux, Y. Ishikawa, K. Keahey, D. Keyes, W. Kramer, J.-F. Lavignon, Y. Lu, S. Matsuoka, B. Mohr, D. Reed, S. Requena, J. Saltz, T. Schulthess, R. Stevens, M. Swany, A. Szalay, W. Tang, G. Varoquaux, J.-P. Vilotte, R. Wisniewski, Z. Xu, I. Zacharov, Big data and extreme-scale computing: Pathways to convergence – Toward a shaping strategy for a future software and data ecosystem for scientific inquiry, The International Journal of High Performance Computing Applications 32 (4) (2018) 435–479.

[942] J. Dean, S. Ghemawat, MapReduce: A flexible data processing tool, Communications of the ACM 53 (1) (2010) 72–77.

[943] M. Zaharia, M. Chowdhury, T. Das, A. Dave, J. Ma, M. McCauly, M.J. Franklin, S. Shenker, I. Stoica, Resilient distributed datasets: A fault-tolerant abstraction for in-memory cluster computing, in: Proceedings of the USENIX Symposium on Networked Systems Design and Implementation (NSDI), 2012.

[944] P. Carbone, A. Katsifodimos, S. Ewen, V. Markl, S. Haridi, K. Tzoumas, Apache Flink: Stream and batch processing in a single engine, Bulletin of the IEEE Computer Society Technical Committee on Data Engineering 38 (4) (2015) 28–38.

[945] J. Kreps, N. Narkhede, J.R. Kafka, A distributed messaging system for log processing, in: Proceedings of the International Workshop on Networking Meets Databases (NetDB), 2011.

[946] A. Knüpfer, C. Rössel, D. an Mey, S. Biersdorff, K. Diethelm, D. Eschweiler, M. Geimer, M. Gerndt, D. Lorenz, A.D. Malony, W.E. Nagel, Y. Oleynik, P. Philippen, P. Saviankou, D. Schmidl, S. Shende, R. Tschüter, M. Wagner, B. Wesarg, F. Wolf, Score-P: A joint performance measurement run-time infrastructure for Periscope, Scalasca, TAU, and Vampir, in: H. Brunst, M.S. Müller, W.E. Nagel, M.M. Resch (Eds.), Tools for High Performance Computing 2011, Springer, 2012, pp. 79–91.

[947] J. Frenzel, K. Feldhoff, R. Jäkel, R. Müller-Pfefferkorn, Tracing of multi-threaded Java applications in Score-P using bytecode instrumentation, in: Proceedings of the GI/ITG International Conference on Architecture of Computing Systems (ARCS), Workshop Proceedings, 2018.

[948] H. Brunst, M. Weber, C. Herold, Performance tracing of heterogeneous exascale applications: Pitfalls and opportunities, in: Proceedings of the International Conference on Exascale Applications and Software (EASC), 2018.

[949] K. Henricksen, J. Indulska, Developing context-aware pervasive computing applications: Models and approach, Pervasive and Mobile Computing 2 (1) (2006) 37–64.

[950] I. Jaouadi, R. Ben Djemaa, H. Ben-Abdallah, A model-driven development approach for context-aware systems, Software and Systems Modeling 17 (4) (2018) 1169–1195.

[951] K. Häussermann, C. Hubig, P. Levi, F. Leymann, O. Siemoneit, M. Wieland, O. Zweigle, Understanding and designing situation-aware mobile and ubiquitous computing systems: An interdisciplinary analysis on the recognition of situations with uncertain data using situation templates, International Journal of Computer and Information Engineering 4 (3) (2010) 562–571.

[952] H.R. Schmidtke, W. Woo, A size-based qualitative approach to the representation of spatial granularity, in: Proceedings of the International Joint Conference on Artificial Intelligence (IJCAI), 2007.

[953] G.M. Kapitsaki, I.S. Venieris, Model-driven development of context-aware web applications based on a web service context management architecture, in: M.R.V. Chaudron (Ed.), Models in Software Engineering, in: Lecture Notes in Computer Science, vol. 5421, Springer, 2009, pp. 343–355.

[954] D. Grzelak, U. Aßmann, Preparatory reflections on safe context-adaptive software (Position paper), in: Proceedings of the International Conference on Internet of Things, Big Data and Security (IoTBDS), 2020.

[955] M. Hennessy, Context-awareness: Models and analysis, talk given at the UK-UbiNet Workshop: Security, trust, privacy and theory for ubiquitous computing, 2004.

[956] D. Broman, E.A. Lee, S. Tripakis, M. Törngren, Viewpoints, formalisms, languages, and tools for cyber-physical systems, in: Proceedings of the International Workshop on Multi-Paradigm Modeling (MPM), 2012.

[957] D. Grzelak, U. Aßmann, Bigraphical meta-modeling of fog computing-based systems, in: Proceedings of the International Conference on Discrete Models of Complex Systems (SOLSTICE), 2019.

[958] G.-C. Roman, C. Julien, J. Payton, A formal treatment of context-awareness, in: M. Wermelinger, T. Margaria-Steffen (Eds.), Fundamental Approaches to Software Engineering, in: Lecture Notes in Computer Science, vol. 2984, Springer, 2004, pp. 12–36.

[959] T. Strang, C. Linnhoff-Popien, A context modeling survey, in: Proceedings of the International Conference on Ubiquitous Computing (UbiComp), Workshop on Advanced Context Modelling, Reasoning and Management (W12), 2004.

[960] C. Bolchini, C.A. Curino, E. Quintarelli, F.A. Schreiber, L. Tanca, A data-oriented survey of context models, ACM SIGMOD Record 36 (4) (2007) 19–26.

[961] I. Cafezeiro, E.H. Haeusler, A. Rademaker, Ontology and context, in: Proceedings of the IEEE International Conference on Pervasive Computing and Communications (PerCom).

[962] S.W. Loke, Incremental awareness and compositionality: A design philosophy for context-aware pervasive systems, Pervasive and Mobile Computing 6 (2) (2010) 239–253.

[963] C. Bettini, O. Brdiczka, K. Henricksen, J. Indulska, D. Nicklas, A. Ranganathan, D. Riboni, A survey of context modelling and reasoning techniques, Pervasive and Mobile Computing 6 (2) (2010) 161–180.

[964] S.W. Loke, Representing and reasoning with the Internet of Things: A modular rule-based model for ensembles of context-aware smart things, EAI Endorsed Transactions on Context-aware Systems and Applications 3 (8) (2016) e1, pp. 1–17.

[965] R. Hirschfeld, P. Costanza, O. Nierstrasz, Context-oriented programming, Journal of Object Technology 7 (3) (2008) 125–151.

[966] H.R. Schmidtke, W. Woo, Towards ontology-based formal verification methods for context aware systems, in: H. Tokuda, M. Beigl, A. Friday, A.J.B. Brush, Y. Tobe (Eds.), Pervasive Computing, in: Lecture Notes in Computer Science, vol. 5538, Springer, 2009, pp. 309–326.

[967] S. Edelkamp, S. Schrödl, Heuristic Search, Morgan Kaufmann, 2012.

[968] R. de Lemos, H. Giese, H.A. Müller, M. Shaw, J. Andersson, L. Baresi, B. Becker, N. Bencomo, Y. Brun, B. Cukic, R.J. Desmarais, S. Dustdar, G. Engels, K. Geihs, K.M. Göschka, A. Gorla, V. Grassi, P. Inverardi, G. Karsai, J. Kramer, M. Litoiu, A. Lopes, J. Magee, S. Malek, S. Mankovskii, R. Mirandola, J. Mylopoulos, O. Nierstrasz, M. Pezzè, C. Prehofer, W. Schäfer, R.D. Schlichting, B.R. Schmerl, D.R. Smith, J.P. Sousa, G. Tamura, L. Tahvildari, N.M. Villegas, T. Vogel, D. Weyns, K. Wong, J. Wuttke, Software engineering for self-adaptive systems: A second research roadmap, in: R. de Lemos, H. Giese, H.A. Müller, M. Shaw (Eds.), Software Engineering for Self-Adaptive Systems II, in: Lecture Notes in Computer Science, vol. 7475, Springer, 2010, pp. 1–32.

[969] V. Grassi, A. Sindico, Towards model driven design of service-based context-aware applications, in: Proceedings of the International Workshop on Engineering of Software Services for Pervasive Environments (ESSPE) in Conjunction with the ESEC/FSE Joint Meeting, 2007.

[970] On the protection of natural persons with regard to the processing of personal data and on the free movement of such data, and repealing Directive 95/46/EC (General Data Protection Regulation), Regulation (EU) 2016/679 of the European Parliament and of the Council, Article 13(2)(f), 2016.

[971] I.J. Good, A theory of causality, British Journal for the Philosophy of Science 9 (36) (1959) 307–310.

[972] E. Eells, Probabilistic Causality, Cambridge University Press, 1991.

[973] J. Pearl, Causality: Models, Reasoning and Inference, Cambridge University Press, 2009.

[974] J. Williamson, Probabilistic theories of causation, in: H. Beebee, C. Hitchcock, P. Menzies (Eds.), The Oxford Handbook of Causation, Oxford Handbooks, Oxford University Press, 2009, pp. 175–197.

[975] D. Lewis, Causation, Journal of Philosophy 70 (17) (1973) 556–567.

[976] J.Y. Halpern, A modification of the Halpern-Pearl definition of causality, in: Proceedings of the International Joint Conference on Artificial Intelligence (IJCAI), 2015.

[977] J.Y. Halpern, J. Pearl, Causes and explanations: A structural-model approach – Part I: Causes, in: Proceedings of the International Joint Conference on Artificial Intelligence (IJCAI), 2001.

[978] J.Y. Halpern, J. Pearl, Causes and explanations: A structural-model approach – Part II: Explanations, in: Proceedings of the International Joint Conference on Artificial Intelligence (IJCAI), 2001.

[979] T. Eiter, T. Lukasiewicz, Complexity results for structure-based causality, Artificial Intelligence 142 (1) (2002) 53–89.

[980] G. Aleksandrowicz, H. Chockler, J.Y. Halpern, A. Ivrii, The computational complexity of structure-based causality, Journal of Artificial Intelligence Research 58 (2017) 431–451.

[981] T. Eiter, T. Lukasiewicz, Causes and explanations in the structural-model approach: Tractable cases, Artificial Intelligence 170 (6–7) (2006) 542–580.

[982] C. Hitchcock, Probabilistic causation, in: E.N. Zalta (Ed.), The Stanford Encyclopedia of Philosophy, spring edition, Metaphysics Research Lab, Stanford University, 2018.

[983] L. Fenton-Glynn, A proposed probabilistic extension of the Halpern and Pearl definition of "actual cause", British Journal for the Philosophy of Science 68 (4) (2017) 1061–1124.

[984] S. Beckers, J. Vennekens, A general framework for defining and extending actual causation using CP-logic, International Journal of Approximate Reasoning 77 (2016) 105–126.

[985] S. Kleinberg, Causality, Probability and Time, Cambridge University Press, 2012.

[986] S. Kleinberg, A logic for causal inference in time series with discrete and continuous variables, in: Proceedings of the International Joint Conference on Artificial Intelligence (IJCAI), 2011.

[987] T. Han, J.-P. Katoen, Counterexamples in probabilistic model checking, in: O. Grumberg, M. Huth (Eds.), Tools and Algorithms for the Construction and Analysis of Systems, in: Lecture Notes in Computer Science, vol. 4424, Springer, 2007, pp. 72–86.

[988] H. Aljazzar, S. Leue, Directed explicit state-space search in the generation of counterexamples for stochastic model checking, IEEE Transactions on Software Engineering 36 (1) (2010) 37–60.

[989] N. Jansen, Counterexamples in probabilistic verification, PhD thesis, RWTH Aachen University, Germany, 2015.

[990] F. Leitner-Fischer, S. Leue, Causality checking for complex system models, in: R. Giacobazzi, J. Berdine, I. Mastroeni (Eds.), Verification, Model Checking, and Abstract Interpretation, in: Lecture Notes in Computer Science, vol. 7737, Springer, 2013, pp. 248–267.

[991] F. Leitner-Fischer, S. Leue, SpinCause: A tool for causality checking, in: Proceedings of the International Symposium on Model Checking of Software (SPIN), 2014.

[992] I. Beer, S. Ben-David, H. Chockler, A. Orni, R.J. Trefler, Explaining counterexamples using causality, Formal Methods in System Design 40 (1) (2012) 20–40.

[993] F. Leitner-Fischer, Causality checking of safety-critical software and systems, PhD thesis, University of Konstanz, Germany, 2015.

[994] M. Blum, S. Kannan, Designing programs that check their work, Journal of the ACM 42 (1) (1995) 269–291.

[995] R.M. McConnell, K. Mehlhorn, S. Näher, P. Schweitzer, Certifying algorithms, Computer Science Review 5 (2) (2011) 119–161.

[996] O. Kupferman, M.Y. Vardi, From complementation to certification, Theoretical Computer Science 345 (1) (2005) 83–100.

[997] N. Jansen, E. Ábrahám, M. Volk, R. Wimmer, J.-P. Katoen, B. Becker, The COMICS tool – Computing minimal counterexamples for DTMCs, in: S. Chakraborty, M. Mukund (Eds.), Automated Technology for Verification and Analysis, in: Lecture Notes in Computer Science, vol. 7561, Springer, 2012, pp. 349–353.

[998] F. Funke, S. Jantsch, C. Baier, Farkas certificates and minimal witnesses for probabilistic reachability constraints, in: Proceedings of the International Conference on Tools and Algorithms for the Construction and Analysis of Systems (TACAS), 2020.

[999] P. Zave, Feature-oriented description, formal methods, and DFC, in: S. Gilmore, M. Ryan (Eds.), Language Constructs for Describing Features: Proceedings of the FIREworks workshop, Springer, 2001, pp. 11–26.

[1000] K.C. Kang, S.G. Cohen, J.A. Hess, W.E. Novak, A.S. Peterson, Feature-oriented domain analysis (FODA) feasibility study, Technical report, Carnegie-Mellon University, Software Engineering Institute, 1990.

[1001] S. Apel, C. Kästner, An overview of feature-oriented software development, Journal of Object Technology 8 (5) (2009) 49–84.

[1002] P. Clements, L. Northrop, Software Product Lines: Practices and Patterns, Addison-Wesley Professional, 2001.

[1003] K. Czarnecki, S. She, A. Wasowski, Sample spaces and feature models: There and back again, in: Proceedings of the International Software Product Line Conference (SPLC), 2008.

[1004] F. Damiani, L. Padovani, I. Schaefer, C. Seidl, A core calculus for dynamic delta-oriented programming, Acta Informatica 55 (4) (2018) 269–307.

[1005] C. Dubslaff, C. Baier, S. Klüppelholz, Probabilistic Model Checking for Feature-Oriented Systems, Lecture Notes in Computer Science, vol. 8989, Springer, 2015, pp. 180–220.

[1006] P. Chrszon, C. Dubslaff, S. Klüppelholz, C. Baier, ProFeat: Feature-oriented engineering for family-based probabilistic model checking, Formal Aspects of Computing 30 (1) (2018) 45–75.

[1007] M. Acher, P. Collet, F. Fleurey, P. Lahire, S. Moisan, J.-P. Rigault, Modeling context and dynamic adaptations with feature models, in: Proceedings of the International Workshop Models@run.time (MRT), 2009.

[1008] C. Dubslaff, P. Koopmann, A.-Y. Turhan, Ontology-mediated probabilistic model checking, in: W. Ahrendt, S.L. Tapia Tarifa (Eds.), Integrated Formal Methods, in: Lecture Notes in Computer Science, vol. 11918, Springer, 2019, pp. 194–211.

[1009] T. Pett, T. Thüm, T. Runge, S. Krieter, M. Lochau, I. Schaefer, Product sampling for product lines: The scalability challenge, in: Proceedings of the International Systems and Software Product Line Conference (SPLC), 2019.

[1010] C. Dubslaff, A. Morozov, C. Baier, K. Janschek, Reduction methods on probabilistic control-flow programs for reliability analysis, in: Proceedings of the European Safety and Reliability Conference (ESREL) and the Probabilistic Safety Assessment and Management Conference (PSAM), 2020.

[1011] A. von Rhein, A. Grebhahn, S. Apel, N. Siegmund, D. Beyer, T. Berger, Presence-condition simplification in highly configurable systems, in: Proceedings of the International Conference on Software Engineering (ICSE), 2015.

[1012] W.V. Quine, The problem of simplifying truth functions, Mathematical Association of America 59 (1952) 521–531.

[1013] E.J. McCluskey Jr., Minimization of Boolean functions, Bell System Technical Journal 35 (6) (1956) 1417–1444.

[1014] P. Chrszon, C. Baier, C. Dubslaff, S. Klüppelholz, From features to roles, in: Proceedings of the International Systems and Software Product Line Conference (SPLC), 2020.

[1015] T. Kühn, M. Leuthäuser, S. Götz, C. Seidl, U. Aßmann, A metamodel family for role-based modeling and programming languages, in: B. Combemale, D.J. Pearce, O. Barais, J.J. Vinju (Eds.), Software Language Engineering, in: Lecture Notes in Computer Science, vol. 8706, Springer, 2014, pp. 141–160.

[1016] T. Thüm, S. Apel, C. Kästner, I. Schäfer, G. Saake, A classification and survey of analysis strategies for software product lines, ACM Computing Surveys 47 (1) (2014) 6, pp. 1–45.

[1017] A. Filieri, G. Tamburrelli, C. Ghezzi, Supporting self-adaptation via quantitative verification and sensitivity analysis at run time, IEEE Transactions on Software Engineering 42 (1) (2016) 75–99.

[1018] A. Aijaz, M. Sooriyabandara, The Tactile Internet for industries – A review, Proceedings of the IEEE 107 (2) (2019) 414–435.

[1019] M. Fiedler, T. Hossfeld, P. Tran-Gia, A generic quantitative relationship between quality of experience and quality of service, IEEE Network 24 (2) (2010) 36–41.

[1020] P. Reichl, B. Tuffin, R. Schatz, Logarithmic laws in service quality perception: Where microeconomics meets psychophysics and quality of experience, Telecommunication Systems 52 (2) (2013) 587–600.

[1021] M. Maier, A. Ebrahimzadeh, Towards immersive Tactile Internet experiences: Low-latency FiWi enhanced mobile networks with edge intelligence, IEEE/OSA Journal of Optical Communications and Networking 11 (4) (2019) B10–B25.

[1022] L. Meli, C. Pacchierotti, D. Prattichizzo, Experimental evaluation of magnified haptic feedback for robot-assisted needle insertion and palpation, International Journal of Medical Robotics and Computer Assisted Surgery 13 (4) (2017) e1809, pp. 1–14.

[1023] E. Wong, M.P.I. Dias, L. Ruan, Predictive resource allocation for Tactile Internet capable passive optical LANs, IEEE/OSA Journal of Lightwave Technology 35 (13) (2017) 2629–2641.

[1024] I.A. Tsokalo, D. Kuß, I. Kharabet, F.H.P. Fitzek, M. Reisslein, Remote robot control with Human-in-the-Loop over long distances using digital twins, in: Proceedings of the IEEE Global Communications Conference (GLOBECOM), 2019.

[1025] S.R. Fiorini, J. Bermejo-Alonso, P. Gonçalves, E. Pignaton de Freitas, A. Olivares Alarcos, J.I. Olszewska, E. Prestes, C. Schlenoff, S.V. Ragavan, S. Redfield, B. Spencer, H. Li, A suite of ontologies for robotics and automation, IEEE Robotics & Automation Magazine 24 (1) (2017) 8–11.

[1026] IEEE standard ontologies for robotics and automation, IEEE Std 1872-2015, 2015.

[1027] R. Drath, A. Lüder, J. Peschke, L. Hundt, AutomationML – The glue for seamless automation engineering, in: Proceedings of the IEEE International Conference on Emerging Technologies and Factory Automation (ETFA), 2008.

[1028] E. Estevez, M. Marcos, Model-based validation of industrial control systems, IEEE Transactions on Industrial Informatics 8 (2) (2012) 302–310.

[1029] R. Khare, A. Rifkin, XML: A door to automated web applications, IEEE Internet Computing 1 (4) (1997) 78–87.

[1030] Ş. Kolozali, M. Bermudez-Edo, N. Farajidavar, P. Barnaghi, F. Gao, M.I. Ali, A. Mileo, M. Fischer, T. Iggena, D. Kümper, R. Tönjes, Observing the pulse of a city: A smart city framework for real-time discovery, federation, and aggregation of data streams, IEEE Internet of Things Journal 6 (2) (2019) 2651–2668.

[1031] K. Gilani, J. Kim, J. Song, D. Seed, C. Wang, Semantic enablement in IoT service layers – Standard progress and challenges, IEEE Internet Computing 22 (4) (2018) 56–63.

[1032] T. Berners-Lee, J. Hendler, O. Lassila, The Semantic Web, Scientific American 284 (5) (2001) 34–43.

[1033] A. Schmidt, C. Martin, T. Dietz, A. Pott, A generic data structure for the specific domain of robotic arc welding, in: L. Wang (Ed.), 51st CIRP Conference on Manufacturing Systems, in: Procedia CIRP, vol. 72, Elsevier, 2018, pp. 322–327.

[1034] S.K. Jensen, T.B. Pedersen, C. Thomsen, Time series management systems: A survey, IEEE Transactions on Knowledge and Data Engineering 29 (11) (2017) 2581–2600.

[1035] K. Goldberg, Robots and the return to collaborative intelligence, Nature Machine Intelligence 1 (1) (2019) 2–4.

[1036] P.J. Rousseeuw, M. Hubert, Robust statistics for outlier detection, Wiley Interdisciplinary Reviews: Data Mining and Knowledge Discovery 1 (1) (2011) 73–79.

[1037] M. Hubert, E. Vandervieren, An adjusted boxplot for skewed distributions, Computational Statistics & Data Analysis 52 (12) (2008) 5186–5201.

[1038] J.W. Tukey, Exploratory Data Analysis, Addison-Wesley, 1977, pp. 39–43, chapter Box-and-whisker plots.

[1039] D. Freedman, P. Diaconis, On the histogram as a density estimator: L_2 theory, Zeitschrift für Wahrscheinlichkeitstheorie und verwandte Gebiete 57 (4) (1981) 453–476.

[1040] A. Bhardwaj, B. Çizmeci, E. Steinbach, Q. Liu, M. Eid, J. Araújo, A. El Saddik, R. Kundu, X. Liu, O. Holland, M.A. Luden, S. Oteafy, V. Prasad, A candidate hardware and software reference setup for kinesthetic codec standardization, in: Proceedings of the IEEE International Symposium on Haptic Audio-Visual Environments and Games (HAVE), 2017.

[1041] J. Kirsch, A. Noll, M. Strese, Q. Liu, E. Steinbach, A low-cost acquisition, display, and evaluation setup for tactile codec development, in: Proceedings of the IEEE International Symposium on Haptic Audio-Visual Environments and Games (HAVE), 2018.

[1042] N. Landin, J.M. Romano, W. McMahan, K.J. Kuchenbecker, Dimensional reduction of high-frequency accelerations for haptic rendering, in: A.M.L. Kappers, J.B.F. van Erp, W.M. Bergmann Tiest, F.C.T. van der Helm (Eds.), Haptics: Generating and Perceiving Tangible Sensations, in: Lecture Notes in Computer Science, vol. 6192, Springer, 2010, pp. 79–86.

[1043] Y. Gao, S.S. Vedula, C.E. Reiley, N. Ahmidi, B. Varadarajan, H.C. Lin, L. Tao, L. Zappella, B. Béjar, D.D. Yuh, C.C.G. Chen, R. Vidal, S. Khudanpur, G.D. Hager, JHU-ISI gesture and skill assessment working set (JIGSAWS): A surgical activity dataset for human motion modeling, in: Proceedings of the Workshop on Modeling and Monitoring of Computer Assisted Interventions (M2CAI), 2014.

[1044] G. Aceto, V. Persico, A. Pescapé, A survey on information and communication technologies for Industry 4.0: State-of-the-art, taxonomies, perspectives, and challenges, IEEE Communications Surveys & Tutorials 21 (4) (2019) 3467–3501.

[1045] L. Lo Bello, W. Steiner, A perspective on IEEE time-sensitive networking for industrial communication and automation systems, Proceedings of the IEEE 107 (6) (2019) 1094–1120.

[1046] F. Engelhardt, M. Güneş, Modeling delay of haptic data in CSMA-based wireless multi-hop networks: A probabilistic approach, in: Proceedings of the IEEE International Symposium on Personal, Indoor and Mobile Radio Communications (PIMRC), Workshop on Wireless Sensor Networks for the Internet of Things (S2), 2019.

[1047] L. Linguaglossa, D. Rossi, S. Pontarelli, D. Barach, D. Marjon, P. Pfister, High-speed data plane and network functions virtualization by vectorizing packet processing, Computer Networks 149 (2019) 187–199.

[1048] Z. Ma, M. Xiao, Y. Xiao, Z. Pang, H.V. Poor, B.S. Vučetić, High-reliability and low-latency wireless communication for Internet of Things: Challenges, fundamentals, and enabling technologies, IEEE Internet of Things Journal 6 (5) (2019) 7946–7970.

[1049] A. Nasrallah, A.S. Thyagaturu, Z. Alharbi, C. Wang, X. Shao, M. Reisslein, H. ElBakoury, Ultra-low latency (ULL) networks: The IEEE TSN and IETF DetNet standards and related 5G ULL research, IEEE Communications Surveys & Tutorials 21 (1) (2019) 88–145.

[1050] A. Nasrallah, A.S. Thyagaturu, Z. Alharbi, C. Wang, X. Shao, M. Reisslein, H. Elbakoury, Performance comparison of IEEE 802.1 TSN time aware shaper (TAS) and asynchronous traffic shaper (ATS), IEEE Access 7 (2019) 44165–44181.

[1051] P. Shantharama, A.S. Thyagaturu, M. Reisslein, Hardware-accelerated platforms and infrastructures for network functions: A survey of enabling technologies and research studies, IEEE Access 8 (2020) 132021–132085.

[1052] S. Vitturi, C. Zunino, T. Sauter, Industrial communication systems and their future challenges: Next-generation Ethernet, IIoT, and 5G, Proceedings of the IEEE 107 (6) (2019) 944–961.

[1053] A.J. Ferrer, J.M. Marquès, J. Jorba, Towards the decentralised cloud: Survey on approaches and challenges for mobile, ad hoc, and edge computing, ACM Computing Surveys 51 (6) (2019) 111, pp. 1–36.

[1054] M. Maier, A. Ebrahimzadeh, M. Chowdhury, The Tactile Internet: Automation or augmentation of the human?, IEEE Access 6 (2018) 41607–41618.

[1055] M. Mehrabi, D. You, V. Latzko, H. Salah, M. Reisslein, F.H.P. Fitzek, Device-enhanced MEC: Multi-access edge computing (MEC) aided by end device computation and caching: A survey, IEEE Access 7 (2019) 166079–166108.

[1056] V. Huang, G. Chen, P. Zhang, H. Li, C. Hu, T. Pan, Q. Fu, A scalable approach to SDN control plane management: High utilization comes with low latency, IEEE Transactions on Network and Service Management 17 (2) (2020) 682–695.

[1057] W. Kellerer, P. Kalmbach, A. Blenk, A. Basta, M. Reisslein, S. Schmid, Adaptable and data-driven softwarized networks: Review, opportunities, and challenges, Proceedings of the IEEE 107 (4) (2019) 711–731.

[1058] Y. Zhao, Y. Li, X. Zhang, G. Geng, W. Zhang, Y. Sun, A survey of networking applications applying the software defined networking concept based on machine learning, IEEE Access 7 (2019) 95385–95405.

[1059] T. Brogårdh, Present and future robot control development – An industrial perspective, Annual Reviews in Control 31 (1) (2007) 69–79.

[1060] A. Kelly, N. Chan, H. Herman, D. Huber, R. Meyers, P. Rander, R. Warner, J. Ziglar, E. Capstick, Real-time photorealistic virtualized reality interface for remote mobile robot control, The International Journal of Robotics Research 30 (3) (2011) 384–404.

[1061] J. Guivant, S. Cossell, M. Whitty, J. Katupitiya, Internet-based operation of autonomous robots: The role of data replication, compression, bandwidth allocation and visualization, Journal of Field Robotics 29 (5) (2012) 793–818.

[1062] M. Liou, Overview of the $p \times 64$ kbit/s video coding standard, Communications of the ACM 34 (4) (1991) 59–63.

[1063] A. Pulipaka, P. Seeling, M. Reisslein, L.J. Karam, Traffic and statistical multiplexing characterization of 3-D video representation formats, IEEE Transactions on Broadcasting 59 (2) (2013) 382–389.

[1064] P. Seeling, M. Reisslein, Video traffic characteristics of modern encoding standards: H.264/AVC with SVC and MVC extensions and H.265/HEVC, The Scientific World Journal 2014 (2014) 189481, pp. 1–16.

[1065] X. Xu, S. Liu, Recent advances in video coding beyond the HEVC standard, APSIPA Transactions on Signal and Information Processing 8 (2019) e18, pp. 1-10.

[1066] T. Zhang, S. Mao, An overview of emerging video coding standards, ACM GetMobile: Mobile Computing and Communications 22 (4) (2019) 13–20.

[1067] G.B. Akar, A.M. Tekalp, C. Fehn, M.R. Civanlar, Transport methods in 3DTV – A survey, IEEE Transactions on Circuits and Systems for Video Technology 17 (11) (2007) 1622–1630.

[1068] J. Kua, G. Armitage, P. Branch, A survey of rate adaptation techniques for dynamic adaptive streaming over HTTP, IEEE Communications Surveys & Tutorials 19 (3) (2017) 1842–1866.

[1069] T.V. Lakshman, A. Ortega, A.R. Reibman, VBR video: Tradeoffs and potentials, Proceedings of the IEEE 86 (5) (1998) 952–973.

[1070] M. Reisslein, K.W. Ross, High-performance prefetching protocols for VBR prerecorded video, IEEE Network 12 (6) (1998) 46–55.

[1071] Y. Li, M. Reisslein, C. Chakrabarti, Energy-efficient video transmission over a wireless link, IEEE Transactions on Vehicular Technology 58 (3) (2008) 1229–1244.

[1072] J.C.V.S. Junior, M.F. Torquato, D.H. Noronha, S.N. Silva, M.A.C. Fernandes, Proposal of the tactile glove device, Sensors 19 (22) (2019) 5029, pp. 1–17.

[1073] A.S. Muhammad Sayem, S.H. Teay, H. Shahariar, P.L. Fink, A. Albarbar, Review on smart electro-clothing systems (SeCSs), Sensors 20 (3) (2020) 587, pp. 1–23.

[1074] M.C. Silva, V.J.P. Amorim, S.P. Ribeiro, R.A.R. Oliveira, Field research cooperative wearable systems: Challenges in requirements, design and validation, Sensors 19 (20) (2019) 4417, pp. 1–24.

[1075] S. Wilson, R. Laing, Fabrics and garments as sensors: A research update, Sensors 19 (16) (2019) 3570, pp. 1–35.

[1076] M. Şimşek, A. Aijaz, M. Dohler, J. Sachs, G.P. Fettweis, 5G-enabled Tactile Internet, IEEE Journal on Selected Areas in Communications 34 (3) (2016) 460–473.

[1077] R. Drath, A. Horch, Industrie 4.0: Hit or hype?, IEEE Industrial Electronics Magazine 8 (2) (2014) 56–58.

[1078] The IEEE Tactile Internet Standards Working Group (IEEE 1918.1), 2017.

[1079] Service requirements for the 5G system, 3rd Generation Partnership Project (3GPP), 2018.

[1080] Study on scenarios and requirements for next generation access technologies, 3rd Generation Partnership Project (3GPP), 2017.

[1081] The IEEE Tactile Internet Standards Working Group (IEEE 1918.1) project authorization request, 2017.

[1082] G.P. Fettweis, H. Boche, T. Wiegand, E. Zielinski, H. Schotten, P. Merz, S. Hirche, A. Festag, W. Häffner, M. Meyer, E. Steinbach, R. Kraemer, R. Steinmetz, F. Hofmann, P. Eisert, R. Scholl, F. Ellinger, E. Weiß, I. Riedel, The Tactile Internet – ITU-T technology watch report, Technical report, ITU-T, 2014.

[1083] Ergonomics of human–system interaction – Part 910: Framework for tactile and haptic interaction, ISO 9241-910:2011. Technical Committee: ISO/TC 159/SC 4, 2011.

[1084] Definition and representation of haptic-tactile essence for broadcast production applications, SMPTE Standard, ST 2100-1:2017, 2017.

[1085] IPv6 based Tactile Internet, ETSI Work Item DGR/IP6-0014, 2017.

[1086] Minimum requirements related to technical performance for IMT-2020 radio interface(s), International Telecommunication Union, Radiocommunication Sector (ITU-R), 2017.

[1087] J. Sachs, G. Wikstrom, T. Dudda, R. Baldemair, K. Kittichokechai, 5G radio network design for ultra-reliable low-latency communication, IEEE Network 32 (2) (2018) 24–31.

[1088] E. Wong, M.P.I. Dias, L. Ruan, Tactile Internet capable passive optical LAN for healthcare, in: Proceedings of the OptoElectronics and Communications Conference (OECC), 2016.

[1089] T.S. Rappaport, Y. Xing, G.R. MacCartney, A.F. Molisch, E. Mellios, J. Zhang, Overview of millimeter wave communications for fifth-generation (5G) wireless networks – with a focus on propagation models, IEEE Transactions on Antennas and Propagation 65 (12) (2017) 6213–6230.

[1090] N. Abbas, H. Hajj, Z. Dawy, K. Jahed, S. Sharafeddine, An optimized approach to video traffic splitting in heterogeneous wireless networks with energy and QoE considerations, Journal of Network and Computer Applications 83 (2017) 72–88.

[1091] N. Abbas, S. Sharafeddine, H. Hajj, Z. Dawy, Price-aware traffic splitting in D2D HetNets with cost-energy-QoE tradeoffs, Computer Networks 172 (2020) 107169, pp. 1–15.

[1092] R. Islambouli, Z. Sweidan, S. Sharafeddine, Dynamic multipath resource management for ultra reliable low latency services, in: Proceedings of the IEEE Symposium on Computers and Communications (ISCC), 2019.

[1093] M. Chen, Y. Qian, Y. Hao, Y. Li, J. Song, Data-driven computing and caching in 5G networks: Architecture and delay analysis, IEEE Wireless Communications 25 (1) (2018) 70–75.

[1094] N. Kherraf, S. Sharafeddine, C.M. Assi, A. Ghrayeb, Latency and reliability-aware workload assignment in IoT networks with mobile edge clouds, IEEE Transactions on Network and Service Management 16 (4) (2019) 1435–1449.

[1095] R. Islambouli, S. Sharafeddine, Optimized 3D deployment of UAV-mounted cloudlets to support latency-sensitive services in IoT networks, IEEE Access 7 (2019) 172860–172870.

[1096] A. Aijaz, Z. Dawy, N. Pappas, M. Şimşek, S. Oteafy, O. Holland, Toward a Tactile Internet reference architecture: Vision and progress of the IEEE P1918.1 standard, CoRR, arXiv:1807.11915, 2018.

[1097] J.J. Nielsen, R. Liu, P. Popovski, Ultra-reliable low latency communication using interface diversity, IEEE Transactions on Communications 66 (3) (2018) 1322–1334.

[1098] C. She, C. Yang, T.Q.S. Quek, Radio resource management for ultra-reliable and low-latency communications, IEEE Communications Magazine 55 (6) (2017) 72–78.

[1099] S.M.A. Oteafy, H.S. Hassanein, IoT in the fog: A roadmap for data-centric IoT development, IEEE Communications Magazine 56 (3) (2018) 157–163.

[1100] B. Zhou, A.V. Dastjerdi, R.N. Calheiros, S.N. Srirama, R. Buyya, mCloud: A context-aware offloading framework for heterogeneous mobile cloud, IEEE Transactions on Services Computing 10 (5) (2017) 797–810.

[1101] S.M.A. Oteafy, H.S. Hassanein, Resilient IoT architectures over dynamic sensor networks with adaptive components, IEEE Internet of Things Journal 4 (2) (2017) 474–483.

[1102] S.M.A. Oteafy, H.S. Hassanein, Leveraging Tactile Internet cognizance and operation via IoT and edge technologies, Proceedings of the IEEE 107 (2) (2019) 364–375.

[1103] H. Wu, Y. Sun, K. Wolter, Energy-efficient decision making for mobile cloud offloading, IEEE Transactions on Cloud Computing 8 (2) (2018) 570–584.

[1104] M. Peng, S. Yan, K. Zhang, C. Wang, Fog-computing-based radio access networks: Issues and challenges, IEEE Network 30 (4) (2016) 46–53.

[1105] O. Dubuisson, P. Fouquart, ASN.1: Communication Between Heterogeneous Systems, Morgan Kaufmann, 2001.

[1106] J. Takeuchi, Requirements for automotive AVB system profiles, White Paper Contributed to AVnu Alliance, 2011.

[1107] J. Baber, J. Kolodko, T. Noël, M. Parent, L.B. Vlačić, Cooperative autonomous driving: Intelligent vehicles sharing city roads, IEEE Robotics & Automation Magazine 12 (1) (2005) 44–49.

[1108] M. During, K. Lemmer, Cooperative maneuver planning for cooperative driving, IEEE Intelligent Transportation Systems Magazine 8 (3) (2016) 8–22.

[1109] The IEEE Haptic Codecs for the Tactile Internet Standards Task Group (IEEE 1918.1.1), 2017.

[1110] The IEEE Haptic Codecs for the Tactile Internet (IEEE 1918.1.1) project authorization request, 2017.

[1111] S. Hirche, M. Buss, P. Hinterseer, E. Steinbach, Network traffic reduction in haptic telepresence systems by deadband control, in: Proceedings of the International Federation of Automatic Control World Congress (IFAC), 2005.

[1112] K. Iiyoshi, M. Tauseef, R. Gebremedhin, V. Gokhale, M. Eid, Towards standardization of haptic handshake for Tactile Internet: A WebRTC-based implementation, in: Proceedings of the IEEE International Symposium on Haptic Audio-Visual Environments and Games (HAVE), 2019.

[1113] D.H. Guston, D. Sarewitz, Real-time technology assessment, Technology in Society 24 (1) (2002) 93–109.

[1114] R. Owen, P. Macnaghten, J. Stilgoe, Responsible research and innovation: From science in society to science for society, with society, Science and Public Policy 39 (6) (2012) 751–760.

[1115] J. Stilgoe, R. Owen, P. Macnaghten, Developing a framework for responsible innovation, Research Policy 42 (9) (2013) 1568–1580.

[1116] R. von Schomberg, A vision of responsible research and innovation: Managing the responsible emergence of science and innovation in society, in: R. Owen, J. Bessant, M. Heintz (Eds.), Responsible Innovation, Wiley, 2013, pp. 51–74.

[1117] G.H. Hadorn, S. Biber-Klemm, W. Grossenbacher-Mansuy, H. Hoffmann-Riem, D. Joye, C. Pohl, U. Wiesmann, E. Zemp, The emergence of transdisciplinarity as a form of research, in: G.H. Hadorn, H. Hoffmann-Riem, S. Biber-Klemm, W. Grossenbacher-Mansuy, D. Joye, C. Pohl, U. Wiesmann, E. Zemp (Eds.), Handbook of Transdisciplinary Research, Springer, 2008, pp. 19–39.

[1118] J.S.B.T. Evans, Dual-processing accounts of reasoning, judgment, and social cognition, Annual Review of Psychology 59 (1) (2008) 255–278.

[1119] P.C. Wason, J.S.B.T. Evans, Dual processes in reasoning?, Cognition 3 (2) (1974) 141–154.

[1120] R.E. Petty, J.T. Cacioppo, The elaboration likelihood model of persuasion, Advances in Experimental Social Psychology 19 (1986) 123–205.

[1121] S. Chaiken, Heuristic versus systematic information processing and the use of source versus message cues in persuasion, Journal of Personality and Social Psychology 39 (5) (1980) 752–766.

[1122] S. Epstein, Integration of the cognitive and the psychodynamic unconscious, American Psychologist 49 (8) (1994) 709–724.

[1123] J.S.B.T. Evans, Intuition and reasoning: A dual-process perspective, Psychological Inquiry 21 (4) (2010) 313–326.

[1124] K.E. Stanovich, Who is Rational? Studies of Individual Differences in Reasoning, Erlbaum, 1999.

[1125] W. De Neys, Dual processing in reasoning: Two systems but one reasoner, Psychological Science 17 (5) (2006) 428–433.

[1126] M.F. Basch, The concept of affect: A re-examination, Journal of the American Psychoanalytic Association 24 (4) (1976) 759–777.

[1127] M.L. Finucane, A. Alhakami, P. Slovic, S.M. Johnson, The affect heuristic in judgments of risks and benefits, Journal of Behavioral Decision Making 13 (1) (2000) 1–17.

[1128] G. Loewenstein, E.U. Weber, C.K. Hsee, N. Welch, Risk as feelings, Psychological Bulletin 127 (2) (2001) 267–286.

[1129] P. Slovic, M.L. Finucane, E. Peters, D.G. MacGregor, Risk as analysis and risk as feelings: Some thoughts about affect, reason, risk, and rationality, Risk Analysis 24 (2) (2004) 311–322.

[1130] M.C. Nisbet, The competition for worldviews: Values, information, and public support for stem cell research, International Journal of Public Opinion Research 17 (1) (2005) 90–112.

[1131] D.A. Scheufele, B.V. Lewenstein, The public and nanotechnology: How citizens make sense of emerging technologies, Journal of Nanoparticle Research 7 (6) (2005) 659–667.

[1132] C.-J. Lee, D.A. Scheufele, B.V. Lewenstein, Public attitudes toward emerging technologies: Examining the interactive effects of cognitions and affect on public attitudes toward nanotechnology, Science Communication 27 (2) (2005) 240–267.

[1133] V. Venkatesh, M.G. Morris, G.B. Davis, F.D. Davis, User acceptance of information technology: Toward a unified view, MIS Quarterly 27 (3) (2003) 425–478.

[1134] F.D. Davis, Perceived usefulness, perceived ease of use, and user acceptance of information technology, MIS Quarterly 13 (3) (1989) 319–340.

[1135] Y.K. Dwivedi, N.P. Rana, A. Jeyaraj, M. Clement, M.D. Williams, Re-examining the unified theory of acceptance and use of technology (UTAUT): Towards a revised theoretical model, Information Systems Frontiers 21 (3) (2019) 719–734.

[1136] TechnikRadar 2018: Was die Deutschen über Technik denken, acatech and Körber-Stiftung, 2018.

[1137] T. Gnambs, M. Appel, Are robots becoming unpopular? Changes in attitudes towards autonomous robotic systems in Europe, Computers in Human Behavior 93 (2019) 53–61.

[1138] J.N. Druckman, T. Bolsen, Framing, motivated reasoning, and opinions about emergent technologies, Journal of Communication 61 (4) (2011) 659–688.

[1139] S.H. Priest, Misplaced faith: Communication variables as predictors of encouragement for biotechnology development, Science Communication 23 (2) (2001) 97–110.

[1140] P. Liu, R. Yang, Z. Xu, Public acceptance of fully automated driving: Effects of social trust and risk/benefit perceptions, Risk Analysis 39 (2) (2019) 326–341.

[1141] A changing population: Assumptions and results of the 14th coordinated population projection, German Federal Statistical Office, 2019.

[1142] M. Pötschke, C. Müller, Erreichbarkeit und Teilnahmebereitschaft in Telefoninterviews: Versuch einer mehrebenenanalytischen Erklärung, ZA-Information / Zentralarchiv für Empirische Sozialforschung 59 (2006) 83–99.

[1143] D.M. Kahan, D. Braman, P. Slovic, J. Gastil, G. Cohen, Cultural cognition of the risks and benefits of nanotechnology, Nature Nanotechnology 4 (2) (2009) 87–90.

[1144] C. Röcker, Social and technological concerns associated with the usage of ubiquitous computing technologies, Issues in Information Systems 11 (1) (2010) 61–68.

[1145] D.G. Mick, S. Fournier, Paradoxes of technology: Consumer cognizance, emotions, and coping strategies, Journal of Consumer Research 25 (2) (1998) 123–143.

[1146] Â. Guimarães Pereira, A. Benessia, P. Curvelo, Agency in the Internet of Things, European Commission, 2013.

[1147] M. Mital, V. Chang, P. Choudhary, A. Papa, A.K. Pani, Adoption of Internet of Things in India: A test of competing models using a structured equation modeling approach, Technological Forecasting and Social Change 136 (2018) 339–346.

[1148] V. Venkatesh, J.Y.L. Thong, X. Xu, Consumer acceptance and use of information technology: Extending the unified theory of acceptance and use of technology, MIS Quarterly 36 (1) (2012) 157–178.

[1149] A. Parasuraman, Technology readiness index (TRI): A multiple-item scale to measure readiness to embrace new technologies, Journal of Service Research 2 (4) (2000) 307–320.

[1150] T. Walsh, Expert and non-expert opinion about technological unemployment, International Journal of Automation and Computing 15 (5) (2018) 637–642.

Index

Printed in the United States
By Bookmasters